26th Heat Treating Society Conference 2011

Cincinnati, Ohio, USA
31 October - 2 November 2011

Editors:

B. Lynn Ferguson
Roger Jones
Dale Weires

ISBN: 978-1-7138-9959-4

Printed from e-media with permission by:

Curran Associates, Inc.
57 Morehouse Lane
Red Hook, NY 12571

Some format issues inherent in the e-media version may also appear in this print version.

Printed with permission by Curran Associates, Inc. (2025)

For permission requests, please contact ASM International
at the address below.

ASM International
9639 Kinsman Road
Materials Park, Ohio 44073-0002
USA

Phone: +1 440.338.5151

memberservicecenter@asminternational.org

Additional copies of this publication are available from:

Curran Associates, Inc.
57 Morehouse Lane
Red Hook, NY 12571 USA
Phone: 845-758-0400
Fax: 845-758-2633
Email: curran@proceedings.com
Web: www.proceedings.com

TABLE OF CONTENTS

Proceedings of the 26th ASM Heat Treating Society Conference
B.L. Ferguson, R. Jones, D.S. MacKenzie, and D. Weires, editors
https://doi.org/10.31399/asm.cp.ht2011fm01

Preface

Heat treating is a core competency and a vital part of the manufacturing community. Nowhere was this more evident than in the gathering of heat treaters, engineers, scientists, managers, and industry leaders who attended the 26th ASM International Heat Treating Conference and Exhibition held in Cincinnati, Ohio, USA on October 31–November 2, 2011.

Heat treating is a vibrant and growth oriented industry driven by the needs of its users to satisfy the demand for increased product performance, new technology developments and innovation. This event gave all those who participated a unique opportunity to network, exchange ideas, and discuss critical issues so as to ensure that Heat Treating remains the most cost effective technology.

The Heat Treating Society Conference was designed around a "Heat Treat University" theme, with sessions for beginners through experts. The topics range from "fundamentals" through "emerging technologies." Special events included the William (Bill) R. Jones Honorary Symposium — celebrating the life and achievements of William R. Jones, founder of Vacuum Furnace Systems Corporation, Solar Atmospheres and Solar Atmospheres Manufacturing. Mr. Jones' entire career has been associated with the development and application of electro-mechanical products since 1962 with the vacuum heat treating furnace. He holds numerous patents in this field and has long been associated with ASM International and related technical organizations.

This Conference was a joint effort and we wish to especially thank the members of the Heat Treating Society Board, the Heat Treating Society Technology and Programming Committee, and the Exposition Committee for their inspiration, guidance and tireless efforts to identify, solicit, and encourage industry experts to share their knowledge and expertise to create the finest program on Heat Treating in the world. We as individuals and as an industry profit from the effort of these volunteers and the hundreds of hours they devote to advancing the science of Heat Treating.

These papers represent the latest state-of-the-art in heat treating materials, processes, products, equipment, and methodology. The topics cover the depth of our industry and will be a valuable reference source for years to come.

The opportunity to Co - Chair this Conference and Exposition as well as to bring you the 2011 Heat Treating Proceedings has truly been our pleasure and privilege. Thank you for allowing us this opportunity.

B. Lynn Ferguson
Founder and President
Deformation Control Technology, Inc.

Roger Jones
Corporate President
Solar Atmospheres Incorporated

Dale Weires
Technical Fellow
The Boeing Company

Proceedings of the 26th ASM Heat Treating Society Conference
B.L. Ferguson, R. Jones, D.S. MacKenzie, and D. Weires, editors
https://doi.org/10.31399/asm.cp.ht2011fm02

Heat Treat 2011 Organizing Committee

Proceedings of the 26th ASM Heat Treating Society Conference
B.L. Ferguson, R. Jones, D.S. MacKenzie, and D. Weires, editors
https://doi.org/10.31399/asm.cp.ht2011p0001

Evaluation of process control methods for nitrogen-hydrocarbon atmospheres

Xiaolan Wang[1]; Zbigniew Zurecki[2]; and Richard D. Sisson, Jr.[1]
[1]Worcester Polytechnic Institute, Worcester, Mass, USA
evawang824@hotmail.com and sisson@wpi.edu, 508-831-5335
[2]Air Products& Chemicals, Inc., Allentown, Pennsylvania, USA
zureckz@airproducts.com,

Abstract

Atmospheric pressure carburizing and neutral carbon potential annealing in nitrogen containing small additions of hydrocarbon gases can offer cost and steel surface quality alternatives to the comparable, endothermic atmosphere or vacuum operations. An experimental program was conducted for refining real-time process control methods in carburizing of AISI 8620 steel under N_2-C_3H_8 blends containing from 1 to 4 vol% of propane at 900°C and 930°C. Multiple types of gas analyzers were used to monitor residual concentrations of H_2, CO, CO_2, H_2O, O_2, CH_4, C_3H_8, and other hydrocarbons inside furnace. A modified shim stock technique and the conventional oxygen probe (mV) were additionally evaluated for correlation with gas analysis and diffusional modeling using measured carbon mass flux values (g/cm^2/s). Results of this evaluation work are presented.

Introduction

Conventional carbon-containing atmospheres used in carburizing are generated in endothermic generators, external to heat-treating furnaces and, frequently, adjusted to match process requirements by mixing hydrocarbon gases (HC) such as methane (CH_4), propane (C_3H_8), propylene (C_3H_6), acetylene (C_2H_2), and/or nitrogen (N_2) with air.[1] Since endothermic gas with air forms hydrogen (H_2), N_2, and carbon monoxide (CO), with minute quantities of water vapor (H_2O), and carbon dioxide (CO_2). The conventional atmospheres have a potential to carburize the main steel and simultaneously oxidize iron and alloying additions, e.g. chromium (Cr), manganese (Mn), silicon (Si) or vanadium (V). And similar oxidizing-carburizing effects are observed in alternative, dissociated alcohol atmospheres, e.g. N_2-methanol and N_2-ethanol. [2-4]

Nitrogen-hydrocarbon atmospheres which applied in carburizing and neutral carbon potential annealing operations, [5-8] hold the promise for matching the quality of work parts processed in vacuum furnace, i.e. eliminating the intergranular oxidation existed in the conventional, endo-generated atmospheres. Moreover, N_2-HC blends are safer, non-toxic/less-flammable atmospheres. Although occasionally used in atmospheric pressure furnaces, the N_2-HC atmospheres are underutilized due to insufficiently developed process control methods and models of secondary reactions with air leaking to typical atmospheric furnaces. However, the process control of nitrogen-hydrocarbon atmosphere is difficult due to the non-equilibrium process and less precise control systems in atmosphere pressure furnace.

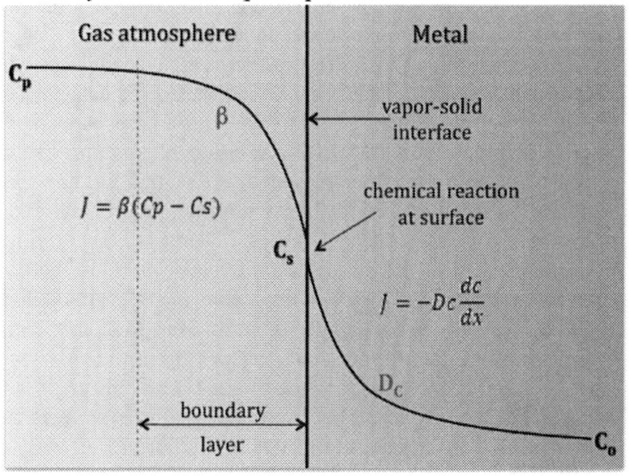

Figure 1 Schematic representation of carbon transport in carburizing

The mass transfer mechanism during gas carburizing/nitriding is a complex phenomenon which involves three stages: (1) carbon/nitrogen transport from the atmosphere to the steel surface, (2) chemical reactions at the surface, and (3) diffusion of the absorbed carbon/nitrogen atoms into the bulk of the steel. Figure 1 schematically shows the mechanisms of carbon transfer during carburizing and the primary control parameters: the mass transfer coefficient (β) defining carbon atoms flux (J) from the atmosphere to the steel surface and the coefficient of carbon diffusion in steel (D) at process temperatures. In this research, by using plasma activation and hydrocarbon gas as the process gas, carbon potential (Cp) is increased from normally 0.8-1.2% (endogas atmosphere)[1] to infinite. And the mass transfer coefficient is also increase by the ionized species which activated steel surface and/or removed oxidation barrier. As the result, carburizing/nitriding can be accelerated. But due to the nature of non-equilibrium N_2-hydrocarbon gas atmosphere, the process control is more challenging compared to conventional endothermic atmospheres. In the former, both the surface carbon concentration and the carburized depth increase simultaneously with the carburizing time. [6] In the latter, the surface carbon concentration is fixed at the level of carbon potential (Cp) so that an increasing carburizing time increases the carburized depth only. [9-10] The process control is even more difficult in the 1-atm-pressure furnaces. It's less precise than vacuum furnaces and involves a number of additional, sometimes uncontrollable processing variables such as air and combustible gas leakage or moisture desorption. [11] The development of heat treating recipes may require more trials than in the case of vacuum furnaces, and the carburizing cycle including carbon boost and diffuse may necessitate real-time, dynamic corrections to the processing parameters using a feedback loop.However, carbon diffusion from the surface into the steel core is based on the same mechanism in both these cases, namely Fick's law,

$$J = -D\ dC/dx \qquad (1)$$

where D is carbon diffusivity in the alloy, cm^2/s, C is carbon concentration and x is depth beneath surface. The carbon diffusivity is controlled by temperature, steel composition, and to a lesser degree by local carbon concentration. [9-10]

For endogas atmosphere process control, metal coupon, metal foil and/or shim stock were widely used for determining Cp in the conventional, equilibrium atmosphere carburizing operations.[11] Since the surface carbon concentration cannot exceed Cp, the method involves a very thin steel foil and long exposure time to saturate the metal throughout and achieve a constant C-concentration profile across the width. Consequently, the measurement of weight gain of the foil directly indicates atmosphere Cp. For N_2-HC atmosphere control, it's more challenge. It normally used the same approach as vacuum atmosphere, by controlling the flowrate and concentration of the process gas. And many in-situ sensors have been developed over the years to address the difficulties of process control in non-equilibrium as well as equilibrium atmospheres by testing the electrical resistance of carburized samples which directly related to carbon concentration. [12-14]

Experimental procedure

Atmosphere carburizing experiments were run in a semi-production scale, electrically heated box furnace (ATS 3350). Process gas analyses were performed by Las gas analyzer (manufactured by ARI, model LGA-4ENAPBT) for CO, CO_2, H_2, HC, by dewpoint meter for H_2O and by ZrO_2 probe for O_2. AISI 8620 coupons, φ1 x 0.5 in (φ2.5 x 1.2 cm) were used in the tests, Table 1 presents the specimen composition used in the test, result obtained by optical emission spectroscopy (OES).

Table 1: AISI 8620 steel composition (wt. %)

C	Mn	Si	Ni	Cr	Mo	Cu	Fe
0.2	0.69	0.194	0.62	0.61	0.212	0.131	Bal

Carburization of these specimens was performed according to the conditions given in Table 2. Tests 1-2 were performed in AC-plasma activated gas atmosphere. The cold plasma, stream-activating injector was used to activate process gas, equipment details were described elsewhere.[15-17] Tests 3-4 were non-activation, thermal carburizing tests. The atmosphere used in the tests were C_3H_8 (< 4 vol%) in N_2-stream with a flowrate of 250 scfh, 3 volumes change per minute in the box furnace. The nitrogen gas used to balance total gas stream has 99.995% purity. Each carburization cycle involved 45min long heating period from room to treatment temperature under pure N_2, 2 or 3-hour carburizing step and quenching in oil at room temperature. The endo-atmosphere coupon which used to compare with T1-4 samples was produced under following condition: the parts were loaded to hot furnace at 900°C with the carbon potential of 0.95 wt%C for 2.5 hours in boost stage, followed by 0.5 hour diffuse stage at Cp of 0.8-0.9 wt% at 843°C, than quenched in the oil and tampered at 180°C for 2hours. The N_2-4.5%CH_4 coupon were carburized at 900°C for 3hours in non-activated methane nitrogen blend gas and quenched from 843°C. More details were described in previous paper. [16]

The specimens were weighted before and after the carburizing cycle with a conventional microbalance, accuracy of 0.1mg. Weight gain, m, were used to determine carbon flux Jt. Microhardness, OES analyses and SEM test were performed on coupons afterwards. Vickers hardness on the cross-sections, 100g@10s, was taken for the fully carburized, quenched, non-tempered or tempered AISI 8620 coupons. Metallographic cross-sections of the coupons were etched with 2% Nital prior to optical microstructural and SEM examination. Residual austenite was measured by XRD on T_3 coupon at 100, 350 and 800μm depth. Carbon profile was tested by OES (SPECTRO MAXx M, manufactured by SPECTRO Analytical Instruments).

Table 2: Carburizing conditions for AISI 8620 test coupons

Test No.	T1	T2	T3	T4
Carburizing				
Temperature (°C)	900	900	900	930
Quenching T (°C)	843	843	843	860
Carburizing Time(hr)	3	3	3	2
Plasma activation	Yes	Yes	No	No
Gas flowrate, scfh (Nm^3/h at 0°C)				
Total gas flowrate	250 (6.7)	250 (6.7)	250 (6.7)	500 (13.4)
N_2-thru-plasma	245 (6.6)	240 (6.4)	0	0
C_3H_8-thru-plasma	5 (0.1)	10 (0.3)	0	0
Furnace inlet CH_4 (vol %)	2	4	2	0.9
Tempering				
Temperature (°C)	180	180	180	-
Time (hr)	2	2	2	-

Result and discussion

Effect of process atmosphere on the carbon flux

The average carbon flux J_t of N_2-C_3H_8 atmospheres for 2 or 3 hours process time, calculated from the weight gain data, was listed in Table 3. It was higher than conventional endogas atmosphere [18] or vacuum furnace carburizing with C_3H_8 [19]. Thus, the plasma activated conditions and higher propane concentration didn't affect carbon uptake much. Within same temperature and process time, T_1 and T_2 with plasma activation, have the similar carbon flux, even when the propane concentration is doubled for T_2 condition. And for T_1 and T_3, the flux is also the same, with or without activation. So, compared to inactive hydrocarbon methane, the active propane gas has enough potential to carburize steel without plasma energy stimulation. The carbon flux was limited by the diffusivity of carbon into austenite, and more related by the process temperature than carburizing potential in the atmosphere. And it also has been noticed, in the semi-scaled furnace, with limited gas circulation system and short residence time, 0.9% C_3H_8+ N_2 atmosphere was not sufficient enough to produce a uniform thickness carburizing layer.

Table 3 Carbon flux and surface hardness for different test conditions

Test No.	T1	T2	T3	T4
Weight gain per unit area due to carburization for AISI 8620 steel coupons, ΔWt, g/cm^2				
ΔWt	0.00336	0.00349	0.00349	0.00341
Time averaged carbon flux calculated from AISI 8620 steel coupons, J_t, g/cm^2/sec				
J_t	3.1E-7	3.2E-7	3.2E-7	4.7E-7
Surface hardness for AISI 8620 coupons, after quenching, before tempering.				
HRC	64.5±0.5	63.3±0.3	64.8±0.3	61.8±0.3

Microstructure, profiles of carbon concentration and microhardness of carburized layers

Cross-section microstructures of carburized surface layers produced at 900°C are shown in Figure 2-4. Microstructural analysis of the carburized test coupons revealed carbides within the first 20μm carburized layer and a mixture of martensite and retained austenite near the surface and a mixture of martensite and bainite in the core. From the SEM-SEI pictures, about 5μm large martensite can be observed along with the carbides at the subsurface area. In the 200-500μm depth area, which has the highest as-quench hardness range, very fine grain (~ 1μm) structure was detected.

While martensite is the desired phase in a carburized case, a large amount of retained austenite, about 53% was also detected at 100μm depth, which resulted from direct oil quenching at 843°C. In the maximum microhardness area, a mid-level of the retained austenite (15-25%) was observed.

The carbon profile is measured from 0-600μm depth, the very surface contained high concentration carbon (>1.6 wt%) for all the conditions. It was expected, using C_3H_8 as carbon source, graphite may generated and with extremely high carburizing potential, cementite may exit at the surface. As shown in Figure 3, carbon content is dropped significantly at 50μm depth to 0.9 wt% C and flattened into the core. By integrated carbon concentration profile, the result of 0.03597g/cm^2 weight gain matched well with weight gain value 0.0349g/cm^2, measured by microbalance. After 3 hours carburizing at 900°C, AISI 8620 steel generated a 520μm case depth (case depth defined as carbon concentration drops to 0.5 wt%).

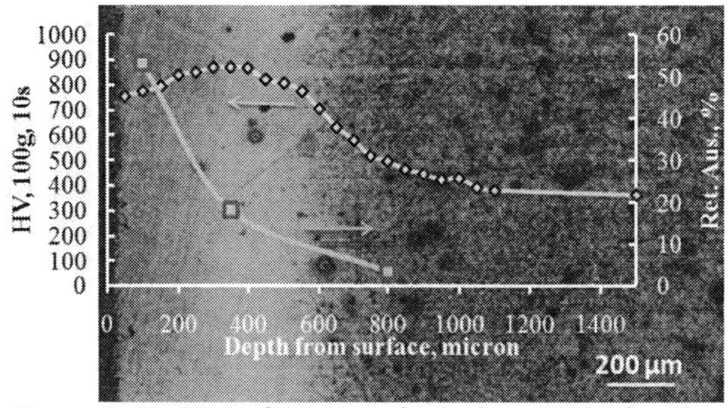

Figure 2 T3 (N_2-2vol%C_3H_8 carburized at 900°C for 3hrs, without plasma activation) oil quenched from 843°C, not tempered AISI 8620 steel microstructure, microhardness and residue stress profile.

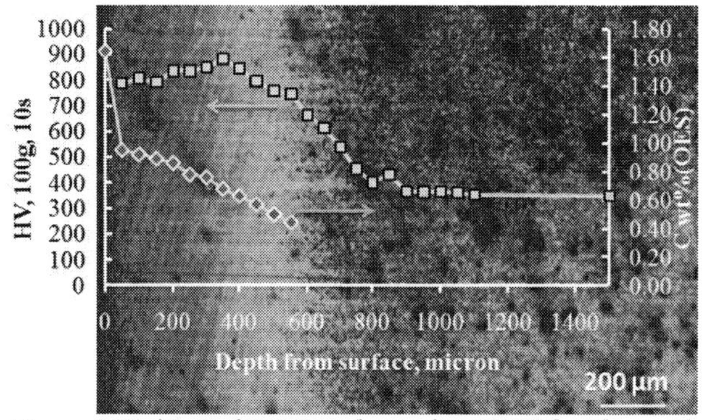

Figure 3 T1 (N_2-2vol%C_3H_8carburized at 900°C for 3hrs, with plasma activation) oil quenched from 843°C, not tempered AISI 8620 steel microstructure, microhardness and carbon profile.

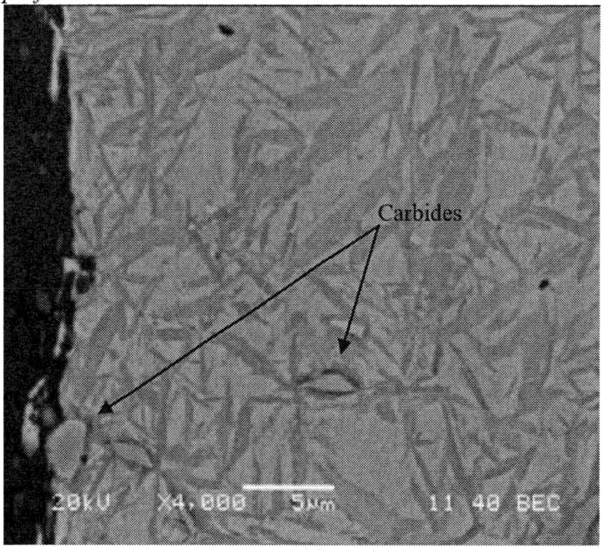

(a)

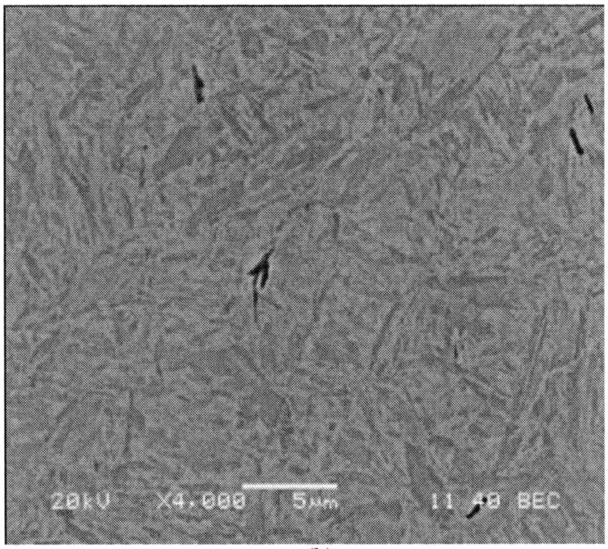

(b)

Figure 4 SEM-SEI cross-sectional images of T1 AISI 8620 coupons after carburizing and quenching cycle. Etched in 2% Nital. (a) sub-surface region; (b) ~350μm depth region.

Comparison of different carburizing atmosphere

The surface hardness HRC of the carburizing parts was measured and found to be identical within a narrow range of measurement error, listed in Table 3. Cross-sectional microhardness was plotted in Figure 5 and also displayed the similarity. As-quenched AISI 8620 sample, started with 700-800 HV surface hardness, and reached the peak at 400μm, then with a sharp drop-off into the core area. The lower hardness at very surface is due to high concentration retained austenite of more than 50%. And these profiles also support carbon flux measurement that 2% C_3H_8 was sufficient to carburize test coupons in the lab furnace. However, during industrial production, the parts per load have enormous surface area compare to lab tests, and atmosphere residence time and circulation also changed, so more than 2% C_3H_8 may be needed depend on different variables.

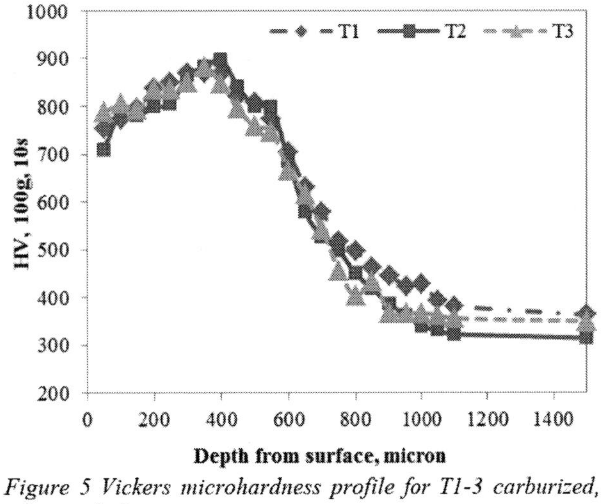

Figure 5 Vickers microhardness profile for T1-3 carburized, oil quenched parts.

The microhardness after tempering at 180°C for 2 hours is presented in Figure 6. From this figure, endogas, 4.5% methane with nitrogen atmosphere samples were compared with T3 (2% propane, thermal run). The same process temperature and schedule were followed for those tests. After tempering, the peak hardness at 400μm for as-quenched condition was dropped from HV 900 to 740. Fine structure martensite turned into tempered martensite, that resulted the hardness dropping and less brittle structure was formed. Compared with methane and endogas samples, the case depth and hardness are all improved by using propane atmosphere, higher microhardness was obtained at 0-400μm working zone, and case depth (defined as hardness drops HV 500) was also increased about 100μm. Overall, the HC atmosphere samples displayed a higher hardness level going deeper into the part with a sharper drop-off in the core area than the endogas samples. This type of hardness profile is desired, in the case of parts requiring an additional surface finishing by machining for restoring dimensional accuracy.

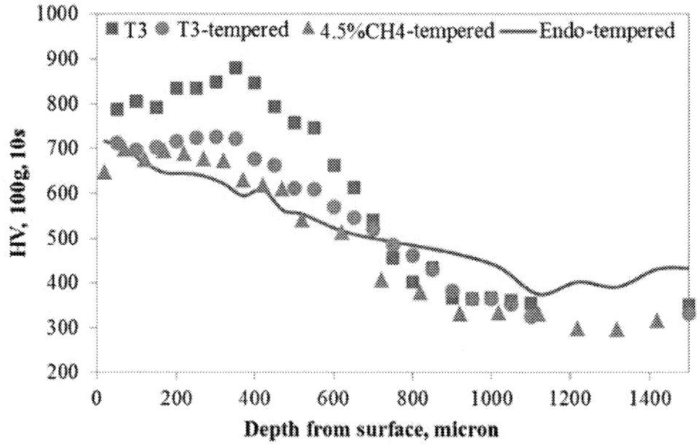

Figure 6 Vickers microhardness profile for T3, 4.5%CH_4 and endogas carburized, oil quenched and tempered and T3 non-tempered parts.

Process control

Numerous gas products are formed in the N_2-C_3H_8 blend in the furnace. They include CH_4, C_2H_2, C_2H_4, C_3H_6, H_2, N_2 and some by-product when HC gas reacts with residual O_2 in the furnace. [20] After preliminary experiments, it was observed that the H_2 concentration in the furnace effluent is the most sensitive real-time process measurement. The changes of H_2O, CO_2, CO or HC are not as directly connected to carbon flux into steel as that of H_2, shown in Figure 7-8. It shows that other process indicators change only within several hundred ppm range or may, like ZrO_2-probe, become affected by catalyzed carbon deposits. Through the correlation between carburizing ability and hydrogen concentration can be changed by process variables. In most case, the H_2 content can still be used to determine the carburizing effectiveness, expect for some rare situations. For example, heavy hydrocarbon impurities in the feed gas will result an increase of carburizing ability, but appear as a decrease for hydrogen concentration in furnace exhaust gas.

During the receipt development, modified shim stock methods (Appendix) can be used, while H_2 readings were recorded for future reference. Then in the following carburizing cycles, H_2 sensor can be used to monitor the atmosphere solely, and by adjusting inlet hydrocarbon concentration to match the pre-recorded H_2 concentration, the carburizing atmosphere can be maintained as same as the previous run to duplicate the results. Conventional ZrO_2-probe can be used in the furnace conditioning stage to monitor the O_2 purge out rate and for safety control during carburizing.

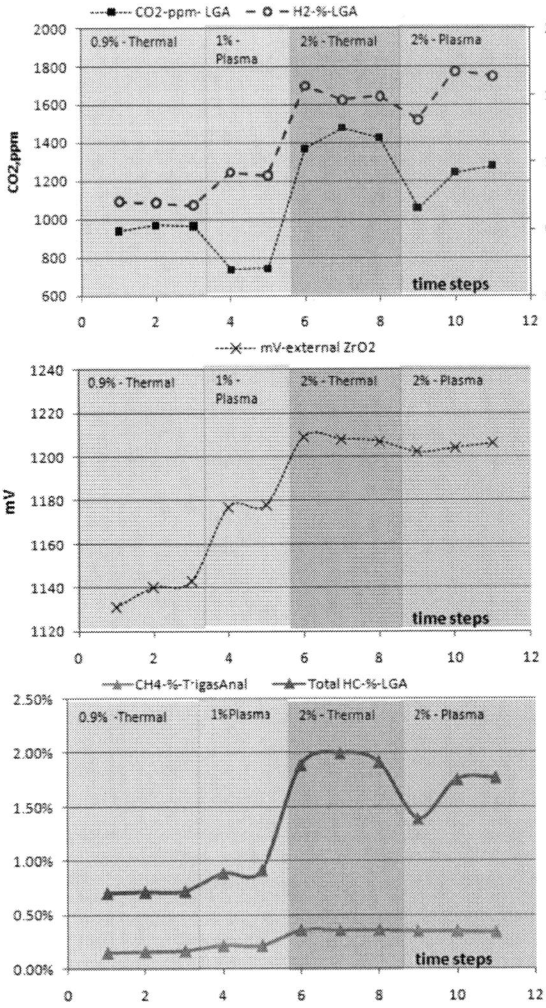

Figure 7 Correlations between effluent gases and external zirconia probe readings during carburizing tests involving N_2-0.9%C_3H_8, N_2-1%C_3H_8, and N_2-2%C_3H_8 atmospheres at 930°C with plasma discharge activation and w/o it, using conventional, thermal-only activation.

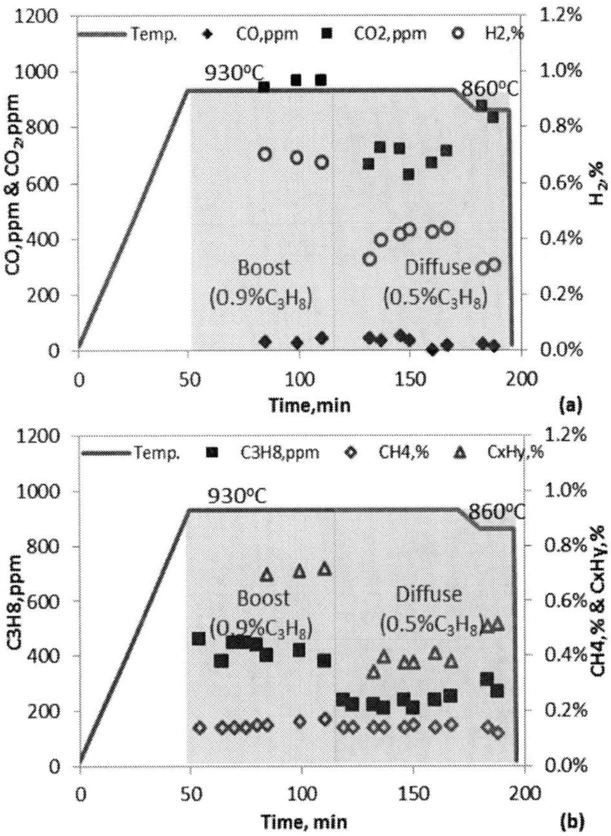

Figure 8 Atmosphere concentrations during T4 process. (a) CO, CO_2 and H_2 gases; (b) Hydrocarbon gases.

Industrial trial

Production scaled test was run at commercial heat treating facility. DIN16MnCr5 steel (equivalent to AISI 5115) parts were used in testing, with the technical target of producing a 0.68 wt%C at the surface and 0.3 wt%C at 0.6 mm case under the surface in a 90 minutes 930°C-boosting/930-860°C diffusing cycle. Due to ingress of oxygen into IQ furnace during carburizing operation, the N_2-C_3H_8 atmosphere becomes 'transitional' between the conventional, non-equilibrium atmosphere characterizing vacuum furnace carburizing and the endothermic gas-based, atmosphere. Although transitional N_2-C_3H_8 atmospheres may, in principle, produce co-existing, intergranular carbides(IGC) and oxides(IGO) during quenching, following carburizing, the actual size of those products is negligible (less than 3-5 μm into surface, shown in Figure 9) because of the equilibrium nature and very limited quantity of the oxygen-containing gases (i.e. CO, H_2O, and CO_2) available for reaction.

(a)

(b)

Figure 9 Optical microstructure graph for coupons, shown (a) IGO/IGC and (b) cementite.

Conclusions

1. The steel carburizing process in atmosphere pressure furnaces involving non-equilibrium atmospheres containing propane gases was evaluated. Measurements of carbon mass flux and calculations of carbon potential in gas phase have shown that the present carburizing rates are comparable to those of low-pressure (vacuum) and endothermic atmosphere carburizing systems.
2. Carburizing effects were compared for the AISI 8620 steel coupons processed with the N_2-C_3H_8, N_2-CH_4 mixture and the conventional endothermic atmosphere using the same heat treatment schedule. The peak hardness and case depth for N_2-C_3H_8 samples were improved compared with N_2-CH_4 sample. The microhardness profile directly under metal surface was relatively flat, similar as by low-pressure carburizing, and beneficial from the post-machining and fatigue strength standpoint.
3. Modified shim stock method and probe can be used for determining carbon flux from atmosphere into metal combined with diffusion calculations for carbon concentration profile at and under metal surface, described in Appendix. Carbon flux measurements can be correlated with H_2 concentration and, optionally, with other gas sensors. Controlling hydrocarbon gas concentration during the subsequent, production operations, where carbon flux measurements are no longer used as long as the HC additions result in the same H_2 during the recipe development run.

Acknowledgments

The authors would like to thank J.L. Green for laboratory support and Air Products for funding and the permission to publish this study.

References

[1] R.L. Davis *et al*, U.S. Patent 4,049,473
[2] An, X. *et al*, "A study of internal oxidation in carburized steels by glow discharge optical emission spectroscopy and scanning electron microscopy", *Spectrochimica Acta* Part B 58 (2003) 689–698
[3] Chatterjee-Fisher, R., "Internal Oxidation During Carburizing and Heat Treating", *Metallurgical Transactions* Vol. 9A, November 1978, pp.1553-1560
[4] Asi, O., *et al*, "The relationship between case depth and bending fatigue strength of gas carburized SAE 8620 steel", *Surface & Coatings Technology* 201 (2007), pp. 5979–5987
[5] Kaspersma, J.H., and Shay, R.H., "Carburization and Gas Reactions of Hydrocarbon-Nitrogen Mixtures at 850°C and 925°C", *Metallurgical Transactions B*, Vol. 13B, June 1982, pp. 267-273.
[6] Estrin, B.M, *et al*, "Carburizing in a nitrogen-based mixture with additives of pure methane", *Metallovedenie i Termicheskaya Obrabotka Metallov*, No. 5, pp. 26-29, May, 1984
[7] Connery, K. and Ho, S., "Optimization of Oxygen-free Heat Treating", *Proc. of the 24th ASM Heat Treating Society Conf.*, September 17-19, 2007, COBO Center, Detroit, Michigan, USA
[8] Baldo *et al*, U.S. Patent 4,992,113
[9] Karabelchtchikova, O. and Sisson, R.D. Jr., "Calculation of Gas Carburizing Kinetics from Carbon Concentration Profiles based on Direct Flux Integration", *Defect and Diffusion Forum Vol. 266*,(2007), pp. 171 - 180.
[10] Karabelchtchikova, O. and Sisson, R.D. Jr., "Carbon diffusion in steels: A numerical analysis based on direct integration of the flux", *Journal of Phase Equilibria and Diffusion*, Volume 27, Number 6,(2006) p598-604
[11] Herring, D.H., "Furnace atmosphere analysis by the shim stock method", *Industrial Heating,* Sept (2004)
[12] P. Beuret, U.S. Patent 5,064,620
[13] L.G. Chedid *et al*, U.S. Patent 7,068,054
[14] Winter, K.M., "A Guide to Better Atmosphere Carburizing Using Both Dynamic and

Equilibrium-Based Measurements", *Industrial Heating*, Oct.(2008)
[15] Z. Zurecki *et al*, U.S. Patent 2008/0283153
[16] Zurecki, Z and Wang, X, "Atmosphere carburizing using electric discharge-activated nitrogen-natural gas mixtures," *Heat Treating Conference and Exposition*, Indianapolis, Indiana, Oct 2009.
[17] Zurecki, Z., "Heat Treating Atmosphere Activation", , *Proc. of the 24th ASM Heat Treating Society Conf.*, Detroit, Michigan, Sept. 2007.
[18] Linde Gas, Special Edition, "Furnace Atmospheres No. 1, Gas Carburizing and Carbonitriding", url: https://b2.boc.com/catweb/CATweb.nsf/noteid/EC84EBA1ADCB86EC802572C1004B3977/$file/SpEd_Carburizing_and_Carbonitriding.pdf, last accessed: March 24, 2009
[19] Altena, H., and Schrank, F., "Low Pressure Carburizing with High Pressure Gas Quenching", *Gear Technology*, March/April 2004, pp.27-32
[20] R.U. Khan *et al*, Pyrolysis of propane under vacuum carburizing conditions: An experimental and modeling study, *J. Anal. Appl. Pyrolysis*, 81 (2008) 148–156

Appendix

The following section describes a procedure for estimating carbon flux into steel during carburizing operations in non-equilibrium atmospheres. Modified shim stock methods can be used to determine the carbon flux into parts in-situ. Tube shape samples were used and only OD was exposed to the carburizing atmosphere. The wall thickness, W, and carburizing exposure time, t, are selected in such a way that the unexposed ID side is not yet carburized by the flux of carbon atoms flowing from the exposed side. Weight gain due to carburizing was measured by using conventional microbalance.

The procedure requires several coupons insertion into the furnace, for different time periods, such as t_1, t_2 and t_3. The formulas for calculating the fluxes from the three weight gain measurements are listed in Table 4, where m_1, m_2, and m_3 are the gains at the end of each exposure time, t' times are the times assigned for specific weight gains, and J are the averaged flux values associated with the t' times.

The following assumptions were made to simplify the procedure and calculation. In a short period of time, carbon flux and time have a liner relationship; during the whole carburizing time, process temperature and atmosphere were maintained the same.

Figure 10 shows the typical weight gains registered by 3 metal coupons exposed to the carburizing atmosphere for t_1, t_2 and t_3. The line connecting the weight gain datapoints measured reflects the decreasing rate in view of increasing carbon concentration at the coupon surface. The m value obtained during the t_1 exposure, i.e. m_1, is the average gain associated with the middle of the exposure time, i.e. $t_1'=1/2t_1$. The same operation can be repeated for the longer exposure times, but it should be noted that the longer the exposure time is, the larger error results from associating the average gain with the half of the exposure time used.

Table 4: Calculation of average fluxes for exposure times

Data used	Time	Carbon flux
$t_1, \Delta m_1$	$t_1' = \frac{t_1}{2}$	$J(t_1') = \frac{\Delta m_1}{A \cdot t_1}$
$t_2, \Delta m_2$	$t_2' = \frac{t_2}{2}$	$J(t_2') = \frac{\Delta m_2}{A \cdot t_2}$
$t_1, t_2, \Delta m_1, \Delta m_2$	$t_{1\sim2}' = \frac{t_1 + t_2}{2}$	$J(t_{1\sim2}') = \frac{\Delta m_2 - \Delta m_1}{A \cdot (t_2 - t_1)}$
$t_3, \Delta m_3$	$t_3' = \frac{t_3}{2}$	$J(t_3') = \frac{\Delta m_3}{A \cdot t_3}$
$t_1, t_3, \Delta m_1, \Delta m_3$	$t_{1\sim3}' = \frac{t_1 + t_3}{2}$	$J(t_{1\sim3}') = \frac{\Delta m_3 - \Delta m_1}{A \cdot (t_3 - t_1)}$
$t_2, t_3, \Delta m_2, \Delta m_3$	$t_{2\sim3}' = \frac{t_2 + t_3}{2}$	$J(t_{2\sim3}') = \frac{\Delta m_3 - \Delta m_2}{A \cdot (t_3 - t_2)}$

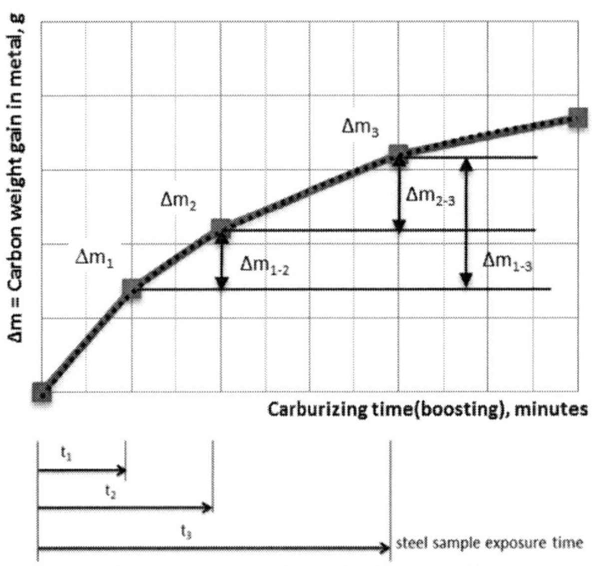

Figure 10 The weight gains of samples due to carburizing.

Carbon mass flux, J, is calculated by dividing weight gain m by coupon surface area exposed to the atmosphere, A, and by the exposure time interval, t, which means that the measured datapoints can be quickly converted into carbon flux values. Figure 11 shows the carbon fluxes recalculated from the weight gains resultant from the three original measurements using the procedure of extracting the additional data described above and the curve fitting obtained. Six J-flux datapoints can be fitted with a power function curve of the general type: $J=at^b$, since carbon flux into metal core typically decays during carburizing and C-saturation according to such a relationship. Here, a and b are constants, and t is running time of the carburizing (boosting) cycle.

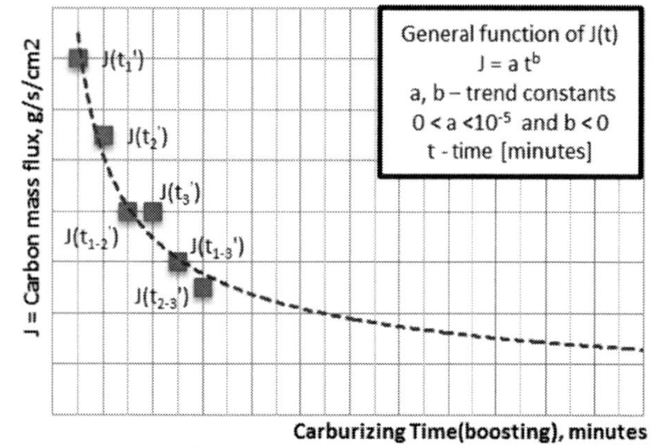

Figure 11 Estimation of average carbon flux using modified steel coupon probes

The fitted curve represents time dependant flux value and, in the next step, can be extrapolated up to the maximum carburizing time of interest, e.g. to 60 minutes, if 60 minutes was the original boosting time intended for the analyzed operation. Thus, in the next step, the average flux for the 60 minute boosting can be calculated using the same, offline computer spreadsheet by averaging the value integrated under this fitted curve.

An offline diffusion software package, e.g. "CarbTool", by Worcester Polytechnic Institute, is needed in the final step to evaluate carbon profile generated by the carburizing process. Carbon profile can be predicted by entering the temperature, time, carbon flux and other parameters. By completing this procedure, the surface carbon and carbon depth of the products can be estimated and evaluated.

Proceedings of the 26th ASM Heat Treating Society Conference
B.L. Ferguson, R. Jones, D.S. MacKenzie, and D. Weires, editors
https://doi.org/10.31399/asm.cp.ht2011p0009

Independently Controlled Carbon and Nitrogen Potential - A New Approach to Carbonitriding Processes -

Karl-Michael Winter
PROCESS-ELECTRONIC GmbH, Heiningen, Germany
km.winter@process-electronic.com

Abstract

Late research projects show that retained austenite, if stabilized by nitrogen, has a positive influence on the fatigue strength of work pieces. The combined diffusion profile of carbon and nitrogen applied in a carbonitriding process plays the major role, besides the process temperature. Yet today, only the carbon potential is somehow controlled and even this is not easy to achieve.

This paper will present a new system able to measure and control both, the carbon potential and the nitrogen potential independently. The knowledge of the activities of nitrogen and carbon in iron and the effect of alloying elements on such activities as well as the solubilities offers an easy to use method to apply the potentials on real steels.

Introduction

Gaseous carbonitriding is a termochemical treatment similar to gas carburizing. The difference consists of an addition of ammonia to the process gas which adds nitrogen to the carburized case. The additional nitriding effect is analog to a nitriding process, where the nascent nitrogen is produced by thermal dissociation of the ammonia on the catalytic surface of the work pieces. The nitrogen then diffuses simultaneously with carbon into the steel.

Nitrogen being present in the interstitial profile of carbon has the advantage of an increased hardenability and wear resistance as well as a higher temperature stability of the heat treated parts /1/. As nitrogen stabilizes the retained austenite it also has a positive effect on form stability and fatigue strength of gears /2/. In addition the increased hardenability enables to oil quench machined parts made from unalloyed steels in order to produce strong and wear resistant parts economically.

As the ammonia will eventually completely dissociate into hydrogen and nitrogen at carburizing temperatures and as on the other hand the nitrogen pick-up is driven by the content of still undissociated ammonia, the fundamental problem in a controlled carbonitriding process is that it cannot be performed properly by applying a certain percentage of ammonia in the inlet gas.

The higher the carbonitriding temperature and the higher the combined catalytic surface of load and furnace walls, the faster the ammonia will dissociate. The lower the temperature, the smaller the reactive surface and the higher the total process gas flow at a given ammonia percentage on the inlet, the less ammonia will dissociate and the higher will be the nitriding effect.

The carbon pick-up is as well affected, as with most of the ammonia decomposing, an addition of ammonia to a carburizing atmosphere has the effect of a dilution with nitrogen and hydrogen further affecting the measurement of the carbon potential. In addition the presence of nitrogen in the steel also influences the carbon activity and therefore the carbon pick-up.

Carburizing and nitriding reactions

Gaseous carbonitriding is performed in an endothermic atmosphere consisting of carbon monoxide, hydrogen and nitrogen and smaller percentages of uncracked hydrocarbons, carbon dioxide and water vapor with an addition of ammonia. Bischoff /3/ presented a graph displaying the concurrently active carburizing and nitriding reactions in this mixture (fig. 1). It can be seen that besides the normal carburizing reactions and the nitriding reaction out of the ammonia dissociation there is also an additional carbonitriding reaction out of cyanide that is created by a reaction of carbon monoxide and ammonia.

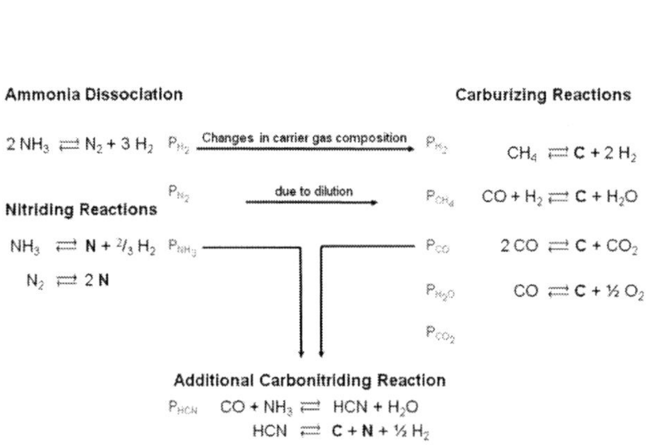

Fig. 1: Carburizing and nitriding reactions /3/

Bischoff /3/ as well calculated the varying gas percentages in a carbonitriding atmosphere in equilibrium as a function of temperature (fig. 2).

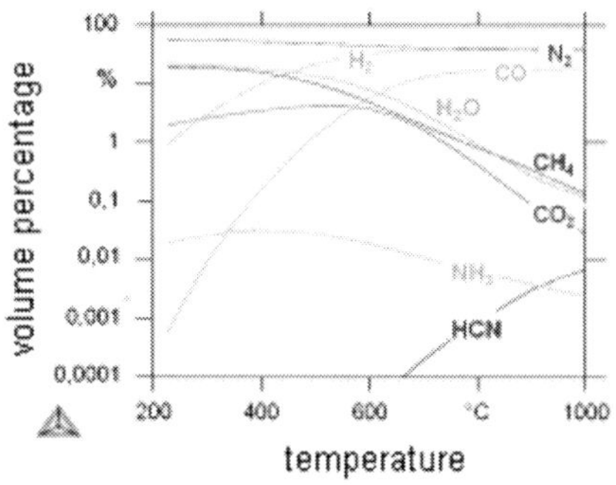

Fig. 2: equilibrium gas percentages in a carrier gas with 20% CO, 40% H_2 and 40% N_2 with an NH_3 addition of 5% /3/

It can be seen that with increasing temperature the partial pressure of ammonia decreases whereas the partial pressure of cyanide increases. If we assume the partial pressures being equal at approximately 1650°C (900°C) we would expect a not negligible nitriding effect out of the cyanide reaction.

$$HCN \rightarrow [C] + [N] + H_2$$

A calculation of the nitrogen activities of NH_3 and HCN at a temperature of 1650°F (900°C) shows that they would give approximately the same nitriding effect. In reality, as in a carburizing furnace there is a comparably rapid gas exchange the dissociation of the ammonia and therefore as well the calculated equilibrium amount of cyanide will never be reached. Typical fractions of residual ammonia measured in such atmospheres will range between 500 and 5000 ppm, well above the calculated ammonia percentages of less than 100 ppm. Therefore the cyanide reaction will not be taken into account in the further document.

If we further assume that the carburizing will mostly be driven by the heterogeneous water gas reaction

$$CO + H_2 \rightarrow [C] + H_2O$$

we can simplify the set of carbonitriding formulas to

$$\lg(a_{Catm}) = \lg(pCO*pH_2/pH_2O) + 7130/T - 7.494 \qquad /4/$$

and

$$\lg(a_{Natm}) = \lg(pNH_3/p^{1.5}H_2) - 2943/T + 6.196 \qquad /4/$$

All partial pressures are given in bar, temperatures in Kelvin.

Alternatively, if we assume CO, CO_2, H_2 and H_2O staying in water gas equilibrium, it is possible to use the carbon activity out of the Boudouard reaction

$$2\,CO \rightarrow [C] + CO_2$$

with

$$\lg(a_{Catm}) = \lg(p^2CO/pCO_2) + 8861/T - 9.107 \qquad /4/$$

Note, that in atmospheres with high residual methane percentages the total carbon activity of the process gas will be affected by the methane reaction and has to be determined by solving the equation

$$J_C = k_{ch4}*(a_{Cch4}-a_{Cfe}) + k_{Boud}*(a_{CBoud}-a_{Cfe}) + k_{het}*(a_{Chet}-a_{Cfe})$$

The factors k_{ch4}, k_{Boud} and k_{het} are the carbon transfer coefficients for the three carburizing reactions. These effects have been described in detail by Collin et all /5/.

Interacting activities of carbon and nitrogen

Once carbon and nitrogen is dissolved in the iron lattice they will both influence each others activities.

$$a_{Cfe} = f\,(a_{Catm}, a_{Nfe})$$

and

$$a_{Nfe} = f\,(a_{Natm}, a_{Cfe})$$

Using interaction parameters given in table 1 we can calculate the equilibrium carbon and nitrogen weight percentages in a carbonitriding atmosphere.

(Re-)Definition of carbon and nitrogen potentials

If we define the carbon potential in a carbonitriding atmosphere as the equilibrium surface carbon weight percentage in iron found in addition to nitrogen and if we further define the nitrogen potential as the equilibrium nitrogen weight percentage in iron found in addition to carbon we can use a regular iron shim to determine the total percentage of nitrogen and carbon using a shim weigh. The carbon percentage can then be determined by the combustion method, the nitrogen percentage would be the difference of both measurements.

As nitrogen acts as an alloying element for carbon, the relation between the atmosphere carbon potential and the surface carbon content can be written as:

$$CP = c_{Cequ} = 10^{-0.081\,*\,\%N} * CP_{atm} \qquad /6/$$

and the relation between the atmosphere nitrogen potential and the surface nitrogen content can be written as:

$$NP = c_{Nequ} = 10^{-0.187\,*\,\%C} * NP_{atm} \qquad /6/$$

The atmosphere potentials CP_{atm} and NP_{atm} can be determined by:

$$\lg(a_{Catm}) = 2300/T - 2.21 + 0.15*CP_{atm} + \lg(CP_{atm}) \ /7/$$

$$lg(NP_{atm}) = lg(pNH_3/p^{1.5}H_2) - 2210/T + 3.91 \quad /8/$$
(converted to bar and wt %)

Example:

In a carbonitriding process at 1562°F (850°C) using Endogas with 20 % CO and 40 % H_2 with a 5% addition of ammonia and aiming for 0.7 wt% carbon and 0.3 wt% nitrogen in an iron surface the atmosphere potentials come to:

$$CP_{atm} = 0.7/0.95 = 0.74$$

and

$$NP_{atm} = 0.3/0.74 = 0.41$$

This converts to a nitriding potential of

$$K_N = pNH_3/p^{1.5}H_2 = 4.65*10^{-3}\ [bar^{-1/2}]$$

and results into the according partial pressures:

$$pNH_3 = 1.32 * 10^{-3}\ bar = 1323\ ppm$$

and

$$pH_2 = 0.432\ bar$$

If we would be aiming for a lower carbon content of 0.6 wt% the NH_3 in the exhaust can be reduced to 1267 ppm.

Solubility and phase transformations

The binary systems for carbon in iron and nitrogen in iron will not reveal the maximum solubility lines of a combined carbon and nitrogen content. The well-known Fe-C or Fe-N phase diagrams typically display carbon and nitrogen in either weight or atomic percentages. Figure 3 combines the two binary systems to a phase diagram using the total nitrogen and carbon contents given in volume percentages, such taking into account that both species will occupy the same interstitial places in the iron lattice.

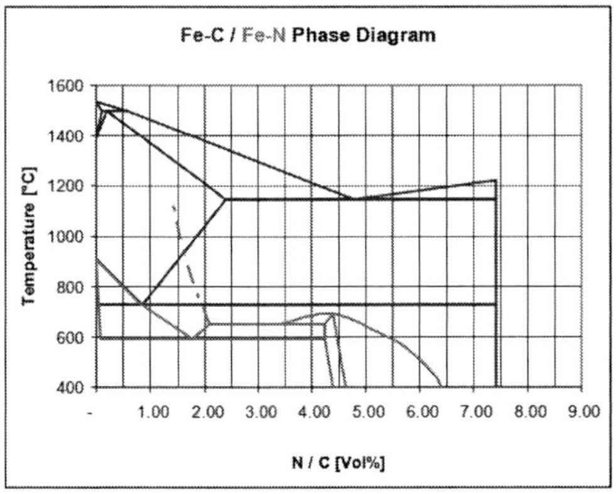

Fig. 3: Combined binary systems Fe-C and Fe-N with C and N given in volume percentages, temperature in centigrade.

It can be derived how the presence of nitrogen expands the existence range of austenite to lower temperatures. The diagram calculates the maximum solubility of nitrogen towards epsilon nitride using the geometric exclusion model given by Slycke and Ericsson /4/. The maximum solubility of carbon towards graphite is taken from standard Fe-C phase diagrams. In absence of nitrogen it can be calculated by setting the carbon activity to 1.

$$lg(a_{Catm}=1) = 2300/T - 2.21 + 0.15 * c_{Cgraphite} + lg(c_{Cgraphite})$$

$c_{Cgraphite}$ is also representing the soot limit in the atmosphere.

As this boundary will be affected by the presence of nitrogen it might approximately be adjusted by:

$$c_{Cmax} = c_{Cgraphite} * 10^{-0.103*\%N}$$

The maximum solubility of nitrogen towards epsilon between 1562°F (850°C) and 1652°F (900°C) can be calculated as:

$$c_{Neps} = 10^{464.28/T - 0.050394} - 0.02$$

and further adjusted to the presence of carbon by:

$$c_{Nmax} = c_{Neps} * 10^{-0.14*\%C}$$

Influence of alloying elements

Both, the activity of carbon and nitrogen will also be affected by the alloying elements being present in regular steel. For control purposes this effect will be taken into account by applying alloying factors for both, carbon and nitrogen. The alloying factors are following the definition:

$$kj = c_{Steel} / c_{Fe}$$

with

$$lg(c_{Fe}/c_{Steel}) = \text{sum of } (c_i * e_{ji})$$

for j = {C, N} and i = {C, N, ... }, c_i in wt%. The according values for e_{ji} can be taken from table 1.

The equilibrium carbon content in alloyed steel exposed to an atmosphere carbon potential will come to:

$$c_{C\text{-}Steel} = k_C * CP_{atm}$$

with

$$lg(k_C) = -(0.010*B\% + 0.081*N\% - 0.014*Al\% \ldots)$$

The equilibrium nitrogen content in alloyed steel exposed to an atmosphere nitrogen potential will come to:

$$c_{N\text{-}Steel} = k_C * NP_{atm}$$

with

$$lg(k_N) = -(0.187*C\% + 0.124*Si\% \ldots)$$

Table 1: Interaction parameters of alloying elements on the activity coefficients for Carbon and Nitrogen in gamma iron at 1652°F (900°C), average values taken from literature /6/.

J	e_{Ci}	e_{Ni}
B	0.070	(unknown)
C	-	0.187
N	0.081	-
Al	-0.014	(unknown)
Si	0.055	0.124
P	0.130	0.050
S	0.140	0.017
Ti	-0.194	(unknown)
V	-0.210	-0.523
Cr	-0.043	-0.158
Mn	-0.016	-0.041
Co	0.012	0.020
Ni	0.014	0.022
Cu	-0.006	(unknown)
Nb	-0.137	-0.358
Mo	-0.015	-0.080
W	-0.023	-0.037

The boundaries towards graphite and cementite formation can be estimated by:

$$c_{Cmax\text{-}Steel} = k_g * c_{Cgraphite} - MC/MB * c_{Boron}$$

with

$$lg(k_g) = sum\ (c_i * e_{Ci}) - c_N * 0.103$$

and i not in {B, N, P, S}, as these elements are interstitials like carbon. Boron is able to replace up to 80% of carbon in cementite, therefore the boron is treated like carbon, taking the different weights into account (MC = 12.011, MB = 10.81).

The boundary towards $Fe_{2\text{-}3}[N+C]$ epsilon can be estimated by:

$$c_{Nmax\text{-}Steel} = k_e * c_{Neps}$$

with

$$lg(k_e) = sum\ (c_i * e_{Ni}) - c_C * 0.14$$

and i not in {C, B, P, S}.

Exceeding this boundary will force the formation of meta stable nitride, causing pores in the parts surface layer /9/ by the possible back reaction:

$$Fe_{2\text{-}3}N \rightarrow Fe + N_2$$

Measuring the atmosphere potentials

According to DIN 17 022 Part 3 /7/ the carbon potential of a carburizing atmosphere can be calculated out of the mV signal of an **OXYGEN PROBE** (following the heterogeneous water gas reaction) by:

$$E=0.0992*T*(lg\ pCO -1.995 -0.15\ CP -lg\ CP) -816.1\ [mV]$$

As the partial pressure of carbon monoxide is typically used as a constant number given by the carrier gas, in a carbonitriding atmosphere either the dilution caused by the ammonia addition has to be taken into account or the actual partial pressure of CO has to be measured using an analyzing instrument. Under regular conditions at atmospheric pressure, where all of the ammonia will be nearly dissociated it is sufficient to use:

$$pCO_{atm} = vol\%\ CO_{carrier} / (100 + 2*NH_{3\text{-}add})$$

One method to determine the nitrogen potential is to measure the partial pressure of ammonia using an analyzer calibrated to a range of 0 – 10000 ppm. The hydrogen partial pressure may then be measured with a second analyzer or estimated as

$$pH_{2\text{-}atm} = (vol\%\ H_{2\text{-}carrier} + 1.5*NH_{3\text{-}add}) / (100 + 2*NH_{3\text{-}add})$$

A second possibility is the direct measurement of carbon and nitrogen in equilibrium in iron using a wire-sensor /10/. This type of sensor exposes a thin iron wire (see figure 4) to the carbonitriding atmosphere. As both, carbon and nitrogen will be interstitially dissolved in the iron lattice, the electrical resistance of the wire changes (see figure 5).

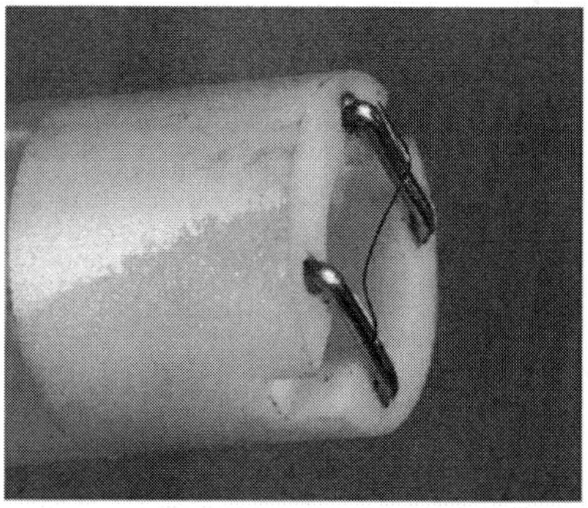

Fig. 4: Active element of the wire sensor

Chatterjee-Fischer /11/ is giving the relation for nitrogen induced increase of resistivity versus carbon induced as:

$$\Delta R_C = 0.7 * \Delta R_N$$

In order to achieve the 0.7 wt% C plus 0.3 wt% N in the example above, the wire would have to be controlled to a [N+C] content of 0.94 wt% virtual carbon. The carbon potential controlled with the oxygen probe has to be adjusted to 0.74 %C and the atmospheric nitrogen potential would be at 0.41 %N.

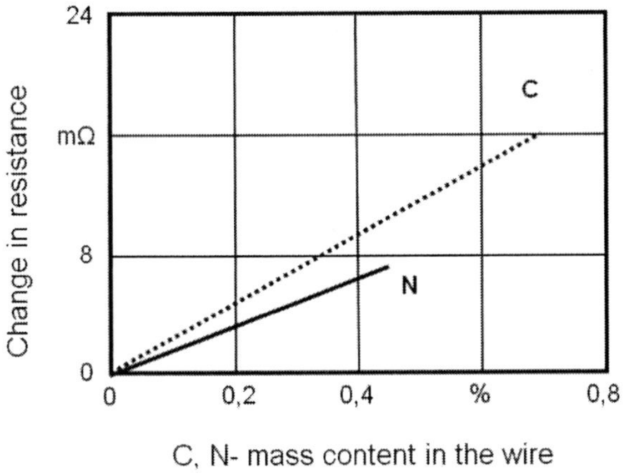

Fig. 5: Induced resistance change in an iron wire due to carbon and nitrogen uptake /11/

Figure 6 shows the computed signals of an oxygen probe and a wire sensor exposed to a carburizing atmosphere. The moment ammonia is added to the process gas the wire's combined carbon and nitrogen potential increases while the carbon potential measured with the oxygen probe and adjusted for the dilution stays level.

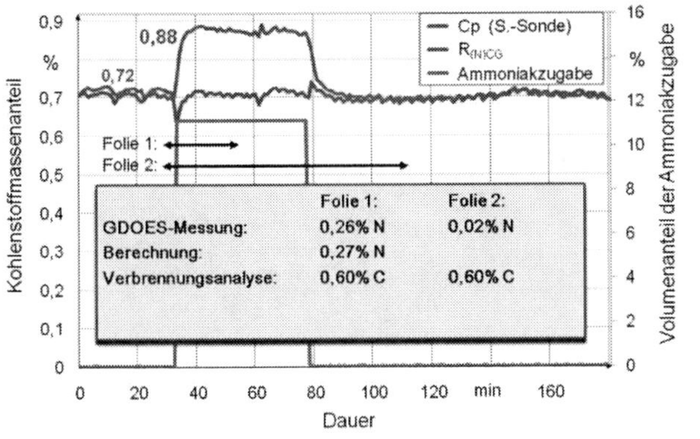

Fig. 6: Nitriding and de-nitriding of wire sensor and iron shims at 1562°F (850°C). The red line indicates the controlled carbon potential (oxygen probe); the blue line displays the increase in virtual carbon [N+C] in the wire. The green line shows the ammonia percentage in the gas inlet /3/

Iron shims (Folie 1 and 2) verified the nitrogen content, the carbon content was lower as expected.

Carbon and nitrogen uptake

Based on the assumption of the heterogeneous water gas reaction being the dominant carburizing reaction, the carbon uptake can be determined by the partial pressures of carbon monoxide and hydrogen in the process atmosphere shown in figure 7. The curve can be approximated by:

$$\beta_C = 0.00018*pCO*pH_2 - (pCO*pH_2/65.78)^2 \text{ [cm/s]}$$

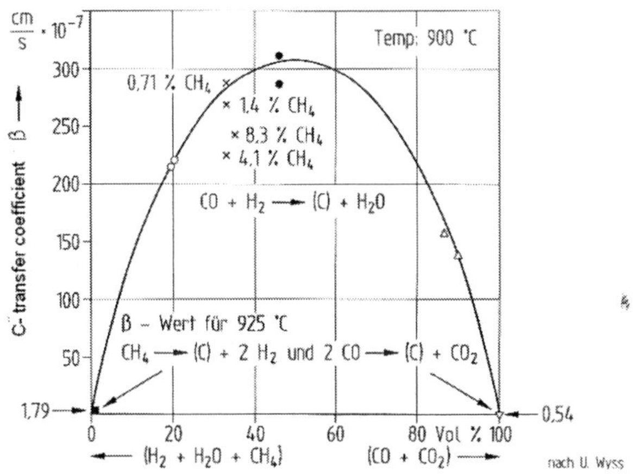

Fig. 7: Carbon transfer coefficient beta as a function of the volume percentages of carbon monoxide and hydrogen at atmospheric pressure and 1652°F (900°C) /12/

While CO will be diluted whereas H_2 will increase by the dissociated ammonia, the transfer coefficient will be affected very little. An ammonia addition of 10 % decreases the beta coefficient only by 4.5 %.

In a carbonitriding process the nitrogen activity given by the ammonia dissociation is greater than the activity given by the partial pressure of nitrogen. Consequently the equilibrium nitrogen content for $pNH_3/p^{1.5}H_2$ is as well much higher. Therefore there has to be a local equilibrium of a dissociation and a formation of ammonia on the parts surface. As the nitrogen uptake is limited by the denitriding reaction it can be determined by the partial pressure of hydrogen:

$$\beta_N = 0.09*e^{-64247/RT}*pH_2 \text{ [cm/s]} \qquad /13/$$

The carbon and nitrogen flux into the surface can be expressed as:

$$J_C = \beta_C * (c_{Cequ} - c_{C\text{-surface}}) \text{ [wt\%*cm/s]}$$

and

$$J_N = \beta_N * (c_{Nequ} - c_{N\text{-surface}}) \text{ [wt\%*cm/s]}$$

Carbon and nitrogen diffusion

The carbon diffusivity in γ-iron is a function of temperature and dissolved carbon.

$$D_C = e^{-(18900/T+0.38)} * e^{(4300/T-2.63)*(\%C\wedge 1.5)} \; [cm^2/s] \qquad /14/$$

As both agents are occupying the same interstitial places one can assume that the diffusivity of carbon and nitrogen will be dependent on the actual amount of nitrogen and carbon /4/ already dissolved and causing the lattice to open. As a simplification, nitrogen will be treated as "virtual carbon", applying a factor taking the different dimensions into account.

$$D_C = e^{-(18900/T+0.38)} * e^{(4300/T-2.63)*((\%C+0.62*\%N)\wedge 1.5)} \; [cm^2/s]$$

Based on a relation given by Slycke /4/ the nitrogen diffusion coefficient of nitrogen between 1562°F (850°C) and 1652°F (950°C) can be adjusted to:

$$D_N = 0.45*e^{-(18900/T+0.38)} * e^{0.62*(4300/T-2.63)*((\%C+0.62*\%N)\wedge 1.5)} \; [cm^2/s]$$

Figure 8 displays the result of a simulation with HT-Tools using the above equations. Carbon and Nitrogen profiles are calculated. Based on the carbon and nitrogen content the software calculates the expected hardness curve after quenching and regular tempering.

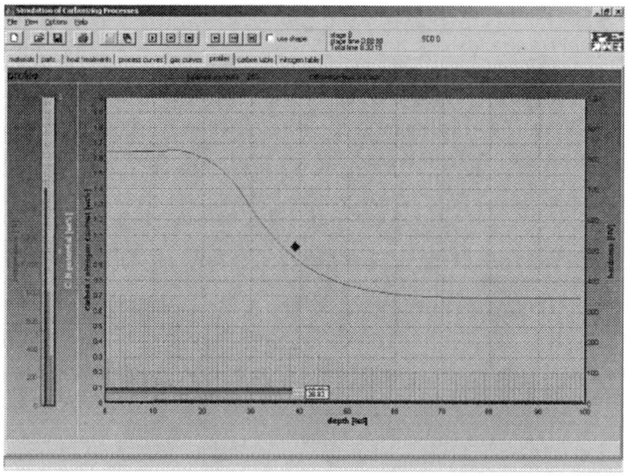

Fig. 8: C- and N- profiles in a steel bar made from AISI 8620 aiming for a case depth of 40 thousands and a surface carbon content of 0.72 %C and a surface nitrogen content of 0.30 %N

Controlling the potentials

In our tests we have been using a PROTHERM 500 programmable controller, able to control temperature and atmosphere.

Like a normal carbon potential controller, the system uses propane or methane as enrichment gas and air for reducing the carbon potential. The nitrogen potential is controlled by increasing or decreasing the ammonia flow.

The controller is able to either calculate the dilution, if the flows are known, or it can use the signals of connected analyzers to adjust carbon monoxide and hydrogen. For determining the nitrogen potential, the controller can be connected to a wire sensor or to an ammonia analyzer.

The recipe holds the composition of the steel to be treated calculates the alloying factors for carbon and nitrogen and calculates the carbon and nitrogen diffusion during the process based on the actual carbon and nitrogen transfer coefficients and the diffusion coefficients at temperature and carbon and nitrogen content throughout the calculated profiles.

Temperature control offers RAMP HEAT, HEAT, HOLD, RAMP COOL, the atmosphere control offers RAMP CP/NP, HOLD CP/NP, and several special functions, SOOT LIMIT, AUTO BOOST, $Fe_{2\text{-}3}N$ LIMIT and SURFACE CONTENT N/C.

In order to achieve short process times, the proposed atmosphere control starts with a boost stage maintaining high potentials. During the AUTO BOOST the controller limits the atmosphere carbon potential to a set percentage of the soot limit (see above).

$$CP_{atm} <= \text{AUTO FACTOR} * c_{Cmax}$$

The moment the calculated surface carbon content in alloyed steels reaches the maximum solubility towards graphite, adjusted to the actual surface nitrogen content, the atmosphere carbon potential will automatically be reduced (see figure 9).

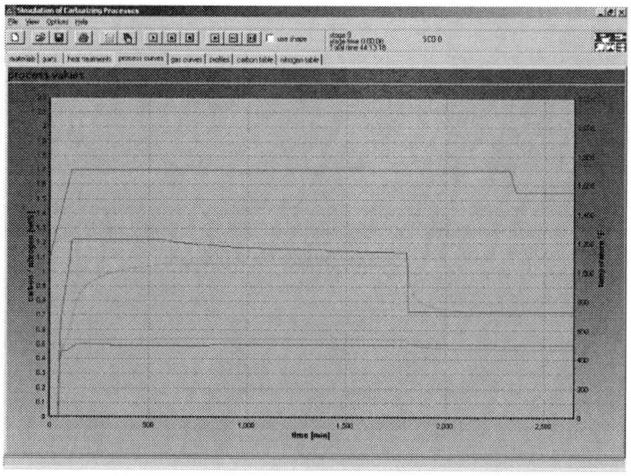

Fig. 9: Chart recorder displaying the controlled atmosphere potentials for carbon (brown) and nitrogen (light green), the temperature (blue) and the calculated surface contents for carbon (orange) and nitrogen (dark green).

The desired nitrogen content in the surface was given as a setpoint of 0.5 wt% N in the steel. The ammonia addition is automatically adjusted to achieve the necessary NP_{atm}.

The end condition for an auto boost stage can be either time or a set percentage of the desired carburizing depth.

In the diffusion stage the controller will set the atmosphere potentials automatically according the DESIRED SURFACE CONTENTS of CARBON and NITROGEN. This stage will typically end once the desired carburizing depth is reached.

It is also possible to use the atmosphere potentials CP_{atm} and NP_{atm} as setpoints, in which case no automatic limitations to soot, cementite or epsilon will be performed. This represents a more classical process variant.

Test Results

Tests done at the IWT and several customers show a good correlation between predicted and measured carbon and nitrogen contents in samples (see figures 10 and 11).

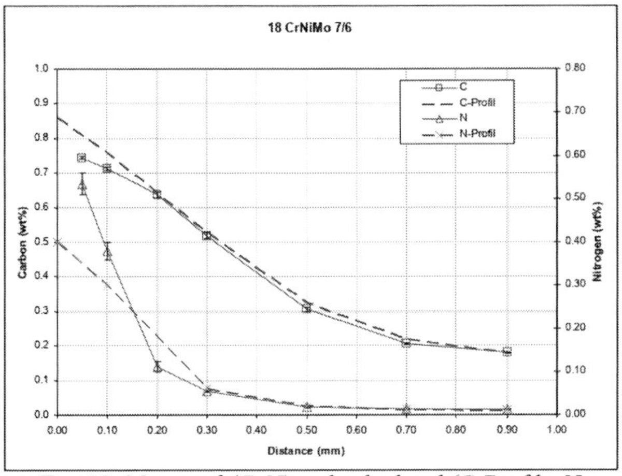

Fig. 10: Measured (C, N) and calculated (C-Profile, N-Profile) carbon and nitrogen contents in an 18 CrNiMo 7/6 test piece, carbonitrided at1544°F (840°C) for 280 minutes.

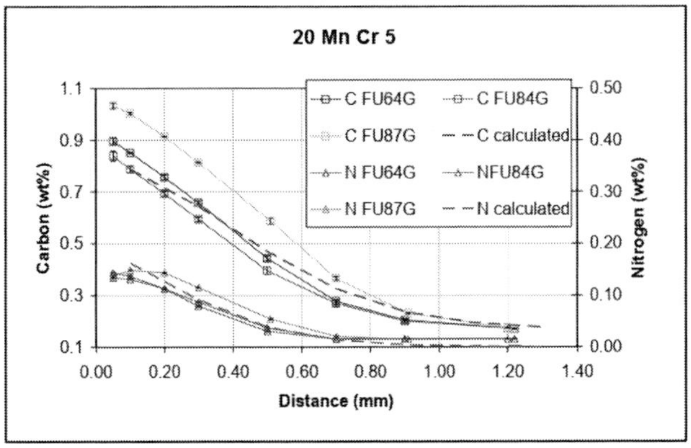

Fig. 11: Measured and calculated carbon and nitrogen contents in 20 MnCr 5 test pieces, carbonitrided at 1706°F (930°C) for 240 minutes.

Conclusions

Incorporating the results of late research projects into a process controller enables a new attempt to performing carbonitriding processes.

The ability to measure and control both, the carbon and nitrogen potential in the atmosphere takes the uncertainty of the ammonia dissociation, caused by changing surface conditions out of the equation. Besides obvious advantages in the process control and the ability to reproduce batch after batch, this will also be valid if furnaces are used for carburizing and carbonitriding processes where the NP control can shorten formation times.

Using the interaction parameters to adjust the atmosphere potentials to the varying steel compositions enables a controlled carbonitriding treatment to desired surface contents of carbon and nitrogen which might open a new field of investigations on working properties of parts.

The inbuilt safety functions to avoid cementite and epsilon formation enables the use of higher potentials during boost stages, aiming for shorter process times while at the same time being safe not to create a porous layer.

The online diffusion calculation for carbon and nitrogen enables a treatment to the specification such as case depth and surface carbon and nitrogen content.

Acknowledgments

The author would like to thank the team of the IWT in Bremen for giving our company the opportunity of testing and modifying our control equipment to their needs and letting us participate in their research.

References

[1] Meinhard, E., Carbonitrieren – warum und wie?, TZ für Metallbearbeitung 1982/10, pp. 23–32.

[2] Hoffmann, F., Clausen, B., Steinbacher M., Carbonitrieren verspricht Verbesserung der Eigenschaften von Zahnrädern, MM Das IndustrieMagazin 30/2008, pp. 32-36.

[3] Bischoff, S., Klümper-Westkamp, H., Hoffmann, F., Zoch, H.-W., Development of a sensor system for gas carbonitriding – Part 1, HTM J. Heat Treatm. Mat. 65 (2010) 3, pp. 141-148

[4] Slycke, J., Ericsson, T., A study of reactions occurring during the carbonitriding process, J. Heat Treating 1981, Vol. 2 (No. 1), pp. 3-19

[5] Collin, R., Gunnarson, S., Thulin, D., Ein mathematisches Model zur Berechnung von Aufkohlungsprofilen bei der Gasaufkohlung, HTM J. Heat Treatm Mat. 25 (1970) 1, pp. 17-21

[6] Literature survey: average interaction parameters for carbon activity based on Houdremont, Gunnarson, Neumann (found in: Schröter, W., Spengler, A., Über den Einfluß von Legierungselementen beim Einsatzhärten, HTM J. Heat Treatm. Mat. 57 (2002) 5, pp. 342-348), AWT (found in: AWT-FA 5 AK 4, Die Prozessregelung beim Gasaufkohlen und Einsatzhärten, Expert Verlag 1997, pp. 12-13) and Kunze, J. (found in: Physikalisch-chemische Grundlagen der Wärmebehandlung und Randschichttechnik, Lehrblätter zur Vorlesung , 2000), average interaction parameters for nitrogen activity based on Zheng (found in: Zheng, X., Nitrogen solubility in iron-base alloys and powder metallurgy of high nitrogen stainless steels, Diss. ETH No. 9488, 1991) and Kunze, J. (see above)

[7] DIN 17 022 Part 3 p. 21, DIN-Taschenbuch 218, Beuth Verlag, 1998

[8] Fromm, E., Gebhardt, E., Gase und Kohlenstoff in Metallen, Springer-Verlag 1976, pp. 581-608

[9] Mittemeijer, E. J., Slycke, J. T., Chemical potentials and activities of nitrogen and carbon imposed by gaseous nitriding and carburizing atmposheres, Surface Engineering 1996 Vol. 12 No. 2, pp. 152-162

[10] Wünning, J., C-Pegel und C-Strommesseung mit dem rechnergesteuerten Widerstandssensor, HTM J. Heat Treatm. Mat. 43 (1988) 5, pp. 266-270

[11] Chatterjee-Fischer, R, Schaaber, O., Einge Betrachtungen zum Karbonitrieren im Gas, HTM J. Heat Treatm. Mat. 26 (1969) 2, pp. 108-110

[12] AWT-FA 5 AK 4, Die Prozessregelung beim Gasaufkohlen und Einsatzhärten, Expert Verlag 1997, p. 31

[13] Grabke, H. J., Reaktionen von NH3, N2, H2 an der Oberfläche von Fe, I. Zur Kinetik der Nitrierung von Eisen mit NH3/H2-Gemischen und der Denitrierung. II. Zur Kinetik der Nitrierung von Fe mit N2 und der Desorption von N2, Berichte Bunsenges. Phys. Chem., 72 (1968), No. 4, pp. 533-548

[14] Leyens, G., Woelk, G., Wünning, J., Berechnung der Aufkohlung nach dem Sättigungs-Ausgleich-Verfahren, Archiv f.d. Eisenhüttenwesen 47 (1976) 6, pp. 385-390

Proceedings of the 26th ASM Heat Treating Society Conference
B.L. Ferguson, R. Jones, D.S. MacKenzie, and D. Weires, editors
https://doi.org/10.31399/asm.cp.ht2011p0017

Capitalizing on Current Technology Used in Data Acquisition

Bob Fincken
Super Systems, Inc., Cincinnati, OH, USA
bfincken@supersystems.com (800) 666-4330

Introduction

Automation can be many different things across many different industries so it's hard to have one definition to cover it in its entirety. In heat treating, automation is approached in categories. The types of processes and equipment may only have certain capabilities when it comes to mechanically automating a process forcing heat treaters to look at non-traditional types of automation. Many heat treaters need flexible equipment when it comes to processes and load sizes because of economic pressures and competition, costs and capital are being scrutinized. Does it take significant capital to automate a process? In the world of heat treating it doesn't have to.

What is Automation in Heat Treating?

An automated process really needs to start with a plan and goals. For most heat treaters the main drivers are obvious: quality, documented productivity and cost reductions. In its simplest form, automation is creating a sequence of steps that flow continuously. Creating a sequence of steps to be automatic doesn't always require equipment. Simple steps for this to happen can be broken into, operational procedures, preventive maintenance programs, shared data and smarter controls.

Proactive vs. Reactive Maintenance

Maintenance is tough and tends to be very reactive. A reactive approach to maintenance bypasses the crucial planning phase of production. The risk is reducing overall equipment utilization as well as any excess capacity thus having opportunities for higher profit margins on the table. Using SCADA (Supervisory Controls and Data Acquisition) and smart process controls can help with this. Quick access to information that can be used by personnel throughout the facility enables a better decision making process. Referencing historical data, questions like these can be answered quickly: Did the load take longer to get to heat than usual? Is the temperature overshooting? Is my carbon percentage oscillating? This information can be valuable when the right person has access to it. The intent is to put yourself in a situation where you have confidence in your ability to produce quality parts.
Traditionally, data may have been available on paper charts but was not used effectively because it was not readily accessible. Today there is more information with the use of open communication standards and plant- wide networks. Access to key data points has to be easy so the right person can analyze it, compare it to historical data, and determine if action is necessary. Rework is double trouble in that it results in twice the furnace time, another notification to the customer, a longer delivery time and reduced margins on the job. To avoid it, put people, processes and equipment in place to identify the problem before it happens.

With a quick glance at a chart, one can identify if a load coming to heat had a problem. Sure this information was available before, but comparing two different runs was not easy. With electronic information and tracking of load times, a quick review periodically can be significant. Often carbon charts can be evaluated and burner tube leaks identified when looking at digital charts and identifying saw tooth trends, (i.e., an indication of air in the work chamber when on high fire and the carbon probe trying to compensate for it.)

It is more common for an out-of-tune burner to be a more reactive maintenance task. With sensor and recording technology today, you can have quick access to current information to allow adjustments to the air-gas ratio that will lead to consistent uniformity and productivity. There are case studies that prove a tuned burner can increase capacity of a furnace tremendously, not to mention the gas savings benefit.

Business Process Optimization

It's no secret good labor in heat treating is tough to find. There is a significant amount of heat treating experience that is retiring from the industry leaving a large gap in knowledge on the floor. Typically key personnel can be categorized into operations, maintenance and quality. Aligning priorities across the organization is critical to one's success. There is no better starting point for making decisions than having information available that can be shared. Whether it is the furnace charts, current customers/jobs running, expected load completed time, or planned maintenance, it all collectively feeds the planning process. This information can be accessible using controls and systems and the data can be used to enable the management team to make more informed decisions that impact resources, equipment, quality and long term planning. When the information is used properly, decision processes become more efficient.

Many heat treating operations have disjointed controls leaving management to continuously monitor furnaces using personnel and chasing down information. Simply put, management can do a better job at planning production and maximizing furnace utilization with the information that is available.

Controls and Data

Process control can be defined as the ability to meet certain parameters over time using inputs from the process and controlling outputs for desired results. Today's control and sensor technology makes much of this automated. With technology enhancements, come more sophisticated controls to make better decisions. The decisions can be on the process itself, productivity, quality, and safety. Traditionally the inputs for a process were disconnected. Today more of these are being brought together to provide a more holistic view of the control decision process.
PLCs, discrete microprocessor controls, and a hybrid of the two is what you see on most heat treating equipment. Microprocessor controllers typically provide single or multiple PID control loops with expanded features such as event and recipe management. In many cases, the controllers are defined for specific processes or can be programmed to provide functionality for a specific process. PLCs can also have PID control loops developed as part of the logic created in the controller.

To get full use of the control technology, it is best to create an opportunity for repeatability and allow the level of flexibility that is necessary for the equipment and operation. A foundation for this is recipe control. A recipe is the steps of the heat treatment process. It can be a simple ramp to temperature and a timed-soak; or a more complex process by which multiple temperature inputs, time, atmosphere, pressure and other variables need to be managed.

Incremental steps can be taken to put more automation in place and provide better access to information. Using recipe control to automate a nitriding, vacuum, carburizing, or even a simple temperature ramp soak, leads to greater chances of running the process correctly leading to repeatability. Automation of a process using programmable controls eliminates variations in process from shift to shift or operator to operator.

A recipe program setup in advance to "step" through the process and automatically advance to the next step upon desired results leaves the controller in charge. Operators can be taken further from the decision process by implementing scanning technology so the recipe is automatically selected based on a Shop Card or Part Number. This reduces the chance for the wrong heat treat cycle and gets the parts heat treated correctly the first time.
Controls today provide user defined audible, visual, and electronic alarms indicating cycle complete or a deviation from the process. Recipes can be setup so upon completion; the heat is turned off; this eliminates unnecessary costs and ensures parts are at heat for the necessary amount of time. With smart alarms it captures the attention of operators when a situation needs to be addressed. Again, simple steps to automating a process which positively impacts labor and utilization.

With more devices and sensors monitoring the process and a system in place to capture information, data is very easy to access remotely. Whether it is a critical job where periodic remote monitoring is necessary or the ability to tweak a cycle, it can be done. Historical information and simulation can lead to more precise control parameters. The more access to these data points, the greater the opportunity for repeatability.

Industry Specifications

The advancements in process control and technology have resulted in more industry requirements. Whether it is CQI-9 requiring an endothermic generator to be continuously controlled and monitored or AMS 2759/12 requiring automatic control to produce a continuous compound layer with controlled phase composition. Both of these requirements can be accomplished using control technology today. Using the AMS 2759/12 as an example, processing equipment is defined in the document. The method of control is not specified but it is pointed out that the furnace should be equipped with gas analyzing equipment capable of measuring the concentration of the reactive gas. This in itself does not specify automation; however, using an integrated process controller that can ensure temperature, time, gas concentration, and maintain a safe operating furnace does.

In many cases, industry specifications drive a certain amount of documentation that is required to provide a customer or auditor for proof of the process. This might be required when the parts are completed or on-demand when a customer requests the information. Well, electronic data and SCADA systems have made this task easier. It is more and more common that information is shared electronically. Whether via email or an attached file, it still gets to the heart of the requirement which is proof of the heat treating process.

Customer Service

There is a whole different level of service that can be provided when putting in place smart controls and data acquisition. Quick access to information allows an efficient response to a customer requesting data or order status. With a system in place, a customer can have an answer while on the phone. Prior to having a system in place personnel were required to find the information on the shop floor by looking through paperwork to determine the time remaining for a part in process.

Conclusions

Today's heat treater depends on the science of the past but will depend on the technology of the future to stay competitive. Technology moves forward every day and requires that we embrace the technology and automation to remain competitive. Those who resist change will be left behind wondering why their business is declining and will be surpassed by the growing global competition. Those who embraces the opportunity automation and digital documentation provides will grow and prosper.

Proceedings of the 26th ASM Heat Treating Society Conference
B.L. Ferguson, R. Jones, D.S. MacKenzie, and D. Weires, editors
https://doi.org/10.31399/asm.cp.ht2011p0020

Transient Liquid Phase Diffusion Bonding of Stainless Steel 304 Using Copper and Aluminium Filler Interlayers

M. Mazar Atabaki[a,b], J. Noor Wati[b], J. Bte Idris[a]
[a]Institute for Materials Research, the School of Process, Environmental and Materials Engineering, Faculty of Engineering, University of Leeds, Leeds, UK
[b] Department of Materials Engineering, Faculty of Mechanical Engineering, Universiti Teknologi Malaysia, 81310, Malaysia

Abstract

Transient liquid phase diffusion bonding is employed to join stainless steel 304 using pure copper and aluminium foils as interlayer. The bonding process was carried out in a vacuum furnace at various temperatures for various diffusion times. The joints were studied with optical and scanning electron microscopy, energy dispersive spectrometry and microhardness measurement. Results showed that the hardness of the bond made by copper interlayer in vacuum were higher than that of made with aluminium interlayer. The poor mechanical properties of the bonds were associated with the formation of intermetallics within the bond region. By holding the parent alloy at the bonding temperature for longer period, complete isothermal solidification was likely to occur. The diffusion of the main elements from the interlayers into the base metal at the bonding temperatures was the main controlling factor pertaining to the microstractural evolution of the joint interface. Selecting an appropriate bonding temperature to achieve a maximum concentration of the melting point depressants was dependent on the time of isothermal solidification. In order to determine the corrosion resistance of the joints, Tafel test was conducted in 3.5% NaCl solution. The presence of eutectoid γFe+ eutectic Cu + Cr and Fe-Al intermetallic was detected at the interface of the joints bonded with copper and aluminium interlayers, respectively. The highest microhardness result was obtained at the diffusion zone and hardness value decreased gradually as the distance from the joint increased. The joints related to stainless steel-copper developed crevice corrosion due to galvanic couple between the stainless steel and copper. Besides, pitting was occurred due to intergranular stress corrosion cracking on the copper surface.

1.0 Introduction

Austenitic stainless steels have been widely used in conventional and nuclear power plants for different applications such as superheaters and heater components and also originally developed for use in cryogenics. This steel also performs well at elevated temperatures and is used extensively for steam pipes and exhaust systems. It is their resistance to elevated temperature, oxidation, and corrosion that makes the alloys the choice for food preparation equipment, including steam-heated boilers and storage tanks. However, there are many difficulties in joining of this material using the fusion welding methods. A concern, when welding the austenitic stainless steels, is the susceptibility to solidification and liquation cracking [1]. Copper does not form brittle intermetallic compounds with iron and the melting point of the copper is lower than Fe and Ni. Thus, the flow-ability will increase at higher temperature and encourage a good contact between the faying surfaces. Besides, it is well known that stainless steel and aluminium react to form intermetallic layers during brazing [2]. In many instances the formation of brittle intermetallic phases in the diffusion zone leads to unfavorable changes in the mechanical and physical properties [3]. The research work presented in this study concern the transient liquid phase diffusion bonding of stainless steel 304 using pure copper and aluminium foil interlayer. The bonding variables and their effect on microstructural changes and mechanical properties of the joint have been investigated using optical and scanning electro microscopy and energy dispersive spectrometry elemental analyses to determine the interface properties of the joints.

2.0 Experimental Procedure

The stainless steel used was received in the form of plate of 2 mm thickness and the chemical composition as shown in Table 1. For the bonding process, the plates were cut using abrasive cutting saw to the dimensions of 2×10×20 mm. The samples were oriented as shown in Fig. 1. The mating surface of the stainless steel was prepared by conventional grinding on 1200 grade silicon carbide paper followed by polishing using diamond paste. The specimens then were cleaned in an ultrasonic bath using acetone for 15 min and dried in air. The copper foil (50 μm thick, 99.95% purity) and aluminium foil (25 μm thick) were used as intermediate material and both the surfaces of the interlayer were polished in the same fashion. A stainless steel fixture (see Fig. 2) was used to fix the sample and hold the sandwich assembly during the bonding process. The bonding process was performed in a vacuum furnace and the substrates and interlayer contact was enhanced by a pressure of 0.5MP at a set of bonding temperatures. The bonding cycle is shown in Fig. 3. Test samples were heated to the bonding temperature, then left in the furnace for a variable holding time and heating rate and finally cooled to the room temperature. The bonding process parameter for copper and aluminium interlayer are shown in Table 2 and 3, respectively. Specimens for the metallographic examination were prepared by grounding on 240 to 2400 grade SiC paper and polished the cross section, then etched using 2g $FeCl_3$, 24 ml distilled water and 6 ml HCl. Microstructure observations were conducted using optical microscope (Nikon microphot-FXL), scanning electron microscope (PHILIPS XL 40) using back-scattered mode (SEM-BSE) and energy dispersive spectrometry (EDS).

Table 1: Chemical composition of the base metal (wt%)

Material	C	Fe	Mn	Si	P	S	Cr	Ni	N
304 SS	0.08	Bal	2.0	0.75	0.045	0.03	20.0	10.5	0.1

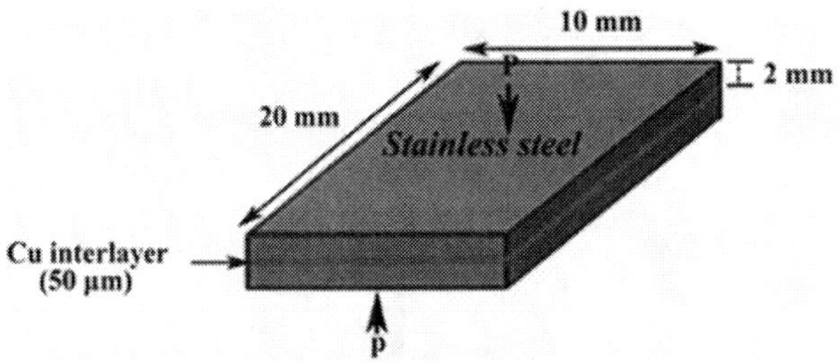

Figure 1: Dimensions of the diffusion brazing bonded specimens (mm).

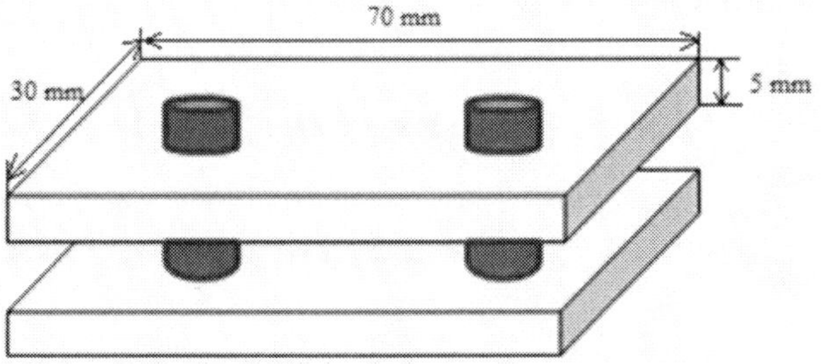

Figure 2: Dimensions of the stainless steel fixture (mm).

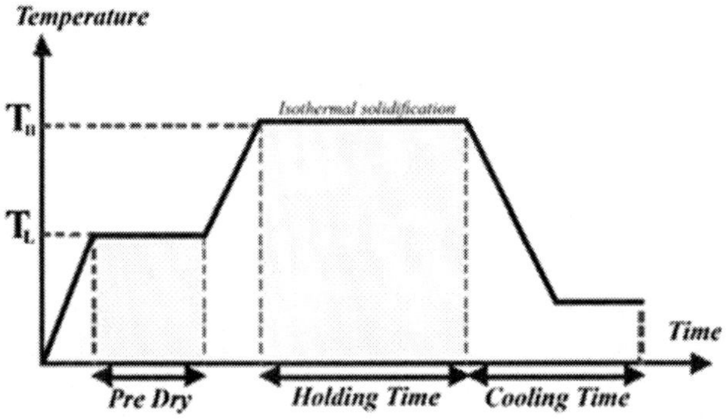

Figure 3: Diffusion brazing cycle.

Table 2: Process parameter for stainless steel with copper interlayer

PROCESS PARAMETERS	
Low Temperature (°C)	900
Pre Dry (minutes)	24
Rate (°C/minutes)	5 and 24
High Temperature (°C)	950
Holding Time (minutes)	16, 20 and 24
Cooling Time (minutes)	10
Vacuum (mm Hg)	740
Vacuum Start (°C)	740
Vacuum Release (°C)	740

Table 3: Process parameter for stainless steel with aluminium interlayer

PROCESS PARAMETERS	
Low Temperature (°C)	900
Pre Dry (minutes)	24
Rate (°C/minutes)	20
High Temperature (°C)	920
Holding Time (minutes)	22
Cooling Time (minutes)	10
Vacuum (mm Hg)	740
Vacuum Start (°C)	740
Vacuum Release (°C)	740

3.0 Results and Discussion

3.1 Microstructure analysis of 304SS-Cu-304SS

Figure 4 (a) and (b) shows the microstructure of the stainless steel before and after the bonding at 900 °C, respectively. As it can be seen, after the bonding process, recrystallization of the grains occurred and chromium carbides precipitated at the grain boundaries of the stainless steel, backing into solution in the austenite. The precipitation leads to relieving stress in the structure and causes a transformation to tough austenitic structure. Fig. 5, 6, and 7 show the optical microstructure of the bonded assemblies at 900, 950 and 1000 °C, respectively. It is observed that certain amount of diffusion occurs between the interlayer and two substrates. A thin diffusion layer was revealed parallel to the bond interface on the stainless steel side for all bonds. The interface of the stainless steel and copper consisted of a continuous reaction layer which is free from voids at both sides. It is observed that the copper side shows absence of any diffusion layer.

SEM-BSE images of the joints at 900, 950 and 1000 °C for various holding time are given in Fig. 8, 9, and 10, respectively. The backscatter images are useful to understand the element distribution in the bond line. In order to determine the chemical composition of the bonding interface, EDS analysis were performed using 'spot analysis' taken from stainless steel and copper interlayer. The interface was revealed at higher magnification. From the observations, various reaction layers were found in the bond zone and base metal. These regions can be classified in terms of the shape and location of each phase. The brazed joint can be divided into three distinct zones, which are base metal, diffusion zone and interlayer as shown in Fig. 11.

Compositional profiles indicating the distribution of the alloying elements from the base metal and interlayer in the centerline shown as AB in Fig. 11 for the specimens bonded at 900, 950 and 1000 °C are presented in Fig. 12, 13 and 14, respectively. It can be clearly seen that the weight percent of Fe and Cr in the interlayer significantly decreases while the weight percent of Cu considerably increases in the base metal. To further analyze the distribution of Fe from the base metal to the interlayer and Cu from the interlayer to the base metal, Cu-Fe binary phase diagram was applied for describing the effect of the melting point depressant

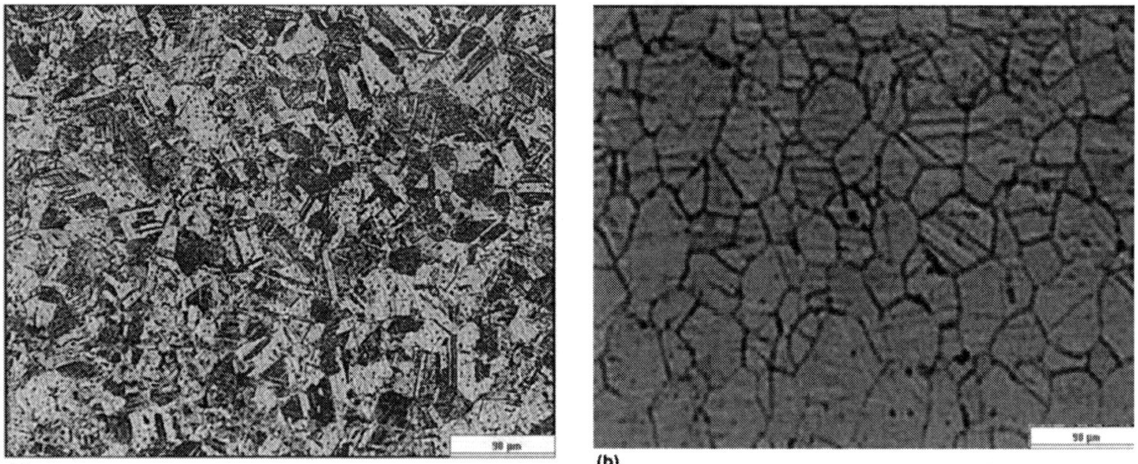

(a) (b)

Figure 4 : Microstructure of the stainless steel (a) before and (b) after bonding at 900°C.

on the mechanism of the bonding process (see Fig. 15). A very low amount of Fe was detected in the copper interlayer due to limited solubility of Fe in copper at the temperature range of 900 to 1000 °C. The copper started to dissolve a little amount of Fe at the stainless steel/copper interlayer interface and continuously diffuses into the interlayer. The diffusion of Cu into the stainless steel produces a solid solution in very limited solubility. In addition, the solubility of Cu in Fe enhanced when the bonding temperature was increased. It can be assumed that the diffusion layer at the bond zone is consisting of γFe + (Cu).

3.1.1 Interface Characterization

For characterizing the diffusion zone of the brazed joints, SEM-BSE images of the transition joints were taken from the diffusion zone at stainless steel/copper interlayer interface as shown in Fig. 16 (a)-(c) for the specimens bonded at 900, 950 and 1000 °C, respectively. Moreover, a quantitative overview of the chemical compositions analyses for different regions (region A to P) in the diffusion zone are listed in Table 4 using EDS analysis.

In the higher magnification, various regions (A-P) were found in the base metal, diffusion zone and interlayer. Fig. 16 (a) shows a dark shaded region (point B) and a light shaded region (point C) at the interface for the joint brazed at 900 °C. Region B and C are enriched with Fe and Cr with small quantity of Ni, Mn and Cu. The shaded differences occur based on the difference in concentration of Fe and Cr into the interlayer and Cu into the base metal. The white island (region A) found in the copper interlayer is enriched with Fe and Cr. Fig. 16 (c) shows significant difference in the bone zone compared with the joints shown in Fig. 16 (a) and (b), presenting a small diffusion layer. Copper is dissolved (region P) due to infiltration of the liquid copper at 1000 °C. The elemental compositions for these regions (A, B, C, O and P) are shown in Fig. 17 based on EDS analysis.

Figure 18 shows schematic diagram describing the possible brazing mechanism and the changes of the phases by increasing the holding time from 0 to 20 min. At the initial stage, during the brazing process with a holding time of less than 1 min (Fig. 18 (a)), the Cu interlayer partially liquefied at the stainless steel and the copper interlayer interface. Stainless steel is dissolved by migrating liquid copper to the base metal. As a result, stainless steel pieces are separated and Fe-Cr from the base metal flows into the diffusion zone. At the same time, Cu elements in the melts diffuse into the dissolved region. This is known as a relationship of solid metal dissolution and liquid metal migration. When liquid interlayer contacts with base metal, the migration of interlayer to base metal governed by volume diffusion of liquid interlayer penetration of liquid interlayer to base metal by grain boundary diffusion [5]. The volume diffusion takes place at the adjacent to interfacial region and the grain boundary diffusion acts beyond interfacial region at the base metal [5]. It is also called as infiltration of copper. Liquid copper changes into Cu phase and reacts with Fe and Cr along the brazed joint. Diffusion coefficients of Fe, Cr and Cu are different. For example, at 920 °C the diffusion coefficients of Cu into Fe and Fe into Cu is equal to 2.2x10-13and 4.63x10-10cm^2/s, respectively [6]. Therefore, the migration of Cu into Fe is smaller than Fe into Cu. By increasing the dissolution of the base metal with applying a longer time (16 min), the phase enriched with Cu and Fe increases in the bond line (Fig. 18 (b)). On the other hand, it is considered that Fe and Cr elements flow from the base metal are transformed into a Fe-Cr phase at this time. With increasing the holding time to 20 min, the amount of Fe-Cr phase shows a tendency to increase (Fig. 18(c)).

3.1.2 Widening of Diffusion Zone

The diffusion zone was widened during the bonding operation. The mechanism of the diffusion zone widening is studied by measuring the thickness of the diffusion layer at both sides of the interlayer with initial thickness 50 μm. The thickness measurement scheme of the layer shown in Fig. 19 is an optical micrograph at 900 °C magnification of 200X of the brazed joint prepared at of 950 °C for 16 min. It can be seen that the diffusion zone advanced into the stainless steel. Table 5 shows

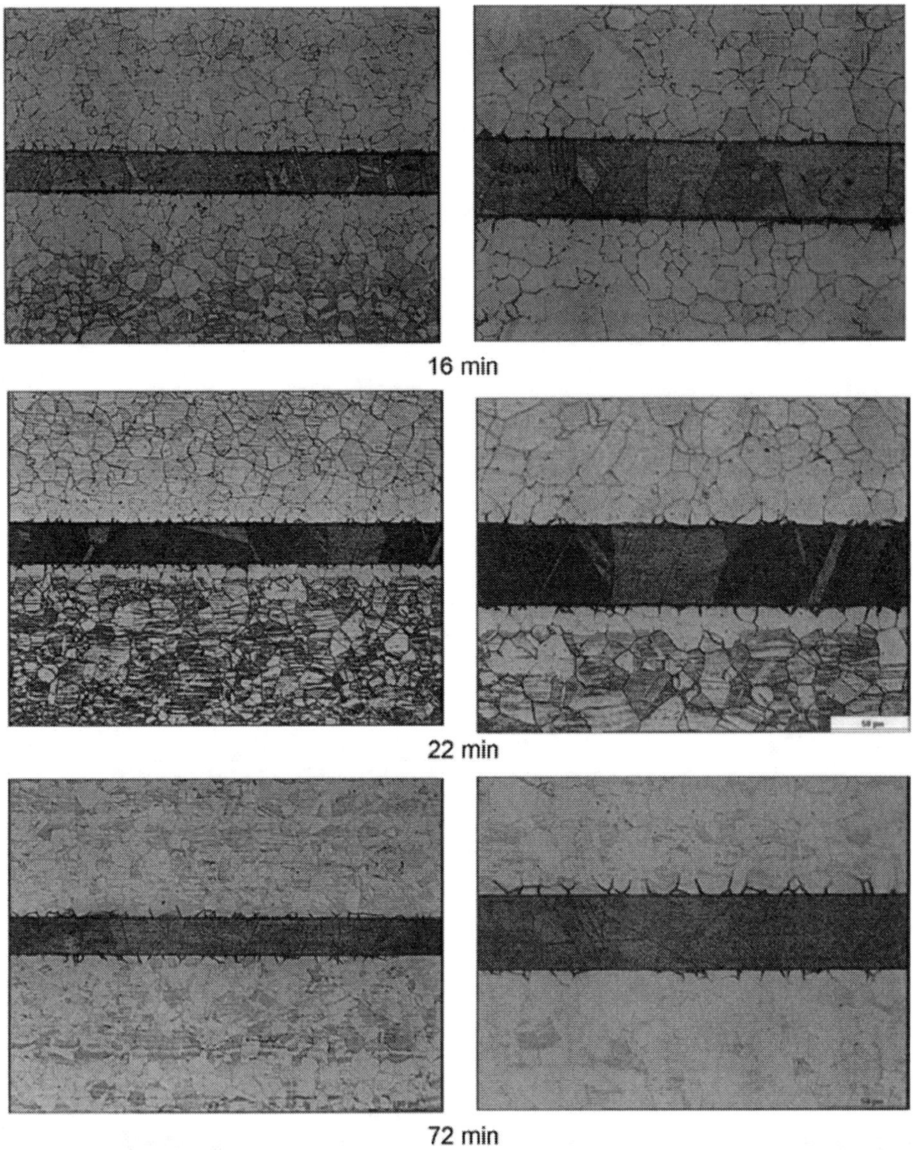

Figure 5: Optical microstructure of the bonds prepared at 900C for three different holding times.

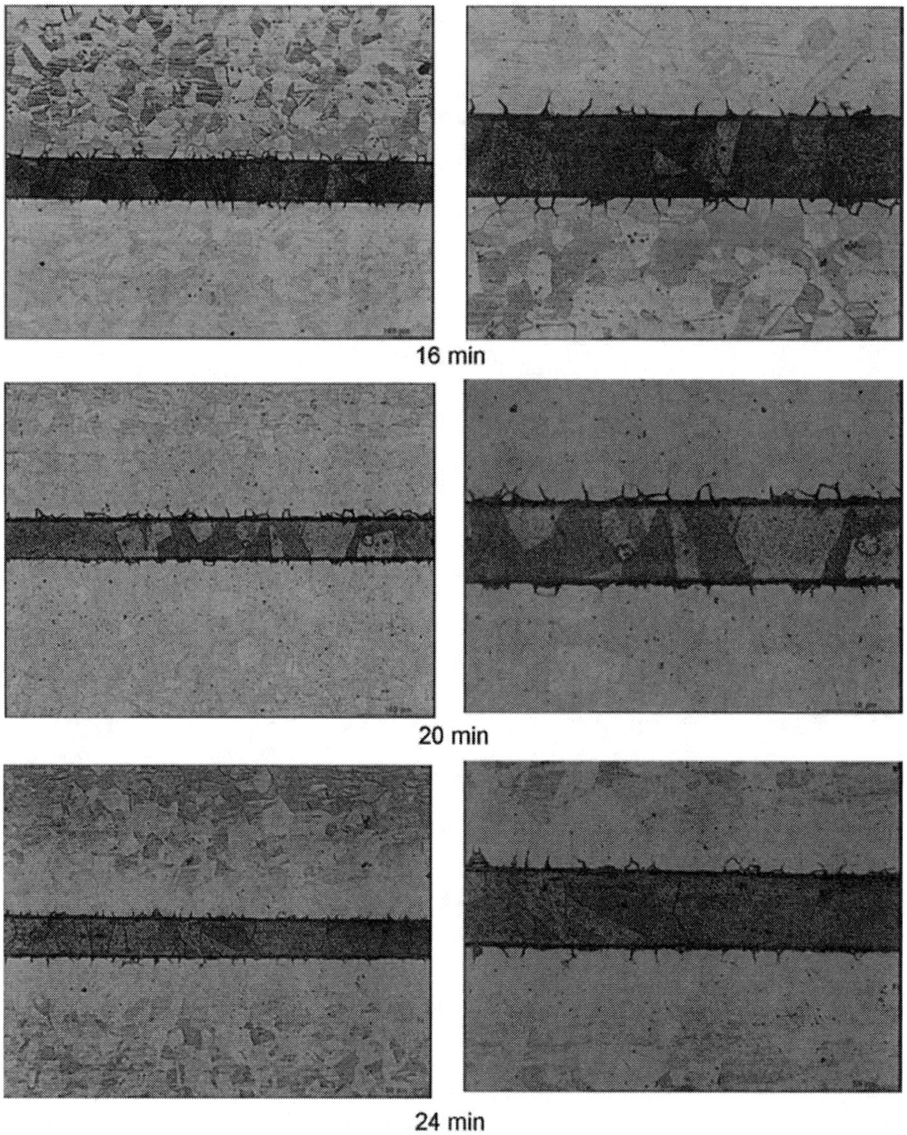

Figure 6: Optical microstructure of the joints prepared at 950C for three different holding times.

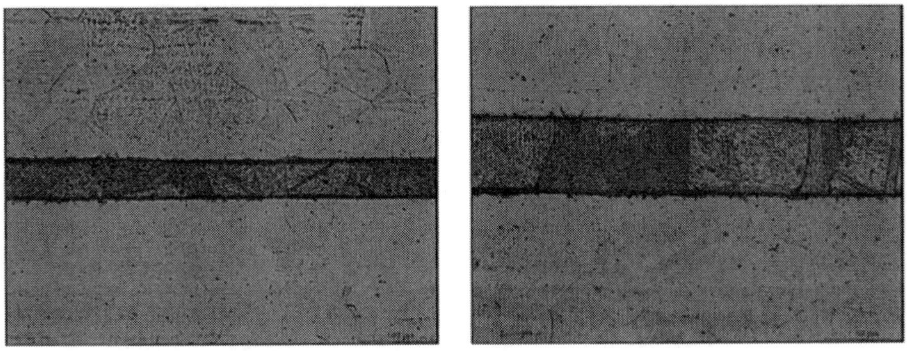

Figure 7: Optical microstructure of the specimens bonded at 1000C for 20 min holding time.

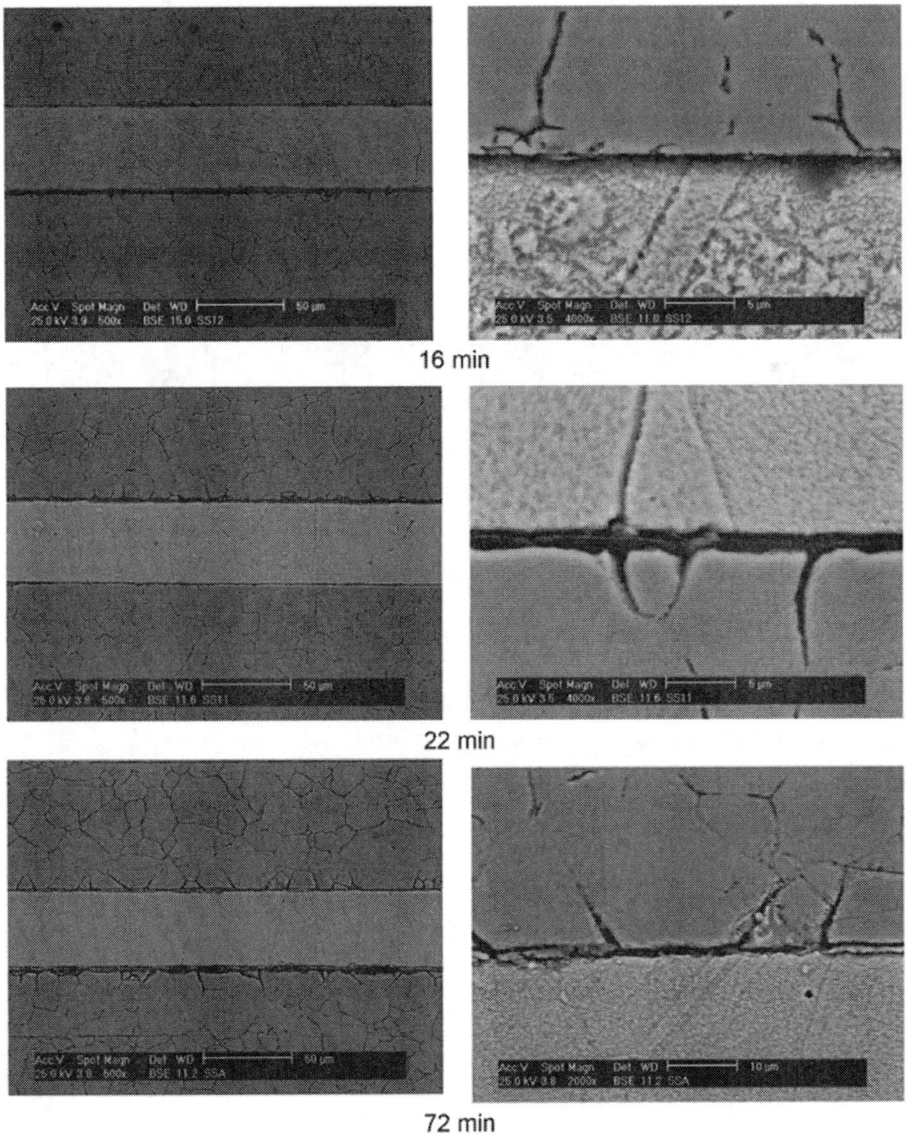

Fig.8: SEM-BSE images of the joints prepared at 900C for three different holding times.

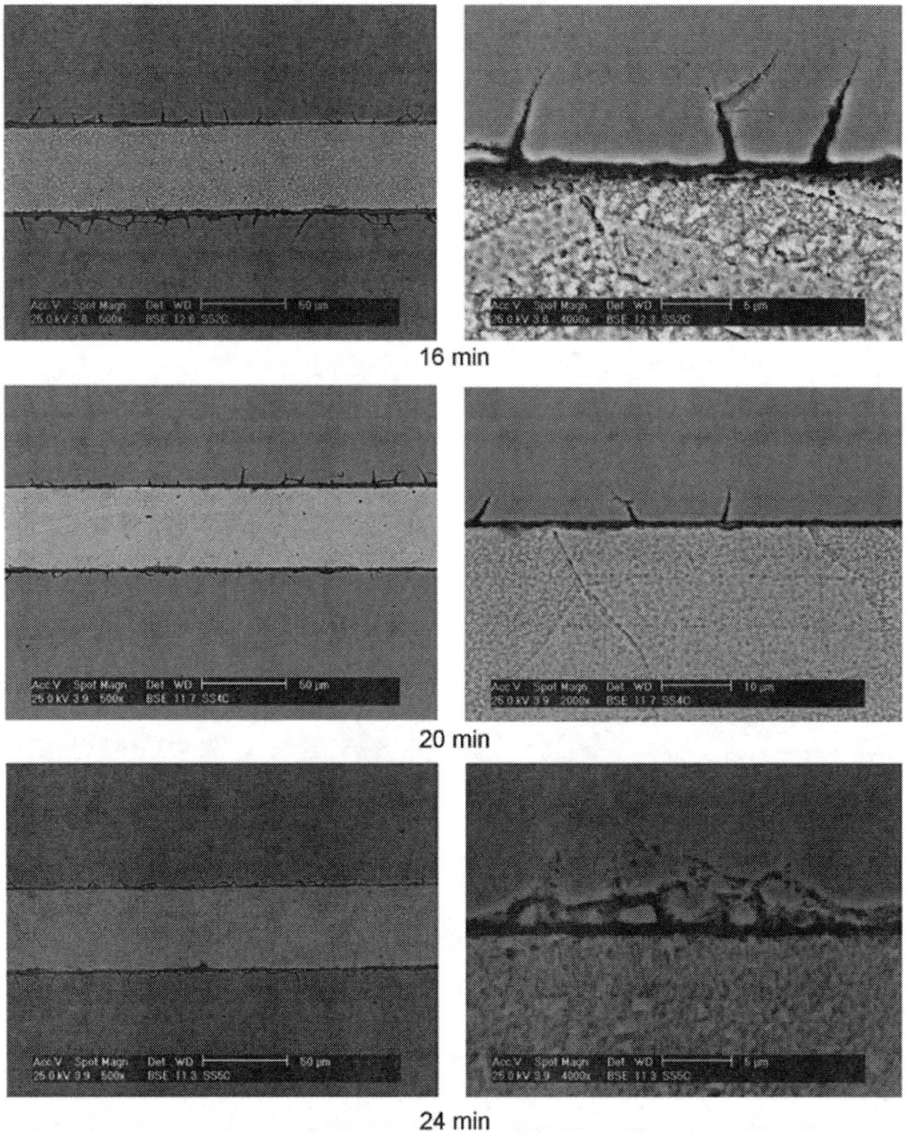

Figure 9: SEM-BSE images of the joints prepared at 950C for three different holding times.

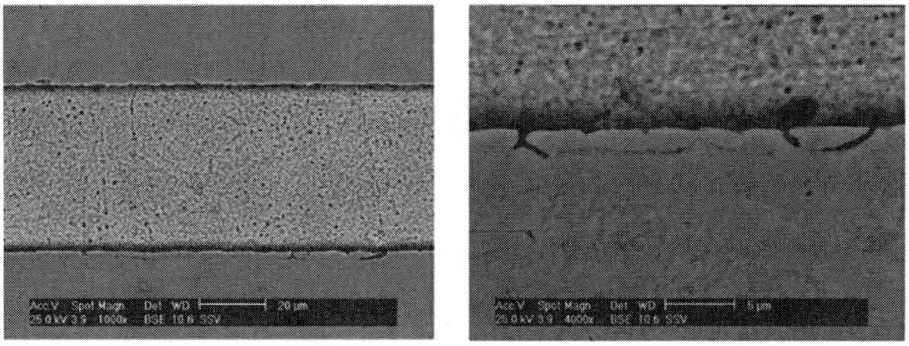

Figure 10: SEM-BSE images of the bonds prepared at 1000C for 20 min holding time.

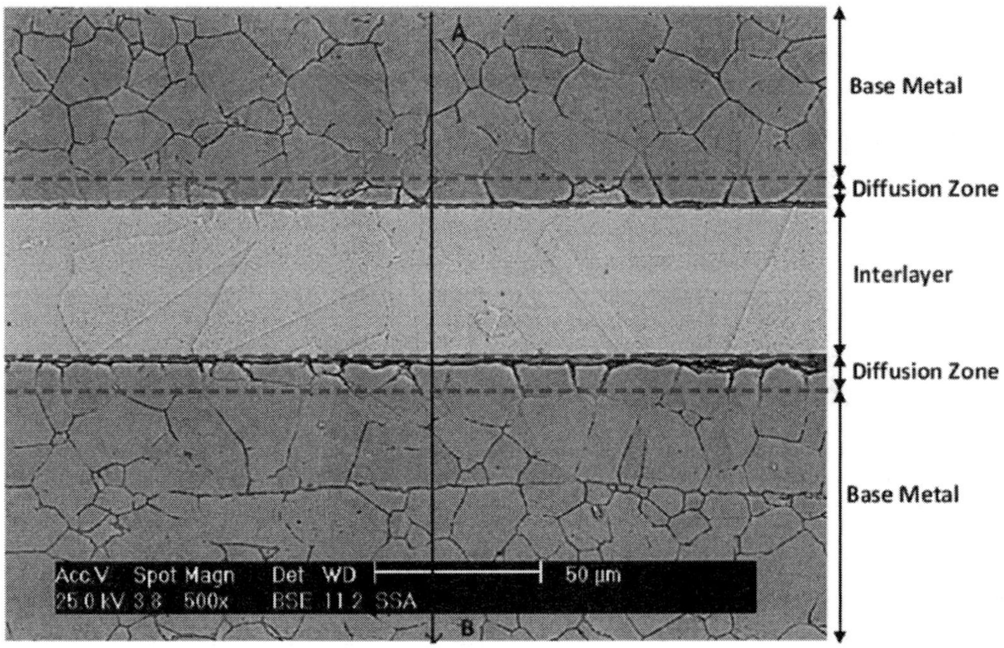

Figure 11: Three distinct zones in the bonds, showing the solute infiltrated to the grain boundries of the base metal in the diffusion zone.

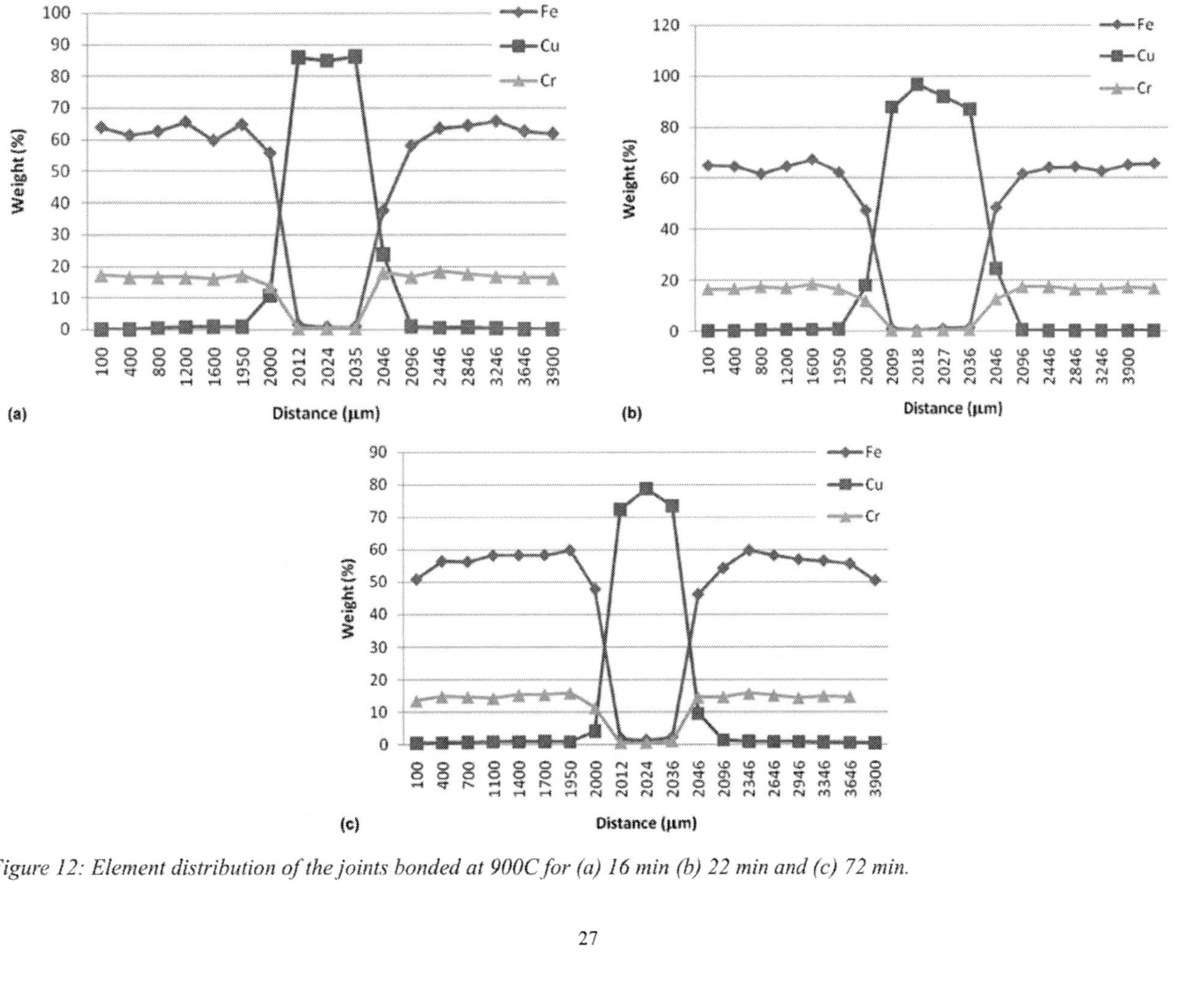

Figure 12: Element distribution of the joints bonded at 900C for (a) 16 min (b) 22 min and (c) 72 min.

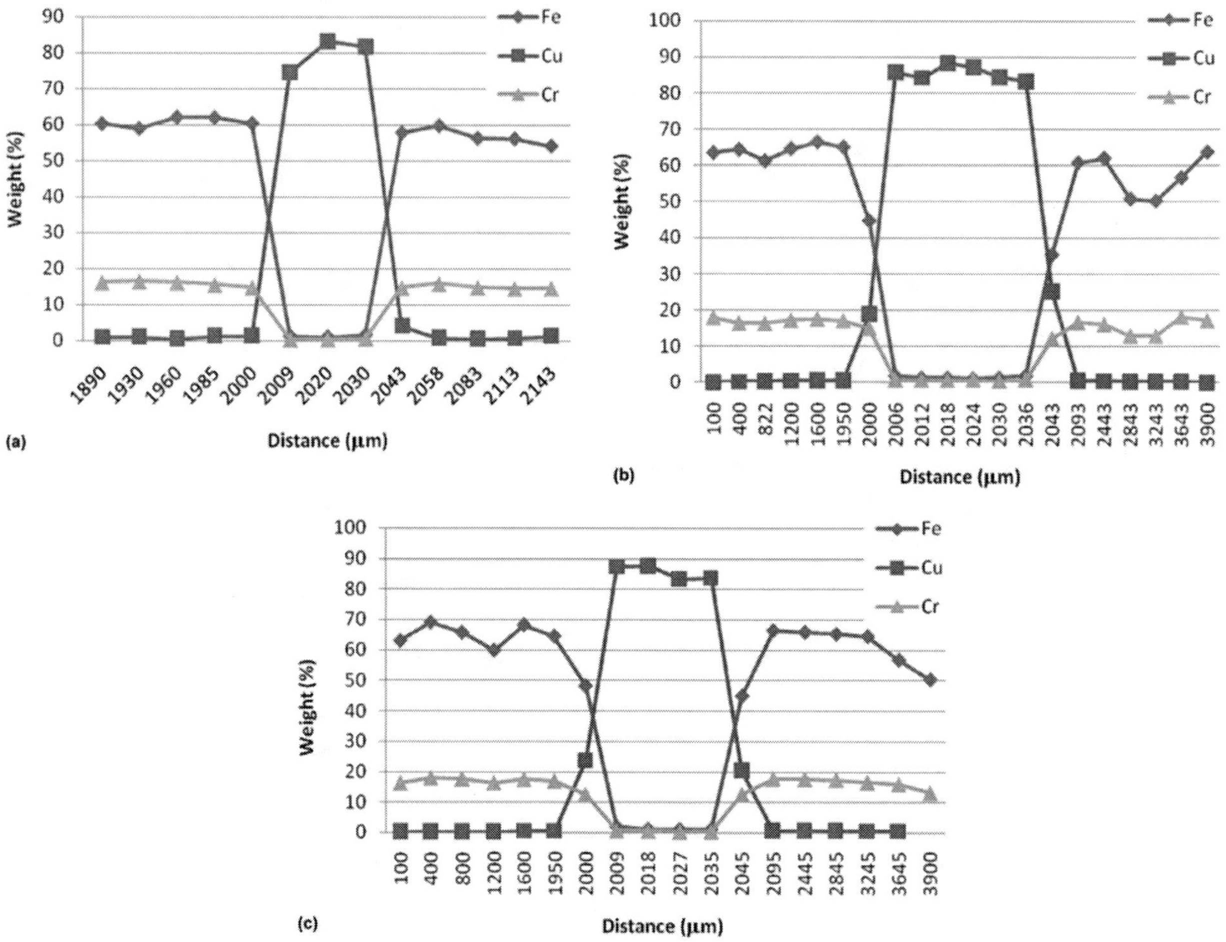

Figure 13: Element distribution of the joints bonded at 950C for (a) 16, (b) 20 and (c) 24 min.

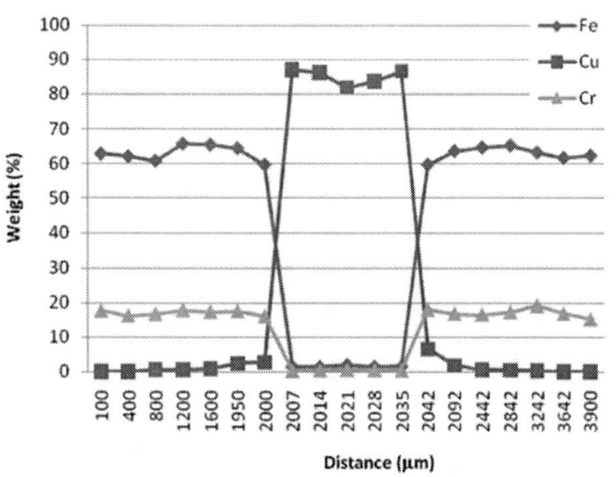

Figure 14: Element distribution of the joints prepared at 1000C for holding time of 20 min.

the average thickness of the diffusion zone in the stainless steel and copper at 900, 950 and 1000 °C for various holding time. The morphology of the diffusion zone was non-uniform therefore thickness measurements were taken at three different locations of the diffusion zone to have an estimation of average thickness.

Obviously, the thickness of the diffusion zones in the stainless steel and copper is increased with increasing the holding time. Generally, mass transfer has to be extended, depending on the bonding temperature. As can be seen, increasing the brazing time and temperature has considerable effect on the interlayer thickness. By increasing the brazing temperature more number of atoms migrated across the interface, hence the reaction layer widened considerably. By approaching to the melting temperature of Cu interlayer, 900-950 °C, Cu atoms were stimulated to move faster and in larger quantity. Therefore, it is enough for Cu atoms to vibrate and give possibility to Fe and Cr atoms to diffuse in the interlayer. Figure 20 shows the percentage of Cu in the interlayer at different temperature as a function of distance from the interlayer. The concentration of Cu decreased by increasing the brazing temperature and as a result the thickness of the interlayer decreased as listed in Table 5.

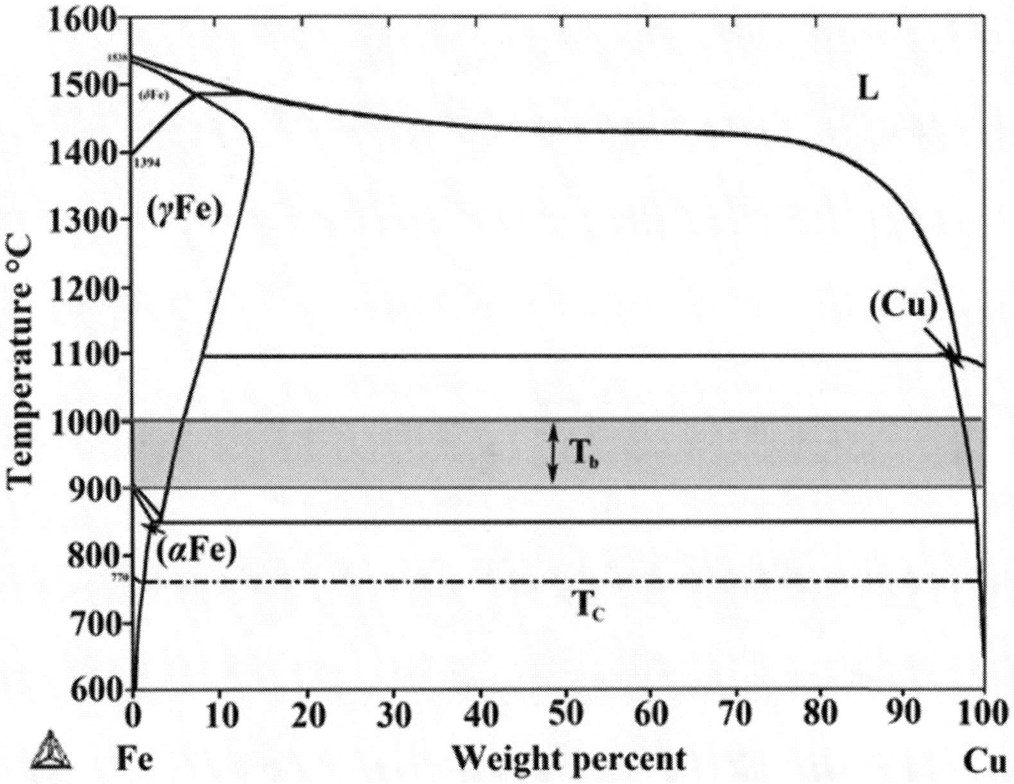

Figure 15: Fe-Cu binary phase diagram [4].

Table 4: EDS analysis at different regions of the brazed joints

	Average				
Region	**Fe**	**Cr**	**Cu**	**Ni**	**Mn**
A	69.11	17.48	0.99	7.33	1.69
B	68.64	18.02	0.78	7.78	2.06
C	65.82	18.32	1.26	8.52	1.83
D	71.47	17.68	1.28	7.58	1.8
E	67.48	18.4	0.46	8.57	2.09
F	67.89	17.98	1.26	7.99	1.49
G	69.06	18.54	0.5	7.48	1.96
H	48.9	14.2	26.31	6.46	1.23
I	20.09	7.07	59.49	2.04	0.29
J	68.29	17.64	1.74	7.87	2
K	68.19	18.51	0.75	7.17	1.64
L	69.55	18.06	4.39	4.09	0.81
M	58.15	17.96	18.64	2	0.54
N	68.31	18.56	0.74	7.44	1.9
O	5.35	1.88	87.32	1.99	0.4
P	7.82	6.97	81.3	1.5	0.45

Table 5: Average layer thickness in the joints prepared at 900, 950 and 1000°C at various holding time

Temperature	**900°C**			**950°C**			**1000°C**
Holding Time (min)	**16**	**22**	**72**	**16**	**20**	**24**	**20**
T_1 (μm)	9.286	9.657	15.15	10.013	14.8	16.385	7.581
T_3 (μm)	7.996	8.664	11.91	6.257	9.748	13.895	5.957
T_2 (μm)	47.129	46.933	46.05	45.113	43.357	43.24	42.355

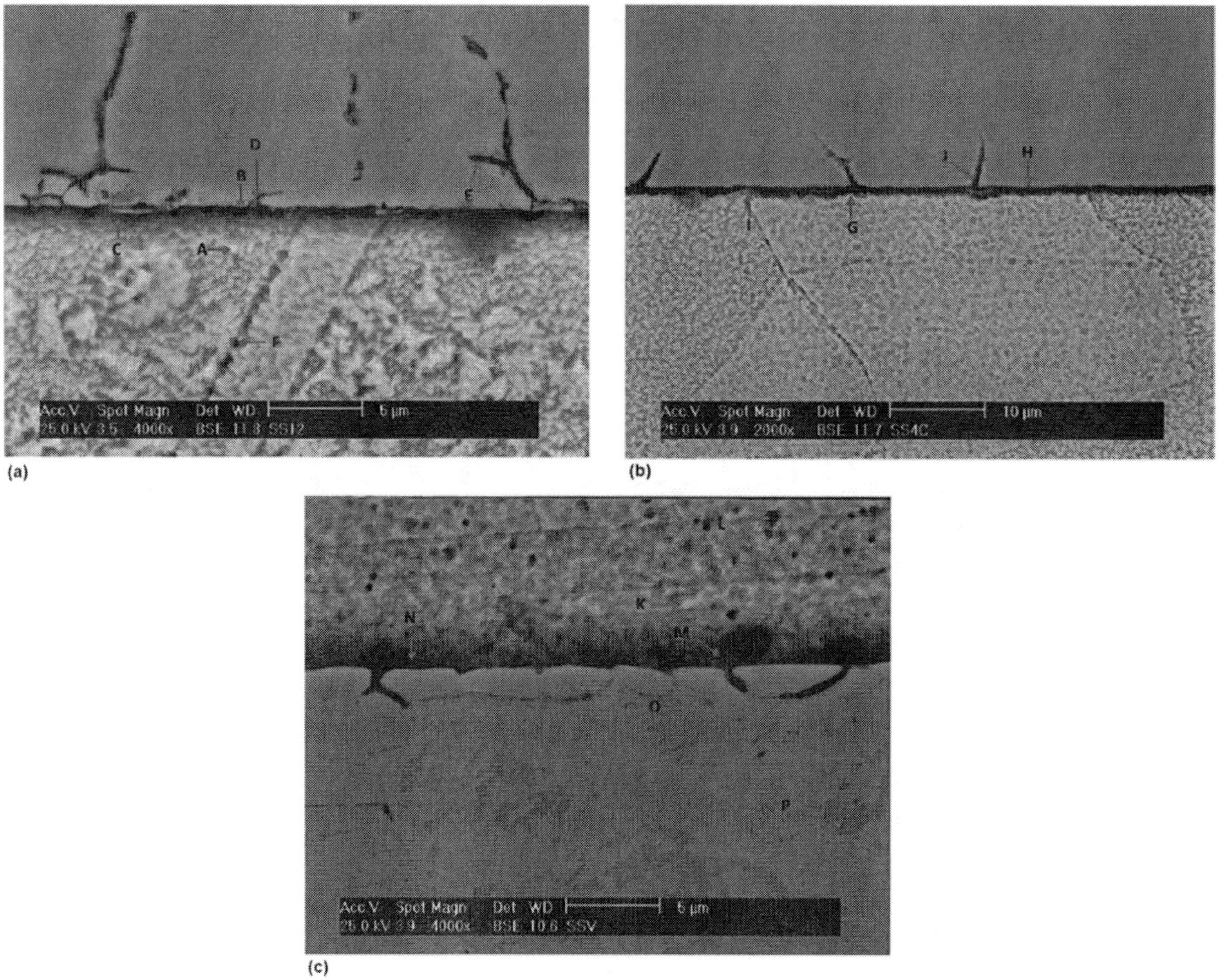

Fig.16: SEM-BSE images of the specimens brazed at (a) 900, (b) 950 and (c) 1000 °C.

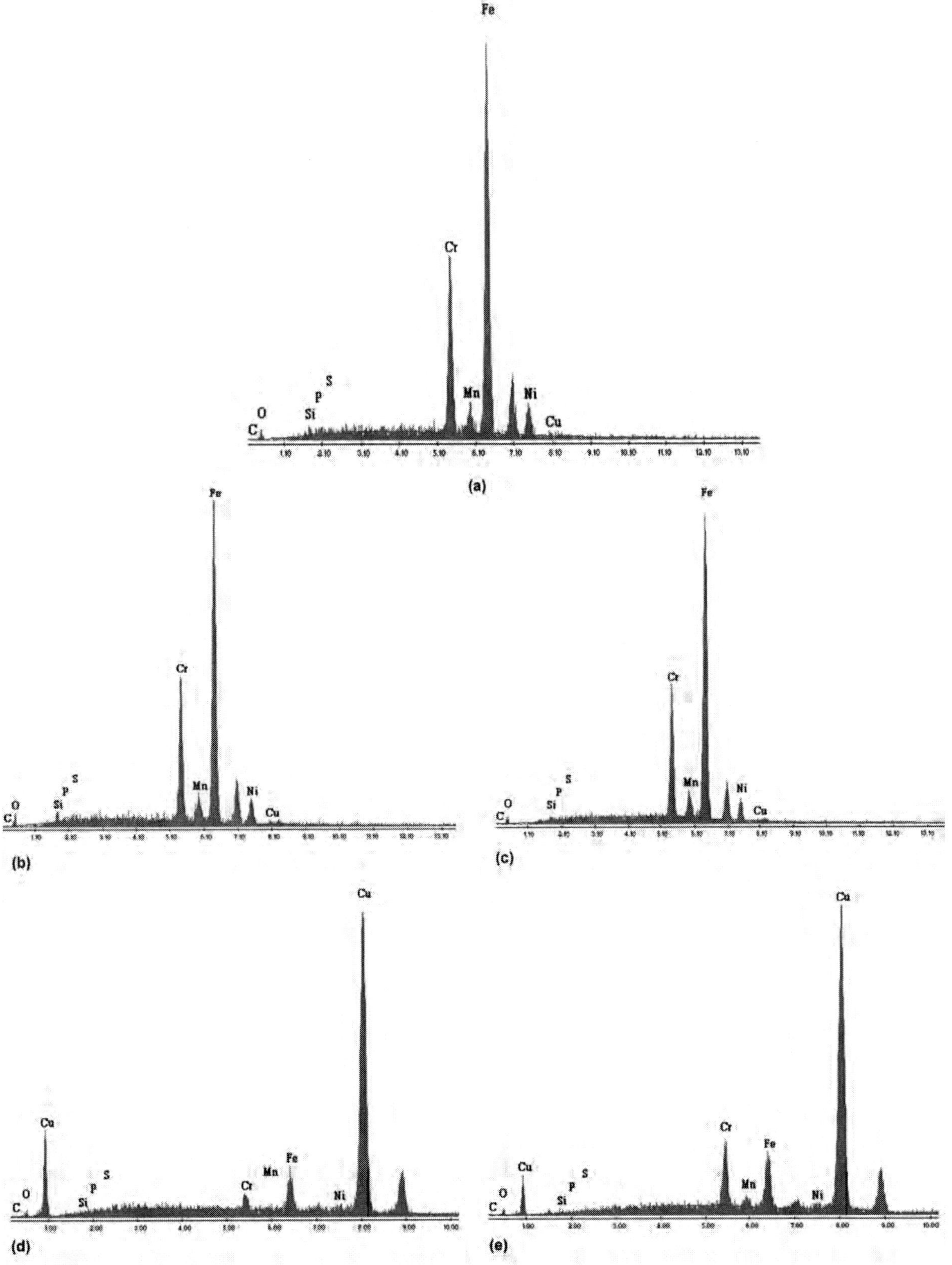

Figure 17: EDS analysis for region (a) A, (b) B, (c) C, (d) O and (e) P.

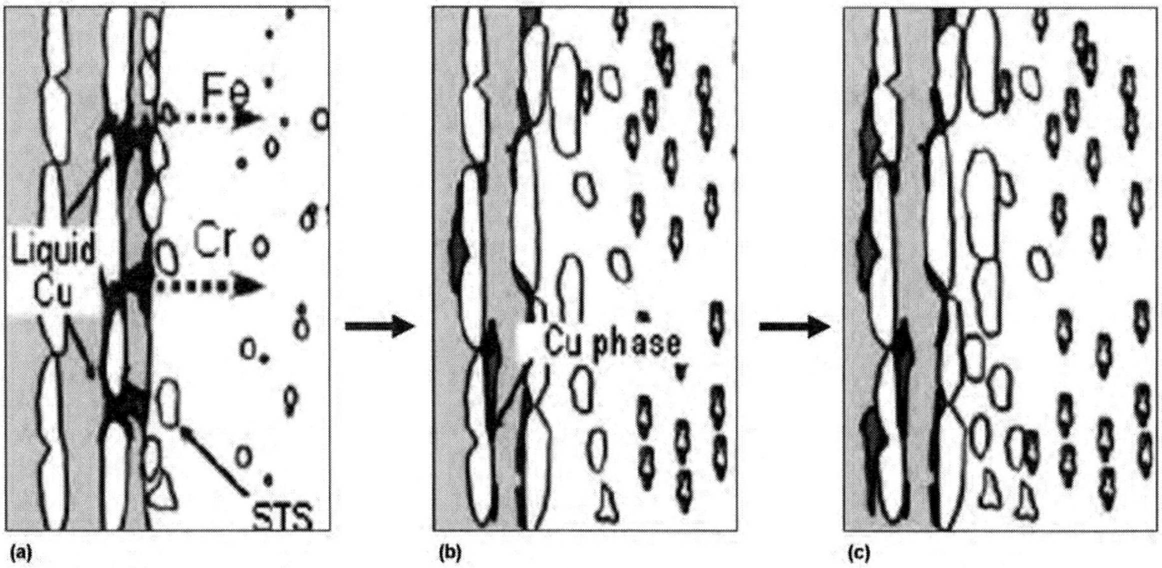

Figure 18: Schematic diagram of the brazing operation.

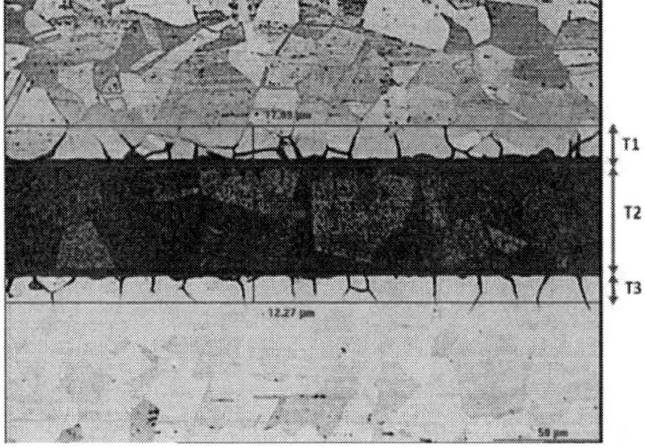

Figure 19: Scheme of layer thickness measurement.

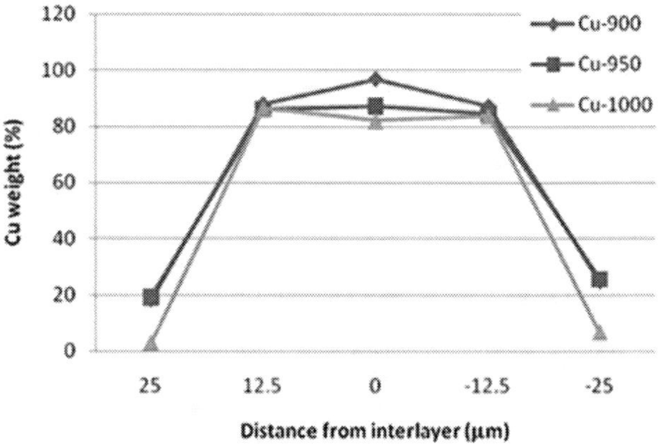

Figure 20: Concentration of Cu in the interlayer at different temperature as a function of distance from the interlayer.

3.1.3 Microhardness

Microhardness was measured in different regions of the base metal, diffusion zone and the interlayer. The micrographs of indentations on each region is shown in Fig. 21 and the microhardness results are listed in Tables 6, 7, and 8 for the joints prepared at 900, 950 and 1000 °C, respectively. Figure 22 (a) and (b) show the microhardness profiles as a function of distance from the interface at 900 and 950 °C, respectively. The microhardness tests were performed under an indentation load of 245.2 mN for 15 s. It is seen that the highest hardness value were obtained from diffusion zone and the hardness decreased gradually as the distance from the diffusion zone increased. On the stainless steel side, there was an increase in hardness near the interface which is due to the strain hardening effect. It is believed that by interdiffusion of the alloying elements from the base metal and the interlayer the new phases are formed at the diffusion zone and hence the microhardness increased. It was observed that γFe + eutectic Cu + Cr accumulated along the diffusion zone.

In addition, it is determined that the hardness taken from the stainless steel region bonded at 900, 950 and 1000 °C are lower than the values taken from the original stainless steel. The decrease in hardness values after bonding can attribute to the grain growth due to high bonding temperature and long diffusion time [7]. The hardness values measured on the interlayer of the bonded specimens were higher due to the presence of Fe and Cr in the vicinity of Cu interlayer. Figure 23 shows the microhardness profile of the joints prepared at different temperature for 20 min. The highest microhardness value was achieved for the specimen brazed at 950 °C. Generally speaking, microhardness value at diffusion zone increased by increasing the bonding temperature. In this case, the joint prepared at 1000 °C has 269 Hv while the joint prepared at 950 °C has 283 Hv. This could be explained owing to the grain growth at the diffusion zone due to high brazing temperature and increasing the ductility of the joint. Grain growth in the brazed joint materials can be attributed to recrystallization process. The tendency for grain growth after recrystallization is related to grain boundary energy.

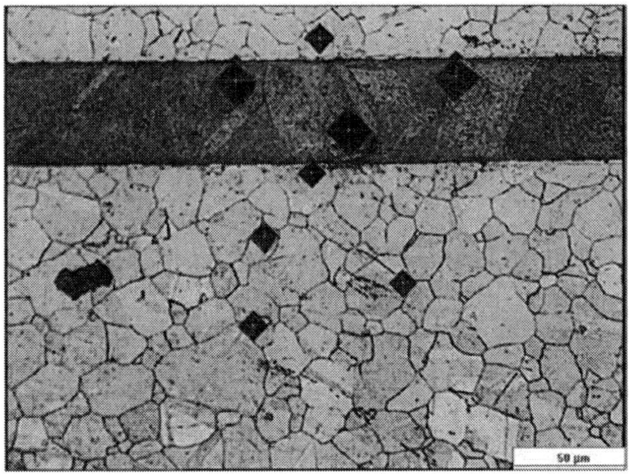

Figure 21: Micrographs of indentations on each region.

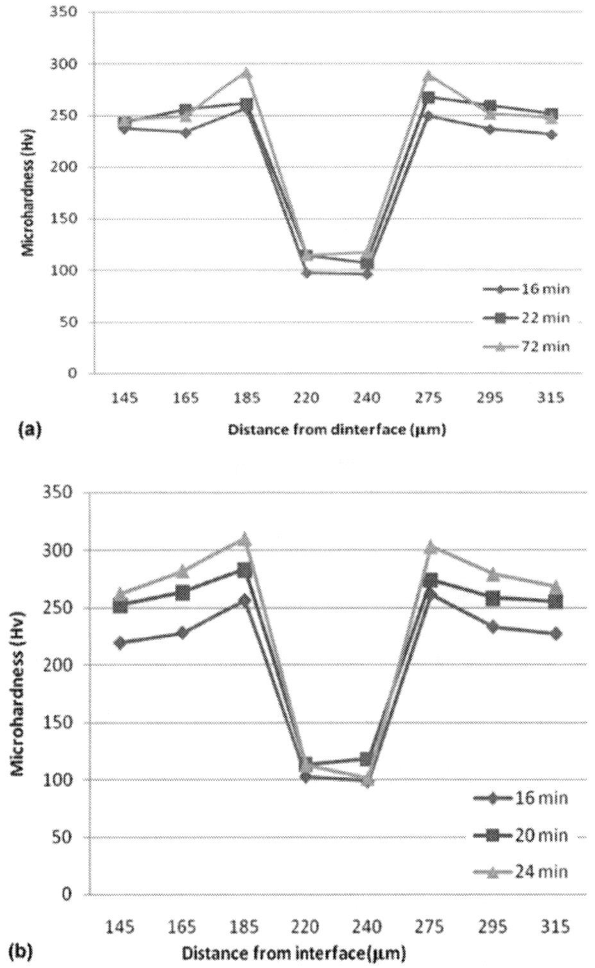

Figure 22: Microhardness profiles as a function of distance from the interface of the joints prepared at (a) 900 °C and (b) 950 °C.

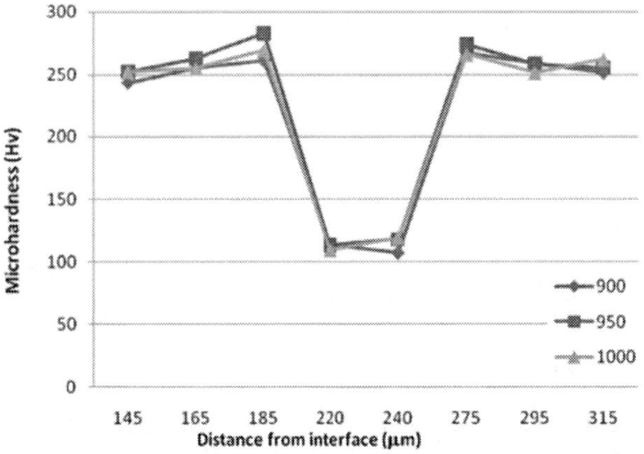

Figure 23: Microhardness profiles as a function of distance from the interface of the bonds prepared at 900, 950 and 1000 °C for 20 min holding time.

Table 6: Microhardness results for brazed joint at 900 °C

	Region							
Time	SS	SS	DZ	Cu	Cu	DZ	SS	SS
16 min	237	233	256	97.2	96	249	236	231
22 min	243	255	261	114	107	267	259	251
72 min	245	249	292	114	117	289	251	247

Table 7: Microhardness results for brazed joint at 950 °C

	Region							
Time	SS	SS	DZ	Cu	Cu	DZ	SS	SS
16 min	219	228	256	103	98.7	261	233	227
20 min	252	263	283	113	118	274	258	255
24 min	262	282	310	113	101	303	279	268

Table 8: Microhardness results for brazed joint at 1000 °C

	Region							
Time	SS	SS	DZ	Cu	Cu	DZ	SS	SS
20 min	252	255	269	109	119	266	251	262

3.1.4 Corrosion Test Result

The corrosion test was carried out on the brazed joints in 3.5% NaCl solution in order to determine the corrosion resistance of the bonds. Figure 24, 25, and 26 show the optical microstructure of the brazed joints after corrosion test at 900, 950 and 1000 °C. Black film deposits were present along the copper surface for all joints. It was seen that chloride ions were very aggressive ions to the copper and its alloys, due to the tendency of the chloride ion to form an unstable film (CuCl) and soluble chloride complexes (CuCl2 and CuCl32-) [8]. Thus, even trace amounts of chloride ions can cause corrosion problem to the interlayer. The most critical parameters such as corrosion potential (Ecorr), current density (Icorr) and corrosion rate (mmpy) values are summarized in Table 9.

Figure 27 shows the polarization curve for stainless steel and Fig. 28 (a) and (b) show polarization curve for the brazed joint samples treated at 900 and 950 °C for various holding time, respectively. Figure 29 shows polarization curve for three different temperatures at 20 min holding time for clear observation of each temperature at constant holding time. From general observation, it is found that all brazed joints show same behavior of polarization curve within each temperature due to uniform and continuously bonding contact. Besides, optical examination indicated corrosion attack along the stainless steel-copper interface and copper side.

From the polarization curve, it was found that the corrosion current density for stainless steel substrate increases quickly at the beginning of the anodic site. Diffusion-controlled anodic current behavior at the end of the curves was occurred due to fast corrosion rate and thus the stainless steel suffered severe attack in NaCl solution [9]. Obviously, cathodic Tafel line at all bonding temperature (900 and 950 °C) seems to significantly exhibit current density range, which is cathodic hydrogen reaction, was reduced by reducing the holding time (see Fig. 28. a-b). Thus, the free corrosion potential (Ecorr) become more negative at lower holding time. Moreover, by reducing the bonding temperature from 1000 to 900 °C, cathodic hydrogen reaction reduces and frees corrosion potential (Ecorr) increases negatively. This statement was shown in detail in Fig. 30 and 31, which are corrosion potential results (Ecorr) as function of the holding time and the bonding temperature, respectively. The corrosion potential of the brazed joints treated at longer holding time and higher temperature is more noble than shorter holding time and lower bonding temperature.

In addition, the Ecorr values of the brazed joints increase positively with increasing diffusion zone thickness. For example, for brazed joint at 950 °C for 16 min (DZ=10.013 μm), it is measured to be ~59 mV which is higher than the brazed joint at 24 min (DZ=16.385 μm) that listed in Table 9. Anodic polarization curve for all brazed joints (see Figure 28 a-b) shows that the current density significantly increases from free corrosion potential (Ecorr) due to dissolution of electrodes at the beginning of the anodic polarization and then the current density decreases due to the passivation of the electrode. At passivation region, the brazed joint surface is covered with passive oxide film. Then, we can see that the current density starts to increase suddenly due to breakdown of the passive layer (transpassive) until the end of the polarization curve. Transpassivation is a process of some chemical species in the original passive film turning into higher valence and more soluble products [10].

In order to further characterize the diffusion zone of the brazed joint after corrosion test, SEM-BSE examination of the transition joints were performed. In this examination, there are two conditions of the brazed joint to be studied. First, brazed joint after Tafel test and second is the brazed joint after immersing in 3.5% NaCl solution for 12 hours at room temperature and then proceed with Tafel test. Immersion of the brazed joint was performed in order to determine the open circuit potential (OCP). Open circuit potential is a steady-state or equilibrium potential of an electrode in the absence of the external current flow from the electrode. Figure 32 shows SEM-BSE images for the joints prepared at 900 °C for 22 min after Tafel test without immersing in the 3.5% NaCl solution. The Tafel test was directly performed after the diffusion brazing process. Figures 33 and 34 show optical microstructure and SEM-BSE images for the joints prepared at 900 °C for 22 min after Tafel test with immersing in 3.5% NaCl solution for 12 hours, respectively. Polarization curve for the brazed joint is shown in Fig. 35.

As mentioned, the corrosion attacks mostly along the stainless steel-copper interface and inside the copper region. At the anodic polarization curve, the current density was increased significantly for all brazed joints which are due to dissolution of some phases (γFe + eutectic Cu + Cr), presenting along the interface which would result in preferential localized corrosion. Figure 32 (b) and 11 (a) show SEM-BSE images that indicated severe corrosion attack with wide opening (trench formation) at the interface which may be attributed to the dissolution of the phases and the galvanic corrosion. The crevice could be formed after dissolution of the phases (γFe + eutectic Cu + Cr) along the stainless steel-copper interface. Crevice corrosion will develop in all brazed joint and preferential dissolution of the copper interface was favored during polarization. Stainless steel-copper joint will promote galvanic between stainless steel and copper induce a sufficiently different galvanic potential with the metal with the more active potential serving as anode by referring the galvanic series in seawater as shown in Figure 36. Corrosion on the anode of the couple usually exceeds its corrosion before the electrical connection of the two metals. On the other hand, galvanic action generally causes a reduction in the corrosion rate of the cathodic member of the couple [10]. In this case, stainless steel is more positive than copper so when stainless steel is in contact with copper, the copper will corrode first. The rate of galvanic attack is governed by the size of the potential difference. As refer to the brazed joint, the surface area of the stainless steel is bigger than copper. Stainless steel has large surface area in contact with the electrolyte, 3.5% NaCl solution, while the sacrificial copper has very small surface area in contact with 3.5% NaCl solution; therefore, the stainless steel will generate a large corrosion current, concentrating a small area of sacrificial copper. Thus, firstly copper corroded and finally stainless steel oxidized. The larger the area of the stainless steel, the greater is the acceleration of the copper corrosion.

At the copper side, chloride ions were very aggressive ions, due to the tendency of the chloride ion to form an unstable film (CuCl) and soluble chloride complexes (CuCl2- and CuCl3 2-). Thus, even trace amounts of chloride ions can cause corrosion problem to the copper. In the aqueous chloride solution and at anodic potentials $\leq$200 mV, copper electrode undergoes dissolution-precipitation mechanism [11], leading to the formation of a diffusive compound CuCl2- and precipitation of CuCl on the surface as the concentration of CuCl2- reached saturation limit.

Figure 32 (b) and 42 (a) shows brazed joint after the Tafel test without the immersion process and with immersing in the 3.5%

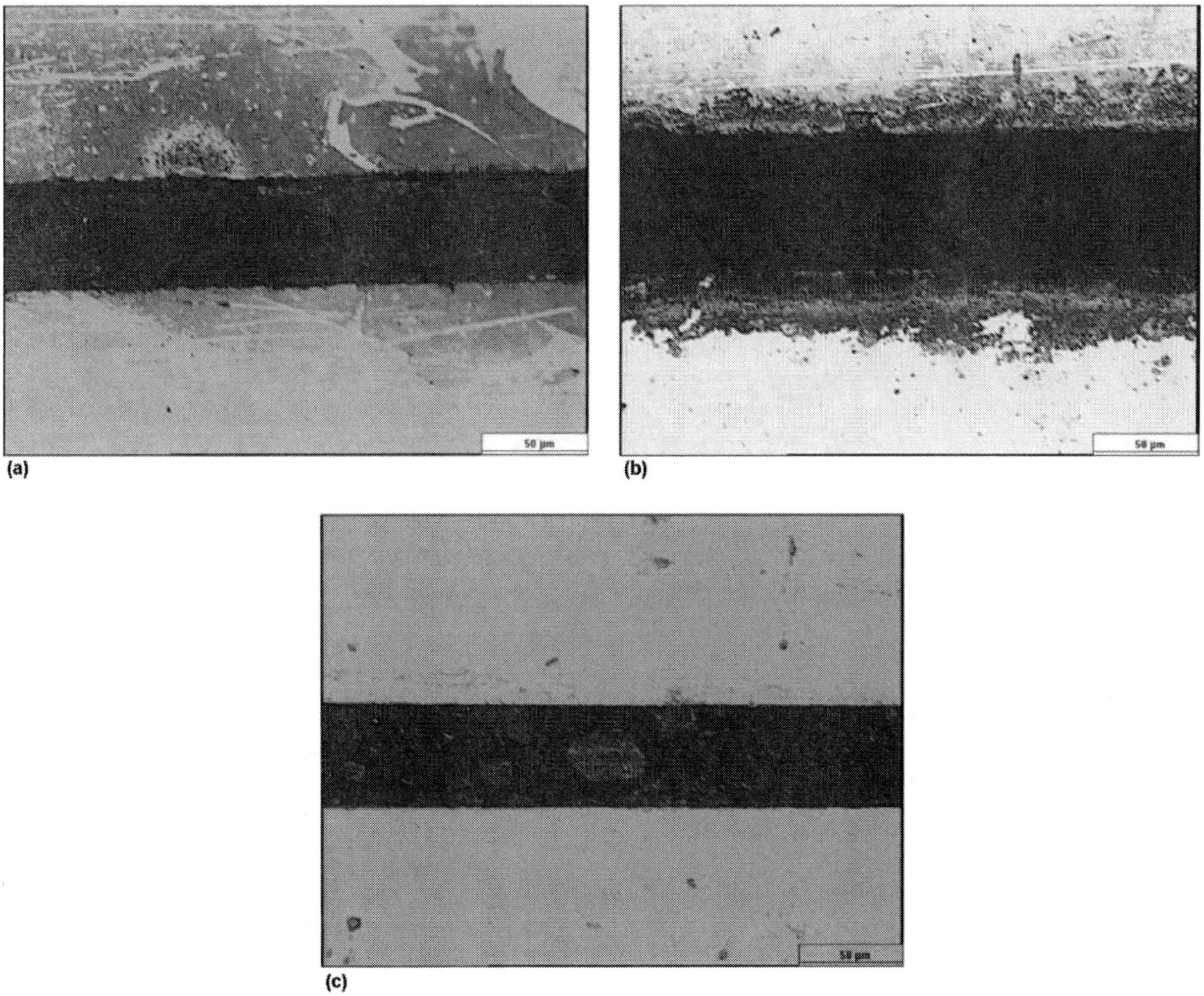

(a) (b) (c)

Figure 24: Optical microstructure of the joints bonded at 900 °C for holding time: (a) 16 min, (b) 22 min and (c) 72 min.

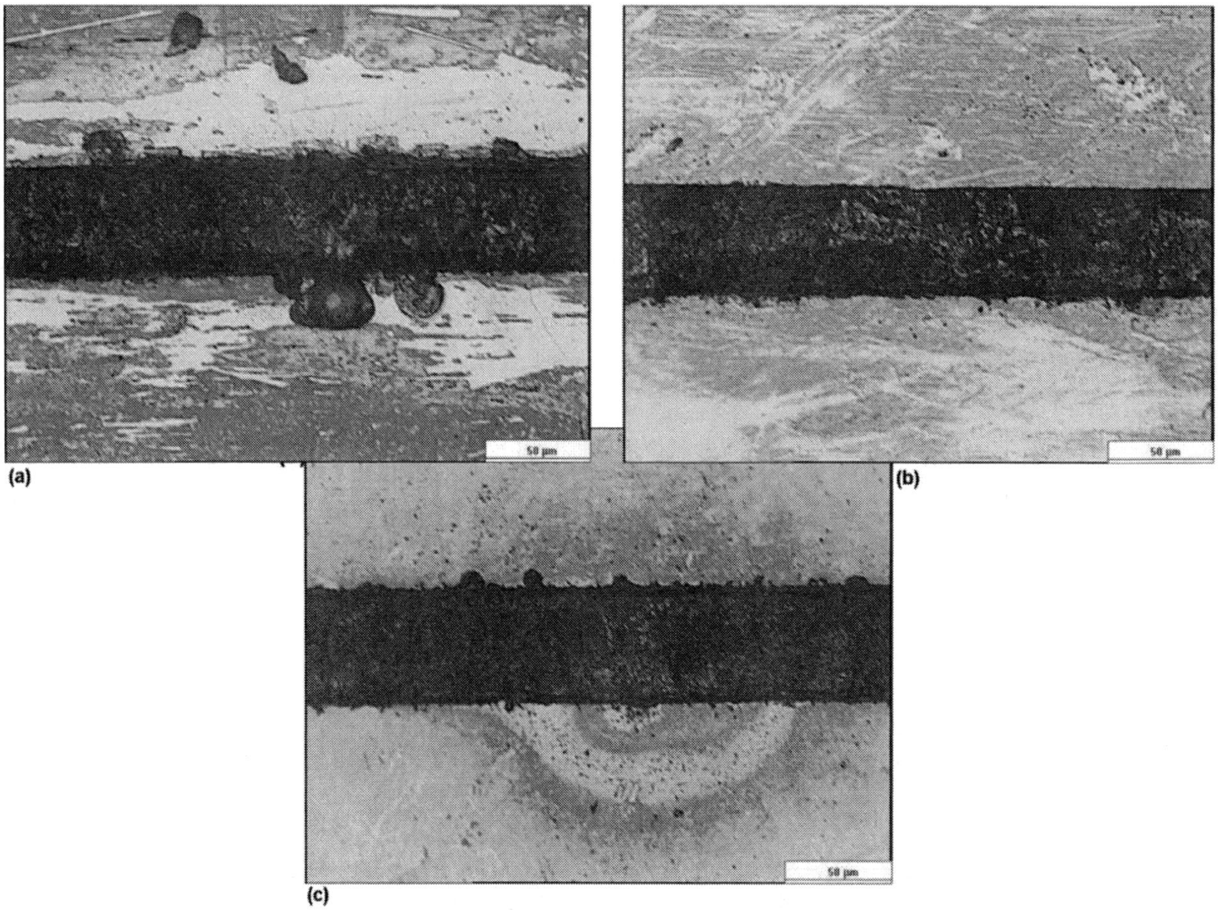

Figure 25: Optical microstructure of the joints bonded at 950 °C for holding time: (a) 16 min, (b) 20 min and (c) 24 min.

Figure 26: Optical microstructure of the specimen brazed at 1000 °C for 20 min holding time.

NaCl solution, respectively. It shows that crevice corrosion located along the interface of stainless steel-copper due to dissolution of (γFe + eutectic Cu + Cr) along the interface. For 316L SS, Oldfield and Sutton [12], reported a corrosion potential range from −350 to −400 mV vs SCE for the case of well-developed crevice corrosion in 1 M NaCl solution. It was shown that stainless steel and copper developed crevice corrosion very early during cathodic polarization. This presented preferential dissolution of the copper interlayer under anodic polarization. Besides, intergranular corrosion cracking also found at the copper surface. It is due to interdiffusion of Fe and Cr element along the copper grain boundary and also formed Fe-Cr intermetallics phase. This phase dissolves into (γFe + eutectic Cu + Cr) along the interface. Figure 34 (b) shows pitting attack on the surface of the copper. Generally, grain boundaries, inclusions such as (sulphides, oxides and nitrides) and local segregation of the alloying elements can act as irregularities, initiating the pitting corrosion. It can be clearly showed that the main alloying elements such as Cr, Ni, Mo and N were segregated into the copper interlayer. Pitting corrosion attacks and propagates easiest to regions where Cr and Mo contents are locally depleted to microsegregation, precipitation of carbides and the formation of intermetallic phases. Meanwhile, for the brazed joint without immersion test, there is no pitting attack found on the copper surface.

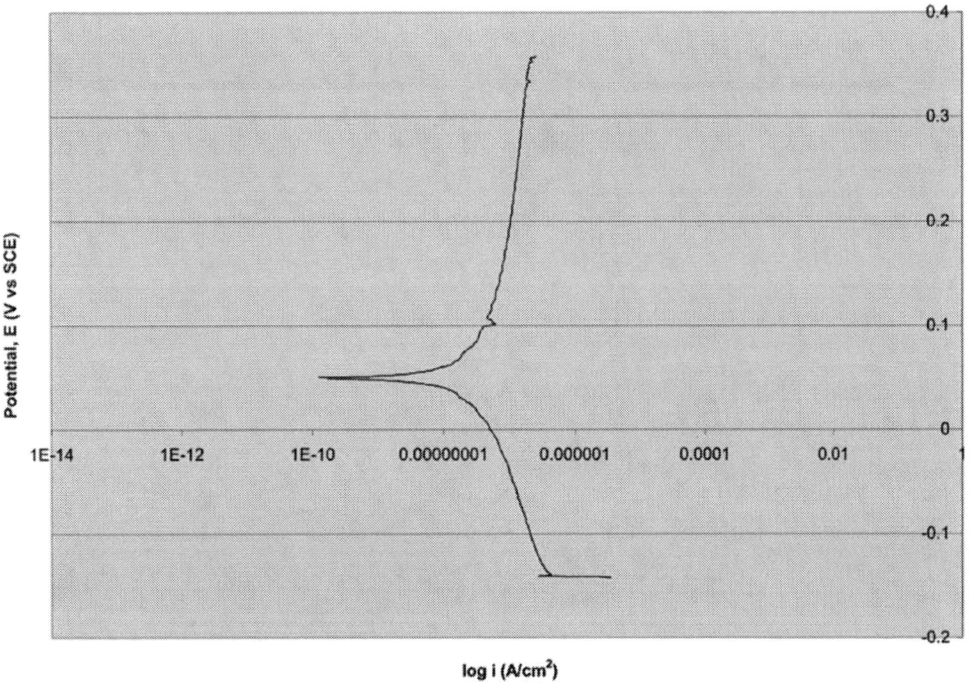

Figure 27: Tafel curve of the original stainless steel

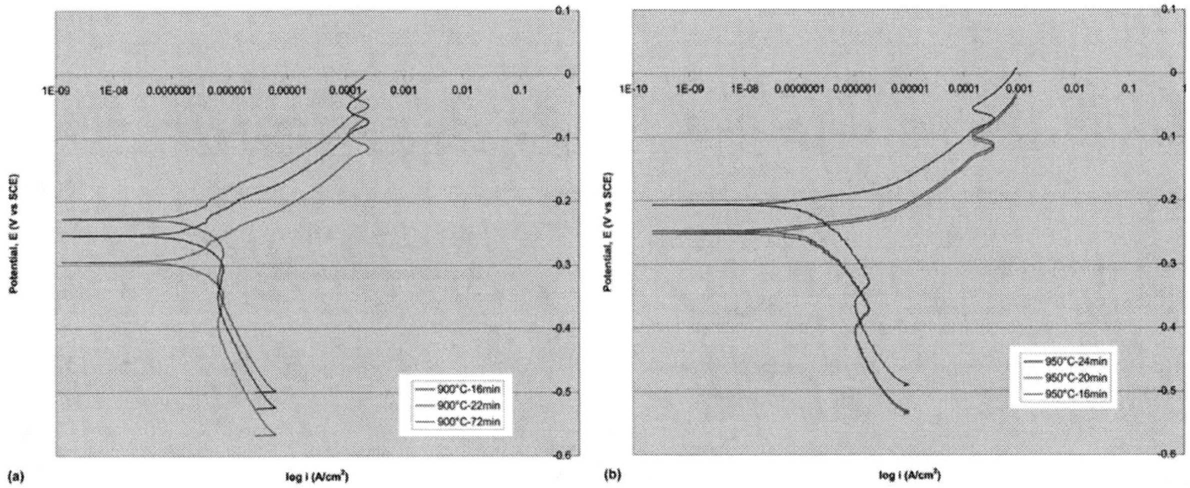

Figure 28: Tafel curve of the joints bonded at (a) 900 and (b) 950 °C for three different holding times.

Table 9: Corrosion potential (E_{corr}), current density (I_{corr}) and corrosion rate (mmpy) values of brazed joint at three different temperatures

Temperature (°C)	900			950			1000
Holding Time (min)	16	22	72	16	20	24	20
Diffusion zone thickness (μm)	9.286	9.657	15.15	10.013	14.8	16.385	7.581
E_{corr} (mV)	−251.782	−320.736	−236.496	−274.332	−267.202	−208.213	−227.98
I_{corr} (□A)	2.49E-1	3.83E+0	1.18E-1	1.25E+0	2.15E-1	3.40E-1	6.49E-2
Corrosion Rate (mmpy)	1.38E-1	2.13E+0	8.74E-2	6.90E-1	1.19E-1	1.89E-1	4.80E-2

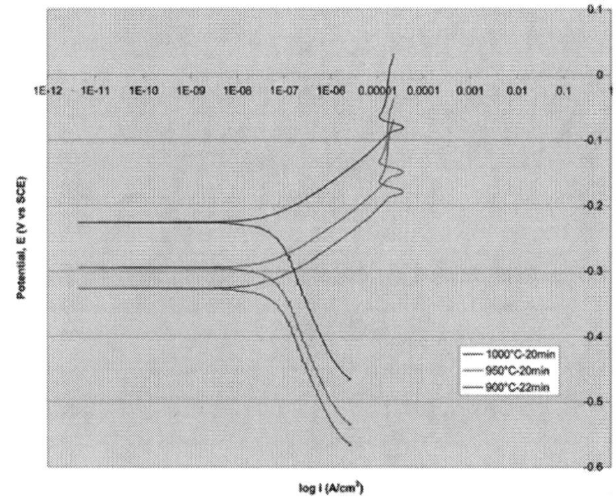

Figure 29: Tafel curve of the joints prepared at three different temperatures for 20 min holding time.

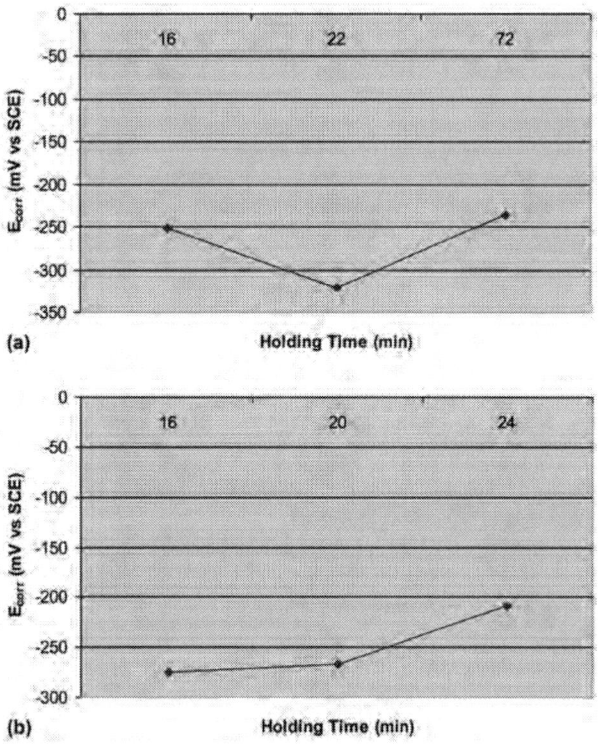

Figure 30: Corrosion potential as a function of holding time for the bonds prepared at: (a) 900 and (b) 950 °C.

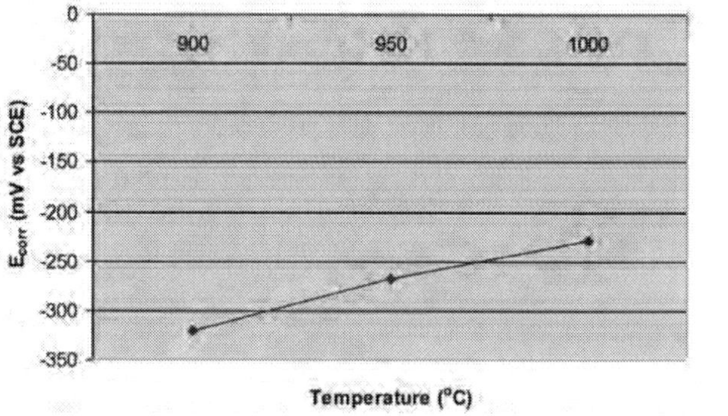

Figure 31: Corrosion potential as a function of different temperature for the specimens bonded for 20 min holding time.

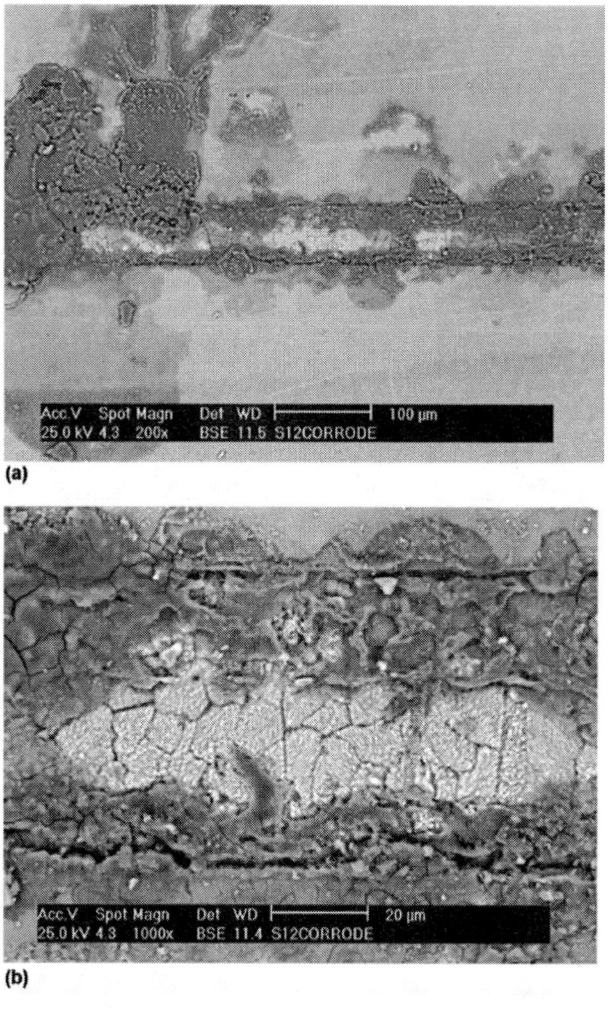

Figure 32: SEM-BSE images for brazed joint after tafel test.

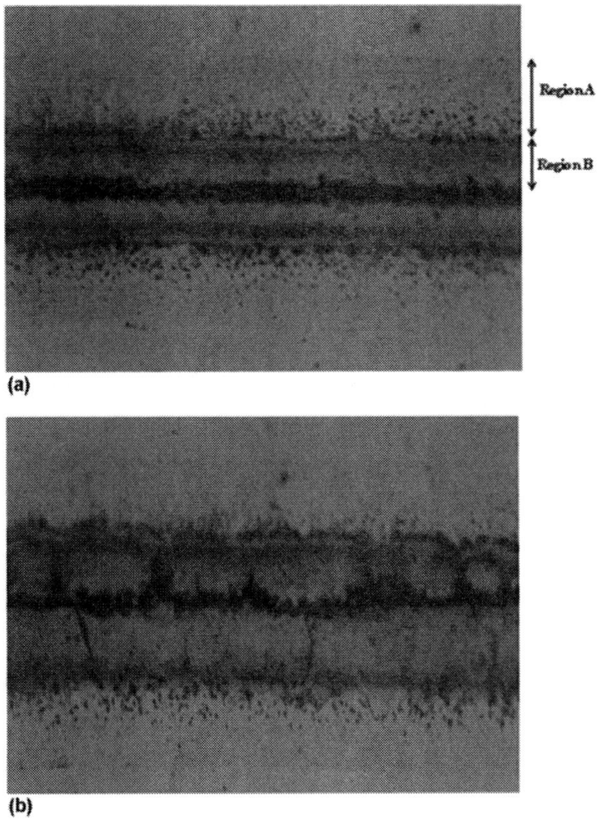

Figure 38: Optical microstructure of diffusion brazing assemblies (a) two distinct regions within the diffusion zone joined (and (b) diffusion zone.

of interlayer, 920 °C, aluminium interlayer was continue to shrink due to chemical reaction at stainless steel and aluminium interface. From the observation, almost the entire aluminium layer is consumed at 22 min holding time and aluminide layers from both sides seems to in contact with each other.

3.2.1 Interface characterization

According to Al-Fe binary phase diagram (Fig. 39), aluminium forms with iron solid solutions, intermetallic compounds and a eutectic composition. The solubility of iron in aluminium is extremely poor. Formation of new phase Fe-Al intermetallic when brazing temperature is 920 °C. During brazing, the solid stainless steel base metal reacts with the molten aluminum filler metal, resulting in the formation of intermetallic layers at the interface between stainless steel and aluminium. Formation of intermetallic compounds is due to the enhanced Al concentration in the diffusion zone of the stainless steel surface.

Figure 40a and b show SEM-BSE images of the transition joints were taken from the diffusion zone at stainless steel/aluminium interlayer interface. A quantitative overview of the chemical compositions analyses for different regions (region a to h) in the diffusion zone are listed in Table 11 using EDS analysis. From the EDS analysis, region A to d actually form in the stainless steel substrate and region g and h form in the aluminide region between stainless steel and aluminium. In addition, its shows that the concentrations of Al element change monotonically with distance accept other elements. It is possibly owing to the random distribution of fine aggregates different phases within the bond zone.

The variation in microhardness across the diffusion zone will provide useful information about the nature of the phases constituting the diffusion zone. Figure 41 shows the hardness distribution from stainless steel, region A, region B to aluminium. The microhardness tests were performed under an indentation load of 20g for 20 s. It is clear that the diffusion zone is harder than either aluminium or stainless steel region. Moreover, region B is harder than region A. However, region A that contains a large number of pores shows some decreasing in hardness value.

4.0 Conclusions

The diffusion brazing was carried out for stainless steel base metal and using copper and aluminium as interlayers. Bonding was carried in the temperature 900, 950 and 1000 °C for 16, 20, 24 and 72 min for stainless steel-copper and 920 °C at 22 min for stainless steel-aluminium in vacuum furnace. The important findings are as follows:

- The microstructure studies revealed that the brazed joint that γFe + eutectic Cu + Cr accumulated along the diffusion zone.
- The diffusion zone for the brazed joint increase with increasing temperature and holding time.
- The brazed joints developed crevice corrosion due to a galvanic couple formed between stainless steel and copper interlayer and presented preferential dissolution of copper interlayer under anodic polarization in 3.5% NaCl solution at room temperature.
- In addition, intergranular stress corrosion cracking also found on the copper region.
- After immersed the brazed joints for 12 hours, pitting attack was concertrated on the copper surface.
- Diffusion brazing of stainless steel using aluminium interlayer produced different layers of intermetallics compounds along the interface between the stainless steel and aluminium. According to Al-Fe binary phase diagram, the new phase is Fe-Al intermetallic.
- At the given temperature, 950 °C, a large number of pores appear along the interface close to aluminium. It is clearly shows that by using high brazing temperature is impractical while it is lead to oxidation and produced pores and void at the interface.
- The highest microhardness result was obtained at the diffusion zone and hardness value decreased gradually as the distance from the joint increased using both interlayer.

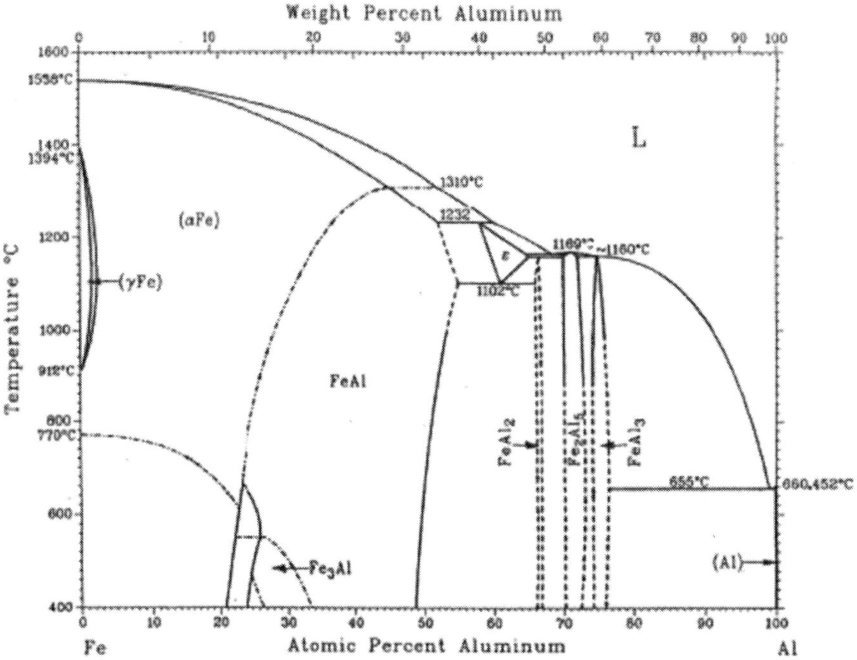

Figure 39: Al –Fe phase diagram.

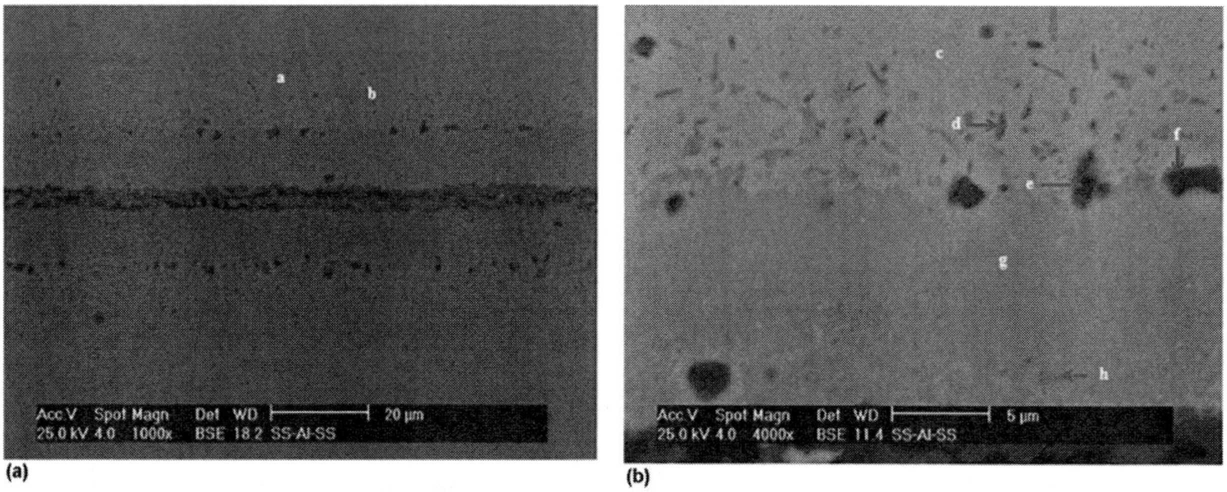

Figure 40: SEM-BSE images of the specimens brazed.

Table 11 : EDS analysis at different regions of the brazed joints

REGION	Al	Fe	Cr	Ni
a (matrix)	1.69	68.38	18.35	6.41
b (precipitate)	4.99	64.9	19.16	4.89
c (matrix)	7.43	64	19.81	5.56
d	5.62	64.08	20.01	6.92
e	15.34	59.41	10.78	4.3
f	12.29	47.56	9.89	3.39
g (matrix)	15.47	53.78	9.83	17.25
h (precipitate)	15.23	62.98	9.17	3.92

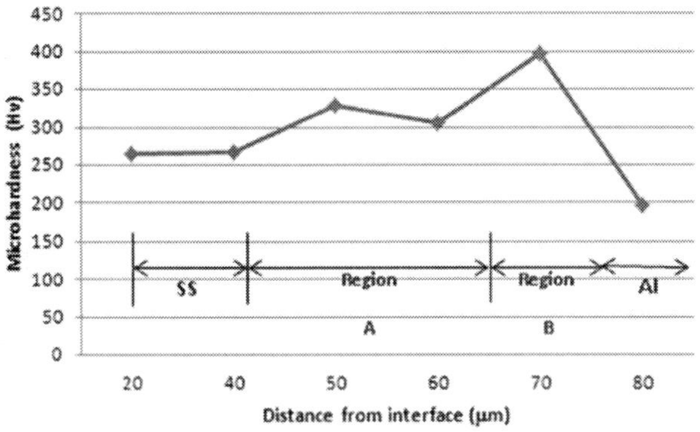

Figure 40: SEM-BSE images of the specimens brazed.

REFERENCES

1. N. Orhan, T.I. Khan, M. Eroglu. Diffusion bonding of a microduplex stainless steel to Ti-6Al-4V, Scripta Mater. 45 (2001) 441-446.
2. Brazing Manual. 1975. American Welding Society, Miami, Fla.
3. S.K. Mannan, V.Seetharaman and V.S.Raghuthan. A study on interdiffusion between AISI type 316 stainless steel and aluminium. Mater. Sci. Eng. 60 (1983) 79-86.
4. Stephen D. Cramer, Bernard S. Covino, ASM handbook, ASM International. Handbook Committee, 2005.
5. Hye Sung Na, Jeong Kil Kim, Bo Young Jeong and Chung Yun Kang. Effect of brazing conditions on microstructure and mechanical properties of duplex stainless steel to Cr-Cu alloy with Cu-base insert metal. Metals and Mater. International, vol. 13, No. 2007), pp. 511-515.
6. E.A. Brandes, Brook G B (Eds.), Smithells Metals Reference Book, 7th ed., Butterworth–Heinemann, Oxford, 1992, 938.
7. A. Elrefaey and W. Tillmann. Solis state diffusion bonding of titanium to steel using copper base alloy as interlayer. Journal Mater. Process. Techn. 209 (2009) 2746-2752.
8. Yeow CW, Hibbert DB and Galvanostatic. Pulse plating of copper and copper (I) halides from acid copper (II) halide solutions. J. Electrochem. Soc. 1983, 130: 786-90.
9. Y.W. Song, D.Y. Shan, E.H. Han, Materials Letters 62 (2008) 3276–3279, Electrodeposition of hydroxyapatite coating on AZ91D magnesium alloy for biomaterial application3. G. Song, Corr. Sci. 47 (2005) 1953.
10. Harvey P. Hack, ASTM Committee G-1 on Corrosion of Metals ASTM International,1988 - Science.
11. Crousier J, Pardessus L, Crousier J-P. Voltammetry study of copper in chloride solution. Electrochim Acta 1988; 33: 1039-42.
12. Al-Abdallah MM. Impedance measurements on Inconel and Monel alloys in dead seawater and sulphate solution. Br Corros J 1996; 31: 213-7.
13. N. Bhuvaneswaran, U. Kamachi Mudali and P. Shankar. Characterization of aluminide coatings formed by diffusion alloying on nitrogen-containing type 316L stainless.
14. H.K.D.H. Bhadeshia, Proceedings of International Conference on Aluminum (INCAL 03), Aluminum Association of India, Bangalore, S.Subramanian, D.H. Sastry (Eds.), 2003, p. 195.

Proceedings of the 26th ASM Heat Treating Society Conference
B.L. Ferguson, R. Jones, D.S. MacKenzie, and D. Weires, editors
https://doi.org/10.31399/asm.cp.ht2011p0044

The Effect Of Cryogenic Processing On The Mechanical Properties Of Austempered Ductile Cast Iron (ADI)

Susil K. Putatunda, Codrick Martis
Department of Chemical Engineering and Materials Science, Wayne State University, Detroit, MI 48202, USA
sputa@eng.wayne.edu

Frederick Diekman
Controlled Thermal Processing Inc. Park City, Illinois 60085, USA
Info@metal-wear.com

Rozalia Papp
Air Liquide US LP, Countryside, Illinois 60525, USA
rozalia.papp@airliquide.com

Abstract

An investigation was carried out to examine the influence of cryogenic processing on the microstructure and mechanical properties of Austempered Ductile Cast Iron (ADI). ADI has emerged as a major engineering material in recent years because of its many attractive properties. These include high yield strength with good ductility, good fatigue strength, fracture toughness and wear resistance. In this investigation, compact tension and cylindrical tensile specimens were prepared from ductile cast iron as per ASTM standards and were austempered at a lower bainitic temperature of 288°C (550°F). These specimens were then cryogenically processed. The mechanical properties and fracture toughness of these samples were evaluated and compared with the non-cryogenically treated samples. The influence of cryogenic heat treatment on the microstructure of these samples was also examined. Test results show that the cryogenic processing can improve the mechanical properties without compromising the fracture resistance of the material.

Introduction

The focus of this investigation was to examine the influence of cryogenic processing on the microstructure and mechanical properties of Austempered Ductile Cast iron (ADI). ADI has emerged as a major engineering material in recent years because of its excellent properties such as high strength, with good ductility [1-4], good fatigue strength [5-10], fracture toughness [11-17] and excellent wear resistance [18-20]. It is now used extensively in many structural applications such as automotive components, locomotive wheels, gears, crankshaft, connecting rods, brake shoes etc. ADI has other advantages such as low production cost because of its good castability and excellent machinability and consequently longer tool life and shorter heat treatment processing cycles. Moreover, the density of ADI is about 10% lower than steel. Thus it has the added advantage of higher specific strength compared to wrought or forged steel. Because of these advantages, the use and application of ADI continues to grow worldwide and its market share is going to exceed more than several billion dollars in coming years.

In recent years there has been significant interest in cryogenic processing of materials. There have been several applications of cryogenic processing of materials for aerospace and other industries. This process has been used for aircraft components like turbine blades, landing gear systems, gears, cutting tools etc. Cryogenic processing has also been used to treat a variety of other materials such as nickel based super alloys, and even nonmetallic materials. Cryogenic processing has been found to improve the mechanical properties of these materials.

The conventional cryogenic processing involves subjecting materials to extremely cold temperature (in the range of -80 to -155°C) and then heating them to above room temperature. The cryogenic heat treatment cycle consists of several steps such as (a) Ramp down, (b) Hold, (c) Ramp up and (d) Tempering. During ramp down, the temperature of the material is cooled down to a very low temperature from ambient temperature very slowly. This slow cooling helps to reduce the temperature gradient within the component and keep stresses to a minimum. During holding period the temperature is held at the cryogenic temperature for a predetermined time period and in the ramp up stage the temperature of the bath is brought back to ambient temperature slowly. Finally the material is tempered at slightly elevated temperatures.

While cryogenic processing has been successfully used to improve the mechanical properties of various materials [21-22], its use in cast iron and ADI has been rather limited. This investigation was undertaken to explore the use of cryogenic processing in ADI and the objective of this investigation was to examine the influence of cryogenic processing on the microstructures and mechanical properties of ADI.

Experimental procedure

Material

The material used in this investigation was an unalloyed nodular cast iron. The chemical composition of the material is reported in Table 1. The material was cast in the form of KEEL blocks and from these cast blocks, cylindrical tensile specimens and compact tension specimens for fracture toughness tests were prepared as per ASTM standards E-8 [23] and E-399 [24] respectively.

Table 1: Chemical composition of the material

Element	wt %
C	3.6
Si	2.5
Mn	0.4
S	0.015
P	0.015
Mg	0.010
Fe	93.46

Heat Treatment

After fabrication the samples were austempered as follows. The sample were initially austenitized at 927°C (1700°F) for 2 hrs and then subsequently quenched in a molten salt bath maintained at 288°C (550°F) .The samples were then austempered at this temperature for 2 hours. After austempering the samples were treated cryogenically as follows;

The samples were initially cooled down to -150°C from ambient temperature very slowly i.e. 8 hour ramp down to -150° C and held at this temperature for twelve hours. After that the temperature of the samples were increased to ambient temperature in 8 hours. The samples were then tempered at +150°C.

Tensile Testing

Tensile testing of these samples was carried out as per ASTM standard E-8 [23]. The tests were carried out at a constant engineering strain rate of $4x10^{-4}$ s^{-1} on a servo hydraulic MTS (Material Test System) test machine at room temperature and ambient atmosphere. Load and displacement plots were obtained on a X-Y recorder and from these load-displacement diagrams, yield strength, ultimate tensile strength and % elongation values were calculated. Four samples were tested from each heat treated condition and the average values are reported in this paper.

Fracture Toughness Testing

After heat treatment, the compact samples were ground and then polished with 600 grit emery paper. The specimens were then pre-cracked in fatigue at ΔK level of 10 MPa√m with a load ratio R = 0.10 to produce a 2 mm sharp crack-front in accordance with the ASTM standard E-399 [24]. After fatigue precracking, the specimens were loaded in tension in a servo-hydraulic MTS test machine and the load-displacement diagrams were obtained with a clip-gauge in the knife edge attachment on the specimens. From these load displacement diagrams, P_Q values were determined using the 5% secant deviation technique as per ASTM standard E-399 [24]. From these P_Q values, K_Q values were determined using the standard stress intensity factor calibration function for the compact tension specimens. Since these K_Q values satisfied all the requirements for a valid K_{IC} test as per ASTM standard E-399, they are all valid K_{IC} values. Four samples were tested from each heat treated conditions and the average values are reported here.

Metallography and X-ray Diffraction Studies

Microstructures of all the samples were examined by optical microscopy after polishing and etching with 5% nital solution. X-ray diffraction (XRD) analysis was performed to estimate the austenite content and the carbon content of austenite following the procedure of Rundman and Klug [25]. XRD was done using a monochromatic copper K_α radiation at 40 kV and 100 mA. A Rigaku rotating head anode diffractometer was used to scan angular 2θ range from 42°-46° at a scanning speed of 0.25° per minute and in 2θ range of 72°-92° at a scanning speed of 1° per minute. The profiles were analyzed using Jade 5 software to obtain the peak positions and the integrated intensities of {111}, {220} and {311} planes of FCC austenite and {110} and {211} planes of BCC ferrite. The volume fractions of ferrite (X_α) and austenite (X_γ) were determined by the direct comparison method using the integrated intensities of the above planes [26]. The carbon content of the austenite was determined by the equation [27]

$$a_\gamma = 0.3548 + 0.00441\ C_\gamma \quad \dots (2)$$

Where a_γ is the lattice parameter of austenite in nanometer and C_γ is the carbon content of austenite in wt%. The {111}, {220} and {311} planes of austenite were used to estimate the lattice parameter. Three samples were examined from each heat treated condition and the data reported is the average from these samples.

Results and Discussion

Influence of Austempering Temperature on the Microstructure

The microstructure of the as-cast sample is reported in the Figure 1. It shows a predominantly pearlitic structure with graphite nodules dispersed in it. The graphite nodules were well rounded with nodularity of about 85%. The microstructure of the austempered samples is reported in Figure 2 while the microstructure of the samples processed by cryogenic heat treatment is reported in Figure 3. The microstructures of Austempered samples show a mixture of bainitic ferrite and austenite. It had the appearance of a lower bainitic microstructure. The ferrite appears as dark needles where as austenite appears as white in the micrograph. The austenite has the appearance of slivers between finite needles. The microstructures of cryogenically heat treated samples consisted of ferrite and tempered martensite. The austenite was almost non- existent in the cryogenically treated samples. Moreover some precipitation of carbide was observed in these samples.

During austempering process, ferrite first nucleates out of austenite by the nucleation process and then grows with austempering time. As the ferrite grows, the remaining austenite becomes more and more enriched with carbon.

Fig 1: Microstructure of as-cast ductile cast iron samples.

Fig 2: Microstructures of samples austempered at 550°F

Fig 3: Microstructure of austmpered and cryogenically treated samples.

Table 2 reports the microstructural constituents of the material in both austempered and cryogenically treated samples as determined by X-ray diffraction technique. While austemepred samples had 8.33% austenite, in the cryogenically treated samples no austenitic peak was observed. This indicates that the austenite had transformed into tempered martensite in these samples. The microstructure in both the austempered and cryogenically treated samples were very fine scale in nature. The ferritic cell size (d) was determined for austempered samples using the well-known Scherrer equation [26]. This is also a measure of the mean free path for the dislocation motion in the material. Table 2 reports the ferritic cell size of the austempered sample. The ferritic cell size in the austempered samples were about 31.54 nm. This fine scale ferrite and austenite has caused significant strength in the austempered samples [28-29]

Table 2: Microstructural Constituents of the Material

Material condition	% Retained Austenite Content, $X\gamma$	Carbon Content of Austenite, $C\gamma$ (%)	$X\gamma C\gamma$	Ferritic cell size, d (nm)
Austempered	8.33	1.5619	0.13008	31.54

Table 3: Mechanical Properties of the Material

Material condition	Hardness (RC)	Yield Strength (MPa)	Tensile Strength (MPa)	% elongation
Austempered samples	37.5	1198.5	1328	1.92
Cryogenically treated samples	41.4	1303	1466	1.4

Influence of Austempering Temperature on Mechanical Properties

Table 3 compares the mechanical properties of Austempered and cryogenically heat treated samples. It is evident that the hardness of the cryogenically treated samples are higher than the austempered samples. The table also shows that cryogenically treated samples have higher yield and tensile strength than the austempered samples. On the other hand, the ductility of the material has decreased as a result of cryogenic heat treatment. This increase in hardness and strength in the cryogenically treated material is due to the absence of austenite content and presence of carbides in these samples. The cryogenic heat treatment has converted the retained austenite of the austempered samples in to a tempered martensitic structure with some precipitation of carbides. Figure 4 shows a micrograph of the sample indicating the presence of carbides. Austenite is a FCC (face centered cubic) and soft and ductile phase. Due to the presence of about 8.33% austenite in austmepered samples, the ductility of these samples was higher. On the other hand in cryogenically treated samples virtually no austenite was left and some carbide precipitation has taken place. Consequently the strength and hardness of these samples were higher and ductility was lower.

The strain hardening exponent of the austempered and the cryogenically treated samples are reported in Table 4. This table shows that the cryogenically treated samples had a lower strain hardening exponent compared to the austempered samples. This is because the austenitic carbon content ($X\gamma C\gamma$) of the Austempered samples were significantly higher. The austenitic carbon content is a very important microstructural parameter. This is a measure of the total carbon content of the austenite. As this parameter increases the strain hardening exponent of the material will increase [20]. Higher austenitic carbon promotes greater interaction between dislocations and solute carbon atoms in ADI. This leads to higher strain hardening exponent of the material [20]. Since cryogenically treated samples did not have any austenite left its strain hardening exponent was lower. Since the wear resistance is directly related to the hardness of the material the cryogenically heat treated ADI will have higher wear resistance also.

While it is now a well-known fact that the cryogenic treatment can improve the mechanical properties of the materials there is still a controversy about its origin. Several investigators [30-32] have suggested that improvement in the mechanical properties occur due to austenite to martensitic phase transformation. However, this improvement in properties has been observed in non-ferrous material also [33], where austenite was absent and hence no austenite to martensitic phase transformation can take place. Other investigators have suggested various other contributing factors for improvement in mechanical properties such as precipitation of very fine scale carbide [34], reduction of crystal defects such as dislocations and vacancies [35] and presence of compressive residual stress [36].

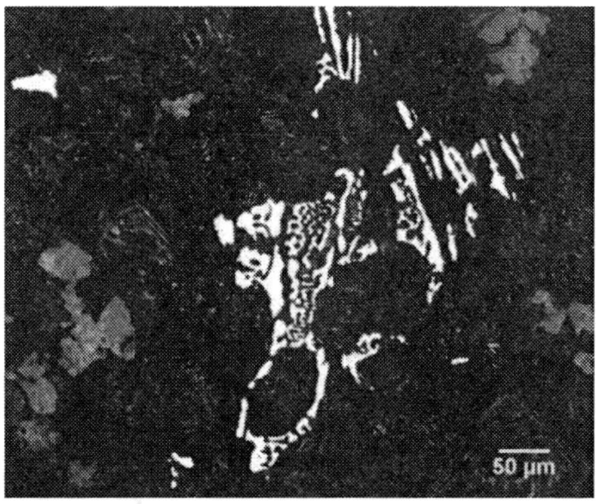

Fig.4: Microstructure of the cryogenically treated sample showing the presence of carbide.

While all these factors need to be investigated thoroughly, the test data of this investigation indicate that at least in ADI transformation of austenite to martensite and precipitation of carbides take place and contributes to improvement of mechanical properties of ADI.

The fracture toughness values of the austempered and cryogenically treated samples are reported in Table 4. It is evident that the fracture toughness values of both these batches of samples are very similar. It is interesting to note that even though the cryogenically treated samples had higher yield and tensile strength (about 8% higher) the fracture toughness of the cryogenically treated samples was only 2% lower. Generally in all structural materials as the yield or the tensile strength increases the fracture toughness decreases. However, it appears that cryogenic treatment can improve mechanical properties of ADI without compromising the fracture toughness of the material.

Table 4: Fracture Toughness of the Material.

Material condition	Fracture Toughness, (MPa√m)	Strain Hardening exponent (n)
Austempered	59.65	0.063
Cryogenically treated	58.4	0.053

Fractograph

The fractographs of the cryogenically treated samples are shown in Figures 5 and 6 respectively. Both these samples show the presence of quasi cleavage type of fracture with occasional dimples. These dimples are mostly at the graphite nodules/matrix interface. Basically no significant difference was observed in this fracture mode of the samples.

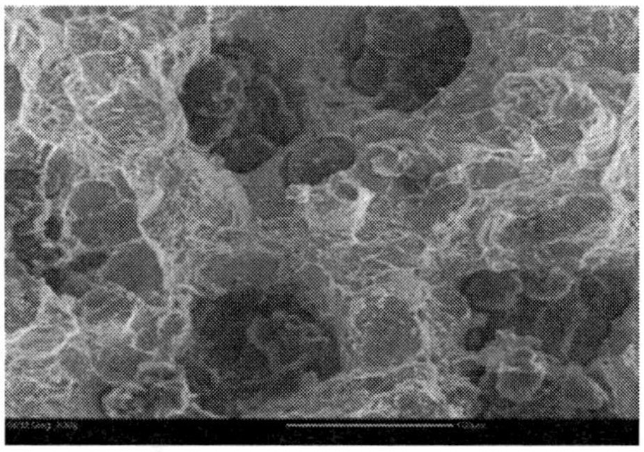

Fig.5: Fractograph of the cryogenically treated samples.

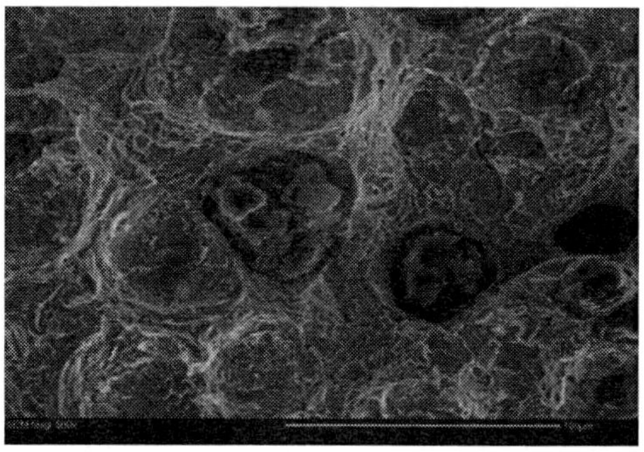

Fig.6: Fractograph of the non-cryogenically treated samples.

Conclusions

Unalloyed ductile cast iron was austempered in the lower bainitic temperature range of 288°C (550°F) and then cryogenically processed. Cryogenic processing has resulted in improvement in mechanical properties such as hardness, yield strength, ultimate tensile strength of the material compared to austempered samples. This improvement in mechanical properties can be attributed to precipitation of some carbides and transformation of austenite to tempered martensitic structure. Moreover, the improvement in mechanical properties occurred without any significant reduction of the fracture toughness of the material. Thus it appears that a unique opportunity exists for material processing for ADI by cryogenic treatment i.e. it can be used to improve the mechanical properties without compromising the fracture resistance of the material.

References

1. R.B. Gundlach and J.F. Janowak, Metal Progress,, 128, 19-26 (1985)
2. J. Dodd, Modern Casting, 68, 60-66 (1978)
3. R.A. Harding, Cast Iron, 2, 65-71, (Feb 1986)
4. D.J. Moore, T.N. Rouns, and K. B. Rundman, AFS Transactions, 93, 705-718, 1985
5. L. Bartosiewicz, I. Singh, F.A. Alberts, A.R. Krause, S.K. Putatunda, Materials Characterization, 30, 221-234 (1993)
6. L. Bartosiewicz, B.V. Kovacs, A.R.Krause, S.K. Putatunda, AFS Transactions, 92, 135-142 (1992)
7. J.Yang and S.K. Putatunda, Materials Science and Engineering A, 382, 265-279 (Oct 2004)
8. P. Shanmugam, P.P. Rao, K.R. Udupa, N. Venkataraman, Journal of Materials Science, 29, 4933-40 (1994)
9. T. Nakamura , K. Kishima, K. Jnibo, Y. Matsuo, F. Nugai, Journal of society of Mechanical Engineers Int. J A, 36, 348-353 (1993)
10. R. Boschen, H. Bomas, P. Mayr, H. Vetters, ASME, Gear Research Institute: Naperville, IL, pp 179-185 (1986)

11. P.P. Rao and S.K. Putatunda, Metallurgical and Materials Transactions, 28, 1457-1470 (1997)
12. P.P. Rao and S.K. Putatunda, Metallurgical and Materials Transactions, 29, 3005-3016 (1998)
13. P.P. Rao and S.K. Putatunda, Materials Science and Technology, 14, 1257-1263 (1998)
14. S.K. Putatunda and I. Singh, Journal of Testing and Evaluation, 23, 325-332 (1995)
15. J.L. Doong and C. Chen, Fatigue and Fracture of Engineering Material Structure, 12, 155-165 (1989)
16. S.K. Putatunda and P.K. Gadicherla, Materials Science and Engineering A, 268, 15-31 (1999)
17. S.K. Putatunda, Materials and Manufacturing Processes, 16 , 245-263 (2001)
18. M. Johansson, American Foundry Society Transaction, 85, 117-122 (1977)
19. Schmidt and A. Schuchert, Z Metal Kunde, 78, 871-875 (1987)
20. J. Yang and S.K. Putatunda, Materials Science and Engineering A, 406, 217-228 (Oct 2005)
21. F. Diekman and R. Papp, Heat Treating Progress, 33-36 (2009)
22. Anthony L. Velocci, Parts life extended through deep freeze technology, AMP news
23. ASTM E-8, Annual Book of ASTM Standards , ASTM, 3.01, 542-566, 1992
24. ASTM E-399, Annual Book of ASTM Standard, PA, 3.01, 745-756, 1992.
25. K.B. Rundman, R.C. Klug, American Foundry Society Transactions, 90, 499-508 (1982)
26. B.D. Cullity, Elements of X-ray diffraction, Addison – Wesley, Reading, MA (1974)
27. C.S. Roberts, Transactions of AIME, 197,.203-204 (1982)
28. R.B. Gundlach and J.F. Janowak, American Foundry Society Transactions, 94, 377-388 (1983)
29. D.J. Moore, T.N. Rouns and K.B. Rundman, Journal of Heat Treating, 41, 7-24 (1985)
30. P. Chen, T. Malone, R. Bond and P. Torer, Technical report, NASA Marshall Space Flight Center, (1997).
31. A. Molinari, M. Pellizzari, S. Gialenella etal, Journal of Materials Processing Technology, (2001).
32. E.N. Ma, Advanced Materials and Process, January (2003).
33. David N. Collins, Advanced Materials and Process, December, (1998).
34. D.N. Collins and J Dormer, Heat treatment of Metals (UK), 24, 71-74 (1997).
35. A.N. Popandopulo and L.T. Zhukova, Met. Sci Heat Treat, 22, 708-710, (1980)
36. A.Bensely, S. Venkatesh, D. mohan Lal, G. Nagarajan, A. Rajadurai and K. Junik, Materials Sci and Engg, 479, 229-235, April (2008).

Proceedings of the 26th ASM Heat Treating Society Conference
B.L. Ferguson, R. Jones, D.S. MacKenzie, and D. Weires, editors
https://doi.org/10.31399/asm.cp.ht2011p0050

Microstructural Characterization and Creep Properties of Cast Nb, Zr-modified HP Steels

Fernando B. Martins
Federal University of São Carlos (UFSCar), São Carlos, São Paulo, Brazil.
fernando.martins@sulzer.com
Marcelo Martins
SULZER BRASIL S/A, Americana, São Paulo, Brazil.
marcelo.martins@sulzer.com
George E. Totten
Portland State University, Portland, OR, USA.
GETotten@aol.com
Frederico A.P. Fernandes, Luiz C. Casteletti
São Carlos School of Engineering (USP), São Carlos, São Paulo, Brazil.
codoico@gmail.com, castelet@sc.usp.br

Abstract

The microstructure of 25Cr-35Ni-0.4C refractory steels consists of an austenitic matrix and eutectic carbides precipitated in the interdendritic regions. In-depth studies of the morphology and chemical composition of these carbides are extremely important for industry, since the microstructural components of these steels are responsible for their hot mechanical properties. In this context, the microstructural characterization of ASTM A297 Grade HP 40 steels modified with niobium and zirconium is using scanning electron microscopy, microanalysis and X-ray diffraction, and determination of the time to rupture at 1100°C under a constant stress of 17 MPa is reported here.

Introduction

Series H, heat-resistant austenitic stainless steels, are widely applicable in the petrochemical industry for the fabrication of columns for water vapor reform furnaces and pyrolysis furnaces. These furnaces are used in refineries to convert process gases (mixtures of hydrocarbons) and water vapor into a hydrogen-rich gas which is employed in water treatment plants to improve fuel quality or in other applications such as the production of ammonia and methanol. Because the pipes and components of reform furnaces must remain in operation for long periods of time (8 to 12 years) under severe conditions (high temperatures, aggressive environments and high internal pressures), a special class of cast, heat-resistant refractory steels has been developed [1].

These special alloys have been studied for many decades and major advances have been achieved with respect to their increased mechanical strength at high temperatures which has prolonged their service life. Among the various developments for these steel is the addition of carbide-forming alloy elements such as niobium, titanium, zirconium, tungsten, rare earth mixtures (mischmetals) and, more recently, the addition of yttrium that are stable at high temperatures [1-3].

The manufacture of components of these alloys is extremely important for the oil industry since the increasing demand for fuels requires increased production. Consequently, furnace operating times are longer and working conditions more severe. Therefore, augmenting the high-temperature strength of these steels offers several advantages, e.g., prolonged service life of the components, longer intervals between repair stoppages and reduction of unprogrammed stoppages, and reduction of component wall thickness thus maximizing the thermal exchange and design efficiency allowing for higher catalyzer loads. All these factors reduce process costs.

The high carbon content of these steels, which is necessary for their high-temperature strength, gives these alloys little mechanical conformability by conventional processes. For this reason, pipes are fabricated by centrifugal casting in which the molten metal is poured into a rotary metal mold under high solidification rates. However, some geometries such as curves and Y-shaped fittings are not manufactured by this process but instead by static casting in sand molds. Few studies have focused on investigating the properties of steels cast under these conditions and the literature offers very little information to date [3-7].

It is therefore necessary to characterize the microstructures of these materials to gain an understanding of the behavior, morphology and composition of precipitated carbides in order to relate these characteristics with the material properties.

Experimental Procedure

Test specimens of ASTM A297, Grade HP 40 steel were cast in the form of 190 mm long and 30 mm diameter cylinders. The molds were made of silica sand agglomerated with air-cured phenol urethane resin. The casting design was constructed in 3-D using the Inventor 2008 software program and the solidification process was simulated with the aid of the Magmasoft program which is based on finite element heat transfer analysis. The test specimens were analyzed chemically in an optical emission spectrometer with 47 channels.

Table 1. Concentration as weight percentage of the alloys elements.

Alloy	C (%)	Cr (%)	Ni (%)	Mn (%)	Si (%)	Nb (%)	Zr (%)	Fe (%)
1	0.42	24.97	36.42	1.42	1.68	1.19		Bal.
2	0.41	24.92	36.08	1.41	1.71	1.2	0.068	Bal.
% Variation	2.38	0.20	0.93	0.70	1.79	0.84	-	-
Standard	0.35-0.75	24-28	33-37	2.00 max	2.00 max	-	-	Bal.

Scanning electron microscopy (SEM) with images generated by secondary electrons and microanalysis by energy dispersive spectroscopy (EDS) were used. The precipitates were characterized by X-ray diffraction using copper radiation; CuKα, with a wavelength of 1.5406 Å and a scanning rate of 2°/min. Creep tests were conducted in an electromechanical machine with a constant load of 17 MPa at 1100°C.

Results and Discussion

Table 1 lists the chemical composition of the materials used in this study which was determined by optical emission spectroscopy. These values followed the ASTM A297 standard except with the addition of the niobium and zirconium elements. Note that the chemical compositions of these alloys are compatible with the ASTM A297 Grade HP standard and that variations between the alloy elements were very low, falling within the operational error of the assay. It can therefore be stated that the difference between Alloys 1 and 2 lies in the zirconium content (680 ppm) which was intentionally added in an attempt to interrupt the continuity of the eutectic carbide network formed during solidification.

Figure 1 depicts the microstructures of the HP 40 refractory steel in the as-cast condition after cooling in sand molds, by optical microscopy. Note the austenitic matrix with large quantities of eutectic carbides located in the interdendritic spaces. This morphology is explained by the solidification process of these alloys in which elements such as Cr, Nb and Zr – due to their saturation in the newly solidified austenite – are rejected to the liquid regions where they combine with free carbon leading to the formation of carbides in the interdendritic spaces, which are the last regions to solidify.

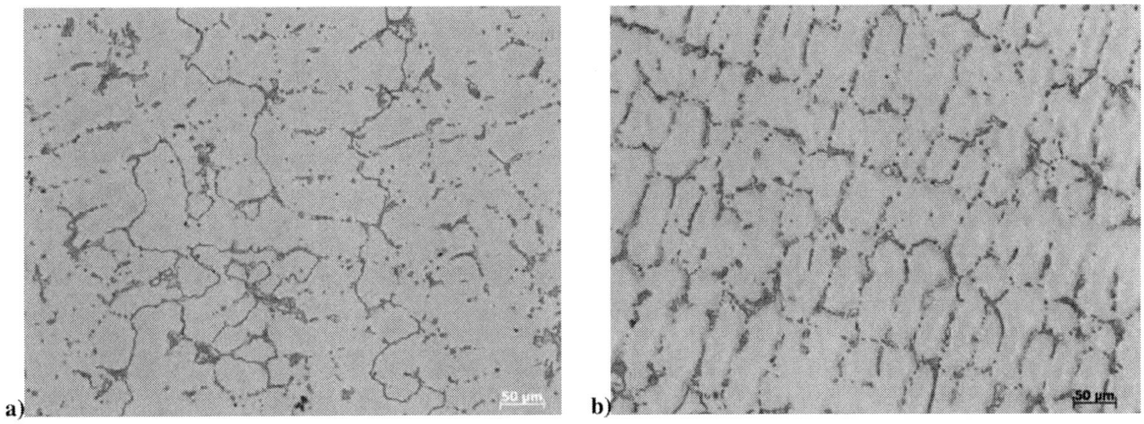

Figure 1. As cast microstructures observed by optical microscopy: (a) Alloy 1; and (b) Alloy 2.

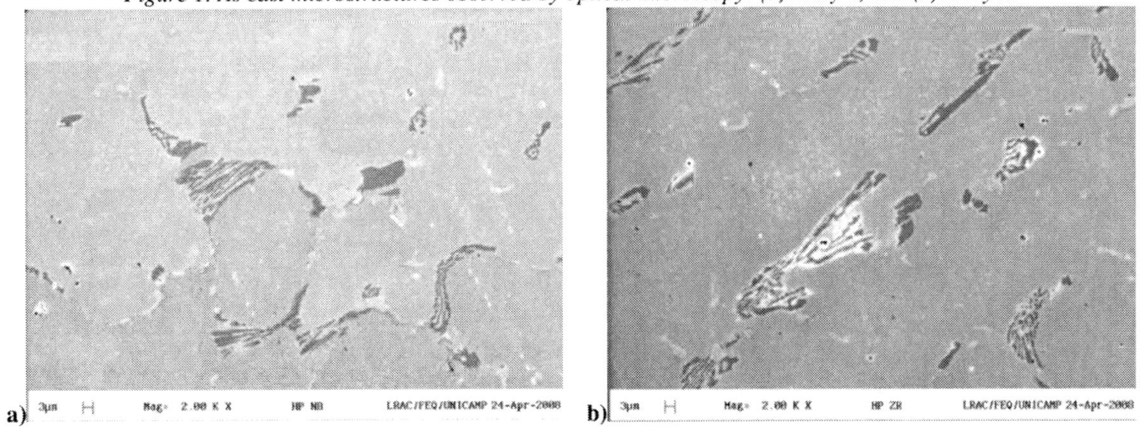

Figure 2. SEM micrographs of: (a) Alloy 1; and (b) Alloy 2.

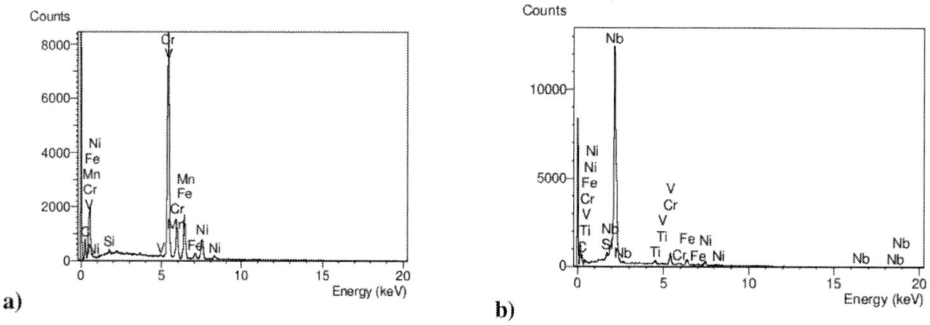

Figure 3. Microanalysis of the carbides found in Alloy 1: (a) Dark, (b) Light.

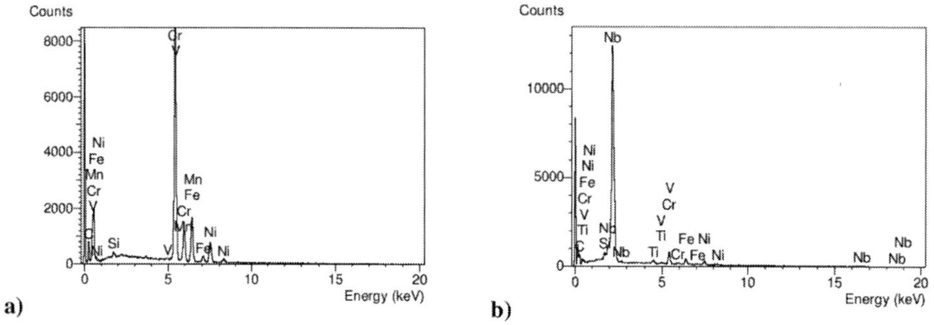

Figure 4. Microanalysis of the carbides found in Alloy 2: (a) Dark, (b) Light.

The fraction of carbides found here is independent of the mold dwell time since they have already precipitated after the passage of the liquidus temperature. The mold dwell time only affects their distribution, i.e., the longer they remain in the mold, the greater the coalescence of these precipitates leading to the formation of large networks.

Another important point is the level of coalescence of the carbide networks. A comparison of Alloys 1 and 2 reveals a visibly greater fragmentation of this network in Alloy 2. Because of their low solubility in austenite, elements such as zirconium are the first to be rejected in the form of complex oxides during solidification. These precipitates act as local substrates for nucleation of the eutectic carbides. This nucleation is random due to the distribution of the substrates which causes discontinuity in the carbide network.

The morphology and the semi-quantitative analysis of the chemical composition of the carbides were determined by scanning electron microscopy and electron microprobe, analysis respectively. Figure 2 shows the morphologies of the carbides detected in Alloys 1 and 2. Even with the use of secondary electron imaging, it is possible to observe carbides in different shades of gray. The lighter colored carbides are rich in niobium while the darker carbides are rich in chromium.

Alloy 1 exhibit dark carbides in the form of alternate layers and light carbides with the appearance of isolated islands [1,2]. Table 2 and Figure 3 show the results of the microanalyses of these carbides and the characteristic spectrum of each carbide. The dark carbides (alternating layer morphology) are chromium-rich while the light carbides (isolated island morphology) are niobium-rich.

The same tests were performed for Alloy 2 (Table 2 and Figure 4), which contained carbides whose morphology differed from that of Alloy 1. The dark carbides showed fewer layers than those characterized in Alloy 1 due to the greater fragmentation of these carbide networks. As for light carbides, no significant differences were detected between Alloys 1 and 2. This microstructure is depicted in Figure 2. Microanalysis by EDS (Figure 4) also revealed no significant differences between the two alloys, as indicated in Table 2.

Table 2 – Approximate chemical composition of the carbides.

	Dark Carbides		Light Carbides	
Element, %	**Alloy 1**	**Alloy 2**	**Alloy 1**	**Alloy 2**
Cr	71.87	70.36	16.07	15.4
Ni	10.49	9.88	17	17.55
Nb	--	1.2	47.71	48.9
Fe	15.84	13.9	16.62	17.4
Si	0.61	0.63	1.31	1.01
Mn	1.19	1.17	1.01	1.2

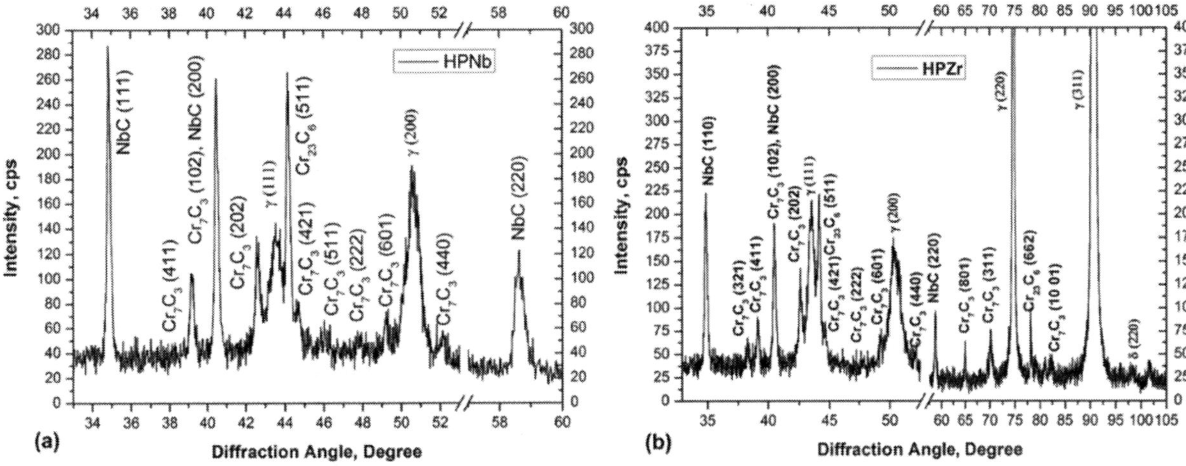

Figure 5. X-ray diffraction patterns of the: (a) Alloy 1; and (b) Alloy 2.

X-ray diffraction was also used to characterize the phases and precipitates. This technique, allied to the microanalytical results and data reported in the literature [2,5,6], provided strong signs of the real stoichiometry of the carbides in the samples. Figure 5a shows the diffractogram of Alloy 1. Note the massive presence of Cr_7C_3 and NbC type precipitates, as well as $Cr_{23}C_6$ type carbides. As suggested in the literature [2, 5], the dark gray carbides are of the Cr_7C_3 type and the lighter areas are carbides of the NbC type. These assumptions were confirmed by X-ray diffraction and electron microprobe tests.

Figure 5b shows the diffractogram of Alloy 2. An important observation is that some peaks indicate more than one possibility. This is because the crystal lattices of these materials are deformed by the saturation of solute atoms, forming substitutional solid solutions that alter the interplanar distances.

It should be noted that some reflections of crystalline planes of a given phase do not appear in the spectrum due to peak overlapping. In other words, the intense reflection of a given crystalline plane of any given phase is superposed over the less intense reflection of another plane belonging to other phases, or even to the same phase with an intense reflection.

In addition to the microstructural characterization, the alloys were subjected to creep tests. The test condition adopted was: Test stress of 17 MPa with a minimum time to rupture of 100 hours at 1100°C. Table 3 lists the results obtained in the creep tests. Each point represents the arithmetic mean of 5 tests.

Alloy 2 presented better results in the creep test than alloy 1. The resistance mechanism is associated principally with the formation of complex eutectic zirconium and niobium carbides. The main characteristics of these carbides are greater structural and chemical stability at temperatures above 950°C, as well as retardation of the precipitation of deleterious phases such as G - phase (the probable preferential site for nucleation of creep holes, besides the grain boundaries) [7].

The function of eutectic carbides during the creep process is to prevent the movement of dislocations by their anchoring the dislocations around the precipitated phases, as well as to maintain the structural integrity by preventing the appearance of holes. Secondary carbides also act in the mechanism of creep resistance. These carbides appear preferentially in the form of fine precipitation at the grain boundaries, acting as barriers and preventing the movement of dislocations by means of a mechanism called pinning. As a consequence of this mechanism, the ductility of the alloy diminishes substantially and its strength increases.

Table 3. Results obtained in the creep tests.

Alloy	Time to rupture, h
1	104
2	116

The difference between the results lies precisely in the presence of complex carbides in the alloys of this study. The composition of Alloy 1 exhibited only Nb additions, forming Cr_7C_3, $Cr_{23}C_6$ and NbC type carbides. Composition 2, besides Nb additions and the presence of the same chromium carbides, also exhibited micro-additions of zirconium that also formed ZrC type carbides, which yielded a 10% increase in the time to rupture of this material.

Conclusions

As a result of this work, it is concluded that:

The microstructure of these materials is composed basically of austenite plus complex carbides precipitated in the interdendritic spaces.

This microstructure is strongly dependent on the cooling rates imposed on the material.

The microanalyses of the carbides revealed two types of precipitates: dark (chromium-rich) and light (niobium-rich).

The diffractometry and microanalyses revealed the presence of dark and light carbides, which were possibly Cr_7C_3 and NbC type carbides, respectively.

The addition of zirconium in HP type stainless steels provides a reasonable increase in these alloys' time to rupture under creep.

References

[1] Almeida, L.H., Ribeiro, A.F., Le May, I., "Microstructural characterization of modified 25Cr-35Ni centrifugally cast steel furnace tubes," Materials Characterization, Vol. 49, (2003), pp. 1-11.

[2] Barbabela, G.D., Almeida, L.H., Silveira, T.L., Le May, I., "Role of Nb in modifying the microstructure of heat-resistant cast HP steel," Materials Characterization, Vol. 26, (1991), pp. 193-197.

[3] Barbabela, G.D., "Estabilidade Estrutural de Aços Resistentes ao Calor do Tipo HP com Adição de Nióbio," Tese de D. Sc. UFRJ. 1990, 132 p.

[4] Eduardo, A.A.G., Ribeiro, R., Papaléo, R., Guimarães, J.R.C., "Microstructure and creep behavior of a niobium alloyed cast heat-resistant 26 pct Cr steel," Metallurgical and Materials Transactions, Vol. 17A, No. 4 (1986), pp. 691-696.

[5] Piekarski, B., "Effect of Nb and Ti additions on microstructure and identification of precipitates in stabilized Ni-Cr cast austenitic steels," Materials Characterization, Vol. 47, (2001), pp. 181-186.

[6] Powell, D.J., Pilkinton, R., Miller D.A., "The precipitation characteristics of 20% Cr/25% Ni-Nb stabilized stainless steel," Acta Metallurgica, Vol. 36, No. 3 (1988), pp. 713-724.

[7] Wen-Tai, H., Honeycombe, R.W.K., "Structure of centrifugally cast austenitic stainless steels: Part 2 - Effects of Nb, Ti and Zr," Materials Science and Technology, Vol. 1, No. 5 (1985), pp. 390-397.

Proceedings of the 26th ASM Heat Treating Society Conference
B.L. Ferguson, R. Jones, D.S. MacKenzie, and D. Weires, editors
https://doi.org/10.31399/asm.cp.ht2011p0055

Quantum Jump in Heat Treating

Mr. Shobhan Paul, Starfire *Technologies LLC, Malibu, CA*

shobhan@aol.com

Abstract

A new heat treatment is discussed wherein Differential Thermocouple Analysis (DTA) is used to give real time feedback of the phase transformations. This results in extraordinary control of the heat treat process. Real world applications are identified The process can be integrated very well with Differential Heat treatment. This process can be used to create mechanical properties that are variable and tailored to end use. For example; high strength in one location and high fatigue resistance in another location. The process can also lead to a new form of thermo-mechanical heat treatment, where first differential heat treatment is done to create variable properties and then laser peening is done to create dislocations which are trapped by subsequent heat treatment. This results in extraordinary fatigue resistance in the core section, while ensuring high strength in the outer surface. This technology results in capabilities that do not exist today in industrial heat treating.

Background

Turbine Engines Are a Critical Military Need - The gas turbine-based jet engine has been in existence over fifty years, and today still remains as the dominant element in aeronautics propulsion. Turbine engines power all high-performance combat aircraft and have revolutionized both military and commercial air transport. The turbine engine's impact is such that nearly every breakthrough in aircraft capability since World War II in some way traces to propulsion advancement.

Improvements Are Needed to Address Future Warfighter Requirements - The U.S. armedforce vision for future aviation is to enable the warfighter to anticipate, find, track, target, engage, and assess any target at anytime and anywhere. To accomplish this, weapon systems are needed that are capable of persistent and responsive Intelligence, Surveillance, and Reconnaissance (ISR), extreme endurance, responsive strike, persistent attack, global reach, multi-role mobility, and responsive space access.

Many mission requirements of these future weapon systems simply cannot be achieved without propulsion advancements. For example, a responsive strike aircraft should have twice the range at half the aircraft unit cost of current systems. Certain Unmanned Combat Air Vehicle (UCAV) concepts require 2.5 times the mission radius or 3 times the mission persistence (loiter time) of today's manned vehicles. For access to space, a fuel-efficient, on-demand turbine engine accelerator up to a speed of Mach 4+ is required. Such capability does not exist today. For multi-role mobility, a future aircraft must be capable of Short Take-Off Vertical Landing (STOVL) with a 2-to-4 times mission radius increase over today's conventional take-off aircraft.

To meet these mission requirements, future turbine engines must feature improved performance as well as reduced cost. Necessary performance improvements include increased thrust, lower fuel consumption, lower weight, improved durability, and lower exhaust noise and pollutant emissions. Cost reductions must be realized at all program phases – development, procurement, and operations and support.

AN EXAMPLE OF A USE OF A TURBOSHAFT ENGINE

State of the art: Operating conditions imposed on components require different mechanical properties at different locations. For example automotive gears may have surface hardening; Samurai swords have had a sharp hard edge and a tough blade. The human bone is not isotropic either. The human bone has different strengths at different locations. The human bone even has different Young's moduli in the x, y and z directions. Hence, in a way this work to produce different microstructures for different mechanical properties copies nature.

The improved performance of the gas turbine engine over the years has been paced by improvements in the elevated temperature mechanical properties of nickel-base superalloys.

These alloys are the materials of choice for most of the components of gas turbine engines exposed to the hottest operating temperatures. Components of gas turbine engines such as, for example, disks, blades, fasteners, cases, and shafts are typically fabricated from nickel-base superalloys and are required to sustain high stresses at very high temperatures for extended periods of time. Nickel base superalloys have been developed for high temperature turbine applications. It is well know that the fatigue crack propagation resistance and the creep resistance of the majority of nickel base superalloys may be improved by increasing the grain size. Nickel base superalloys are composed of two principal phases, a gamma matrix and an ordered strengthening gamma prime phase (Ni_3 Al/Ti). At the gamma prime solvus temperature of the alloy, the gamma prime phase is taken completely into solution in the gamma matrix. Typical gamma prime solvus temperatures of nickel based superalloys are 1120° C. to 1190° C. The gamma prime phase exists as two principal sizes, the primary gamma prime and the secondary gamma prime (there is also a tertiary gamma prime in small quantities as per some researchers it's size is small about 8 nanometers).The primary gamma prime is the larger of the two and is located on the grain boundaries. The primary gamma prime is retained throughout the manufacturing process to prevent the migration of the grain boundaries and hence to control grain size. If the primary gamma prime volume fraction is reduced the grain size is increased, even at temperatures below the gamma prime solvus temperature. The secondary gamma prime is precipitated uniformly throughout the gamma matrix on cooling during heat treatment processes.

Heat treatment at temperatures greater than the gamma prime solvus temperature, super solvus heat treatment, usually results in non-uniform grain growth, and thus it is difficult to generate reproducible structures using supersolvus heat treatments. Heat treatments at a temperature near, but less than, the gamma prime solvus temperature may be used to generate controlled and reproducible uniform grain growth.

In earlier work in this area, Pratt & Whitney has implemented one process to manufacture dual-microstructure disks for production disks within the F119 engine for the USAF F-22 Raptor. This approach relies on induction heating to selectively heat the outer rim to a high temperature to coarsen the grain structure.

NASA GLENN developed the DMHT process, in this process, here a dual microstructure was obtained using Thermal masses and thermal insulators to differentially heat the disk, this resulted in different temperatures, subsequent quenching in oil and aging at 815 C for eight hours was done. The paper concluded that it may be better to pre or post heat DMHT (sub-solvus). It also spoke about optimizing the cooling rates for optimum mechanical properties. Ref 1.

Alloy 10 is a high strength Nickel based alloy with a gamma prime strength of 55%. A study of the DMHT process using Alloy 10 was done. "Since tensile, creep, and fatigue properties are strongly influenced by alloy grain size, the effect of heat treatment on grain size and the attendant properties were studied in detail. It was observed that a fine grain microstructure offered the best tensile and fatigue properties, whereas a coarse grain microstructure offered the best creep resistance at high temperatures. Therefore, a disk with a dual microstructure, consisting of a fine-grained bore and a coarse-grained rim, should have a high potential for optimal performance". Ref 2

Reference 3 was a good paper that showed how simple metal thermal sinks were used to create a temperature differential. The paper went into some degree of detail as to how the dual microstructure process could be implemented in the shop. It showed in a revision of the DMHT process where the furnace temperature was 2170 and the temperature in the upper thermal block being 2100 a band over 2" wide on the outer periphery of the disk was converted to a coarse grain microstructure while the interior of the disk retained a fine grain microstructure. Further, no quench cracks were observed in the disk. The transfer time to the quench tank was under one minute.

Reference 4 reaffirmed that the DMHT process resulted in benefits from; a higher strength fine-grain bore and an enhanced creep resistance of a coarse grain rim. The paper went onto say that "these tests have confirmed the strength benefits of a fine-grain bore and the enhanced creep resistance of a coarse-grain rim. More recently, the fatigue performance of DMHT disks was studied using test coupons machined from disk forgings and cyclic spin testing of DMHTdisks. The coupon testing confirmed the enhanced fatigue resistance associated with a fine-grain bore. Cyclic spin testing of DMHT disks was performed with two goals in mind. First, the grain-size transition zone was intentionally loaded to high stress levels by employing web holes in the disk that bisected the grain-size transition zone. Second, the fatigue lives of the DMHT disks were estimated using advanced elastic-plastic finite-element techniques. The results of these tests and the analyses showed that the grain-size transition zone of the DMHT disk could withstand significant loads that exceeded the predicted fatigue lives".

In a fairly recent publication, Fatigue Resistance of the Grain Size Transition Zone in a Dual Microstructure Superalloy Disk, Ref 5, pretty much the same authors of the above mentioned references, studied the transition zone, between the bore and the rim. They came to the following conclusions, that "failures often initiated at relatively large grains, which failed on crystallographic facets. Grain size distributions were characterized in the specimens, and related to the grains initiating failures as well as location within the transition zone. Fatigue life decreased with increasing maximum grain size. Correspondingly, mean fatigue resistance of the transition zone was slightly higher than that of the rim, but lower than that of the bore. The grain facet failure initiations resulted in substantially lower fatigue lives. The grain facets initiating failure were generally larger than the mean grain size, extending to near ALA grain size. The lives of specimens located mid way in the transition zone, with near bimodal grain size, appeared to still be limited by these large grains, with no additional complications due to wider grain-size variations." Numerous patents that impact the scope of the activities of this SBIR proposal are shown below.

Rolls-Royce had published a paper on cooling from above the γ' solvus and determined that "The interrelationships between composition, γ' morphology, hardness and γ–γ' mismatch were identified for a set of advanced nickel-base polycrystalline superalloys. When cooling from a temperature above the γ' solvus, i.e. a supersaturated γ matrix, it was found that the cooling rate determines the γ' morphology and particle distribution that develops. This in turn dictates the hardness of the alloy system and the mismatch in lattice parameters between the γ and γ' phases. This work shows that the development of these parameters is relatively insensitive to alloy composition. Moreover, the thermal history, especially cooling rate through the γ' solvus temperature is found to be the critical factor in determining alloy response. See Ref 6.
Further Rolls-Royce has applied for a patent , see Ref 7"Method of heat treating a superalloy component and an alloy component" therein it states a method of heat treating a superalloy component comprises solution heat treating the component at a temperature below the gamma prime solvus temperature to produce a fine grain structure in the component. Insulation is placed over a first area of the component to form an insulated assembly. The insulated assembly is placed in a furnace at a temperature below the solvus temperature and maintained at that temperature for a predetermined time to achieve a uniform temperature in the component. The temperature is increased at a predetermined rate to a temperature above the solvus temperature to maintain a fine grain structure in a first region, to produce a coarse grain structure in a second region and to produce a transitional structure in a third region between the first and second regions of the component. The insulated assembly is removed from the furnace when the second region of the component has been above the solvus temperature for a predetermined time and/or the first region of the component has reached a predetermined temperature." Ref 7 .

The effect of the hub of the disc obtaining a temperature significantly lower than the gamma prime solvus is to rapidly coarsen the gamma prime precipitates if the temperature is too low or to dissolve the gamma prime precipitates if the temperature is too high for ageing and too low for solution heat treatment. This would result in a disc with an overaged bore and a significant reduction in mechanical properties, thus negating the benefit of the dual microstructure heat treatment.
Higher cooling rates may be used to provide improved tensile properties and on the contrary lower cooling rates may be used to provide improved fatigue crack propagation resistance

GENERAL ELECTRIC has also applied for a patent for an ageing heat treatment, see Ref 8, US patent application 20070169860, The method comprises providing a gas turbine engine superalloy article having a gamma prime solvus temperature; and locally over aging only a selected portion of the article to locally improve fatigue resistance at the selected portion of the article, wherein the local over age cycle includes heating at about 843.degree C for about 3 to 4 hours. Higher operating temperatures for gas turbine engines are continually sought in order to increase efficiency. However, as operating temperatures increase, the high temperature durability of the components within the engine must correspondingly increase. Material processing for durability in high temperature applications yields a coarser grain microstructure with less fatigue resistance than fine grain structures at temperatures below about 649.degree. C. There are numerous instances where operating conditions experienced by a component place differing materials property requirements on different portions of the component. A turbine disk for a gas turbine engine is an example of a type of component where tailored mechanical behavior in various portions of the article is preferred. Such disks are typically made from nickel-base superalloys, because of the temperatures and stresses involved in the gas turbine cycle. In the hub portion where the operating temperature is somewhat lower, the limiting material properties are often tensile strength and low-cycle fatigue resistance, which are superior in the fine grain condition up to about 649.degree. C. In the rim portion where the operating temperature is higher because of the proximity to combustion gases, resistance to creep and hold time fatigue crack growth (HTFCG) is often the limiting material property. HTFCG is the propensity in a material for a crack to grow under cyclic loading conditions where the peak tensile strain is maintained at a constant value for an extended period of time. Therefore, processing the entire article to the damage-tolerant, coarse grain structure necessary for the high temperature rim location can result in a sacrifice of fatigue life at conditions encountered in the relatively cool-running bore. In accordance with one embodiment of the invention, a method for locally heat treating a gas turbine engine superalloy article to improve resistance to strain-induced fatigue of the article is disclosed. The method comprises providing a gas turbine engine superalloy article having a gamma prime solvus temperature; and locally over aging only a selected portion of the article to locally improve fatigue resistance at the selected portion of the article, wherein the local over age cycle includes heating at about 843 degree C for about 3 to 4 hours.
The above was a review of the DMHT state of the art. New tools are now available, that can lead to improvement in the state of the art.

SINGLE SENSOR DIFFERENTIAL THERMAL ANALYSIS (Ref 9): Most of the available phase transition data in engineering materials is generated by the traditional techniques of differential thermal analysis (DTA), differential scanning calorimetry (DSC), and dilatometry (DA). DTA and DSC are limited to relatively low heating and cooling rates and are not applicable to simulate most of the actual materials processing conditions. DA is insensitive to transformations that cause small (or no) volume changes and is inapplicable for solid-liquid phase transformations. The limitations of these methods restrict the availability of phase transition data for the response of engineering materials to non-equilibrium thermal processing.A technique for single sensor differential thermal analysis (SS DTA) that determines the phase transitions in engineering materials has been developed at the Ohio State University. SS DTA utilizes a single thermocouple to measure the thermal history in the tested material and a calculated reference curve. The reference curve is generated from the measured thermal history by specially developed software and does not reflect the enthalpy variations in the tested material. The phase transformation temperatures are determined by comparing the measured thermal history to the "Reference curve".

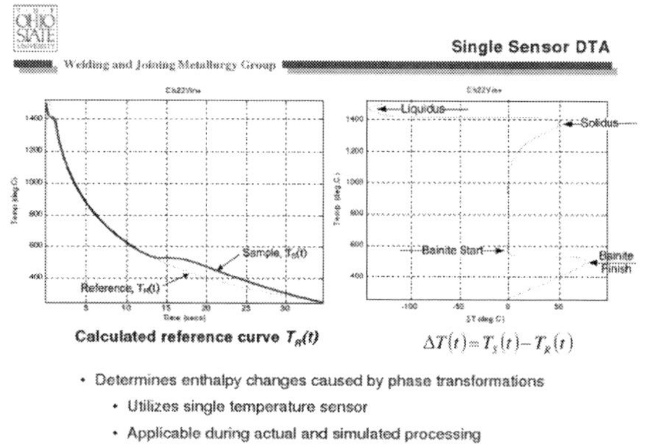

The utilization of a single temperature sensor provides the SS DTA technique with the unique capability to study the entire range of solid-liquid and solid-state non-equilibrium phase transitions that are associated with enthalpy changes. Compared to DTA, DSC, and DA, the SS DTA has a wider application range of heating/cooling rates and temperatures. The sensitivity, accuracy, and precision of the SS DTA measurements have been validated by direct comparison to DTA and DA, and by utilizing the Curie temperature as a reference point.

The SS DTA technique has been applied to determine a number of phase transitions including melting and solidification, liquation and formation of eutectics, dissolution and precipitation, solid-state transformation, recrystallization, and ferromagnetic-to-paramagnetic transition. The range of metallic alloys and weld filler metals tested by SS DTA includes steels (high strength low alloy, advanced high strength, creep strength enhanced ferritic, and stainless), non-ferrous alloys (Ni-base, titanium, and aluminum), and brazing filler metals.

The SS DTA has been implemented to study non-equilibrium phase transformations during actual and simulated processing by fusion welding, friction stir welding and processing, brazing, casting, heat treatment, and PWHT. The SS DTA provides a powerful tool for development of non-equilibrium phase transformation diagrams and databases that has been used for alloy development, for identification and solving of weldability/fabricability problems, and for development and testing of materials processing procedures.

Applications

Components with different mechanical properties can now be manufactured. This would be based on a combination of unsteady state heat transfer analysis and Phase Transformation diagrams generated using by a SSDTA. For example the component would be heated in a furnace, the usual way, however, where different mechanical properties are required, the component would be subject to differential heat treatment, an SS DTA would identify the enthalpy changes based upon the phase transformation, the operator would use this feedback to control the process.

Use the same DMHT process, that has been invented by NASA and adapt it to the geometry of the impeller disk. Use the SSDTA (Single Sensor Differential Thermal analysis for feedback control). The SSDTA can identify gamma prime dissolution beginning and dissolution finishing. The SSDTA will also allow the gamma prime start and gamma prime finish during quenching to be identified. SSDTA in a way will allow to control the process as if looking into the microstructure!!!! Hopefully the SSDTA will also allow to identify the precipitation of gamma prime secondary. This will enable the design of a heat treat process that gives variable microstructure.

Thermo-mechanical Heat treatment: Do laser peening of the bore after the DMHT and quenching. Then age the bore material only. This ageing heat treatment should trap the dislocations due to laser peening. In a way this will be giving the material a thermomechanical heat treatment!! This should greatly improve fatigue properties. FIB/SEM tomography will be done to see the microstructure. Fatigue tests should indicate a huge improvement in low cycle fatigue.

Conclusion

It is possible today to generate variable micro-structure using the methodologies discussed above. This will lead to components with mechanical properties that are tailored to their performance. This is turn will lead to production of turbines/aircraft and other machinery that will have far greater capability than possible today.

References

1. Gayda, John, and David Furrer. "Dual-Microstructure Heat Treatment." Heat Treating Progress 3 (2003): 85-89. ASM International, The Materials Information Society, 01 Oct. 2003. Web. 7 Jan. 2010
2. Gayda, John. "Dual Microstructure Heat Treatment of a Nickel-Base Disk Alloy Assessed"
3. Gayda, John, Tim Gabb, and Pete Kantzos. "Low Cost Heat Treatment Process for Production of Dual Microstructure Superalloy Disks." NASA Technical Reports Server. NASA, 07 July 2009. Web. 7 Jan. 2010.

4. Gayda, John, Timothy Gabb, and Pete Kantzos. "Advanced Heat Treatment Technology for Superalloy Disks Verified." NASA Technical Reports Server. NASA, 29 July 2009. Web. 7 Jan. 2010.
5. Gabb, T., P.; Kantzos, P., T.; Telesman, J.; Gayda Fatigue Resistance of the Grain Size Transition Zone in a Dual Microstructure Superalloy Disk, NASA/TM—2010-216369. April 2010
6. Mitchell, R.J.[1]; Preuss, M., Inter-Relationships between Composition, γ' Morphology, Hardness, and γ-γ' Mismatch in Advanced Polycrystalline Nickel-Base Superalloys during Aging at 800 °C , Metallurgical and Materials Transactions A, Volume 38, Number 3, March 2007 , pp. 615-627(13)
7. Rolls Royce Patent Application , "Method of heat treating a superalloy component and an alloy component," Inventors Robert J. Mitchell David U. Furrer Joseph A. Lemsky Mark C. Hardy
8. US patent application 20070169860 General Electric B. Alexandrov, J. Lippold, J. Tatman, G. Murray, Non-Equilibrium Phase Transformation Diagrams in Engineering Alloys,The Ohio State University, Columbus,Trends in Welding Research, Proceedings of the 8th International Conference, pages 467-476

Proceedings of the 26th ASM Heat Treating Society Conference
B.L. Ferguson, R. Jones, D.S. MacKenzie, and D. Weires, editors
https://doi.org/10.31399/asm.cp.ht2011p0060

Method for accurate surface temperature measurements during fast induction heating

Benjamin Larregain, Nicolas Vanderesse, Florent Bridier, Philippe Bocher
École de technologie supérieure, Montreal, Quebec, Canada
benjamin.larregain@etsmtl.ca

Patrick Arkinson
Pratt & Whitney Canada, Longueuil, Quebec, Canada

Abstract

A robust method is proposed for the measurement of surface temperature fields during induction heating. It is based on the original coupling of temperature indicating lacquers and a high-speed camera system. Image analysis tools have been implemented to automatically extract the temporal evolution of isotherms. This method was applied to the fast induction treatment of a 4340 steel spur gear, allowing the full history of surface isotherms to be accurately documented for a sequential heating, i.e., a medium frequency pre-heating followed by a high frequency final heating. Three isotherms, 704°C, 816°C, and 927°C, were acquired every 0.3 ms with a spatial resolution of 0.04 mm per pixel. The information provided by the method is described and discussed. Finally, the transformation temperature Ac1 is linked to the temperature on specific locations of the gear tooth.

Introduction

As a green manufacturing process, induction heating has become a very interesting alternative for the contour hardening of gears in the aeronautical industry, in replacement of the currently used thermochemical processes. Induction contour hardening equipments are designed to harden locally a part with a coil using medium (MF) and/or high frequency (HF) generators. The depth of the hardened layer depends on numerous factors, including the induction heating parameters, the workpiece geometry, and the material metallurgical and physical properties [1]. Even if the use of designs of experiments [2] can provide pertinent results to develop a specific treatment, it remains time consuming and costly as the parts are already in an advanced manufacturing state before the surface hardening stage. Consequently, the development of computer simulation tools are required to allow designers to develop and optimize induction processes more efficiently. Complex multiphysical numerical approaches have to be used due the nature of the process encompassing electro-magnetic, thermal, metallurgical, and mechanical phenomena [3]. The final state, after both induction heating and quenching phases, are usually considered for the validation of the numerical simulations. Little or no transient information is taken into account. Four final responses can be typically used, i.e., the hardness profile, the microstructure transformation, the residual stresses profile, and the level of distortions [4, 5]. However, the microstructure gradient and the residual stresses profile are typical features which depend on the thermal path of the heated and quenched part. Getting accurate information on the temperature field evolution versus time within a induction heated part is a critical way to not only validate the numerical models but also optimize the process itself [6].

However, the measurement of temperature evolution during induction heating faces several issues. First, the austenitic transformation temperatures of usual hardened steels are between 750 and 1000°C. The temperature measurement technique employed must be able to provide data from ambient temperature to these high temperature levels. On the other hand, current induction heating processes use powerful generators able to heat the part in a fraction of a second [7-9] and the measurement medium must provide very high acquisition rates, i.e., a high frequency of signal acquisition and the shorter response time possible. Finally, the thermal gradients may be highly localized and there is a need to have a high spatial resolution for the measurement.

Four temperature measurement methods were tested. Table 1 summarizes their characteristics relative to the induction heating application, i.e., the allowed temperature range, the location and the accuracy the temperature measurement provided by each method.

The impossibility of temperature measurement during a rotating movement, which is common in induction heating of spur gear, is the major inconvenient of thermocouples. Moreover, thermocouples only provide information on local temperature and also impose a compromise between maximal allowed temperature and acquisition speed. For a local measurement at a 0.5 mm diameter spot, the pyrometer is a good alternative but its accuracy strongly depends on the calibration of the material emissivity which evolves with temperature and the environment reflectivity. For a surface measurement, an alternative is the full-field infrared camera measurement but this technique is also very sensitive to the material emissivity versus its temperature. Furthermore, concerning these three temperature measurement media, the response time and acquisition frequency of the data logger have to be added to the tool response time, providing long system response times. On the other hand, temperature indicating lacquers are usually used as a static measurement to localize where a temperature has been reached or not at the surface of a part.

The method developed in the present work consists in filming at a high speed rate gear teeth covered with different temperature indicating lacquers. It allows a direct and full field measure of the corresponding isotherm during the induction heating. From the acquisition and analysis of three different isotherms during the induction heating of a spur gear, a range of the Ac_1 transformation temperature will be defined versus the local heating rate at the gear tooth surface and compared with literature data.

Table 1: Characteristics of temperature measurements means in the context of induction heating.

	Range and area	Accuracy
Thermocouple (K-type)	[0-1250°C] A spot on the surface or within the part	≈ 0.5 to 4 s of response time 2.2°C or 2% of the temperature
Infrared pyrometer	[0-1650°C] 0.5 mm diameter spot on the surface	± 1% of the temperature ≈ 10 to 100 ms of response time
Infrared camera	[0-1200°C] Full field surface	± 1% of the temperature ≈ 10 to 100 ms of response time
Temperature indicating lacquers	Discrete values from 25°C to 1050°C Full field surface	± 1% of the indicated temperature ≈ 1 ms of response time

Experimental procedure

The induction heating treatments were performed using a single turn coil on a dual simultaneous MF and HF frequencies generator designed by EFD Induction Sa. The coil has a 110-mm inner diameter and a 6-mm thickness. The heating procedure consisted in a 4.5 s MF preheating followed by a 0.5 s HF heating and a final air cooling (Fig. 1) without rotating the gear. The times t0 and t1 correspond to the beginning and the end of the HF heating phase, respectively. The hardened spur gear is made of a quenched and tempered SAE 4340 martensitic steel. The gear's geometry is described in Fig. 2.

The experimental setting is shown in Fig. 3. Prior to each test, the upper side of the spur gear was coated with a specific temperature indicating lacquer. Such lacquers contain chemicals in suspension that transform or evaporate at given temperatures, with an accuracy of ± 1% of the indicated temperature. A high-speed camera is set 25 cm above the gear by means of a tripod. The magnification is adjusted so that two teeth are observed within the camera's field of view. The gear is lightened up enough for the lacquer coating to appear uniform. The camera records 512*352 pixels images at a frame rate of 3000 images per second. A total duration of 4 s (12000 images) is recorded for each test. The resulting images have a spatial resolution of 40 µm/pixel and a temporal resolution of 0.3 ms between two frames. A typical image is shown in Fig. 4.

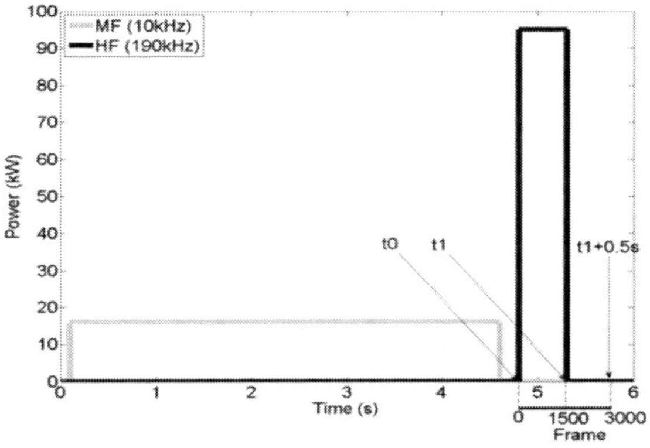

Fig. 1: Induction heating process parameters.

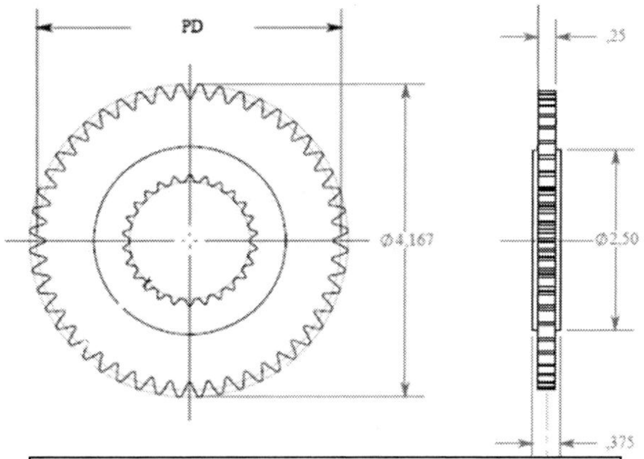

Spur gear data (heat treated part)	
Number of teeth	48
Diametral pitch	12.00
Pressure angle	25.00°
Outside diameter	105.84 mm
Base diameter	92.08 mm
Root diameter	95.38-95.50 mm
Pitch diameter	101.6 mm
Form diameter	97.88 mm
Arc tooth thickness	3.20-3.25 mm
Tooth length	6.35 mm

Fig. 2: Spur gear geometry.

The interface between the evaporated and the non-evaporated lacquer indicates the isotherm specific to this lacquer. The goal of a test is then to determine the exact location of this curve and follow this location during the heating phase. In order to perform automatically the data treatment, image analysis operations have been implemented and carried out using either the free software Fiji, based on ImageJ [10] or the image processing toolbox of Matlab®.

The different steps required for extracting an isotherm from the raw images recorded by the camera system are summarized in Fig. 5, where the following convention, derived from object-oriented programming, is used: a rectangle stands for an image or a group of images, and an oval stands for an operation. An operation takes one or several images as input and returns one or several images as output. Operations and images are related by arrows. An arrow pointing from an image to an operation means that the image is used as an input by the operation, while an arrow pointing from an operation to an image identifies the image as the output of the operation. Only the main output/input images are indicated, along with references to corresponding figures.

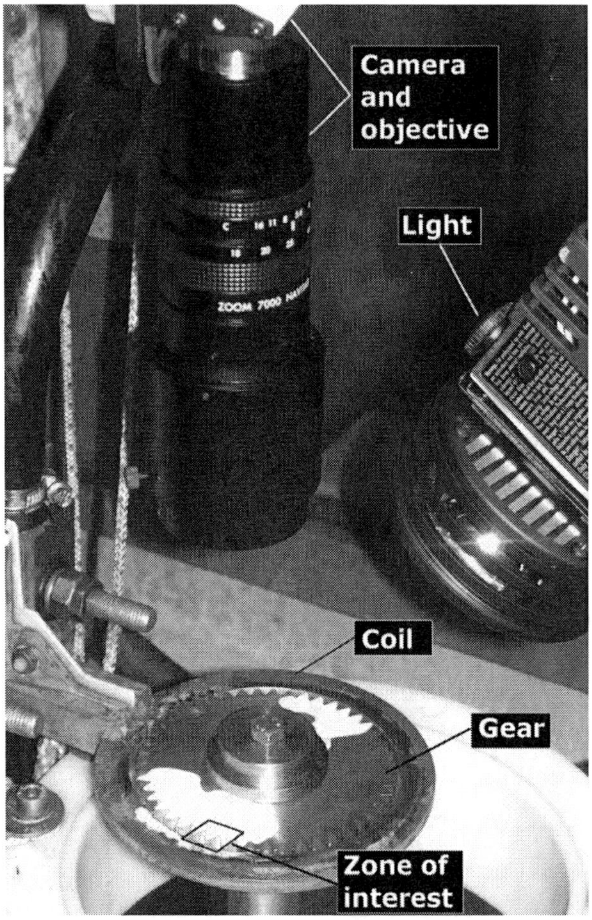

Fig. 3: Experimental setting.

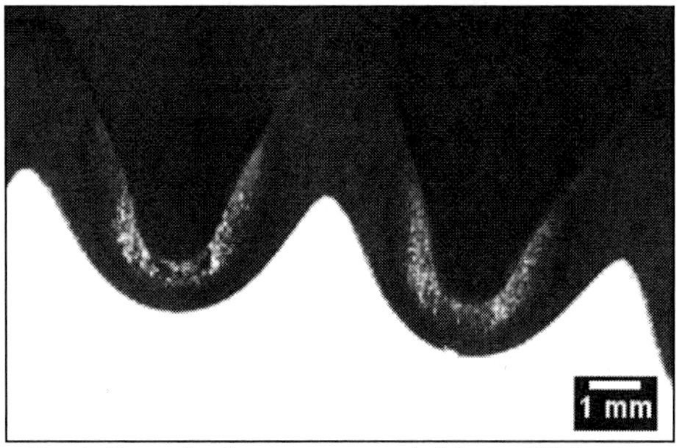

Fig. 4: Example of an image recorded during HF heating.

1. Conversion to gray:
The 12000 images composing the movie are first converted into 8-bit images (255 gray levels).
2. Temporal adjustment:
This operation aims at synchronizing all tests with each other. Indeed, a test is limited to the determination of a single isotherm. The reproducibility of the heating of each of these tests has been thoroughly verified. In order to fully assess the temperature evolution of the part, it is necessary to gather all isotherms within a single, common temporal frame. Images taken for different isotherms, i.e., during different but fully identical tests, are not synchronized. A temporal adjustment is thus needed. It is carried out at the beginning of the treatments and consists in defining a timing reference to which all tests will be referred to. By convention, this temporal reference is defined as the end of the HF heating treatment. It is automatically determined by identifying the frame in which the intensity at the root between the teeth is maximal. 1500 images before this moment and 1500 images after are then selected. This yields a time span of 3000 images (i.e. 1 s) that covers the 0.5 s HF heating treatment and the following air cooling for each test.
3. Stabilization:
The stabilization operation intends to eliminate the vibrations which introduce small periodic translations between successive images. An implementation of the Lukas-Kanade algorithm for ImageJ is used for this task [11].
4. Segmentation:
The goal of the segmentation step is to divide each image into two regions, namely the non-evaporated lacquer and the rest of the image. This step involves distinct sub-operations described in Fig. 6. Choosing a reliable strategy of segmentation adapted to the images and to the information that must be derived from them is a general issue in image processing. In this work, a simple threshold could be considered, i.e. a transformation which assigns the maximal value to bright pixels above a given value and sets the others the minimal value. Nevertheless, such a method is actually error-prone since the intensity of successive images may undergo global and local variations due to external perturbations such as, for the present experiments, the smoke produced by evaporation of lacquers or the light emitted by the material at high temperatures. A more robust approach is needed. While there exists no general solution for efficiently segmenting any given type of image, a so-called "marker-controlled watershed" on the initial image or its gradient proves to be reliable in most cases [12]. This approach is considered as the general paradigm proposed by mathematical morphology. The principle of the watershed relies on the equivalence between a gray valued image and a topographic surface, each gray level corresponding to a certain height. The surface can be described in terms of hills and basins that visually relate to bright and dark zones, respectively. The watershed operation consists in identifying the basins by determining the positions of the ridges that separate them. Each basin thus corresponds to a region that will be segmented. The interested reader is referred to [13, 14] for an in-depth description of this operation.

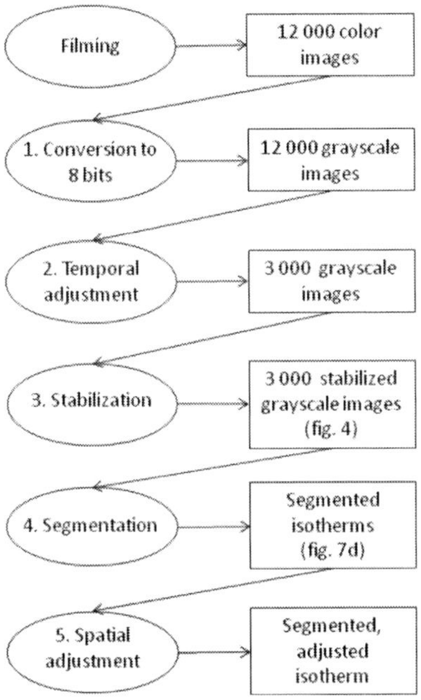

Fig. 5: Image processing operations for all images corresponding to one heat treatment. The segmentation step is detailed in Fig. 6.

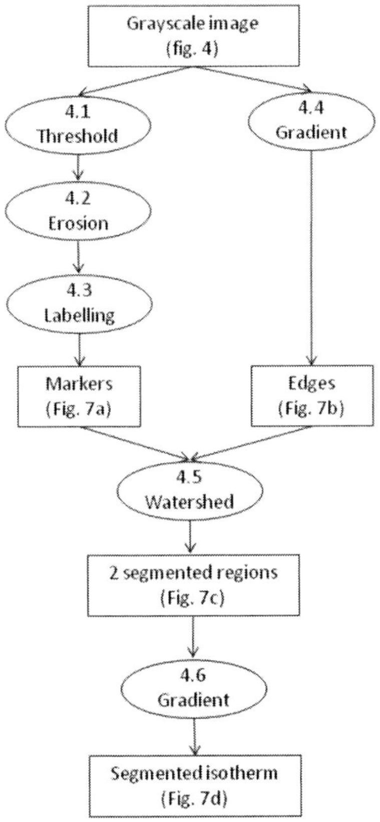

Fig. 6: Image segmentation operations for a single image.

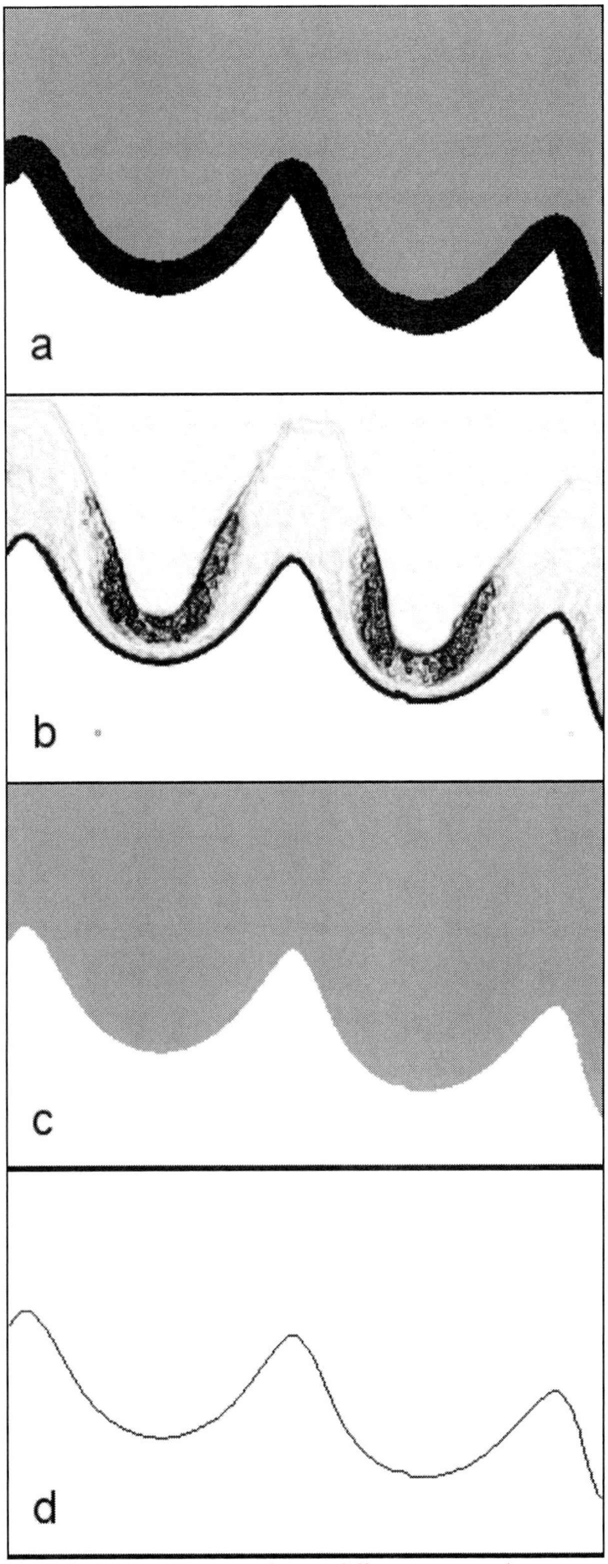

Fig. 7: Images corresponding to different steps in the image processing of Fig. 4. a) Markers, b) Gradient, c) Watershed output, d) Isotherm.

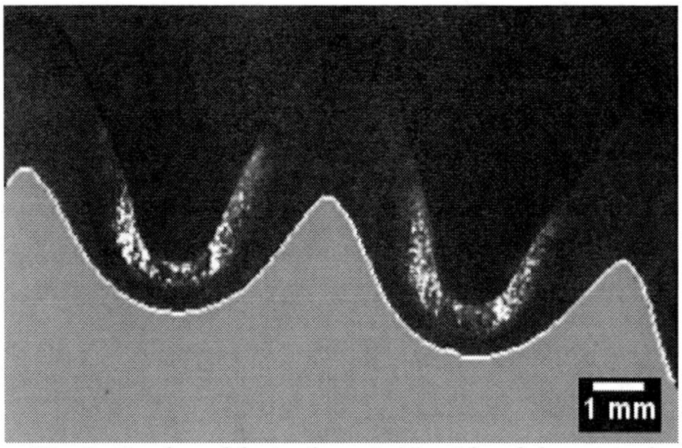

Fig. 8: Overlay of the isotherm in Fig. 7d and the original image in Fig. 4.

In the marker-controlled version of the algorithm, only basins identified by so-called *markers* are taken into account, the others being ignored. Visually, the markers are represented by homogeneous zones, each identified by a distinctive gray level. They are separated by black layers which indicate the approximate location of the ridges between the basins. Two main steps are thus required: first, the different regions to be segmented must be identified by markers (operations 4.1 to 4.3 in Fig. 6); second, the markers are used with the original image or its gradient (operation 4.4) by the watershed algorithm itself (operation 4.5).

4.1, 4.2, 4.3. Threshold, erosion and labelling:

There are various ways to create markers from an image. In this work, they are derived from a threshold of the original image. Two markers are created, one for the intact lacquer, the other one for the evaporated lacquer and the exterior of the part. They appear as two complementary regions that cover the whole image. They then must be separated by a black layer that approximately locates the position of the border between the regions to be segmented. The black layer is created by a grayscale erosion which reduces the surface covered by non-zero pixels. The markers are then each assigned a distinct graylevel (labelling operation). The resulting image is shown in Fig. 7a.

4.4. Gradient of the original image:

The gradient, or spatial derivative, highlights the edges between the different homogeneous zones of the original image and thus facilitates the detection of ridges (Fig. 7b, inverted for the ease of visualization).

4.5. Marker-controlled watershed:

The image of the markers is then used in combination with the gradient of the initial image (Fig. 7b) by the watershed algorithm. The output is an image composed of two complementary regions (Fig. 7c).

4.6. Gradient of the segmented image:

The isotherm is determined by computing the border between both regions (Fig. 7d).

5. Spatial adjustment:

In spite of the care given to the experimental setting, the images produced by different tests may have slightly different orientation and dimensions. The last step consists in the spatial adjustment of the batch of segmented images in order to have a common spatial reference frame for all isotherms issued from different tests. This is achieved by a series of rotations, rescaling and translations. For illustration purpose, the final isotherm is superimposed on the original image in Fig. 8, where the original image has been slightly darkened for the ease of visualization.

Results

A full characterization of the temperature evolution versus time has been conducted at the surface of 2 teeth of the gear heated by MF preheating and 0.5 s HF heating and then cooled in air. Three isotherms, i.e., 704°C, 816°C, and 927°C, are depicted in Fig. 9 for different times of the HF heating phase. A qualitative analysis of the HF heating phase highlights the fact that the temperature rises first at the tooth root and then at the tip, due to the temperature initial state of the surface after the MF preheating phase. Indeed, during the MF preheating phase, Eddy currents are concentrated at the tooth root and the resulting Joule effect make the gear more heated in this area. Consequently, despite the fact that HF heating is known to heat the nearest parts of the tooth to the coil, i.e., the tooth tip, the present method allowed to originally identify three heating steps of temperature evolution during HF heating: temperature increases first at the tooth root, then along the edge of the active profile, and finally in the rest of the tooth surface. Note that the evaporation of the temperature indicating lacquers is by nature non-reversible. It is then only possible to document if a point at the surface reached a certain temperature at a certain time. Therefore, if potential temperature diffusion can be documented during the cooling phase, any temperature decrease cannot be characterized. During the air cooling phase of the present experiment, isotherms slightly evolved during 0.1 s after the HF heating phase and did not evolve anymore after. Therefore, very little temperature diffusion is observed at the surface of the gear after the end of induction heating.

According to the ASTM E384–99 standard, local microhardness measurements were automatically performed on the whole tooth surface with a 300 g load during 13 s. The grid of microhardness indents is defined by a 0.2 mm step in the vertical direction and a 0.3 mm step in the horizontal direction. As shown in Fig. 10, the tooth surface hardness is clearly defined by a hardened layer around 685 HV and a core hardness around 350 HV.

The etched microstructure, result of the induction hardening and air cooling, is also shown in Fig. 11. It can be observed that the hardened layer and the core are separated with a brighter line area corresponding to the transition zone. The final isotherms positions are superimposed with the etched microstructure in order to locate the interface between hardened and non-hardened regions relative to them. The comparison of the distances between the transition line and the isotherms makes it possible to evaluate the transformation temperature locally. The evolution of the 816°C isotherm is also plotted versus time in. Fig. 12 for the tooth root and the tooth tip and compared with the hardened depths at these two locations.

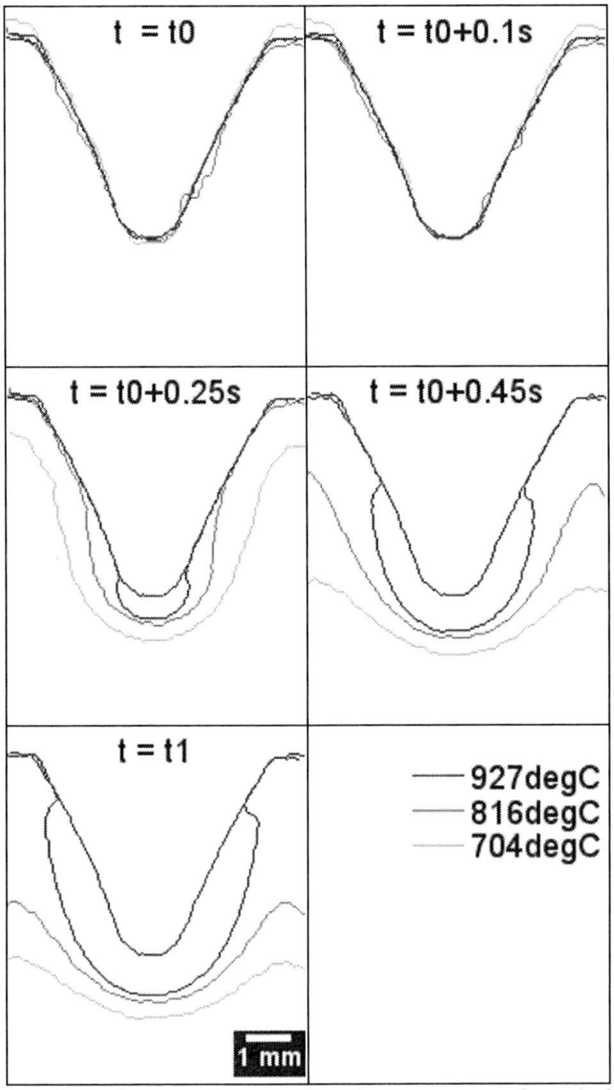

Fig. 9: Isotherms at five steps from the beginning to the end of the HF heating.

As shown in Fig. 11 and Fig. 12, at the tooth root, the hardened depth corresponds to the final position reached by the 816°C isotherm whereas it corresponds to a temperature included into the range 704°C - 816°C at the tooth tip. Thus, the hardness frontier may not be associated with a single isotherm. This result can either suggest very different temperature gradient or that the material experienced different transformation temperatures at different locations of the tooth. This latter is thought by the authors to be more plausible since transformation phenomena are known to be not only temperature-dependent, but are also heating rate dependent [15-17]. From the measurements obtained via the present method, the temperature rate can actually be estimated by plotting the evolution of temperature versus time. In Fig. 13, temperature evolution is given at two arbitrary points within the hardened layer, one located at 0.72 mm from the root and the other at 3.12 mm from the tip (points A and B in Fig. 11, respectively). The result is shown in Fig. 13. It can be observed that the heating rate is much higher at the tooth root than at the tooth tip, i.e., 1650°C/s and 835°C/s, respectively. This significant difference could result in a higher Ac1 at the tooth root than at the tooth tip. This trend is in agreement with the results of [15-17].

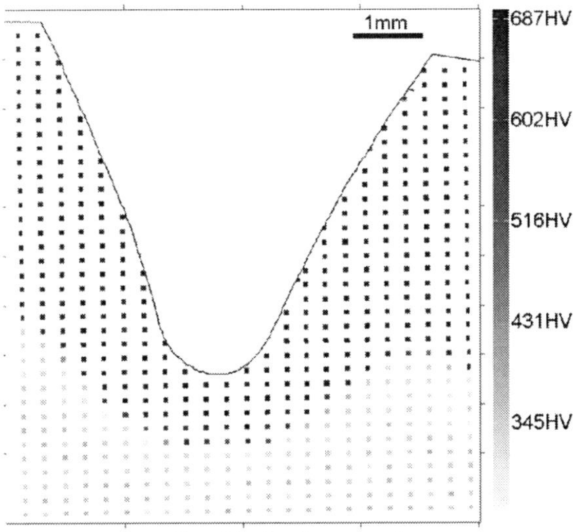

Fig. 10: Hardness results.

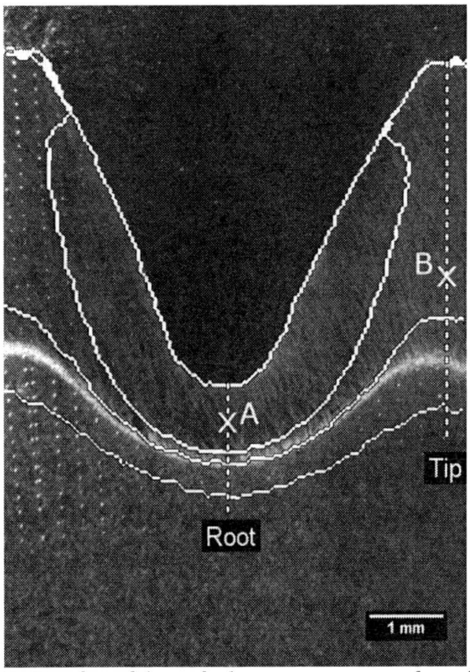

Fig. 11: Overlay of the etched microstructure with isotherms at 927, 816 and 704°C (white lines, from top to bottom). The dashed lines correspond to profiles presented in Fig. 12.

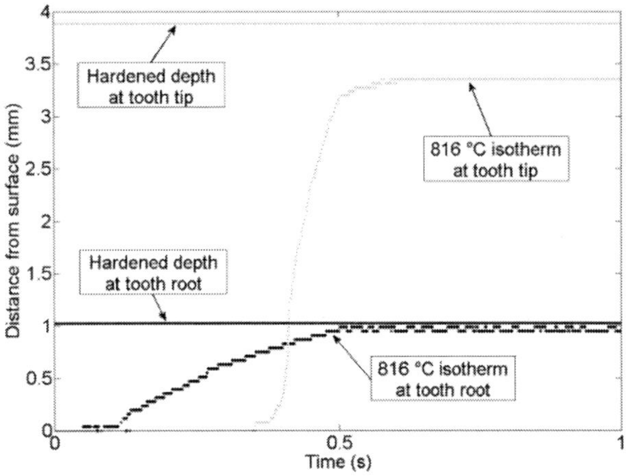

Fig. 12: Temporal profiles of the 816 °C isotherm at the root and tip of a tooth (white dashed lines in Fig. 11).

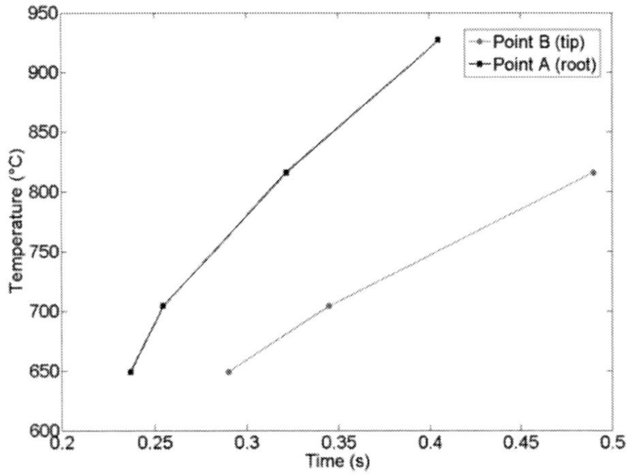

Fig. 13: Temperature versus time at points A (root) and B (tip) in Fig. 11.

Conclusions

A method for full-field surface temperature measurement during fast heating conditions has been developed. It makes use of a high-speed camera system combined with temperature indicating lacquers. In this work, a spatial resolution of 0.04 mm/pixel and a temporal resolution of 0.3 ms between two images were achieved. The thermal history at the side of a spur gear during HF heating was accurately measured by means of an image analysis procedure. In particular, one can document where and when a given temperature is reached at the surface of a whole gear tooth hardened by fast induction heating.

The final position of isotherms was compared to the hardness transition line. The results suggest that the transformation temperature Ac1 varied with the position at the surface of the part. The heating rates were estimated at two locations, showing that the temperature at the root increased faster than at the tip. This method thus allowed confirming the correlation between the Ac1 temperature and the heating rate.

Further work shall focus on determining more isotherms in order to accurately evaluate the variation of Ac1 at very high heating rates. The full history of surface isotherms will also be studied for different heating parameters, i.e., with or without pre-heating, through either simultaneous or sequential high frequency/medium frequency final heating. Finally, such temperature data can be of great interest for validation or calibration of computer simulations of induction heating.

References

[1] Grum, J., *Induction Hardening*, in *Handbook of residual stress and deformation of steel*, G. Totten, M. Howes, and T. Inoue, Editors. 2002, ASM International: New-York. pp. 220-223.

[2] Kristoffersen, H. and P. Vomacka, "Influence of process parameters for induction hardening on residual stresses."*Materials & Design*, Vol. 22, No. 8 (2001), pp. 637-644.

[3] Nemkov, V., *Modeling of induction hardening processes*, in *Handbook of Thermal process Modeling Steels*, C.H. Gür and J. Pan, Editors. 2008, Taylor & Francis group: New-York. pp. 428-431.

[4] Savaria, V., et al., *On the measurement of residual stress gradients in induction hardened parts*, in *8th European Conference on Residual Stresses,* Riva del Garda, Italy, 2010.

[5] Ducassy, C., et al., *Prediction of over-tempering during rapid induction heating*, in *1st Mediterranean Conference on Heat Treatment and Surface Engineering*, Sharm-el-Sheick, Egypt, 2009.

[6] Rudnev, V., et al., *Handbook of induction heating*. 2002, New-York: Taylor & Francis group.

[7] Misaka, Y., Kiyosawa, Y., Kawasaki, K., Yamazaki, T. and Silverthorne, W.O., *Gear Contour Hardening by Micropulse Induction Heating System*, S.o.A. Engineers, Editor. 1997. pp. 121-130.

[8] Komotori, J., et al., "Fatigue strength and fracture mechanism of steel modified by super-rapid induction heating and quenching."*International Journal of Fatigue*, Vol. 23, No. Supplement 1 (2001), pp. 225-230.

[9] Inoue, T., et al., *Simulation of dual frequency induction hardening process of a gear wheel*, in *3rd International conference on quenching and control of distortion*, Prague, Czech Republic, 1999, pp. 243-250.

[10] Abramoff, M.D., Magelhaes, P.J., Ram, S.J., "Image Processing with ImageJ."*Biophotonics International*, Vol. 11, No. 7 (2004), pp. 36-44.

[11] Li, K., Kang, S. *The image stabilizer plugin for ImageJ*. 2008 Last accessed 11-05-18; Available from: http://www.cs.cmu.edu/~kangli/code/Image_Stabilizer.html.

[12] Beucher, S., Meyer, F., *The morphological approach of segmentation: The watershed transformation*, in *Mathematical Morphology in Image Processing*, E.

Dougherty, Editor. 1992, Marcel Dekker: New-York. pp. 433-481.

[13] Beucher, S., *Mathematical Morphology and geology: when image analysis uses the vocabulary of earth science: a review of some applications*, in *Geovision '99: International Symposium on Imaging Applications in Geology*, D. Jongmans, E. Pirard, and P. Trefois, Editors. Université de Liège: Liège, Belgium, 1999, pp. 13-16.

[14] Roerdink, J.B.T.M., Meijster, A., "The watershed transform: definitions, algorithms, and parallellization strategies."*Fundamenta Informaticae,* Vol. 41 (2001), pp. 187-228.

[15] Rudnev, V. (2005) *Be aware of the 'fine print' in the science of metallurgy of induction hardening*. Industrial heating, pp. 1-5.

[16] Feuerstein, W.J. and W.K. Smith, "Elevation of critical temperatures in steels by high heating rates."*Trans. ASM*, Vol. 46 (1953), pp. 1271-1284.

[17] Clarke, K., et al., "Induction Hardening 5150 Steel: Effects of Initial Microstructure and Heating Rate."*Journal of Materials Engineering and Performance*, Vol. 20, No. 2 (2011), pp. 161-168.

Proceedings of the 26th ASM Heat Treating Society Conference
B.L. Ferguson, R. Jones, D.S. MacKenzie, and D. Weires, editors
https://doi.org/10.31399/asm.cp.ht2011p0068

The Aerodynamic Furnaces for Heat Treatment

Dr. Alexey Sverdlin, Dr. Mattew A. Panhans
Milwaukee School of Engineering, Milwaukee, WI, USA
Dr. Yury Sokolov
PO Box 2368, Rzhev, Russia
Dr. Arnold Ness
Bradley University, Peoria, IL, USA

Abstract

AHTF (Aerodynamic Heat Treating Furnaces) furnaces, in which air or gas is heated to 600-700°C without electrical or other special heaters, have been developed and placed in operation in a number of plants for heat treating aluminum, magnesium, and titanium alloys, and also steels. The AHTF chamber furnace is thermally insulated without the use of firebricks. It has a centrifugal fan with vanes having a special contour. The fan, operating in a closed system, converts, into heat, almost all the energy used to turn it; the heat is transferred to the parts by convection. In most machine-building plants aluminum alloys are heat treated in ERF furnaces (electric resistance furnaces with forced air circulation) or in salt baths. This research deals with an investigation of the heating conditions for various semi-finished products of aluminum alloys in the AHTF-3 in comparison with the ERF-2 (Electrical Recirculation Furnace) furnace and a potassium nitrate bath of approximately the same working volume.

Keywords: Heat Treatment, Furnace, Aluminum Alloys, Aerodynamic Heating

1. Introduction

The invention of aerodynamic heating process in the former Soviet Union in 1963-1964 resulted in the introduction of a new class of industrial heating devices. The principle of aerodynamic heating is a transfer of gas flow energy generated by a centrifugal fan into heat. The rotor of the centrifugal fan serves as a compressor, and as a heat generator. It induces circulation in the furnace atmosphere and generates heat at the same time. The first industrial furnaces were developed by a design team headed by Dr. P. I. Tevis and others [1-4].

There is no uniform, widely accepted terminology on the subject. For practical applications the most widespread term is "aerodynamic heat treat furnace," or "aerodynamic losses furnace." We must emphasize the difference between the meaning of the term "aerodynamic heating" in our case, and its usage in the aerodynamics of hypersonic flows. In the latter instance, this term is used to describe the heat generation caused by the friction between a surface of a solid object and high-speed gas flow. This effect is negligible in AHTF.

The AHTF has a number of similar yet different configurations. The differences are generally in the geometry of the heating chamber and in the duct system for air flow and for supply and return. Figures (1-5) show some of these variations.

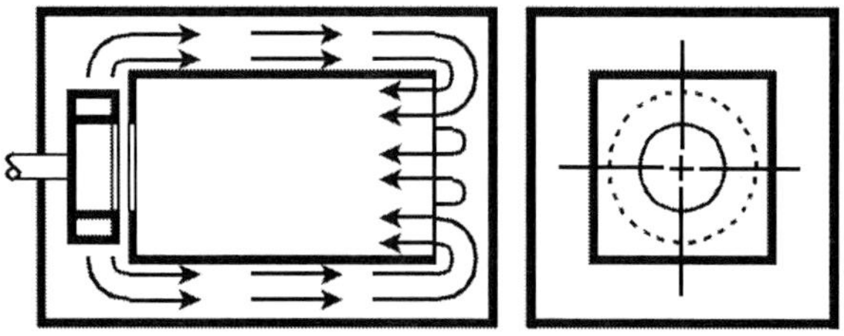

Figure 1. A chamber with a four screen duct system

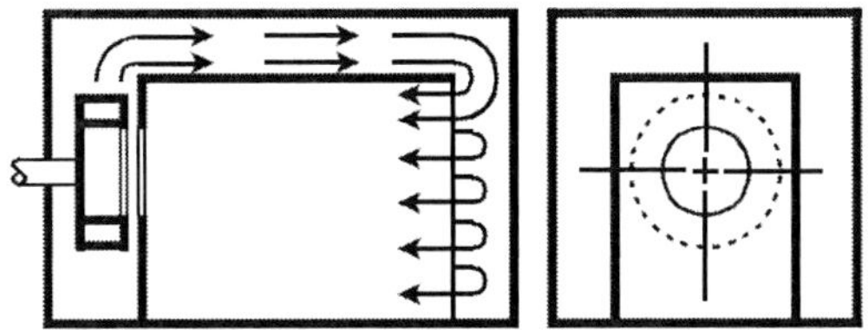

Figure 2. A chamber with a three screen duct system

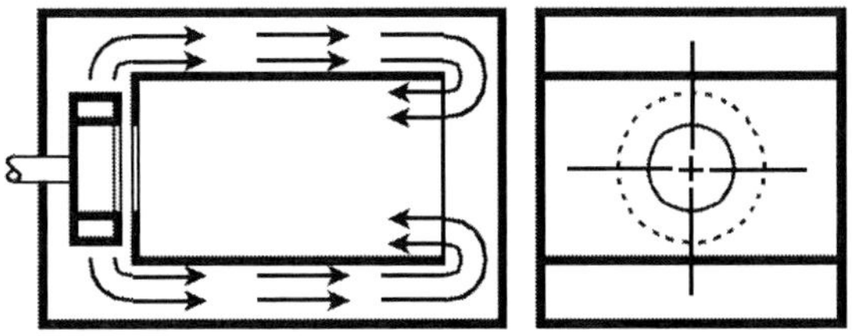

Figure 3. A chamber with a two screen duct system

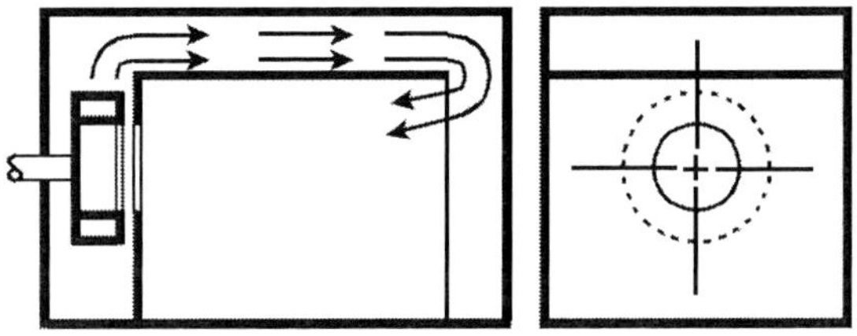

Figure 4. A chamber with a single screen duct system

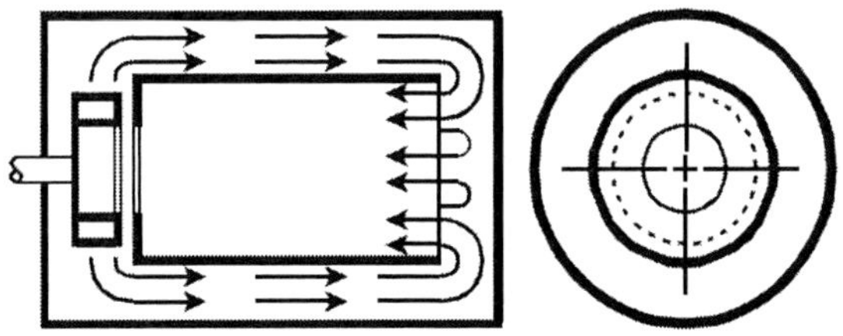

Figure 5. A chamber with a ring duct system

2. Heating Test

A test to compare the heating ability of three different systems was conducted. An AHTF system, an ERF system and a salt bath system were compared.

The duration and uniformity of heating in air and potassium nitrate were recorded with 12 thermocouples equally spaced in the working area. The time for heating various semifinished products to 500-525°C was determined with thermocouples in the center of the samples of the 2024 type alloy (sheets 0.8 - 10 mm thick, pipes with a wall thickness of 1 to 4 mm, bar 40-200 mm in diameter) and the AK6 alloy (forgings 40 - 100 mm thick). The weight of the metal loaded in the furnace and the salt bath was 80-350 kg.

It was found that the AHTF-3 produces a more uniform temperature field in the working space than the ERF-2, the temperature difference in the working space approaching that in the salt bath (Table 1).

Table 1. General Specification of Heating System

Parameters	**AHTF-3**			**ERF-2**	**Salt Bath**
Power, kW	55			129	100
Working Dimensions, mm	1100 x 3100 x 1600			1200 x 3200 x 1500	1000 x 3200 x 1100
Max Operating Temperature, °C	550			550	540
Rate of Air Circulation, m/sec	16-18			6	-
Heating Temperature, °C	160	410	500	500	500
Temperature Uniformity, °C	± 1.5	± 2.0	± 3.0	± 5.0	± 2.0

The air is heated from room temperature to 500°C twice as rapidly in the AHTF-3 (4 hr) as in the ERF-2 (8 hr). The higher heating rate was observed for a furnace of the same size even though the ERF-2 had twice the power.

Despite a larger drop in temperature in the AHTF-3 in comparison to the ERF-2 when metal weighing 180-350 kg is placed in the furnaces, the AHTF-3 heats the metal to the given temperature 1.5 to 2 times as rapidly (Figure 6). However, when the load of metal is 80-160 kg the ERF-2 heats the air to the given temperature 1.5 times as fast as the AHTF-3 (Figure 7). This is due to the large reserve of heat in the ERF-2 in the mass of the heated brick wall, which also leads to a smaller reduction of the furnace temperature when the metal is loaded into a hot furnace.

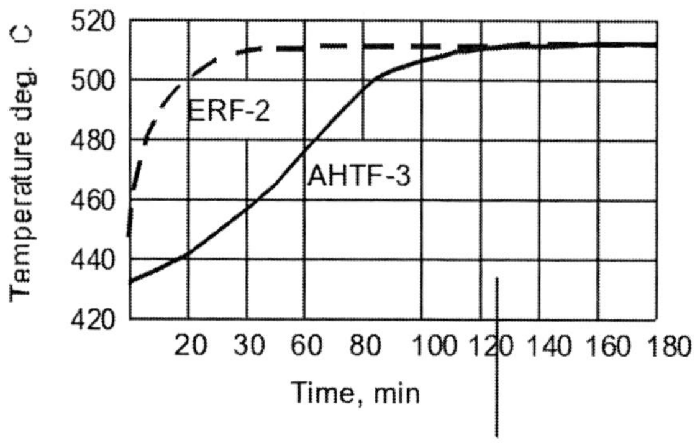

Figure 6. Heating curve of metal from moment of loading the furnace to a temperature of 515°C for parts weighing 350 kg.

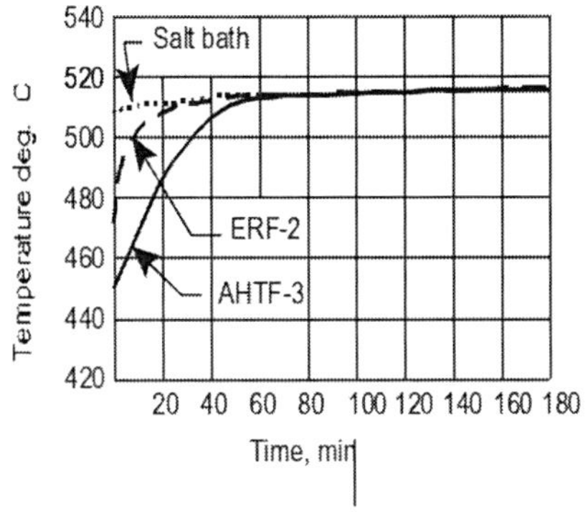

Figure 7. Heating curve of metal from moment of loading the furnace to a temperature of 515 °C for parts weighing 160 kg.

The temperature drop of the salt bath when the metal is loaded is minimal because of the large thermal inertia, and the time of heating to a given temperature is maximum by comparison with the ERF-2 and AHTF-3 (Figure 6).

Bars and forgings weighing 160-350 kg are heated to 515 °C and sheet and pipe weighing 80-180 kg are heated to 500 °C 1.8-2.5 times faster in the AHTF-3 than in the ERF-2 furnace in spite of the more rapid heating of the air in the ERF-2 when the load is small.

The heating time is smallest in the salt bath as compared with the AHTF-3 and ERF-2 (Figure 8). When bars and forgings weighing 160 kg and 40-200 mm thick were heated to 515 °C the heating time was 5-20 min in the salt bath, 40-100 min in the AHTF-2, and 60-180 min in the ERF-2. Then the heating rate of the samples in the salt bath declined sharply and the difference in time for heating to 515°C decreased: the heating time in the AHTF-3 was only 1-2.4 times more, and in the ERF-2 2.5-5 times more, than in the salt bath.

The advantage of the AHTF-3 in the uniformity of the temperature field in the working space and the rate of heating the metal in spite of its lower power (one-half) by comparison with the ERF-2 is explained by the high rate of circulation of the air, the low thermal inertia, the existence of a single thermal zone and the complete absence of heat transfer by radiation. ERF-2 has three independently controlled thermal zones; heat transfer is by radiation and convection.

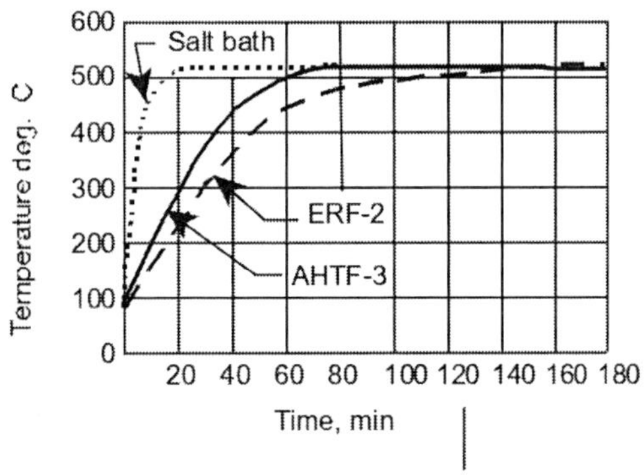

Figure 8. Heating curve of metal from moment of loading the furnace to a temperature of 515 °C for parts 100 mm thick and weighing 160 kg.

3. Technological Efficiency AHTF

The advantages of AHTF are uniform and intensive heating. This results in good control of process, with resulting high quality thermal processing (thermo-chemical). Also these units produce a combination of thermal profitability and productivity of the unit, with simplicity of design reliability. An additional advantage of AHTF is that these units give good gas tightness of the chamber (in contrast to, for example, electric furnaces) eliminating the necessity to support a small positive pressure in the working volume. With normal atmospheric air without protective gasses, when the tightness of the chamber is not required, the pressure is constantly above atmospheric in the unit. As air is always being let out from chamber, inflows are excluded, reducing interim cooling of the load, and increasing uniformity of the heating. All this simplifies the problem of the maintenance of the stable temperature in the air in the workspace.

It is important to emphasize the special requirements for uniformity of a temperature field in the heat treatment of materials with space - time uniformity in the development of structural transformations. As a rule the discrete control of the temperature on the surface of a material is used in the heat treatment processes so for a given temperature diagram only points of control are obtained, but phase transitions occur throughout the entire volume of the material. Temperatures at each interior point are not controlled.

4. Heat Treatment of Aluminum Alloys in AHTF

The results of research and experience with industrial applications testify, not only to the high quality of heat treatment in AHTF, but also the opportunity in a number of cases for significant reduction process cycle times and simplification of procedures with significant decrease in the cost of the process as a whole, and with improvement and increased stability of the material's properties.

Such effects are realized because of the uniformity of heating, accuracy of regulation and maintenance of a temperature mode. The processing with AHTF allows, in a number of cases, one to realize optimum modes unattainable with other methods of heating. For example, the heat treatment of magnesium alloy ML-5 requires heating up to temperature of 420°C with an accuracy of ± 2°C. This furnace allows one to raise the temperature of the metal up to the top limit of hardening temperatures without risk of burning. This has generated a time reduction by one half (from 16 to 8 hours).

The special interest from the point of view of an estimation of quality of heat treatment is represented by aluminum and magnesium alloys, as they are most sensitive to temperature fluctuations: The ability to obtain of quantitative parameters on these alloys guarantees the quality of heat treatment of all other materials.

The quality of heat treatment of aluminum deformed alloys was determined by comparison of properties of specimens from different semifinished products (alloys D16, AK6, V65) (Table 2) after complete strengthening heat treatment with heating to the quench temperature with a AHTF and a saltpeter bath. Specimens from D16 were treated with natural age hardening, from alloys AK6 and V65 - with artificial aging, from alloys AK6 and V65 - with artificial aging using the AHTF.

Table 2. Composition of Aluminum Alloys.

	Chemical Composition, %				
	AK6	**D16**	**V65**	**V95**	**AK4-1**
Cu	2.2	4.3	4.2	1.70	2.40
Mg	0.65	1.5	0.2	2.30	1.60
Zn	-	0.3	-	6.00	0.30
Si	0.9	0.5	-	-	0.35
Mn	0.6	0.6	0.4	0.40	0.20
Cr	0.7	0.5	-	1.18	-
Fe	-	-	-	1.50	1.10
Ni	-	-	-	-	1.10

Specimens were cut from rolled plates D16, extruded tube and bar D16, forged AK6 billets and pins from V95. The results for tensile strength (TS), yield strength (YS), percent of elongation, shear strength, τ_{sh} for rivets, fracture toughness K_{Ic}, electric conductivity of specimens, heat treated in AHTF and saltpeter bath were obtained (Tables 3-5).

Losses of mechanical properties due to stress corrosion of plates D16 with thickness 0.8 and 1.5 mm, quenched in water after heating with the AHTF, are less than after heating in the saltpeter bath (Table 2); during the heating to the quench temperature with an AHTF the scattering of properties of specimens (TS, percent elongation) was smaller. The depth of diffusion of copper and magnesium into the cladding of alloy D16 specimens with thickness from 0.8 to 10 mm and tube D16 (30 by 15 mm), heat treated in the AHTF and in salt bath were the same.

Table 3. Fracture toughness K_{IC} (kJ/m^2) of the specimens after heat treating in the furnaces.

	Sheet, D16			
		Saltpeter bath		
Thickness of a sheet, *mm*	**AHTF**	**3**	**7**	**20**
1.5	107.90	107.90	-	95.16
3.0	101.00	94.18	-	91.23
5.0	100.00	96.14	-	88.30
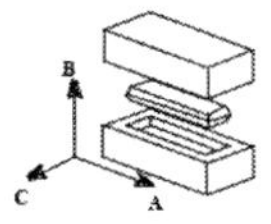				
	Forging AK6			
		Saltpeter bath		
Orientation of specimens	**AHTF**	**3**	**7**	**20**
B	71.61	-	65.65	-
C	117.72	-	107.90	-
A	196.20	-	166.77	-

Table 4. Electrical conductivity of D16H aluminum alloy, $m/(ohm \cdot mm^2)$

		After heat treatment in the furnaces:	
Thickness of the sheet, *mm*	**As shipped**	**AHTF-3**	**Saltpeter bath**
3	16.8	16.8	17.1
5	19.0	19.2	19.4

Table 5. Properties of specimens after heat treatment by different heating methods before and after corrosion tests

Ultimate Tensile Strength UTS, MPa and elongation, %					
			Before	After	Change, %
Sheet D16S, 0.8mm	AHTF	UTS	447.34	302.15	33.9
		Elongation	22.5	4.40	80.5
	Salt bath	UTS	429.7	340.4	20.8
		Elongation	20.2	6.70	67.0
Sheet D16H, 0.8mm	AHTF	UTS	454.2	253.0	54.7
		Elongation	21.9	4.400	80.0
	Salt bath	UTS	449.3	255.0	43.3
		Elongation	21.1	4.00	80.0
Sheet D16S, 1.5 mm	AHTF	UTS	456.16	426.73	6.5
		Elongation	19.7	11.40	42.0
	Salt bath	UTS	435.6	395.34	10.0
		Elongation	18.9	9.20	51.0
Sheet D16H, 1.5 mm	AHTF	UTS	458.13	415.94	9.3
		Elongation	19.3	9.80	49.0
	Salt bath	UTS	451.26	364.93	19.0

Investigation of the processes of heating in the AHTF of forgings from alloy AK-6, produced from preliminary deformed semifinished products, indicated the possibility of a drastic reduction of the heat treatment cycle times. According to a study of mechanical properties of specimens, cut from forging in three directions (σ_b, $\sigma_{0.2}$, δ) were the practically the same after the heat treatment with a reduced holding time of hold, t = 100 min as compared to the recommended holding time, t = 150 min (Table 4).

The heat treatment (artificial aging) without preliminary quenching, which needs close control on the temperature, of automobile pistons from heat-resistant alloyed AL-30 with the AHTF produced good quality of parts. Optimum processing, such as age hardening with at 185°C for 8 hours and cooling in the air, generated better mechanical and manufacturing properties. The cycle of heat treatment was reduced by 4 to 6 hours, compared with previous technology. Further, for manufacturing conditions, with the heat treatment of big groups (more than 2000) pistons without special placement, positive results were obtain too.

Aging hardening of alloy AK4-1 was also done with an AHTF. The basic problem of these processes is a requirement of high uniformity of heating with a gas temperature variation not more than ± 1 to 1.5°C and at the metal ± 3°C. Optimum temperature of artificial age hardening : for plates AK4-1 is t = 195 ± 5°C with holding time of τ = 12 h.; for as quenched, t = 192 ± 2°C, τ = 24 h.

AHTF satisfy the tight requirements, which are necessary for the process of heat treatment, and exceed special electro-thermal equipment for medium- and low-temperature heating with regard to quality of treatment, productivity and efficiency of process. AHTFs do not concede to the technology parameters of other methods and sometimes they exceed the most effective heating technologies available.

5. Conclusions

The AHTF is more economical, simpler to manufacture, and safer in operation than electric air furnaces or salt baths.

The AHTF produces a more uniform temperature in the working space and heats the air from room temperature to the given temperature twice as rapidly as ERFs.

Heating of different semifinished products with sections 40-200 mm thick in loads of 80-350 kg to 500-515°C is 1.8-2.5 times more rapid in the AHTF-3 than in the ERF-2 with similar dimensions in spite of the fact that the ERF has twice the power.

The mechanical properties, resistance to corrosion and stress corrosion, and electrical conductivity of various semifinished aluminum alloy products are the same after heat treatment in the AHTF-3, ERF-2, and the salt bath.

References

[1] Kolobnev N., *et al*, *Metallovedenie I Termicheskaya Obrabotka Metallov,* No. 1, January 1968, pp. 49-51.

[2] Lemke R. G., Troubleshooting of Electrical Machines, Moscow-Leningrad, Gosenergoizdat, 1963, 176 pages.

[3] Loitsyansky L.G. , Fluid Mechanics, Moscow, Nauka, Fifth Edition 1978, 736 pp.

[4] Solomahova T.C., Tchbisheva K.V., Centrifugal Fans, Moscow, Mashinostroenie, 1980, 176 pages.

[5] A. Sverdlin, A.R. Ness, A. Drits, *Aerodynamic Furnaces, Advanced Materials & Processes*, April, 1996, pp. 40-44.

[6] P. Tevis, V.A. Ananjev, E.G. Shadek, Recirculating Aerodynamic Heating Furnaces, Moscow, Mashinostroenie, 1986, 208 pp.

Proceedings of the 26th ASM Heat Treating Society Conference
B.L. Ferguson, R. Jones, D.S. MacKenzie, and D. Weires, editors
https://doi.org/10.31399/asm.cp.ht2011p0076

Operational Efficiency Improvements Resulting from Monitoring and Trim of Industrial Combustion Systems

Damian Bratcher, Director International Operations
Super Systems Inc, Cincinnati, Ohio, USA
dbratcher@supersystems.com, +1 513 772 0060, +1 513 772 9466 (fax)

Introduction

Combustion is the exothermic chemical reaction (a reaction in which heat is given off) of hydrogen and carbon atoms contained in fuels with oxygen. In this reaction, the carbon is oxidized to form carbon dioxide (CO_2) or, if insufficient oxygen is available, carbon monoxide (CO). If there is an excess of oxygen present, non-reacted O_2 will be in the products of combustion. Excess O_2 makes heating inefficient, thus requiring more gas for the same results. In addition, excess air also allows for the formation of pollutants such as Nitrous Oxide (NO) and Nitrogen Dioxide (NO_2). Together, NO and NO_2 make up what is referred to as NO_x.

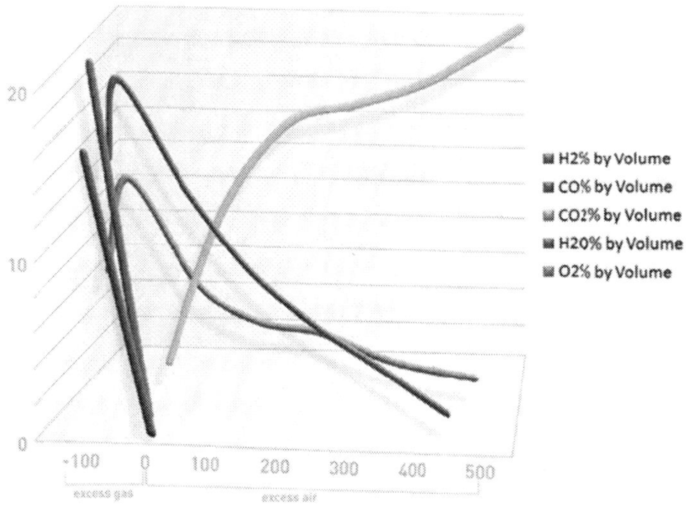

According to the Department of Energy, most high temperature direct-fired furnaces, radiant tubes and boilers operate with about 10 to 20% excess combustion air at high fire to prevent the formation of dangerous CO and "soot" deposits. It is estimated that precise control of air to fuel ratio will yield 5 to 25% or more savings in heat generation. The air gas ratio can be determined by analyzing the flue gas and, with this information, the mixture for combustion can be altered to produce the most clean and efficient heat for the process. Our studies have shown that burners are typically running with excess O_2 greater than 4% in the flue gas.

Optimizing operational efficiency, minimizing production costs and maximizing utilization, is a competitive advantage in good economic conditions. In leaner times, it is a basic necessity. Periodic checking and resetting of air-fuel ratios is one of the simplest ways to get maximum efficiency out of fuel-fired process heating equipment. In heat treatment facilities, the customer would find potential efficiency improvements on generators, radiant tubes, furnaces, ovens, heaters, and boilers.

The two main areas where heat treatment facilities benefit from combustion optimization are fuel savings and throughput improvements. Combustion optimization will be reviewed first. Next, the impact these improvements have on throughput and utilization will be explored.

Combustion Efficiency

Most high temperature direct-fired furnaces, radiant tubes, and boilers are designed to operate with 10 to 20 percent excess combustion air at high fire. This excess air helps prevent the formation of carbon monoxide and soot deposits which can affect heat transfer surfaces and radiant tubes.

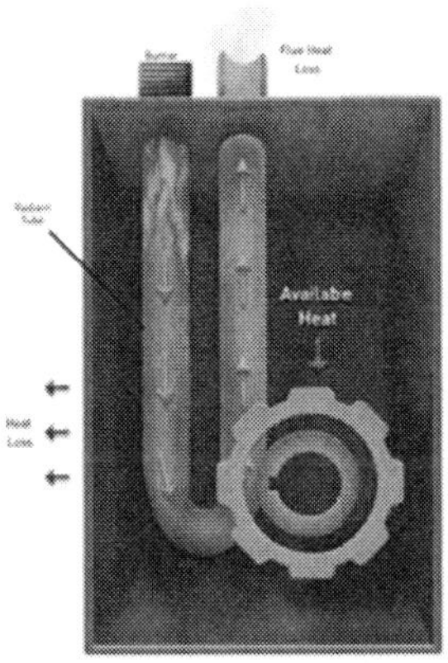

For the fuels most commonly used in the US, including natural gas, propane, and fuel oils, approximately one cubic foot of air is required to release 100 British Thermal Units (BTUs) in complete combustion. Process heating efficiency is reduced considerably if the air supply is significantly higher or lower than the theoretically required air.

In a September 1997 ***Process Heating*** magazine, Mr. Richard Bennett provided calculations for an available heat chart which was republished in May 2002 by the Department of Energy. This chart is an excellent basis to determine potential savings in a combustion process. To determine the potential savings, you will need the following information:

- Exhaust gas temperature as it exits the furnace, tube, etc.
- % excess air or oxygen in the flue gas (actual)
- % excess air or oxygen in the flue gas (target)

The available heat chart is shown in Figure 1.

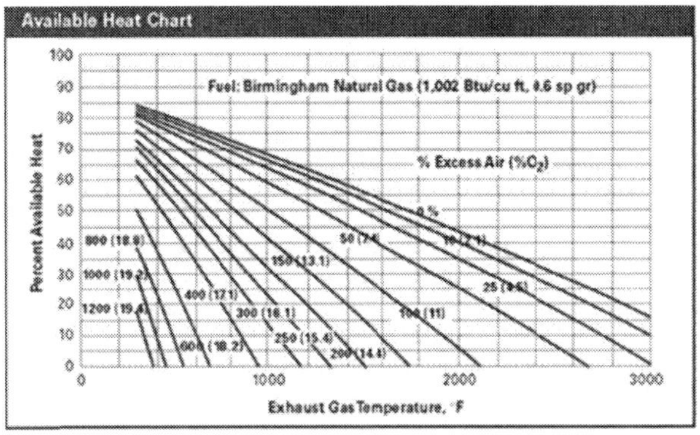

Figure 1- Available Heat Chart

Using the chart, determine the percent available heat under actual and target conditions. The intersection of the measured exhaust gas temperature and % excess air ($\%O_2$) curves provides these values. The potential fuel savings would be calculated as follows:

% Fuel Saving = 100 x ((%AH Target - %AH Actual) / %AH Target)

Documented Savings

To illustrate the value of combustion optimization, two case studies will be presented.

Forge Heat Furnace

A 6 mmbtu/hr open die forge reheat car bottom furnace was equipped with a high-temperature SuperOX oxygen sensor, reference air system and 9120 oxygen controller. Baseline readings of excess O_2 and fuel consumption were collected over a three month period. Based upon this data, monthly fuel consumption was determined as was the average high-temperature O_2 readings (6.5% average at 2200° F).

Picture 2 - Open Die Forge Heat Furnace

The controls and operation's personnel were concerned about over-trimming the excess O_2 level. Lowering O_2 levels can lead to reduced uniformity on the heated ingot. Thus, the O_2 levels were lowered incrementally to ensure that no impact occurred to product quality.

At the end of the first incremental change and after process verification, the customer has lowered his excess O_2 from 6.5% to 5.5%. After numerous runs at this O_2 level, the customer documented a 20.5% reduction in metered gas consumption. Using the data shown in figure 1, the % available heat at 5.5% O_2 and 2200F is ~25%. Similarly, 6.5% and 2200F, it is ~20%. The potential savings is ~20% [100 x (25-20)/25]. The actual results are very similar to the expected results.

For this customer, the 20.5% reduction in fuel cost corresponds to a $15,000 per year savings for this single furnace based upon its current utilization rate. At full utilization, the savings would be $53,874.

The customer has a goal of reducing the excess % O_2 several percent. At his target level, he would reduce his fuel costs by an estimated 37%. At the current utilization, the savings would be $27,750 per year. At full utilization, the savings reach $98,550 for this furnace. For all 14 furnaces in the facility, the fuel savings have the potential to exceed $1,000,000.

A side benefit to the fuel savings is a documented CO_2 reduction. For each MCF CH4 burned completely, 117 pounds of CO_2 is produced. In this particular case, the customer was able to document a reduction of 175,500 pounds or 87.75 tons of CO_2. At full utilization on this one furnace with a 1% reduction in excess O_2, the reduction would be 630,006 pounds or 315 tons. If the customer has similar success on other furnaces and is able to achieve the O_2 target, his potential CO_2 reduction is 8000+ tons.

Batch Furnace Utilization and Fuel Savings

The initial R&D on batch furnaces was initiated with John Keough at his Applied Process' Wisconsin and Kentucky facilities. Mr. Keough recognized the value in operating burners at an optimum level to save on fuel. He also recognized the even greater value in creating operational efficiencies by increasing load throughput based on increasing the available heat produced during high fire with the optimal gas/air ratio. Mr. Keough and Applied Process challenged Super Systems with deriving a system that would monitor high fire air/gas ratios and provide operators with alarms and trending to monitor the burner performance. The two test sites demonstrated and proved out the sensor and control technology and provided the initial data regarding combustion efficiency and utilization improvement. These results led us to further testing at Queen City Steel Treating.

Queen City Steel Treating in Cincinnati, Ohio worked with Super Systems to document savings relative to varying O_2 levels in combustion exhaust gases. The tests were conducted on a batch furnace with 4 radiant tubes using the same load density with identical initial conditions. Each burner is rated 250,000 BTU/hr.

Four tests were conducted with excess O_2 levels ranging from 2 to 5%. The test results are shown in Table 1.

Table 1 – Furnace trial test results

-	Test 1	Test 2	Test 3	Test 4
In (customer supplied)	4/8/2009 11:30	4/9/2009 13:40	4/22/2009 10:50	4/24/2009 12:25
Out (customer supplied)	4/8/2009 13:13	4/9/2009 15:40	4/22/2009 12:55	4/24/2009 14:35
Target % O_2	5	4	3	2
Min Temp	1137	1131	1015	1143
Target Temp	1580	1580	1580	1580
Heat up rate of change, °/min	6.75	7.31	7.54	8.82
Time to Heat, 5% or 6.75 °/min	65.67	66.56	83.75	64.78
Time to heat, 4% or 7.31 °/min	60.64	61.46	77.34	59.82
Time to heat, 3% or 7.54 °/min	58.78	59.58	74.97	57.99
Time to heat, 2% or 8.82 °/min	50.21	50.89	64.04	49.53

Denote Actual Results

Theoretical Time to Heat

Note: Test 3 come to heat longer due to lower minimum load temperature

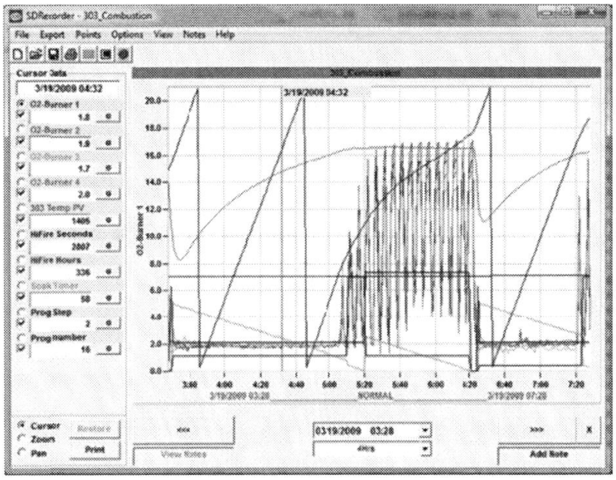

Figure 2 – Test Data

The two significant highlights evidenced by the data are the significant improvement in ramp rate (8.82 vs. 6.75 °/min) and the reduction in the amount of high-fire time. The improved heating rate shortens the time required for the load to reach heat and shortens cycle time by 15 minutes per load. This come to heat time was calculated based upon the 2% and 5% excess O_2 rate and was consistently more than 15 minutes.

The reduction in high-fire time reduces fuel and operating costs along with minimizing CO_2 emissions. Table 2 provides a summary of CO_2 and fuel savings for the reduction in high-fire time. The burner's total demand on high fire is 1,000,000 BTU or one (1) dekatherm. The calculations are based upon a dekatherm cost of $5 and a 90% uptime availability.

Table 2 – CO_2 production and fuel cost

Excess O_2 level	5%	4%	3%	2%
Soak Cost per hour	$4.00	$3.48	$3.20	$2.81
CO_2, lbs per hour	97.60	84.79	77.96	68.44
Soak cost per day	$96.00	$83.40	$76.68	$67.32
CO_2, lbs per day	2342.4	2035.0	1871.0	1642.6
Soak cost per year	$31,536	$27,396	$25,189	$22,114
CO_2, lbs per year	769478	668484	614621	539597

The cost to maintain temperature is reduced by 30% as are the CO_2 emissions. Over the course of one year, the savings will exceed $9,000 and 200,000 pounds of CO_2 by reducing the excess O_2 from 5% to 2% in the combustion process.

Table 3 provides a summary of the improved utilization that is achieved by reducing the excess O_2 in the radiant tubes for various cycle times. The calculations are based upon a 15-minute savings in come to heat time. Cycle times will impact the improvement in utilization and the number of additional loads that can be pushed through the furnace on an annual basis.

Table 3 – Utilization improvement

Cycle time (in hours)	3	4	5	8
15 minute savings, % of cycle	91.67%	93.75%	95.00%	96.88%
utilization improvement	109.09%	106.67%	105.26%	103.23%
Optimal Loads per year	2920	2190	1752	1095
Max increase loads per year	265	146	92	35

As the cycle times decrease, the utilization improvements become more significant. For a typical one hour come to heat and 3 hour soak (4 hour total cycle), The improvement is 6.67% and 146 additional loads per year.

Conclusions

Continuous monitoring and adjustment of excess O_2 levels in combustion applications provides significant fuel savings, reduced emissions and improved utilization. The savings and improvements will vary from facility to facility and from furnace to furnace depending upon how the combustion system is currently tuned and maintained. As process temperatures increase, the fuel and emissions savings rise exponentially. Several state governments currently offer grants and credits that help further reduce the cost of O_2 monitoring and reduce the payback time. Even without these grants and credits, customers should see paybacks from fuel savings in less than 12 months along with gains in utilization. By optimizing combustion efficiency, companies will minimize production costs and maximize utilization and have a competitive advantage over those who overlook this part of their process.

Proceedings of the 26th ASM Heat Treating Society Conference
B.L. Ferguson, R. Jones, D.S. MacKenzie, and D. Weires, editors
https://doi.org/10.31399/asm.cp.ht2011p0081

Automatic Vickers Case Depth Measurement

Arnold Horsch
[1]AHOTEC® e.K., Werkstoffprüftechnik, 42859 Remscheid, Germany
info@ahotec.biz

Abstract

Full automatic Vickers measurement with image analysing systems, is one of the common used measurement systems since more then 20 years, for Case depth measurement of Carburizing – Nitriding – Surface hardening. One of the most discussed points is the reproducibility by existing systems and between different systems. On examples from Industrial users will be shown the challenge which is still happen on used systems and the main points which are happen for discussions. The examples will shown the difficulties which existing, if systems are working in the accepted accuracy range of existing standards. Also will discuss the challenge for the future increase of Automatic Hardness Test Systems.

History

The first fully automatic hardness test machines for case depth examination were prepared from the beginning of the 1980 years in Japan by the companies like Akashi (Fig. 1) and Matsuzawa. Beginning in 1988 Leco ® Instruments built the first on PC based semi-automatic hardness test machines (Fig. 2), these machines are manufactured in Europe in high quantities, this was the first successful commercially system.

From about 1990 the first well functioning fully automatic systems on PC based image analysis systems are available. These systems are running from various manufacturers. Modern systems like the ecoHARD® VA (Fig. 3) have been working reliably in various industrial applications.

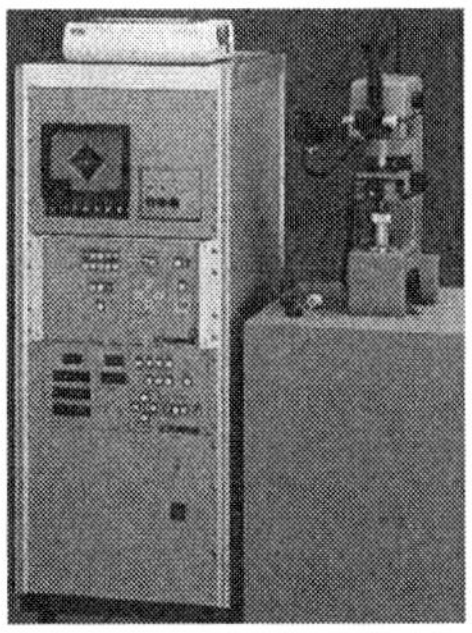

Fig. 1: Automatic Vickers machine from Akashi / Japan

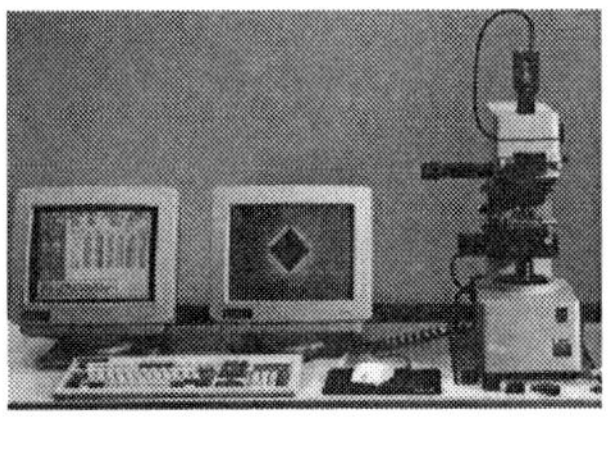

Fig. 2: Semi automatic Case Depth Measurement System from Leco® Instruments 1990

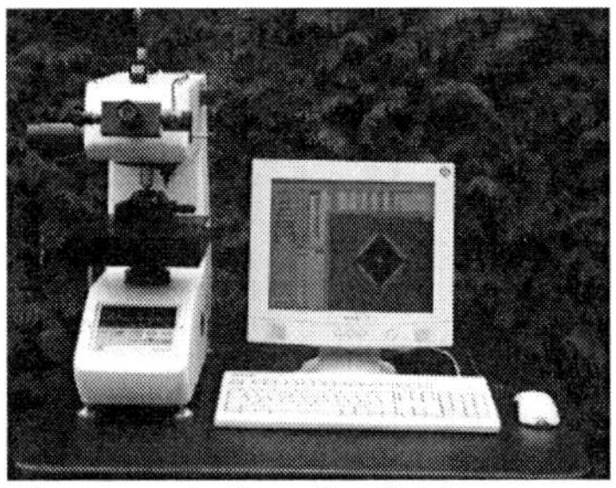

Fig. 3: Full automatic Case Depth Measurement System from AHOTEC® e.K.,2001

The most common applications is the test of case depth measurements and welding:

- Carburizing depth = CHD (Fig. 4)
- Nitriding depth = NHD
- Surface hardness depth = SHD
- Welding tests (Fig. 5)

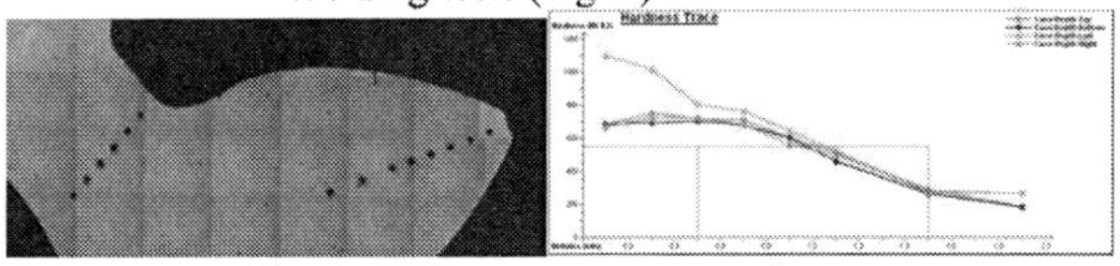

Fig. 4: CHD measurement on Gears, with evaluation curve

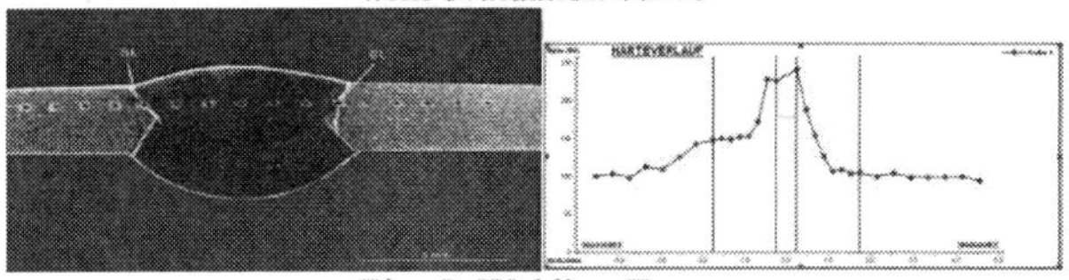

Fig. 5: Welding Test, with evaluation curve

Actual situation of automatic Vickers test

Actual are a lot of different full automatic Vickers systems on the market, all of them use different solutions for the evaluation of indentations. There is no standard which regulate the evaluation with image analysis systems. This sometimes leads to different results on the same samples

In the case of the most used application, automatic Vickers case depth measurement of Carburizing – Nitriding – Surface Hardening, are often discussions about the results. These discussions often lead to no clear result, cause considerable costs. There are no clear statements about the reproducibility / accuracy of automatic measurements with automatic Vickers systems, by the test procedure Vickers HV0,5 and HV1.

Iso standard 6507, general rules for evaluation

Accuracy of the optical measuring device

In the ISO Standard 6507, Part 2, the general rules for the evaluation of Vickers indentations are presented.

This means for HV 0,5:

	Resolution of the measuring device	Repeatability of the measuring device
Indentations < 0,040 mm	0,002 mm	0,004 mm
Indentations > 0,040 mm	0,5 % of d	1 % of d

Maximum permissible deviation by HV 0,5

Method HV 0,5	Maximum permissible deviation of the hardness testing machine in ± in %							
	Hardness HV							
	50	100	150	200	250	300	350	400
		4		5		5		6
	450	500	600	700	800	900	1000	1500
		6	7	7	8	8	9	11

Repeatability of the testing machine

	Repeatability of testing machine max.		
	đ ¹)	HV 0,2 to < HV 5	
Testblock value	HV 0,2 to < HV 5	Testblock value	HV
> 225 HV	0,04 đ	350 600 **750**	28 48 **60**
¹) đ is the average of 5 diagonal measurements			

This shows, that f.e. a Vickers value from 750 HV0,5 has a possible scattering from 694 up to 806 HV, this is a total range from 112 Vickers, by a possible repeatability from 60 HV in this range.

Example and problem

By diesel injectors (Fig.6) we've a very strong tolerance for the Vacuum carburizing process. That means we have to measure, on certain areas 0,05 mm under the surface a Vickers hardness from 750^{+100} HV0,5.

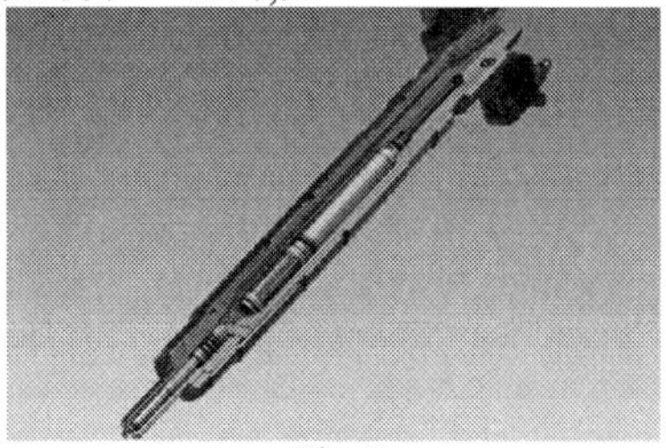

Fig.6
Diesel injector

The question is, is this tolerance acceptable and what is the real accuracy and reliability from measurements in this range.

Task

To obtain accurate and reproducible results the following experiment was conducted. Here are several important questions to answer:

1. Is there a different between manual and automatic measurement with image analysis
2. Is there a different between the 3 considered machines
3. The repeatability of the testing machines
4. The repeatability of the image analysis system
5. The repeatability of manual evaluation

With three different hardness testers, from different age, automated indentations were performed on a calibrated test block (Fig. 7).

Fig. 7: MPA calibrated test block, 760 HV0,5

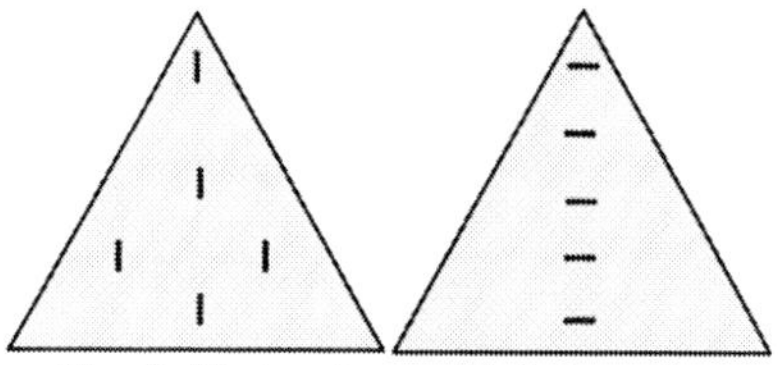

Fig. 8: Test pattern with indentation on Test block

The following test procedure was performed:

1. Set 25 indentations in 5 certain areas of the Test block, Fig. 8
2. Manual measuring of the test block on video screen, with 1 operator, all measurements with new focus
3. Automatic evaluation with ecoHARD® VA software, iteration loop of this measurement 10 time of each indentation, to se the repeatability, all measurements with new auto focus
4. Repeat the measurement consecutively
5. Measurement under Industrial environment in a 3 shift operation

Results

Different between manual measurement and measurement using image analysis

On one hardness testing machine automatically set hardness indentations were evaluated manually (Fig. 9 - 11) and automatically (Fig. 12 + 14), we analyzed the same hardness indentations. This experiment was repeated several times, always with same reproducible results.

Measured on machine AHOTEC® ecoHARD® XV 1270 P.

Indentation	Measurement 1	Measurement 2	Measurement 3	Measurement 4	Measurement 5	Measurement 6	Measurement 7	Measurement 8	Measurement 9	Measurement 10	Range	Average	MIN	MAX
1	750	752	749	748	751	750	751	750	750	751	4	750	748	752
2	747	742	742	743	744	746	748	748	748	745	6	745	742	748
3	765	764	765	767	768	768	767	766	768	769	5	767	764	769
4	759	759	759	759	759	758	757	763	761	761	6	760	757	763
5	758	762	760	759	764	762	760	761	761	764	6	761	758	764
6	746	746	746	745	744	746	748	744	735	737	13	744	735	748
7	763	762	765	765	762	766	766	765	768	768	10	764	758	768
8	770	772	769	771	774	770	773	773	765	764	10	770	764	774
9	750	752	753	752	754	752	753	741	746	746	13	750	741	754
10	745	745	745	746	748	746	746	747	747	748	3	746	745	748
11	765	766	766	767	769	769	768	771	768	770	6	768	765	771
12	776	763	761	761	766	765	765	763	764	766	14	765	761	775
13	766	765	771	771	771	773	772	771	776	774	11	771	765	776
14	741	744	744	743	744	747	745	746	746	747	6	745	741	747
15	762	766	762	764	768	765	763	771	767	768	11	765	760	771
16	750	752	749	748	751	750	751	750	750	751	4	750	748	752
17	747	742	742	743	744	746	748	748	748	745	6	745	742	748
18	765	764	765	767	768	766	767	766	769	769	5	767	764	769
19	759	759	759	759	759	758	757	763	761	761	6	760	757	763
20	758	762	760	759	764	762	760	761	761	764	6	761	758	764
21	746	746	746	745	744	746	748	744	735	737	13	744	735	748
22	763	762	765	765	762	766	766	765	768	768	10	764	758	768
23	770	772	769	771	774	770	773	773	765	764	10	770	764	774
24	750	752	753	752	754	752	753	741	746	745	13	750	741	754
25	745	745	745	746	748	746	746	747	747	748	3	746	745	748
Range	14	30	29	28	30	27	28	32	41	37				
Average	757	756	756	757	758	758	758	758	757	756				
MIN	741	742	742	743	744	746	745	741	735	737				
MAX	775	772	771	771	774	773	773	773	776	774				

Fig. 9: manual evaluation

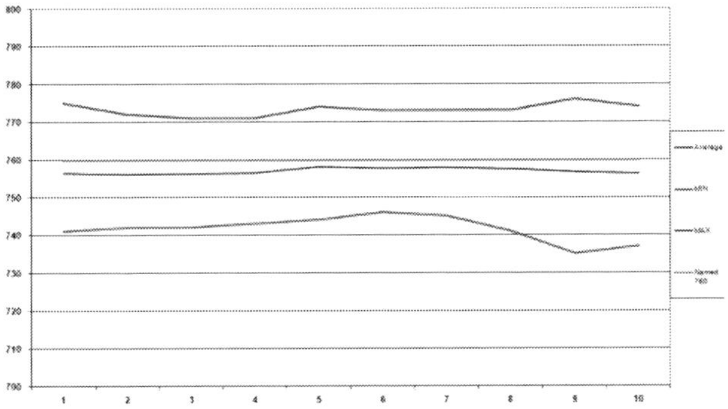

Fig. 10: manual evaluation
Influence Software
average of the 25 indentations 10 times measured

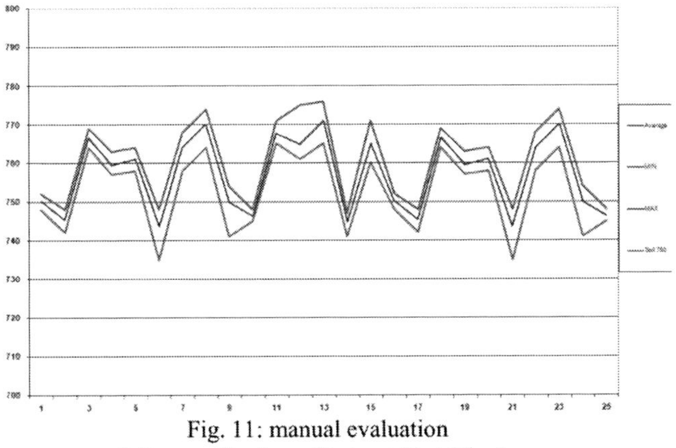

Fig. 11: manual evaluation
Influence Hardness Tester + Test Block
spread of 25 indentations, 10 times measured

We see that the manual evaluation, a larger variation than the automatic analysis has. The difference is not huge but noticeable. Cause might be the problem of pixel resolution at higher magnifications with a camera, this leads to the manual evaluation of the measurement errors measuring lines.

Indentation	Measurement 1	Measurement 2	Measurement 3	Measurement 4	Measurement 5	Measurement 6	Measurement 7	Measurement 8	Measurement 9	Measurement 10	Range	Average	MIN	MAX
1	772	772	765	772	778	765	778	772	765	772	13	771	765	778
2	772	778	772	772	772	765	772	778	765	778	13	772	765	778
3	765	765	765	765	765	759	772	759	765	765	13	765	759	772
4	772	765	758	765	759	765	765	765	765	765	13	765	759	772
5	765	772	765	765	765	765	765	765	772	765	7	766	765	772
6	765	765	765	765	765	765	772	765	765	765	7	766	765	772
7	765	759	759	765	765	759	765	765	765	765	6	763	759	765
8	759	759	759	759	765	759	759	765	765	759	6	761	759	765
9	753	753	765	753	765	759	759	759	753	753	12	757	753	765
10	759	765	759	759	765	765	759	759	765	753	12	761	753	765
11	772	772	765	765	765	765	765	772	765	772	7	768	765	772
12	765	759	765	765	759	765	765	772	765	765	13	765	759	772
13	765	772	772	772	778	778	772	765	778	772	13	772	765	778
14	765	759	765	765	765	759	765	765	772	765	13	765	759	772
15	765	765	765	771	772	765	765	772	772	772	7	768	765	772
16	759	765	759	759	759	765	765	765	765	759	6	762	759	765
17	765	759	765	759	765	765	765	765	759	765	6	763	759	765
18	765	772	772	765	765	772	765	765	765	759	13	767	759	772
19	765	765	765	765	765	759	772	759	765	765	13	765	759	772
20	759	759	759	765	765	772	765	765	765	765	13	764	759	772
21	772	784	778	772	771	772	778	778	778	778	13	776	771	784
22	765	765	772	772	765	772	765	772	772	772	7	769	765	772
23	772	772	778	778	772	772	772	772	772	778	6	774	772	778
24	778	772	772	778	772	778	778	778	778	778	6	776	772	778
25	753	759	759	765	759	759	759	759	759	759	12	759	753	765
Range	25	31	19	25	19	19	19	19	25	25				
Average	765	766	766	766	766	766	767	767	767	767				
MIN	753	753	759	753	759	759	759	759	753	753				
MAX	778	784	778	778	778	778	778	778	778	778				

Fig. 12:automatic evaluation

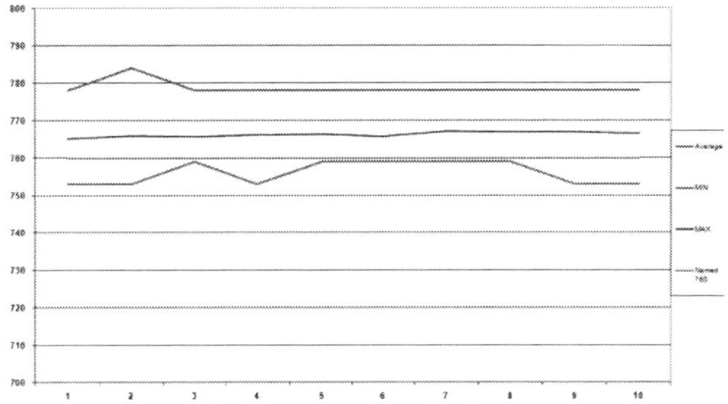

Fig. 13: automatic evaluation
Influence Software
average of the 25 indentations 10 times measured

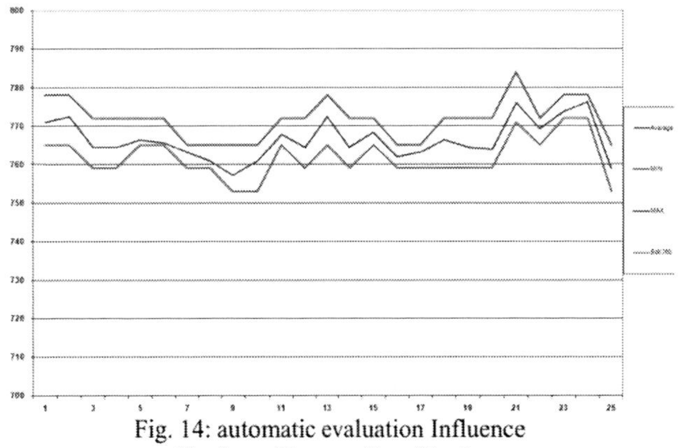

Fig. 14: automatic evaluation Influence
Hardness Tester + Test Block
spread of 25 indentations, 10 times measured

Different between 3 machines

On 3 different machines we set 25 indentations on the same test block, on all three machines we set new indentations (Fig. 15-19).

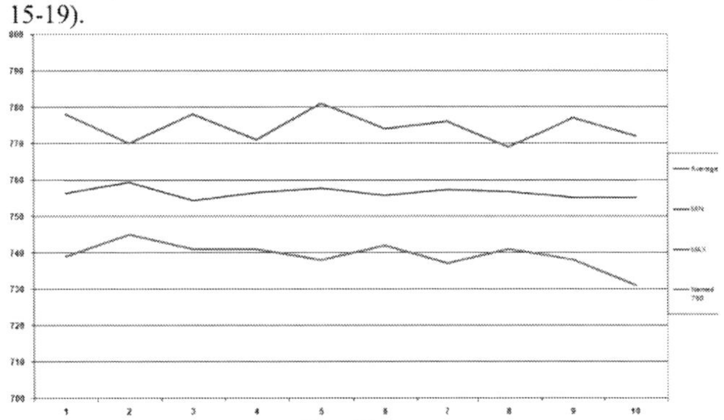

Fig. 15:
Future Tech LM 300, from 2001
Influence Software
average of the 25 indentations 10 times measured.

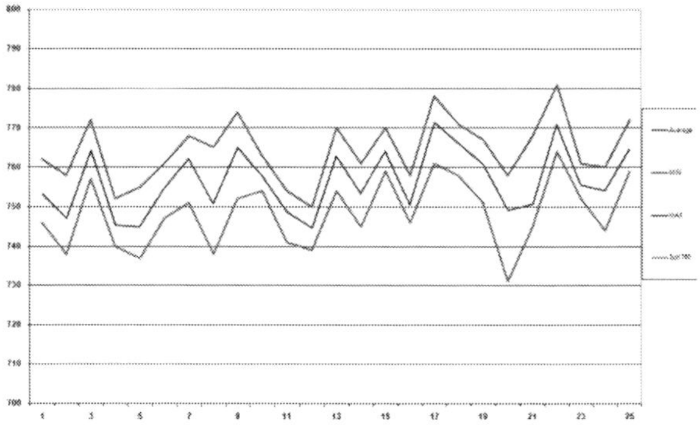

Fig. 16:
Future Tech LM 300, from 2001
Influence Hardness Tester + Test Block
spread of 25 indentations, 10 times measured

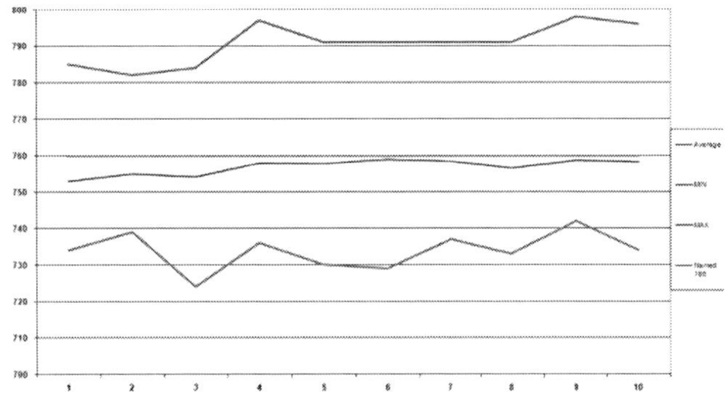

Fig. 17:
Future Tech LM 700, from 2003
Influence Software
average of the 25 indentations 10 times measured

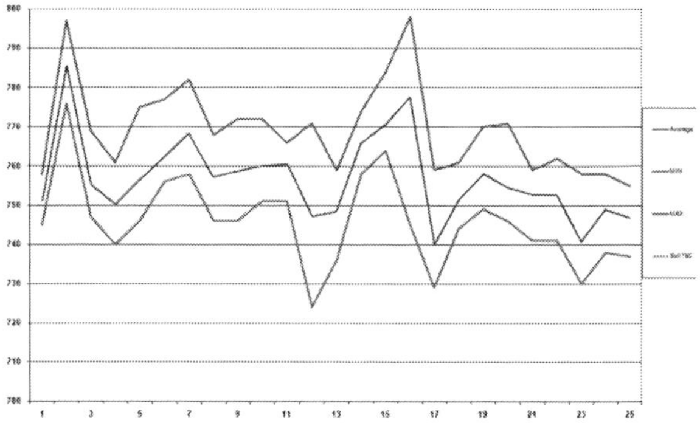

Fig. 18:
Future Tech LM 700, from 2003
Influence Hardness Tester + Test Block
spread of 25 indentations, 10 times measured

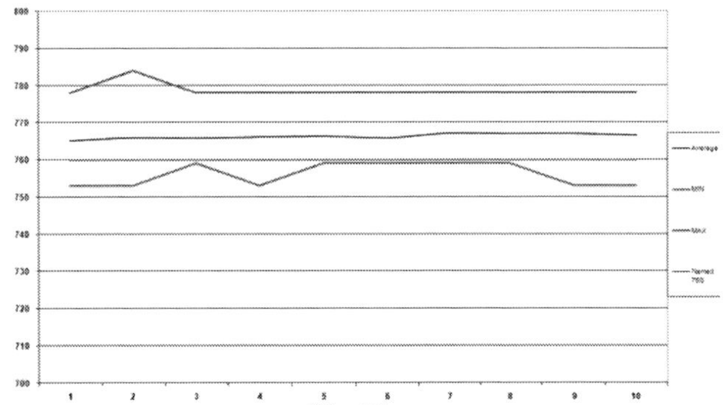

Fig. 19:
AHOTEC® ecoHARD® XV 1270 P, from 2008
Influence Software
average of the 25 indentations 10 times measured

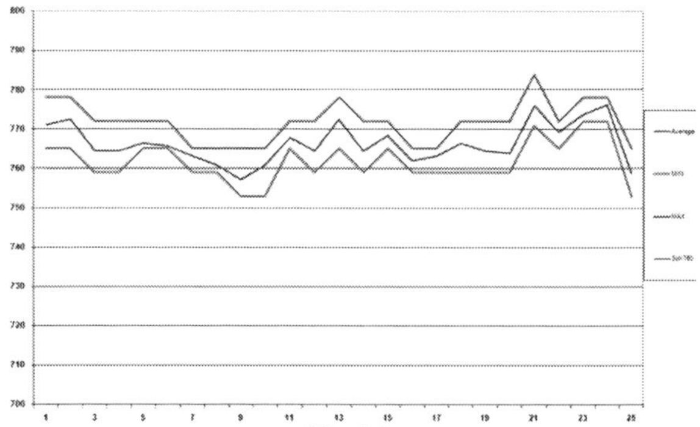

Fig. 20:
AHOTEC® ecoHARD® XV 1270 P, from 2008
Influence Hardness Tester + Test Block
spread of 25 indentations, 10 times measured

Image analysis system

Difficulty of determining whether an image analysis system is working properly. A glimpse of how the image analysis is, the measurement carried out a large number of indentations with the applied measurement lines was observed and documented (Fig. 21). This also gives us an accurate statement about the quality of autofocus.

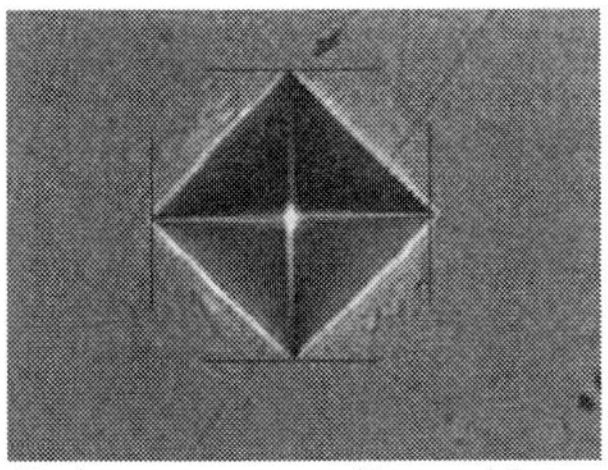

Fig.21: Applied measurement lines with image analysis

Conclusions

In considering all used machines, we see clearly differences in the spread of the measured values; all values are in the required values of the ISO 6507. Looking at the averages of the measurements is clearly seen that the mean of the measurements take very well the value of the test block. Equally visible is that the automatic evaluation with image analysis has a good reliability and accuracy. Apparently, the measured scattering is depending on the hardness testing system and the test block.

The analysis manual. vs. automatic measurement shows no significant differences in the results. Automatic evaluations have obviously a slightly better repeatability and thereby they achieve more reproducible results. Basically, it was found that the actually required values in the ISO 6507, necessary to make sure that the machine is working under the asked accuracy and repeatability.

The results found of the test block shows clearly that the specified tolerances of 750^{+100} HV0,5 by diesel injectors will not safely observed. Where it doesn't matter whether automatic or manual analysis is performed.

All measurements were over a period of 1 year, repeated several times and confirmed

Outlook

Currently no test system is available on the market which is much better than the in the ISO 6507 asked reliability and reproducibility. More measurements with other testing systems must be carried out to see what is really possible over a wide range of systems.

References

[1] ISO 6507-Part 2, 2006

[2] Arnold Horsch, Fortschritte in der Metallographie 22, Dr. Riederer Verlag GmbH – Stuttgart, M400PC3 - Automatisiertes DIN Konformes Mikrohärteprüfgerät für die Metallographie, Page 303-312

Proceedings of the 26th ASM Heat Treating Society Conference
B.L. Ferguson, R. Jones, D.S. MacKenzie, and D. Weires, editors
https://doi.org/10.31399/asm.cp.ht2011p0086

New Heat Treatment Temperatures for HR-120™ Alloy

Octavio Covarrubias [1,2]***, Rafael Colas*** [2]
[1]*Frisa Forjados SA de CV, Santa Catarina, NL, Mexico*
E-mail:ocovarrubias@frisa.com, Phone:+52 81 81243637
[2]*Universidad Autonoma de Nuevo Leon, San Nicolas, NL, Mexico*
E-mail: colas.rafael@gmail.com, Phone: +52 81 83294020

Abstract

HR-120™ alloy is a Ni-Fe-Cr alloy designed to have high strength at elevated temperature and resistance to attack in carburizing and sulfidizing environments. Applications for this alloy include components to be assembled in combustion turbines for power generation. Some of these components can be manufactured by ring-rolling procedures followed by heat-treating operations. Typically, HR-120™ alloy is heat-treated in a temperature range of 1175 to 1230°C in order to promote microstructure homogenization and dissolution of heavy precipitates. An inconvenient of such heat-treating temperature range is related to excessive grain-size coarsening of wrought parts. This work presents the results of a series of heat-treatment procedures performed on seamless rings produced by ring-rolling considering industrial conditions. Lower heat-treatment temperatures are considered for evaluation of mechanical and microstructural properties. It is reported that an excellent combination of mechanical properties and microstructural characteristics is obtained when alloy is exposed to 1050°C for soaking periods above 30 minutes.

Introduction

Developed during the late 1970s, HR-120™ (UNS N08120) is a Ni-Fe-Cr heat-resisting alloy designed to provide strength at temperatures up to 1095°C, combined with excellent resistance to carburizing and sulfidizing environments. Applications for this alloy include industrial heating and gas turbine components[1,2].

In the field of power generation, HR-120™ components for combustion turbines can be fabricated by ring-rolling procedures[3,4,5], where parameters as temperature and deformation ratio shall be controlled to promote adequate microstructural characteristics as an adequate grain size and microstructure uniformity[6].

It must be considered that HR-120™ is not hardenable by heat treatment, and then imparted mechanical properties are consequence mainly of hot-working and cold-working operations.

Heat-treatment procedures recommended for HR-120™ include annealing at temperatures above 1100°C[7], which main objective is the promotion of microstructure recrystallization.

Standard practices for ring-rolling of HR-120™ seamless rings can include hot-working operations up to 1100°C. Such practices are related to promotion of uniform grain size from 31.8 micron to 89.8 micron (ASTM 7.0 to ASTM 4.0) required by industrial specifications. When this promoted microstructure is exposed to annealing temperatures above 1100°C, excessive grain size coarsening can be developed. This behavior is particularly intense when heat treating temperature is significantly above the last hot-working temperature and can lead to non-conforming microstructures despite appropriate hot-working procedures to promote the required grain-size and uniformity.

Since limited information is available for heat treating of HR-120™, the aim of this work is to present the results of a series of heat treatment procedures performed below 1100°C temperature on hot-worked seamless rings to evaluate their effect on grain-size and mechanical properties.

The evaluation was conducted to optimize the heat treatment parameters that will allow for the consistent microstructure and mechanical properties required by industrial specifications. It is reported that HR-120™ alloy seamless rings forged/rolled between 1200°C to 1000°C are capable of fulfill standard industrial requirements when they are annealed at 1050°C temperature for a minimum period of 30 minutes.

Experimentation

An EAF/AOD + ESR 216 mm round-corner-square billet was selected to extract several increments for the manufacture of seamless rings with a weight of 73.0 kg and dimensions: 643.0 mm outer diameter, 510.0 mm and 75.0 mm height. Table 1 summarizes the chemical composition of the starting material.

Table 1: Alloy composition in weight percent.

Ni	Fe	Cr	Mn	Nb+Ta	Si
37.25	34.64	25.27	0.72	0.68	0.54
Mo	Co	Al	C	Cu	Ti
0.17	0.13	0.09	0.053	0.04	<0.01

Mentioned increments were upset and pierced by open-die forging and rolled considering standard production parameters. Considered temperatures for hot-working operations were between 1000°C and 1200°C; ring-rolling stages were designed to promote an average grain-size of 31.8 micron (ASTM 7.0). Figure 1 shows a seamless ring during ring-rolling operation.

Figure 1: Ring-rolling operation for manufacture of a HR-120^{TM} seamless ring. Courtesy of Frisa Forjados SA de CV.

Seamless rings were cooled down to room temperature in still air and inspected to detect for any type of surface defect. Segments from seamless rings were cut and exposed to 1000°C, 1050°C and 1121°C temperatures for soaking periods from 30 minutes to 90 minutes and quenched in water at room temperature.

Tangential samples from segments exposed to listed heat-treating conditions were extracted for mechanical testing and microstructural evaluation. Mechanical testing included tensile at room temperature per ASTM E8 and stress-rupture testing per ASTM E139 according standard industrial requirements.

Microstructural evaluation by optical microscopy considered grain-size measurement per ASTM E112 and ASTM 1181 on specimens extracted from non-stressed end of tensile samples. Scanning electronic microscopy (SEM) observation complements microstructural analysis.

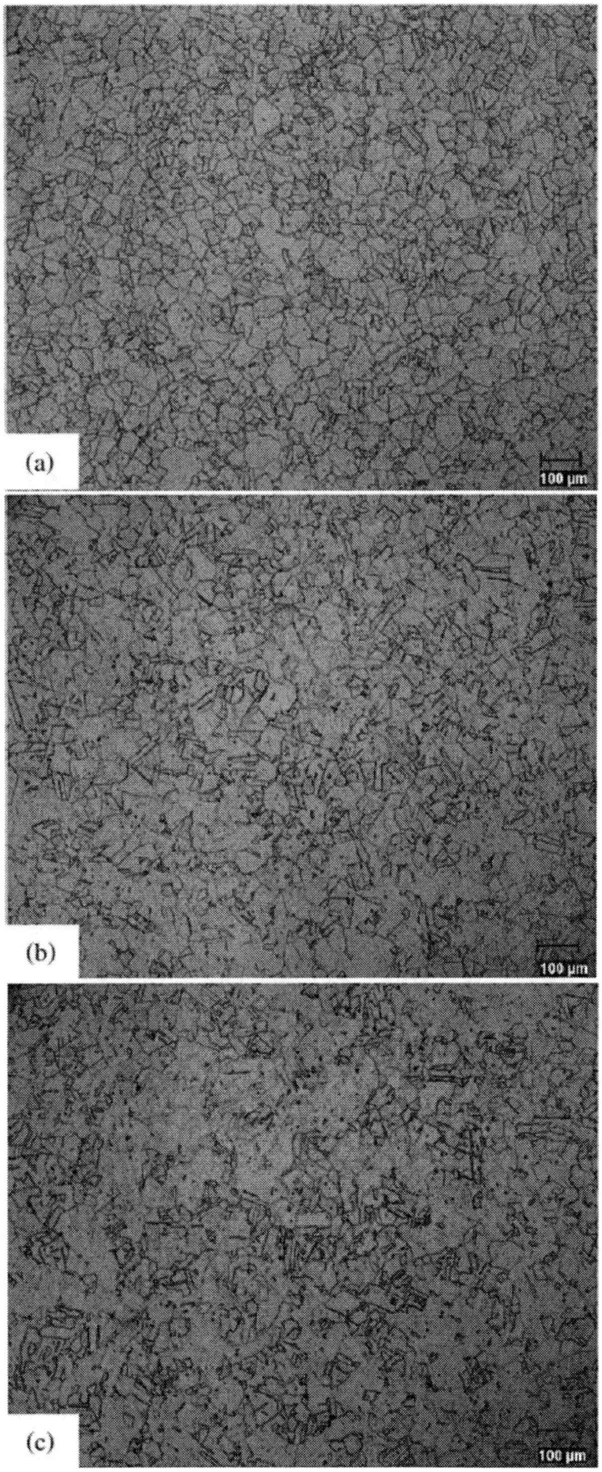

Figure2: Light optical micrographs showing effect of temperature in HR-120TM microstructure: (a) Sample exposed to 1000°C for 30 minutes. A duplex bimodal microstructure is observed; (b) Sample exposed to 1050°C during 30 minutes. A uniform microstructure is promoted; (c) Sample exposed to 1121°C during 30 minutes, initial grain coarsening can be observed.

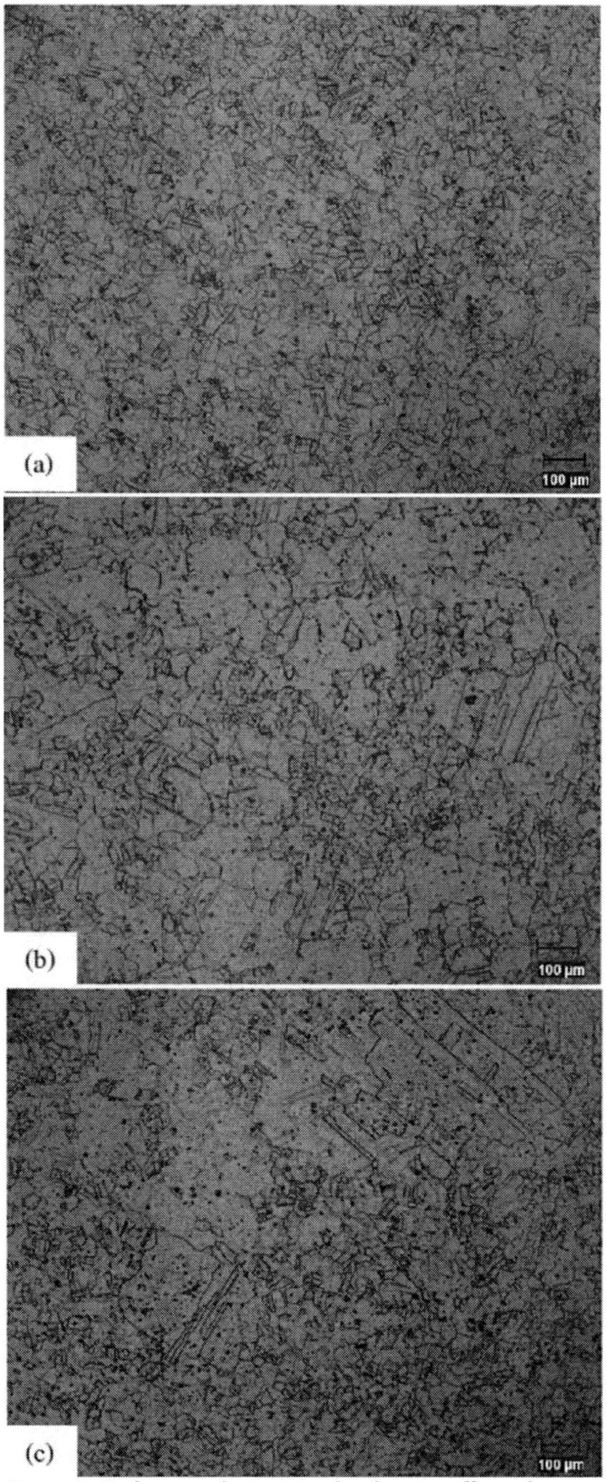

Figure3: Light optical micrographs showing effect of time in HR-120TM microstructure: a) Sample exposed to 1121°C for 30 minutes. A uniform microstructure is observed, some grain coarsening can be observed; b) Sample exposed to 1121°C during 60 minutes. Grain coarsening is evident, leading to a non-uniform microstructure; c) Sample exposed to 1121°C during 90 minutes, more grain coarsening is promoted and a non-conformant microstructure is generated.

Results

Microstructural Characteristics

Reported effect of annealing procedures in HR-120™ microstructural characteristics as grain-size, can be described as follows:

Related to annealing temperature Figure 2 shows its effect in HR-120™ grain-size and uniformity. It is evident how increment on temperature promotes grain-coarsening as it is increased from 1000°C to 1121°C when samples are soaked for 30 minutes.

Promoted microstructure after hot-working operations revealed a duplex-bimodal distribution of grains with an average diameter of 22.5 micron (ASTM 8.0) and grains with an average diameter of 63.5 micron (ASTM 5.0). Such microstructure can be considered as non-conformant and annealing operations are required to promote controlled recovery phenomenon and make a uniform grain size. Limited presence of twinning was reported.

When alloy is exposed to 1000°C during 30 minutes, minimal effect is observed in microstructure since energy provided to the alloy is insufficient to activate boundary mobility. As consequence, the duplex-bimodal microstructure promoted by hot-working operations remains, leading to a non-conformant microstructure according industrial specifications.

Increment of annealing temperature to 1050°C leaded to development of a more uniform microstructure with an average grain size of 63.5 micron (ASTM 5.0). Resulting microstructure developed ALA grains (*As Large As*) with a grain-size up to 63.5 micron (ASTM 5.0). Increase in the number of observed twins suggests recovery mechanism is governing microstructural changes in grain-size, leading to grain growth.

The phenomenon is more evident when alloy is exposed to 1121°C. Number of coarsened grains is increased and presence of twinning is more intense. Resulting microstructure developed occasional ALA grains with a grain-size up to 254.0 micron (ASTM 0), resulting in a non-conforming microstructure. Such ALA grains are consequence of abnormal grain growth, which is consequence of activation energy provided by this temperature (that is higher than the last deformation temperature above 1000°C). Despite of this, grain coarsening is not more intense since soaking time is insufficient to promote further grain growth.

These results show that if uniform microstructure is required for HR-120™ seamless rings with an average grain-size of 63.5 micron (ASTM 5.0), it can be promoted exposing alloy to 1050°C during 30 minutes as minimum time.

Since no second-phase particles are developed by HR-120™ during annealing, main factor affecting grain-growth is related to temperature and time. When this second factor is increased in samples exposed to 1121°C, grain coarsening is stimulated.

Figure 3 present representative microstructures exposed to 1121°C exposed to incremented soaking periods. When alloy is exposed for a time period of 60 minutes more zones of the microstructure develop abnormal grain growth, leading to ALA grains with a grain-size up to 359.2 micron (ASTM 0).

The phenomenon is reinforced when samples of HR-120™ are exposed at this temperature during 90 minutes, developing abnormal grain growth that lead to 508.0 micron ALA grains (ASTM 00). Both microstructures can be considered as non-conformant and must be avoided in seamless rings by appropriate manufacturing processing.

SEM analysis of samples revealed presence of Ni-Fe-Cr fine particles at grain boundaries. Heavy particles with similar chemical composition are reported. It was observed that as annealing temperature is increased, presence of fine particles at grain boundaries diminished. It is not clear if such diminution of fine particles at grain boundaries is related to their dissolution or to the grain boundaries mobilization during recovery phenomenon leading to grain coarsening and abnormal grain coarsening. Representative SEM micrographs are shown in Figure 4.

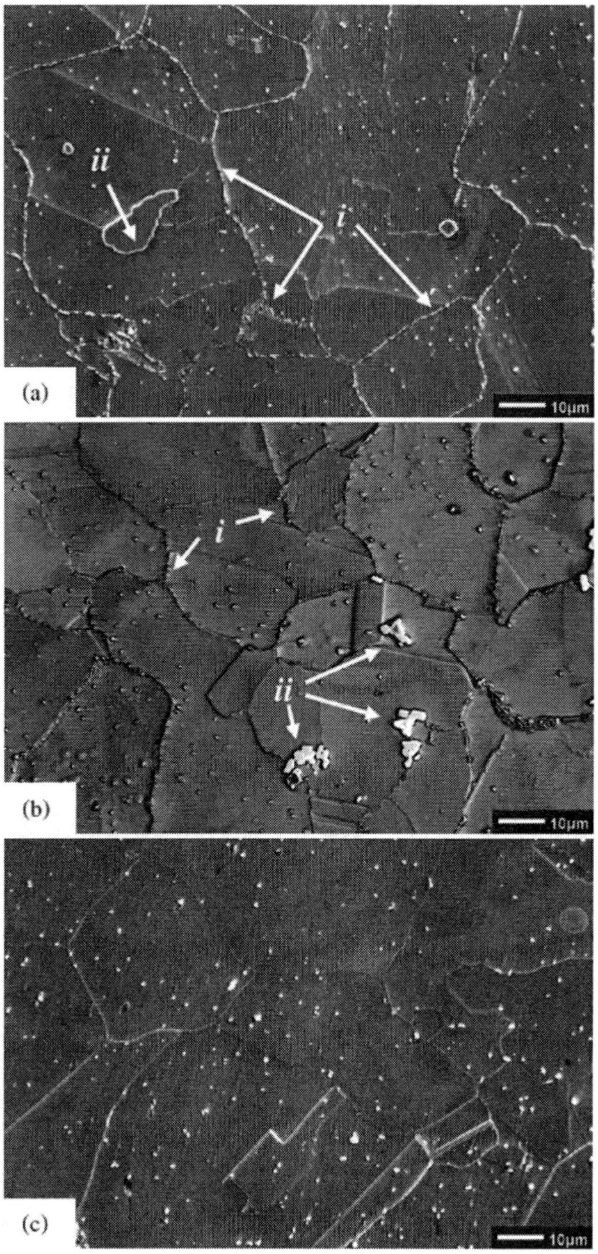

Figure 4: SEM micrographs showing representative zones of samples exposed to annealing conditions: (a) Sample exposed to 1121°C for 30 minutes; a network of Ni-Fe-Cr particles (i) is observed at boundaries, heavy particles of same composition are observed (ii). (b) Sample exposed to 1121°C during 60 minutes exhibits a diminution of Ni-Fe-Cr particles at grain boundaries (i), but some heavy particles of similar composition are present (ii); (c) Sample exposed to 1121°C during 90 minutes reveals particles at grain boundaries are not present; Ni-Fe-Cr heavy particles (not shown) were observed.

Mechanical Properties

As other non-hardenable Nickel-base alloys, HR-120™ mechanical properties are consequence of microstructural properties, mainly grain-size. Lack of hardening elements as Nb (Cb), Al or Ti, necessary for the formation of second phase particles as gamma-prime (γ'), makes mechanical properties of this alloy dependent of other processing parameters as deformation at hot-working o cold-working (if performed) operations.

After deformation processing, annealing processes will promote final mechanical characteristics. Tested samples extracted from involved seamless rings on this research reflect this situation as follows.

Yield Strength (YS) is a property affected heavily by HR-120™ microstructure. It is reported that as grain size is coarsened as consequence of annealing temperature increase, YS values tend to diminish. This effect is intensified when alloy is exposed to 1121°C for extended soaking periods. Mentioned behavior is presented as Figure 5.

Considering Ultimate Tensile Strength (UTS), a similar behavior but less severe effect is observed as expected. Figure 6 shows reported values from tested samples for this property. Figure 7 presents values reported for Elongation property (E): reported behavior is that as tensile properties diminish, elongation values are incremented.

Related to stress-rupture properties, it was reported that no single sample failed the test: all tested samples exposed to temperatures between 850°C to 900°C at selected applied stresses passed the test. Stress-rupture testing was interrupted after testing periods up to 120 hours.

Considering presence of Ni-Fe-Cr fine particles at grain boundaries, further analysis must be considered to determine if they have any influence on mechanical properties of evaluated samples.

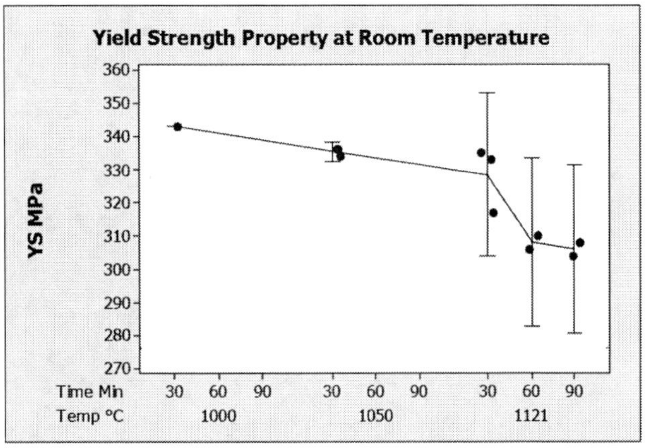

Figure 5: Tensile strength as a function of the YS of the specimens tested in tension.

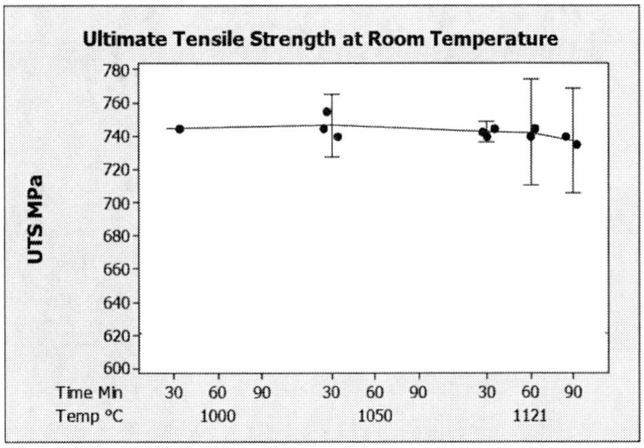

Figure 6: Tensile strength as a function of the UTS of the specimens tested in tension.

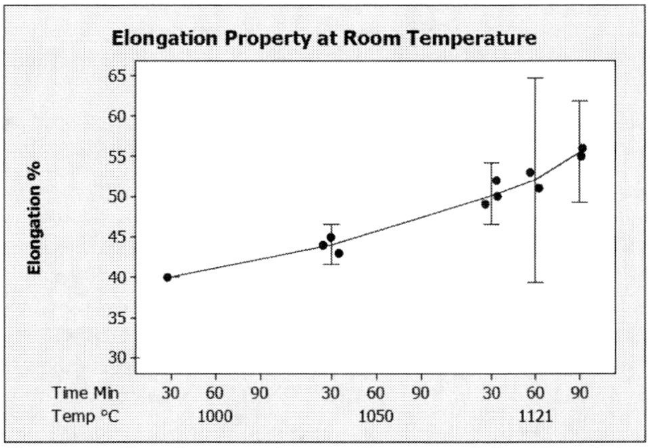

Figure 7: Tensile strength as a function of the E% of the specimens tested in tension.

Conclusions

The mechanical properties and microstructure of HR-120™ alloy was evaluated when it is exposed to annealing conditions below recommended 1121°C temperature: required properties per industrial specifications were fulfilled.

Best combination in mechanical properties and control in grain-size for rolled seamless rings is achieved when involved material is exposed to 1050°C temperature for a minimum of 30 minutes.

Wrought HR-120™ alloy develop abnormal grain growth when annealing temperature is above last hot-working temperature; effect is increased as soaking period is extended.

Reported mechanical properties shows tensile properties tend to diminish as grain size is coarsened as consequence of increasing annealing temperature and soaking period.

References

[1] Davis, J.R *et al*, Nickel, Cobalt and Their Alloys, ASM International, (USA, 2000), pp. 16, 172, 174.

[2] ASM International, ASM Handbook Volume 4 Heat Treating, ASM International, (USA, 2002), pp. 1129-1130.

[3] ASM International, ASM Handbook Volume 14 Forming and Forging, ASM International, (USA, 1998), pp. 224-268.

[4] Altan, T. *et al*, Cold and Hot Forging Fundamental and Applications, ASM International, (USA, 2007), pp. 144-146.

[5] Forging Industry Association, Open Die Forging Technology, Forging Industry Association, (USA 1997), pp. 24-28.

[6] ASM International, Alloy Digest Haynes HR-120 Alloy, ASM International, (USA, 2002), pp. 1-2.

[7] Haynes® HR-120™, Haynes International Inc., (Kokomo, Indiana, 1992)

Proceedings of the 26th ASM Heat Treating Society Conference
B.L. Ferguson, R. Jones, D.S. MacKenzie, and D. Weires, editors
https://doi.org/10.31399/asm.cp.ht2011p0091

Intelligent Heat Treating: Simulation of Carburization Process

Y. Wei, G. Wang, R.D. Sisson, Jr.,
Center for Heat Treating Excellence, Worcester Polytechnic Institute, Worcester, MA, USA
yingyingwei@wpi.edu

B. Bernard, R. Poor
Surface Combustion, Inc. Maumme, OH, USA

Keywords: Modeling, Gas carburizing, Vacuum carburizing, CarbTool©

Abstract

Heat treaters need an effective simulation tool to predict the carburization performance of a variety of steels. The tool is needed not only to predict the carbon profile but also to optimize the process in terms of the cycle time and the cost. CarbTool© has been developed to meet these needs for gas and vacuum carburization. In this paper, CarbTool© predictions were compared with industrial experimental results of four types of steels, heat treated by both gas and vacuum carburizing processes. Based on the excellent agreement of model predictions and experimental results, CarbTool© may be used to predict the carbon concentration profile for a variety of alloys in both gas and vacuum carburizing processes.

Introduction

Carburizing media are divided into three kinds: liquid, solid carbonaceous compounds, and atmospheric. In both gas and vacuum carburization processes, the steel part is subjected to an atmosphere with high carbon potential.

The optimization of the industrial carburizing parameters is typically pursued by a trial and error method. In addition to being time consuming and expensive, this approach yields suboptimal results at best [1]. In this paper, effective gas and vacuum carburizing models are presented. The models may also be used to optimize the processes.

Carburizing Modeling

In this section carburization models for gas carburizing and vacuum carburizing processes developed at CHTE (Center for Heat Treatment Excellence) are presented and discussed.

CarbTool© – Carburizing Simulation Tool

A simulation program has been developed by CHTE for the calculation of the carbon concentration profile during the carburizing processes.

The solution algorithm used in CarbTool© is based on the finite difference method (FDM), and the code is developed using Microsoft Visual C++ in Window OS. The tool has been built so that the carbon flux at the surface between the gas and steel is to be specified by the user [2]. The outputs of CarbTool© are the carbon concentration distribution profile inside the steel part and the surface carbon concentration profile as a function of time.

Users input parameters about the carburization process, such as carburization temperature, carburizing time, and carbon potential or carbon flux. After a quick simulation the carbon profile along the distance below the surface can be plotted with the case depth labeled according to a user defined value.

CarbTool© has two modules – Gas carburizing and vacuum carburizing. The functions for each module are listed below:

1) Gas Carburization

Variable operating temperature;

Constant mass transfer coefficient;

Variable carbon potential;

Single boost-diffuse process;

Data export of carbon profile at certain interval and final time;

Effective case depth indication at 0.35 wt.% carbon or other user defined condition.

2) Vacuum Carburization

Variable operating temperature;

Constant flux process design;

Cementite transformation indication;

Cyclic and non-cyclic boost-diffuse process design;

Data export of carbon profile at certain interval and final time;

Effective case depth indication at 0.35 wt.% carbon or other user defined condition.

Gas Carburizing Model

Gas carburizing is a complex phenomenon which involves three distinct stages [3]: 1) carbon transport from the atmosphere to the steel surface, 2) surface reductive chemical reactions, and 3) diffusion of the absorbed carbon atoms towards the bulk of the steel down the chemical potential gradient. Total carbon transfer from the atmosphere to the steel is thus determined by the limiting process, which kinetically becomes the rate controlling stage of carburizing. Figure 1 shows the mechanisms of carbon transfer during carburizing and the primary control parameters: the mass transfer coefficient (β) defining carbon atoms flux (J) from the atmosphere to the steel surface and the coefficient of carbon diffusion in steel (D) at Austenizing temperatures [2].

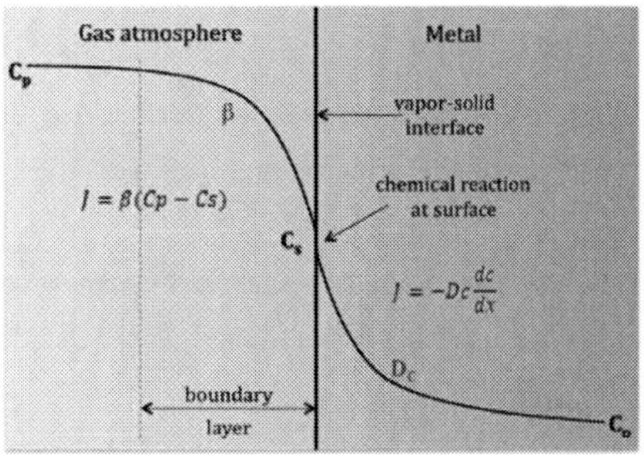

Fig. 1 Schematic of gas carburization process[2].

Mass Transfer Coefficient

A constant value for the mass transfer coefficient is applicable for most cases since once the carbon potential approaches the near-solubility limit in Austenite with carbon content greater than 0.5 wt.%, the value of β becomes consistent with temperature and has little relation with gas compositions [5].

The carbon potential of the carburizing atmosphere is set as the boundary condition to define the physical problem. The mass balance of the steel is

$$\beta(C_P - C_S) = -D_C \frac{dc}{dx} \qquad (1)$$

Vacuum Carburization Model

Boundary Conditions

In vacuum carburizing, a small quality of hydrocarbon gas (i.e. C_2H_2, C_6H_{12}) is added to a vacuum furnace. The addition creates a very high carbon potential. In fact, in some cases soot may form. Therefore, the mass balance boundary condition is not applicable for this process. However, a constant surface flux can be used. It is assumed that the amount of carbon produced by the surface reactions is equal to the mass flow rate for diffusion described by Fick's first law.

$$J|_{x=0} = -D|_{x=0} \cdot \frac{dC}{dx}|_{x=0} \qquad (2)$$

where x=0 at the gas-steel surface.

Process Control – Surface Carbon Saturation

During vacuum carburizing, the surface concentration of carbon can exceed the Austenite solubility limit due to the high carbon flux and might form carbides [5,6]. In plain carbon steel, cementite will start forming at 1.3 wt.% carbon (point A) at 925 °C, as shown in Figure 2 (a). In alloy steel, the saturation point is reached at 1.0 wt.% carbon (point B) at the same temperature, as shown in Figure 2 (b).

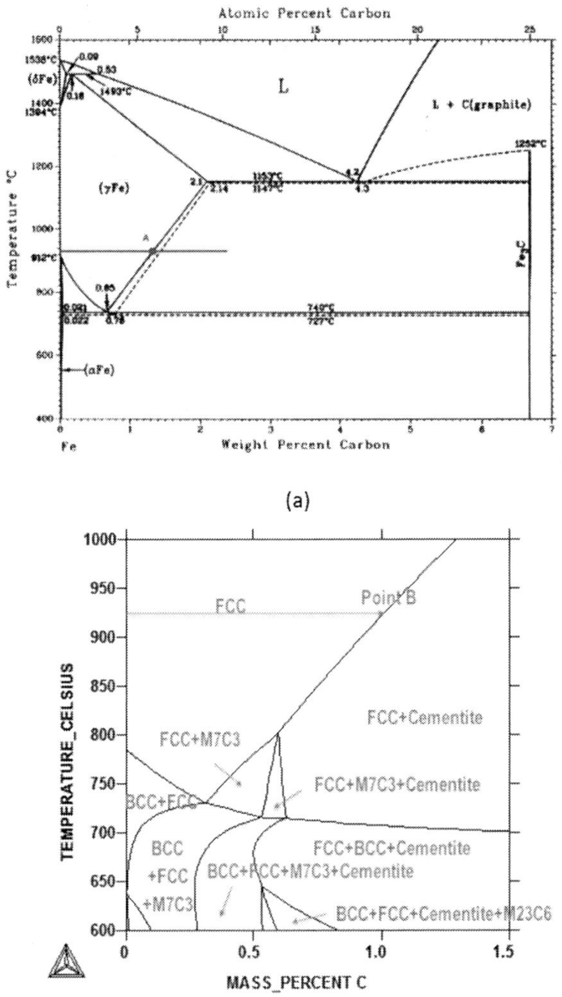

(a)

(b)

Fig. 2 Fe-Fe3C phase diagram (a) and 93xx isopleth diagram (b).

Current commercial carburization simulation software does not consider the formation of carbides on the steel surface [8]. When surface carbon concentration of 93xx steel is plotted as a function of time, the surface carbon concentration will increase beyond the solubility limit, as seen in Figure 3. But in reality, the steel will start to form cementite at 1.03 wt.% carbon (as seen in Figure 4). Therefore in simulation it is necessary to consider the maximum solubility of carbon in Austenite, otherwise, the formation of carbides will not be identified. CarbTool© indicates at which carbon concentration cementite will start to form, and shows the cementite volume fraction along with increasing total carbon content. In this case, cementite formation in processes can be controlled.

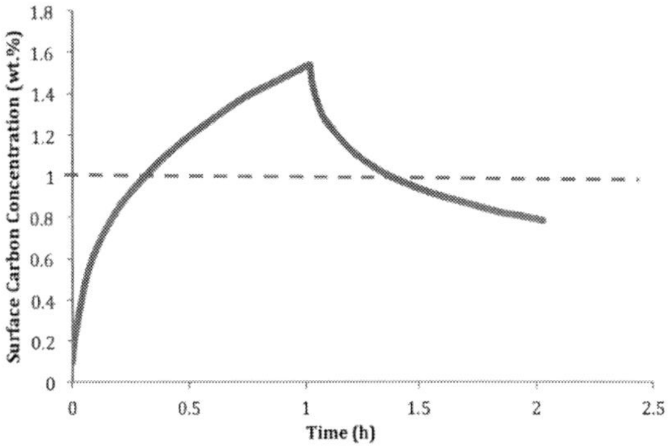

Fig. 3 Output of carbon concentration as the function of time of single boost and diffuse vacuum carburizing

To determine the Austenite maximum carbon solubility, ThermoCalc was utilized to draw the diagram of mole fraction of cementite for alloys. The diagram for 93xx, 86xx, 51xx, 43xx and 10xx was drawn and presented in Figure 4. As seen in Figure 4, the carbon solubility of 93xx is the least, approximately at 1.0 wt. % carbon, while the limits of other alloys are larger.

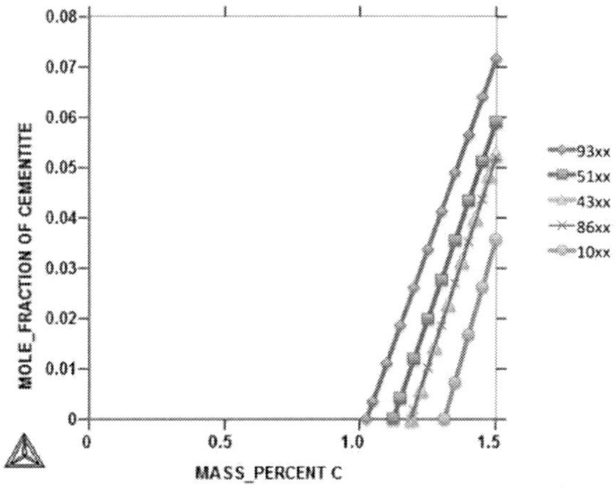

Fig. 3 Mole fraction diagram of AISI 93xx, 86xx, 51xx, 43xx and 10xx.

Case Study

In a current case, identical carbon profiles for a series of steels are required for fatigue testing. These profiles need to be achieved by gas carburizing as well as vacuum carburizing in cyclohexane. The modeling was used to revise the processes to achieve the same surface carbon concentration and effective case depth.

Carburizing Samples

Chemistry

Four materials were chosen, which need to obtain the same carburized properties after both gas and vacuum carburizing. Table 1 is the chemistry (AISI and UNS) of each steel.

Table 1 Chemical Composition (wt %) at room temperature

	C	Cr	Fe	Mn	Mo	Ni	P	S	Si
9310	0.08 – 0.13	1.00 – 1.40	Bal.	0.45 – 0.65	0.08 –0.15	3.00 – 3.50	0 – 0.025	0 – 0.035	0.15 –0.30
8620	0.18 – 0.23	0.40 – 0.60	Bal.	0.70 – 0.90	0.15 – 0.25	0.40 – 0.70	0 – 0.035	0 – 0.040	0.15 – 0.30
5120	0.17 – 0.22	0.70 – 0.90	Bal.	0.70 – 0.90	0	0	0 – 0.035	0 – 0.040	0.15 – 0.30
4320H	0.17 – 0.23	0.35 – 0.65	Bal.	0.40 – 0.70	0.20 – 0.30	1.55 – 2.00	0 – 0.035	0 – 0.040	0.15 – 0.30

Carburizing Objective

Case depth: 0.035" (0.9mm) at C=0.35 wt. %.
Surface carbon: 0.80 ± 0.05 for 8620 and 5120, 0.70 ± 0.05 for 4320 and 9310

Geometry

The geometry is shown in Figure 5.

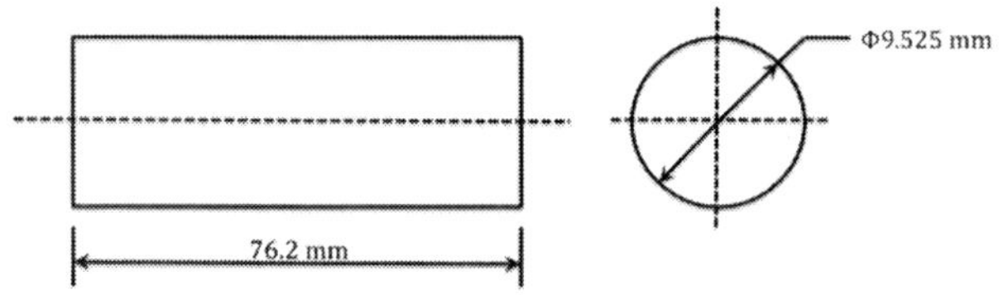

Fig. 4 Geometry of sample

Gas Carburizing Process Development

Test samples were gas carburized in an industrial furnace using a boost and diffuse method. The heat treating process is to heat up to 1700 °F and hold 3.5 hours in endothermic gas at carbon potential of 0.95%, then diffuse at 1550 °F for 1 hour in carbon potential of 0.8%, quench in oil at 140 °F and temper at 350 °F for 2 hours. Figure 6 presents the process schematic.

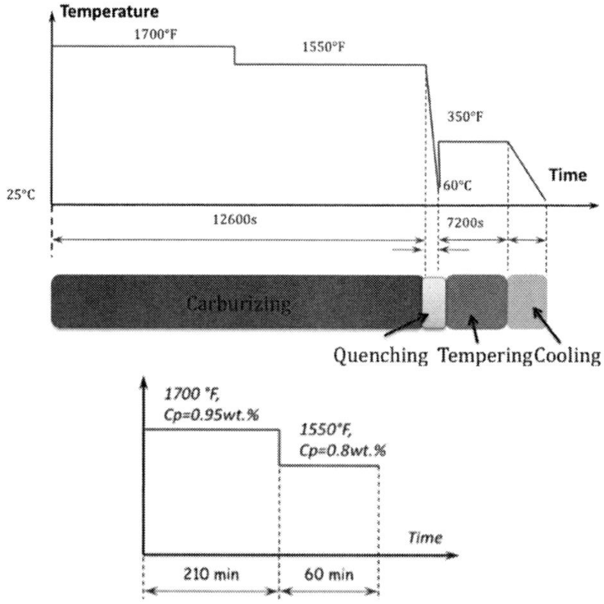

Fig. 5 Process schematic of gas carburizing

The carbon-depth profile measurements of the carburized parts are performed using the Optical Emission Spectrometer (OES). The model was calculated by inputting the parameters of boost and diffuse cycles. Figure 7 presents the measured results and model predictions from CarbTool©. These results agreed very well, which verified the effectiveness of CarbTool© on predicting gas carburizing.

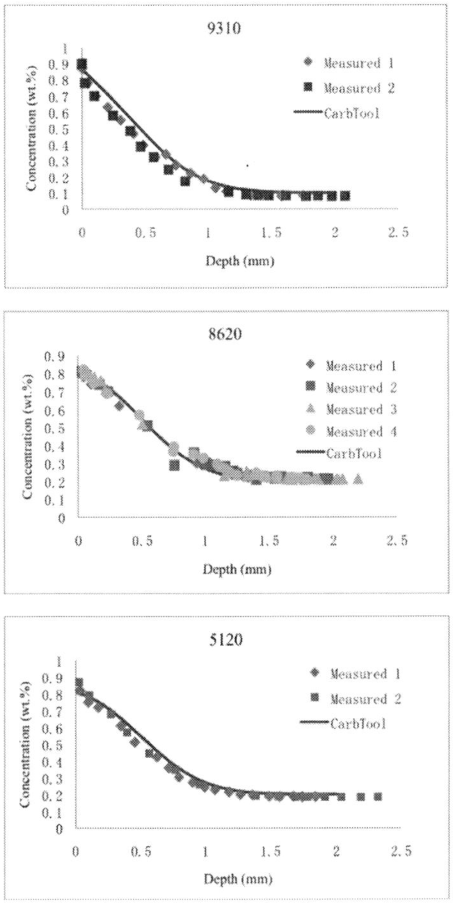

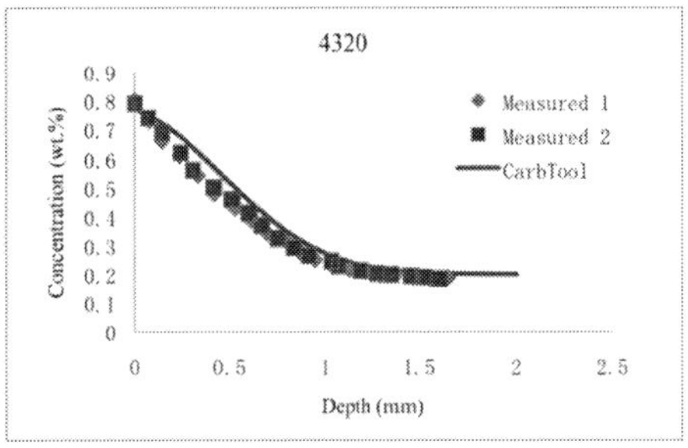

Fig. 6 Comparison between Measured and Predicted Carbon Profile of Four Alloys.

The accuracy of CarbTool© was demonstrated in Table 2; CarbTool©'s calculated surface concentration is compared with the measured result.

Table 2 Comparison of Surface Carbon Concentration between Measured Results and Prediction

	9310	8620	5120	4320
Measured (wt. %)	0.885	0.81	0.85	0.80
CarbTool© (wt. %)	0.86	0.81	0.84	0.77

Table 3 presents a difference between the experiments and specifications. The data shows CarbTool© is able to predict the experimental carbon profile; it was used to revise the industry processes. After changing the inputs in CarbTool©, the corresponding results were obtained in Table 4.

Table 3 Comparison between objective and experimental results

Materials	Surface concentration (wt. %)		Case depth (mm)		Retained Austenite (%)	
	Specification	Experimental	Specification	Experimental	Specification	Experimental
9310	0.70 ± 0.05	0.885(higher)	0.889	0.65(shallower)	20% - 30%	33.08
4320		0.8(higher)		0.67(shallower)		37.21
8620	0.80 ± 0.05	0.81		0.80(shallower)		19.04
5120		0.85		0.74(shallower)		23.54

Table 4 Gas carburizing predictions from CarbTool©

		9310	4320	8620	5120
Specification	Surface concentration (wt.%)	0.7±0.05		0.8±0.05	
	Depth at 0.35 wt% carbon (mm)	0.889		0.889	
Modeling Results	Surface concentration (wt.%)	0.74	0.74	0.84	0.84
	Depth at 0.35 wt% carbon (mm)	0.888	0.889	0.889	0.889

Process Determination of Vacuum Carburizing in Cyclohexane (C_6H_6)

9310 and 8620 were used for process development. Processes of the four alloys were set up by industry based on experience and their own databases. In 9310 process, 9310 and 8620 were used. In the other three processes, 8620 coupons with different dimensions were used. Table 5 demonstrates the trial tests.

From Figure 8, the flux of each process was calculated.

Table 5 Tentative test cycles

	Process	Test Material	Diameter (mm)	Calculated flux
Coupon 1	9310	9310	25.4	Flux 1
Coupon 2	9310	8620	12.7	Flux 2
Coupon 3	8620/5120	8620	25.4	Flux 3
Coupon 4	8620/5120	8620	12.7	Flux 4
Coupon 5	4320	8620	25.4	Flux 5

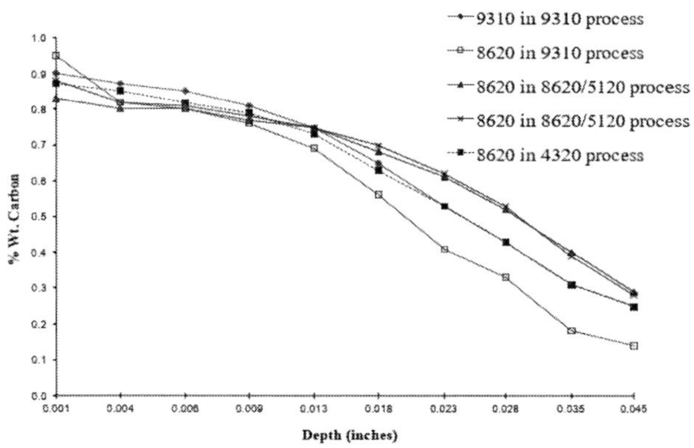

Fig. 7 Carbon profiles of test coupons.

The calculated flux was an input into CarbTool© as a boundary condition; other carburizing parameters (temperature, time and dimension) were kept the same with industry process. The model then predicted carbon profiles of test coupons.

Although we already verified the effectiveness of CarbTool© in gas carburizing, this process has new feature in terms of temperature. Thus, here we verify CarbTool© again. As seen in Figure 9, the modeling results and experimental results agreed with each other very well.

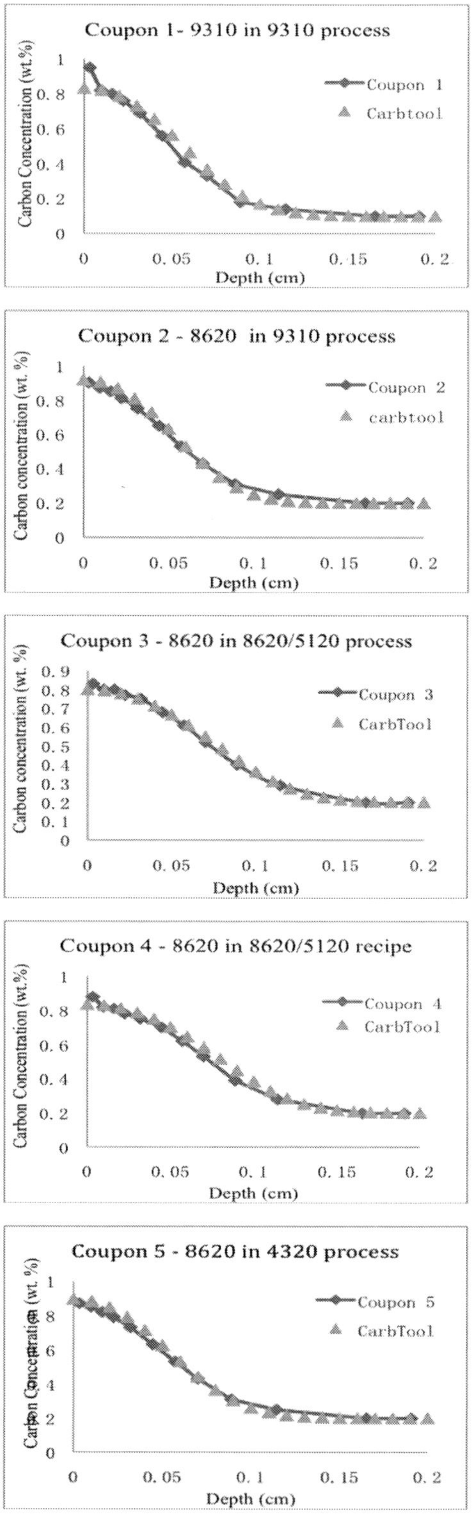

Fig. 8 Carbon profiles of test coupons and their modeling results.

Since the effectiveness of CarbTool© has been proved in this process, the software was utilized to model the fatigue test samples in their corresponding processes. In each process, the calculated flux from the trial test was used as the boundary condition. Meanwhile, diffusivity was changed from the test material (9310 or 8620) to the corresponding material, and the diameter was changed to the correct size (9.525 mm diameter). At first, the industry processes were conducted, the results were shown in "Modeling 1" column in Table 6. The results do not match specification. Then carburized parameters were tentatively changed in CarbTool©, based on the basic knowledge of diffusion. Once the results (shown in "Modeling 2" column in Table 6) matched the objective, the final processes were obtained.

Table 6 Comparison between objective and different modeling

Process with right material	Flux	Surface concentration (wt. %)			Case depth (mm)		
		Specification	Modeling 1	Modeling 2	Specification	Modeling 1	Modeling 2
9310	Flux 1	0.70 ± 0.05	0.82(higher)	0.72	0.889	0.717(shallower)	0.878
	Flux 2		0.84(higher)	0.73		0.727(shallower)	0.891
4320	Flux 5		0.94(higher)	0.69		0.829(shallower)	0.886
8620	Flux 3	0.80 ± 0.05	0.83(higher)	0.78		1.063(deeper)	0.883
	Flux 4		0.85	0.8		1.079(deeper)	0.896
5120	Flux 3		0.84	0.8		1.040(deeper)	0.866
	Flux 4		0.86(higher)	0.82		1.057(deeper)	0.878

Conclusion

The modeling tool, CarbTool©, is effective in predicting the carbon profile for gas carburizing and vacuum carburizing.

Carburization modeling will assist heat treaters to better understand the effects of process parameters on the diffusion process, distribution of carbon concentration, and effective case depth.

Effective modeling will save time and money over a variety of experimental trial.

Finally, the model will give engineers the possibility of optimizing the material, process and design for achieving the optimum performance.

Acknowledgments

The support of the Center for Heat Treating Excellence (CHTE) at Worcester Polytechnic Institute and the member companies is gratefully acknowledged. The author would thank Timken Inc. for providing the raw materials and supporting in this work. The carburization heat treatments were performed at Bodycote Inc. and Surface Combustion Inc. Their support through facilities and experimental work is greatly appreciated.

References

[1] Sahay, S.S., "Cost Model Based Optimisation of Carburising Operation", *Surf. Eng.*, 2004, Vol. 20 No. 5, pp. 379.

[2] Karabelchtchikova, O., "Fundamentals of Mass Transfer in Gas Carburizing", Ph.D. Dissertation, Worcester Polytechnic Institute, 2007.

[3] Karabelchtchikova, O., "Carburization Process Modeling and Sensitivity Analysis using Numerical Simulation", *MS&T '06: Materials Science and Technology 2006 Conference and Exhibition*, 2007.

[4] Munts, V.A. et al, "Rate of Carburizing of Steel", *Met. Sci. Heat Treat. (English Translation of Metallovedenie i Termicheskaya Obrabotka Metallov),* Vol. 22, No. 5-6 (1980), pp. 358-360.

[5] Baran, Ya. et al, "Gyulihandanov, Vysokotemperaturnaya nitrotsementatsiya spietchenykh staliv vysokoy plotnosti", *MITOM*, Vol. 9, (1986), pp. 8-10 (in Russian).
[6] Grafem, W. et al, "Acetylene low–pressure carburizing – a novel and superior carburizing technology", *Heat Treatment of Metals*, Vol. 4, (1999), pp. 79-85.
[7] Kula, P. et al, "Vacuum carburizing - process optimization", *13th International Scientific Conference on Achievement in Mechanical and Materials Engineering,* 2005, Poland.

Proceedings of the 26th ASM Heat Treating Society Conference
B.L. Ferguson, R. Jones, D.S. MacKenzie, and D. Weires, editors
https://doi.org/10.31399/asm.cp.ht2011p0099

Austempered Materials for Powertrain Applications

Justin Lefevre and Kathy L. Hayrynen, PhD
Applied Process Technologies Div, Livonia, MI USA
JLefevre@appliedprocess.com,KHayrynen@appliedprocess.com, (734)464-8000

Abstract

Austempered irons and steels offer the design engineer alternatives to conventional material/process combinations. Depending on the material and the application, Austempering may provide the following benefits to producers of powertrain components like gears and shafts: ease of manufacturing, increased bending and/or contact fatigue strength, better wear resistance and enhanced dampening characteristics resulting in lower noise. Austempered materials have been used to improve the performance of powertrain components in numerous applications for a wide range of industries, from gears and shafts to clutch plates and crankshafts. This paper focuses on Austempered solutions for powertrain applications with an emphasis on gear and shaft solutions.

Introduction

Austempering is an isothermal heat treatment process that can be applied to ferrous materials to increase strength and toughness. Fig. 1 shows a schematic isothermal (I-T) diagram with both the Austempering (line 1) and the quench and tempering (Q&T) (line 2) processes outlined. Austempering consists of austenitizing followed by rapidly quenching to temperatures in the range of 260-385°C (500-725°F) where the material is then transformed isothermally to form either Ausferrite, (acicular ferrite and carbon stabilized austenite), in cast iron, or Bainite, (acicular ferrite and carbide), in steel. The Q&T process consists of austenitizing and then rapidly quenching below the Martensite start line. The Martensite that forms is very hard and brittle; and subsequently must undergo a tempering step to acquire the desired combination of strength and toughness.

Because Austempering is an isothermal process, it offers advantages over Q&T. Since the formation of Bainite or Ausferrite occurs over minutes or hours at a single temperature, distortion is minimized and cracking does not occur. In contrast, the formation of Martensite occurs immediately as the metal temperature drops below the Martensite start temperature. This cooling is non-uniform as it is achieved at different rates in various sections, resulting in non-uniform transformation, which can produce significant distortion and/or cracking.

Carbo-Austempering™ is a heat treat process used on certain steels where the surface of the part is carburized, followed by an isothermal quench at a temperature required to produce a high carbon, Bainitic case. When this process is applied to low carbon steels, it results in the formation of a Bainitic case and a low carbon, tempered Martensite core. For medium carbon steels, Bainite is formed throughout the cross-section of the part.

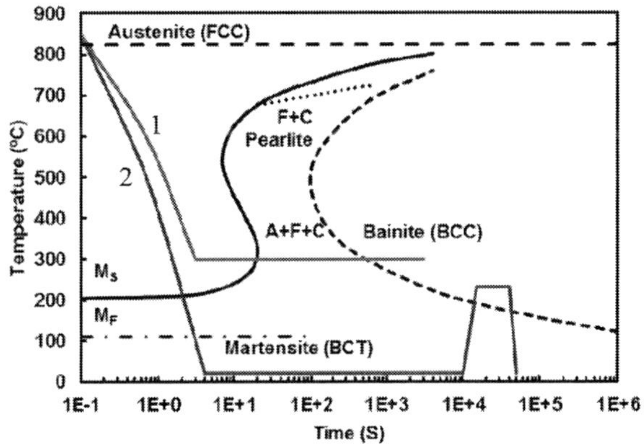

Figure 1: Schematic I-T Diagram illustrating the Austempering (line 1) and Quench & Tempering (line 2) Process. The basic crystal structures are in parentheses.

Austempered Irons

Austempering can be applied to both ductile iron and gray iron to produce beneficial properties. Austempered Ductile Iron (ADI) and Austempered Gray Iron (AGI) components exhibit increased wear resistance, higher strength, and better noise damping properties than their as-cast states. Fig. 2 shows the relationships between Brinell hardness and tensile strength, yield strength, % elongation, and un-notched Charpy Impact energy for ADI. With an increase in hardness, the strength of ADI increases while elongation and impact strength decrease. The standard ASTM grades of ADI and their properties are listed in Table 1.

Fig. 3 shows the tensile strength and yield strength of AGI as a function of hardness including the as-cast condition for Classes 20, 40, and 60 gray irons. Elongation and Charpy impact energy are excluded from this figure because they are not relevant for gray iron or AGI gear applications.

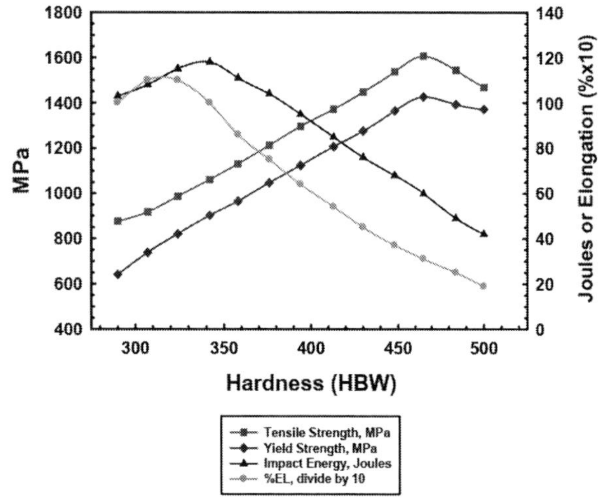

Figure 2: Typical properties of ADI as a function of Brinell Hardness (HBW). Note % Elongation values as plotted should be divided by 10.

Table 1: Properties of ADI per ASTM A897/A897M-06 [1]

Grade	Tensile Strength (MPa/Ksi)	Yield Strength (MPa/Ksi)	Elong. (%)	Impact Energy (J/lb-ft)	Typical Hardness (HBW)
750-500-11 (110-70-11)	750 / 110	500 / 70	11	110 / 80	241 - 302
900-650-09 (130-90-09)	900 / 130	650 / 90	9	100 / 75	269 – 341
1050-750-07 (150-110-07)	1050 / 150	750 / 110	7	80 / 60	302 – 375
1200-850-04 (175-125-04)	1200 / 175	850 / 125	4	60 / 45	341 – 444
1400-1100-02 (200-155-02)	1400 / 200	1100 / 155	2	35 / 25	388 – 477
1600-1300-01 (230-180-01)	1600 / 230	1300 / 185	1	20 / 15	402 - 512

CONTACT FATIGUE – The data for the wear properties of Austempered irons are presented for three wear conditions specific to gear wear: high stress abrasive wear, galling, and gear tooth pitting.

Fig. 4 shows pin abrasion (high stress) test results for ADI and competitive materials. In general, ADI exhibits a lower volume loss for a given hardness level than Q&T steel or ductile iron. This occurs as a result of the Austenite component in the Ausferrite microstructure of ADI. When a high normal force is applied, this Austenite can undergo a strain transformation to Martensite which is an excellent wear material. The depth of this hardened layer (≈5μm) is shown in Fig. 5.

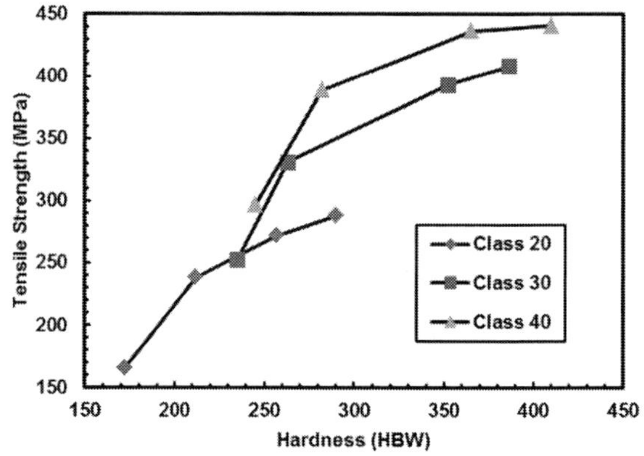

Figure 3: Tensile Strengths of Austempered Gray Iron (AGI) as a function of Brinell Hardness (HBW)

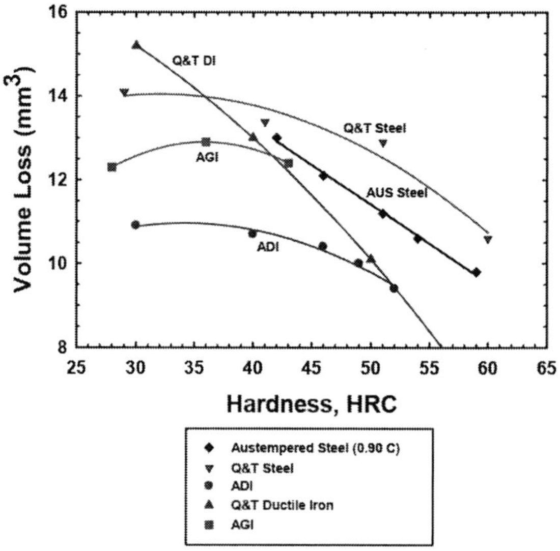

Figure 4: Pin abrasion test results, comparing volume loss at equivalent hardnesses.

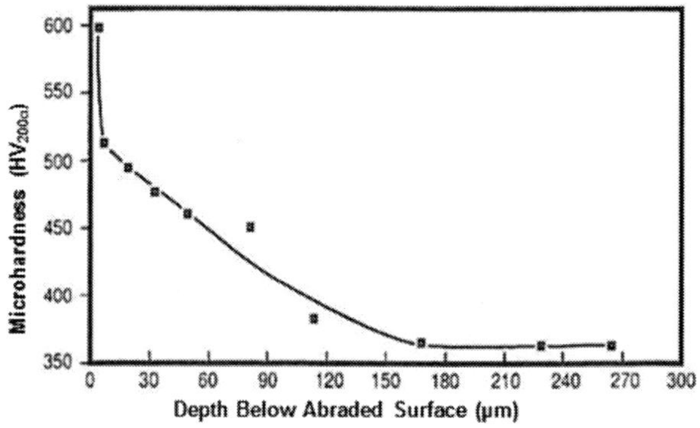

Figure 5: Vickers microhardness profile versus the depth below the surface for Grade 1050-750-07 ADI [2]

Gear failure in the low cycle contact fatigue regime is essentially a galling phenomenon. Fig. 6 shows self-mated galling test results for ADI, Carbo-Austempered™ (C/A) 8620, and carburized quenched and tempered (C/H) 8620. In this experimental work, a spike in the coefficient of friction indicates that galling has occurred. Examination of the curves in Fig. 6 indicates that Grade 900-650-09 did not gall in this self-mated test. It is theorized by the authors that the graphite nodules in ductile iron provide a source of lubrication and, thus, increase galling resistance. [3]

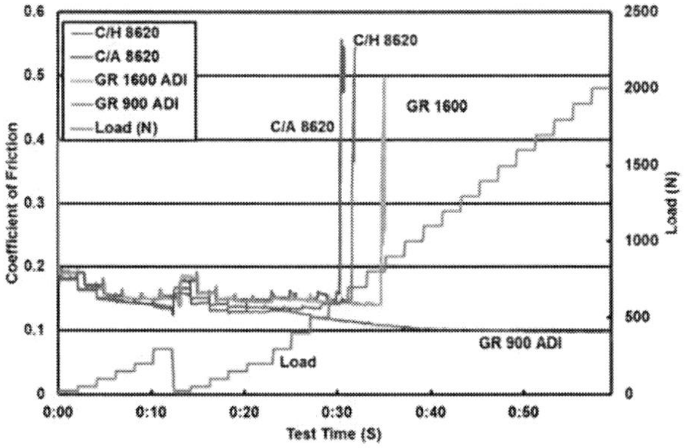

Figure 6: Coefficient of friction versus load and time showing the galling threshold for various materials.

ADI has a Young's Modulus that is 20% lower than that of steel; therefore, ADI gear teeth conform more than steel ones. The elevated conformance of the teeth increases the area of contact and decreases the Hertzian contact stress for a given load. Fig. 7 shows the allowable contact fatigue strength versus hardness for ADI and competitive steel heat treat options for gear design. It shows that ADI competes favorably with through hardened, nitrided, and induction hardened steels.

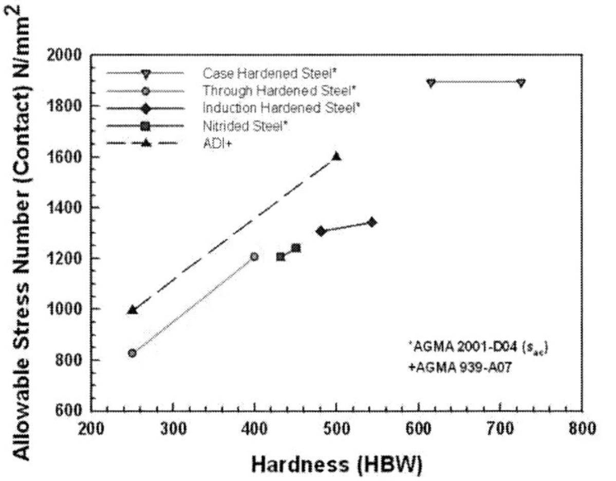

Figure 7: Comparison of the contact fatigue strengths of ADI to various steels used for gear applications. [4]

BENDING FATIGUE- Fig. 8 shows the allowable tooth root bending fatigue strength of ADI and conventionally heat treated steels for gear applications. ADI is competitive with cast and through-hardened steels in tooth root bending. Furthermore, when shot-peened, the fatigue strength of ADI is markedly improved so it can compete favorably with gas-nitrided and case-carburized steels. Shot peening can improve the allowable bending fatigue of carburized quenched and tempered steels by 30%, up to 75% for ADI.

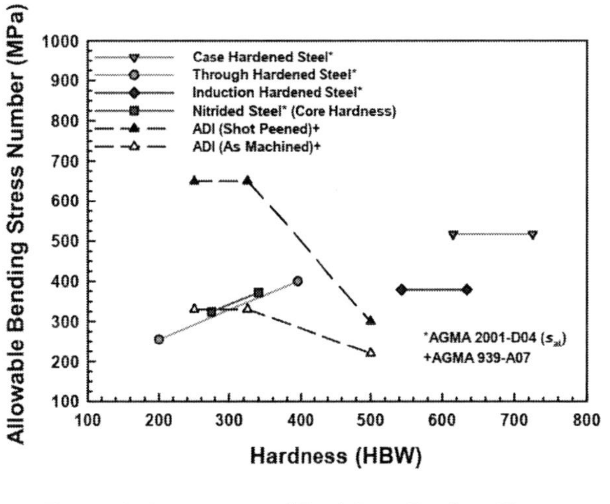

Figure 8: Comparison of Tooth Root Bending-Fatigue Strength of ADI with those of competitive steels used for gear applications.[4]

In Fig. 9, several shot peening combinations are measured for their effect on residual compressive stresses for Grade 1400-1100-02 ADI. The as-Austempered surface compressive stress observed was less than 207 MPa while the maximum shot peened surface compression was more than 896 MPa. Thus, the bending performance of ADI can be greatly enhanced by shot peening and other processes, such as fillet rolling, that impart a residual compressive stress on the surface of a component.

NOISE REDUCTION - Cast iron is inherently quieter than steel alternatives due to the presence of graphite. Fig. 10 is a schematic showing the relative damping of gray iron, ductile iron, and steel.

The finer microstructural scale of Ausferrite (vs. tempered Martensite) further enhances the damping properties of Austempered irons. This allows for the possibility that gears machined to lower AGMA (American Gear Manufacturers Association) classes could be as quiet as those machined to more precise grades. Fig. 11 is a schematic comparing the gear noise of AGI, ADI, and steel over the AGMA gear classes. The red lines illustrate how an ADI gear could be machined to a lower AGMA grade and still be as quiet as a higher grade steel gear.

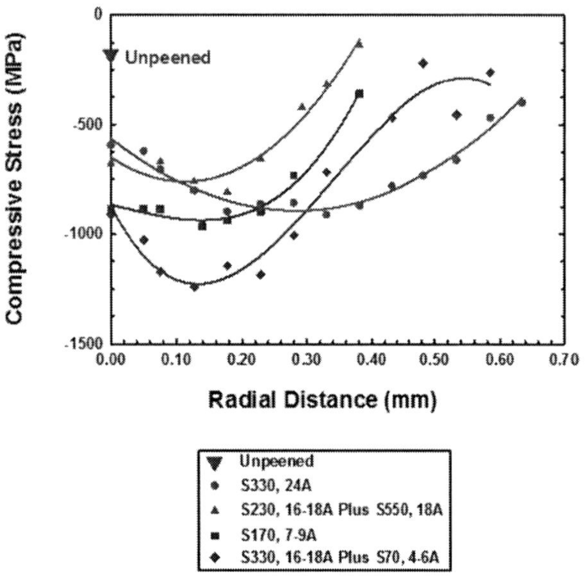

Figure 9: Effect of Various Shot Peening Schemes on the Compressive Stresses of Grade 1400-1100-02 ADI. [5]

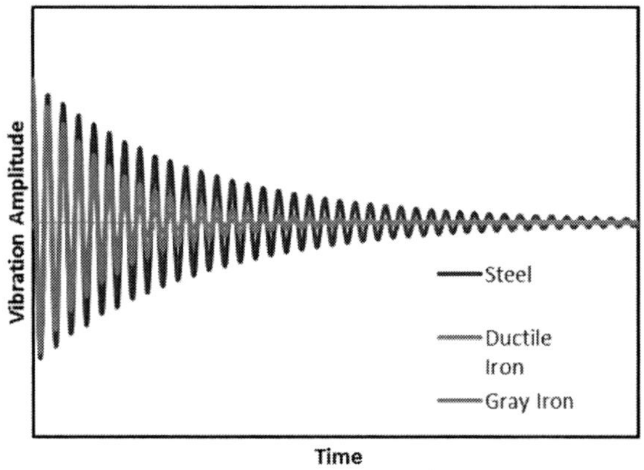

Figure 10: Schematic of Vibration Characteristics of Gray Iron, Ductile Iron and Steel.

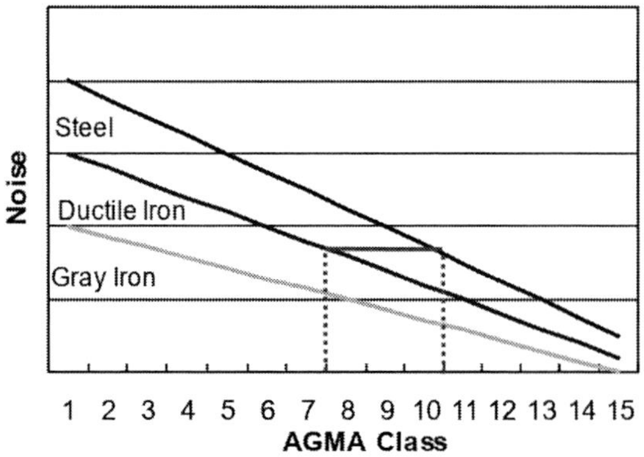

Figure 11: This schematic shows the relative Comparison of Noise Reduction as a Function of Material and AGMA Class (courtesy of Wells Manufacturing, Dura-Bar Division)

A study completed on hypoid gear sets, shown in Fig. 12, compares the noise when using steel or ADI. The curves in the figure represent a steel gear set and an ADI gear set. This study found that a larger noise reduction could be attained with the ADI gear set.

Austempering of gray iron also increases the noise reduction capabilities of gray iron. As shown in Fig. 13, the damping characteristics of gray iron are increased when Austempered, giving the higher strength AGI better noise reduction characteristics than its as-cast counterparts. In fact, an AGI with a tensile strength of nearly 414 MPa can have the noise dampening capabilities of a fully damped, Class 20 Gray Iron (with a UTS of 138 MPa).

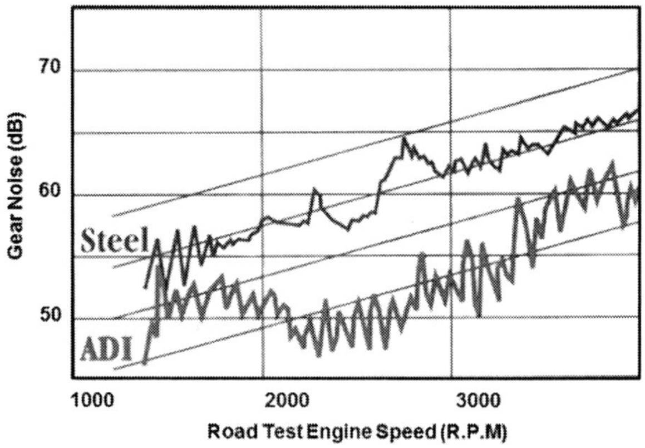

Figure 12: Comparison of Noise in Hypoid Gears During Vehicle Road Tests, from the ASME Gear Research Institute Report A4001 .[5]

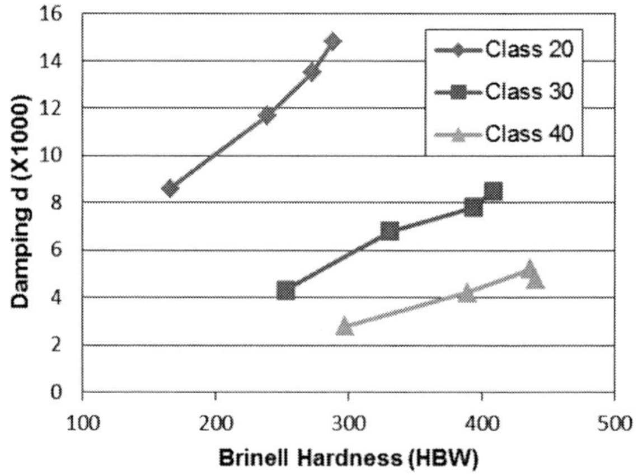

Figure 13: Damping of Austempered Gray Iron versus Brinell hardness. Each curve represents a different base gray iron with the lowest hardness data point for each grade being the as-cast condition.

MANUFACTURABILITY – ADI and AGI offer an opportunity for increased manufacturability of a component, because rough machining can be done prior to heat treatment. In the as-cast condition, the material is much easier to machine reducing the total manufacturing cost. Though many applications can be heat treated after final machining, finish machining after heat treatment increases the strength characteristics of ADI and AGI, giving them superior fatigue strength than prior to finish machining. Fig. 14 compares the relative machinability of several ferrous materials. Note that ductile iron in a ferritic or pearlitic condition is easier to machine than 4140 steel or ADI. Furthermore, machining of ductile iron, gray iron, ADI and AGI results in a compact, discontinuous chip that is easily handled and is fully recyclable. Dry machining techniques can be easily applied to as-cast gray and ductile irons.

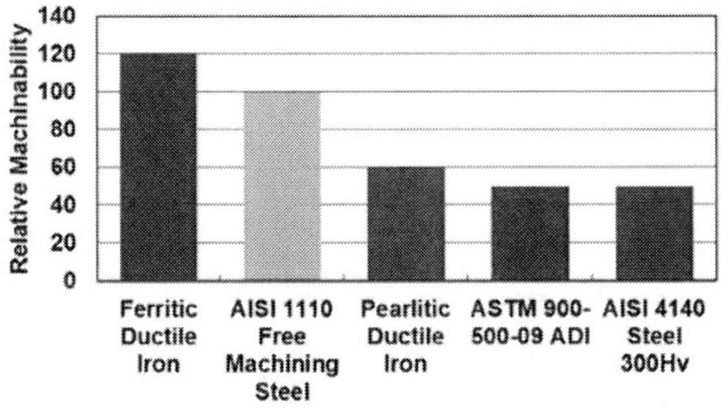

Figure 14: Relative Machinability of Several Ferrous Materials

In addition to the increased ease of manufacturability and improved machinability, iron castings are generally nearer net shape than steel forgings and castings. Fig. 15 shows a comparison of relative material cost of different materials per unit of yield strength. When all material and processing costs are taken into account, ADI and AGI are relatively less expensive to manufacture than other commonly used materials.

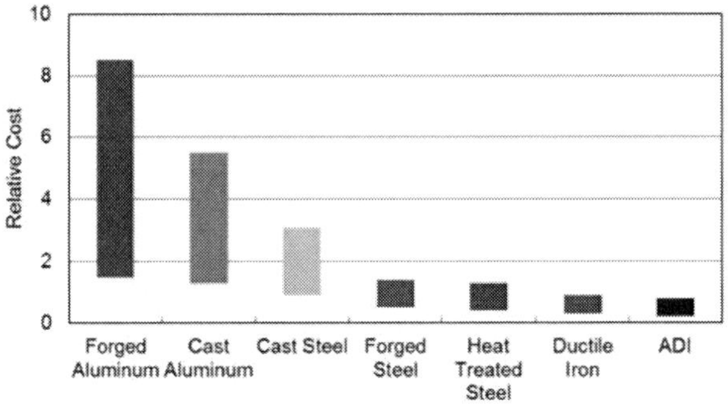

Figure 15: Cost per Unit of Yield Strength of Various Materials. Comparison based on results for the forged steel normalized to 1.

The Austempering process also results in reduced distortion and eliminates the occurrence of quench cracking. When General Motors switched to ADI hypoid differential gears from traditional Carburized quenched and tempered 8620 steel process in the 1970s, they were able to eliminate the need for press quenching.APPLICATIONS OF AUSTEMPERED IRONS- Figs. 16 through 21 show ADI and AGI powertrain components that deliver decreased cost, comparable or improved mechanical properties, and increased design flexibility.

Figure 16: ADI Diesel engine timing gears have replaced carburized Q&T gears at a cost savings for many years.

Figure 17: ADI hypoid differential gears and pinions are common conversions to Grades 1200-850-04 or 1400-1100-02 ADI.

Figure 18: A one piece ADI gear and gxle for commercial lawnmower drives that replaced a three piece carburized steel assembly .

Figure 19: ADI mill gears are produced in segments and assembled post heat treatment.

Figure 20: ADI crankshafts have been used in several notable sports cars for increased fatigue strength and reduced weight and cost. [6]

Figure 21: AGI gear for timing on a light vehicle engine.

Austempered Steels

The Austempering heat treatment can be applied to steels provided the steel has a time-temperature-transformation diagram with the following characteristics:

1. A Martensite start temperature low enough to allow for the formation of Bainite.
2. Sufficient hardenability to avoid the formation of pearlite on quenching to the Austempering temperature.
3. A reasonable Bainite transformation time.

Powdered metal steels can also be Austempered if the alloy meets all of the above criteria and is near full density.

Table 2 shows the tensile and yield strength, hardness, elongation, percent reduction in area, and impact energy of Austempered 1074 and 4340 steel versus their Q&T counterparts. In this data, the impact strengths of the Austempered steels are higher than those of Q&T steels at the hardness levels tested for each alloy. Bainite exhibits higher toughness and equivalent or increased strength in certain high hardness regimes when compared to tempered Martensite. Bainitic toughness surpasses that of Martensite between 40 and 50 HRC depending on the steel alloy.

Table 2: Mechanical Properties of Q&T and Austempered 1074 and 4340 Steels

Material	Quench & Tempered 1074	Austempered 1074	Quench & Tempered 4340	Austempered 4340
Rc Hardness	50	50	45	47
UTS (MPa / ksi)	1701 / 246.7	1949 / 282.7	1465/212.5	1605/232.8
Yield Strength (MPa / ksi)	839 / 121.7	1043 / 151.3	1340 / 194.3	1340 / 194.3
% Elongation	0.3	1.9	12.8	14.2
%RA	0.7	34.5	50.4	56.7
Impact* (J / ft-lbs)	3.9 / 2.9	47.9 / 35.3	27.8/20.5	31.6/23.3

Austempered steels also exhibit higher ductility for high hardness values when compared to their Q&T counterparts. In a bend test of these two heat treatments on 1050 steel at 49 HRC, only the Austempered test piece survived. [7]

Austempered steels also exhibit improved fatigue properties. At high hardness, Bainite does not exhibit a decrease in fatigue strength. In contrast, above a certain hardness level, the fatigue strength of conventional quench & tempered steel drops significantly due to a susceptibility to hydrogen embrittlement as illustrated in Fig. 22.

Improved abrasive wear resistance over tempered Martensite is an additional advantage of Bainite at equivalent hardness. This can be seen in the pin abrasion test results in Fig. 4.

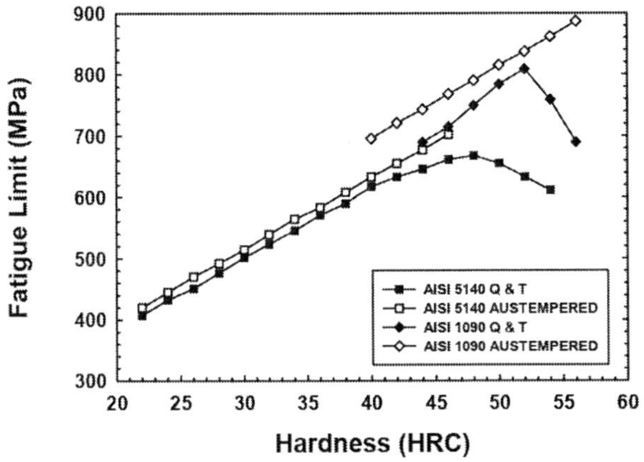

Figure 22: Fatigue limit as a function of hardness for two different steels that have been Austempered and quench & tempered. [8]

In the finite life fatigue regime, Austempered steels excel over Q&T steels. Fig. 23 shows stress life fatigue data for a fully reversed axial load on Austempered and Q&T 4340 steel tensile specimens. The Austempered 4340 has higher finite life strength than the Q&T 4340. The increased finite life strength makes Austempered steel output shafts ideal for occasional overload applications.

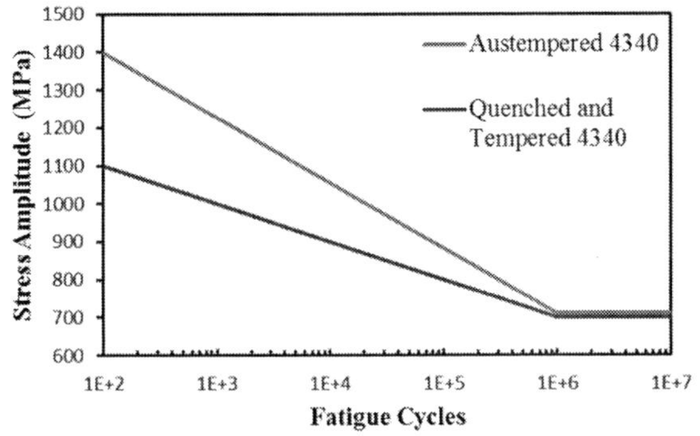

Figure 23: Stress amplitude versus fatigue life for Austempered and Q&T 4340 steel at equivalent hardness.[9]

CONTACT FATIGUE – Documentation of the contact fatigue properties of lower Bainite has been largely confined to high carbon, chromium alloyed steels that are suitable for bearing races. Certainly, this is an area that warrants further investigation.

APPLICATIONS OF AUSTEMPERED STEEL- Figs. 24 through 26 show Austempered steel powertrain components.

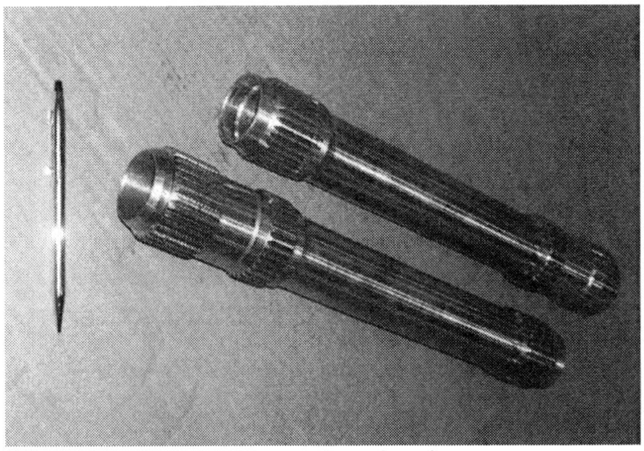

Figure 24: Austempered steel agricultural transmission output shafts.

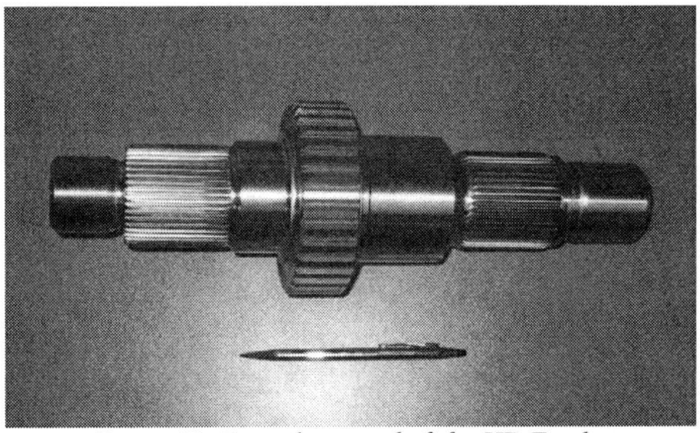

Figure 25: Geared output shaft for HD Trucks.

Figure 26: Austempered steel reverse gears.

Carbo-Austempered Steels™

Low to medium carbon steels are good candidates for Carbo-Austempering™. Typically a high carbon, Bainitic case (HRC 50 - 60) is produced on a component with a lower carbon, tempered Martensite core (HRC < 40). In some instances, advantages have been realized in medium carbon alloy steels with a high carbon, Bainitic case (45-55 HRC) on a medium carbon, Bainitic core (45-50 HRC).

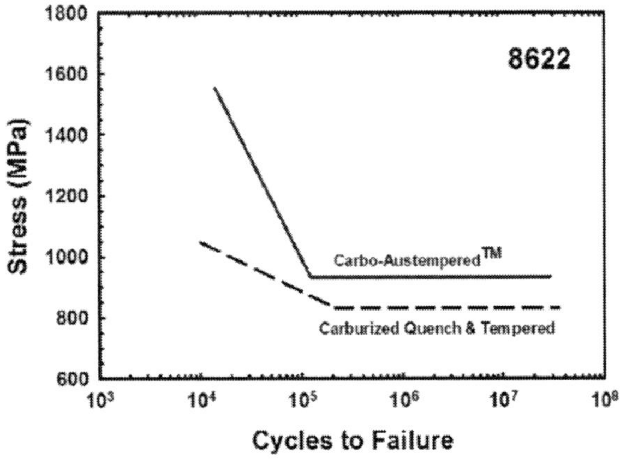

Figure 27: Rotating bending fatigue comparison of 8622 steel that has been Carbo-Austempered™ and Quench & Tempered.

BENDING FATIGUE- Carbo-Austempering™, like Austempering, is a low distortion heat treatment process when compared to conventional carburize quench and temper heat treatments. During Carbo-Austempering™, the transformation begins in the center or core of the part. This results in the formation of compressive stresses as the outside layer or case transforms last during the heat treat process. The residual compressive stresses on the surface of a Carbo-Austempered™ steel result in improved high load, low cycle fatigue properties versus conventional Carburized quenched and tempered steel. This is illustrated in Fig. 27, which contains rotating bending fatigue curves for both Carbo-Austempered™ and conventionally carburized quenched and tempered 8622 steel. The surface hardness for these specimens was 58 HRC with an effective case depth of 0.76 mm (0.03 inches). Note the superior performance of the Carbo-Austempered™ steel in the low cycle regime ($<10^5$ cycles) where improvements in fatigue strength of up to 40% can be realized. This trend also occurs in tooth root bending fatigue testing. Fig. 28 shows the tooth root bending fatigue life of Carbo-Austempered™ and Carburized quenched and tempered 8620 steel.

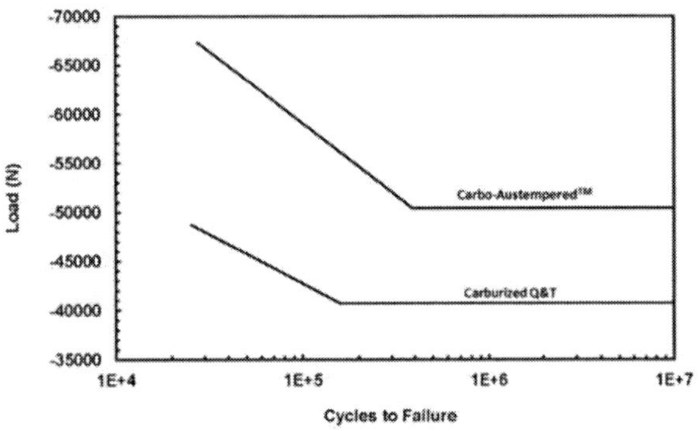

Figure 28: Load versus Cycles to Failure for Single Tooth Bending Fatigue Testing of Carbo-Austempered™ and Carburized Q&T 8620 Gears

APPLICATIONS OF CARBO-AUSTEMPERED™ STEEL- Figs. 29 through 23 show Carbo-Austempered™ steel parts.

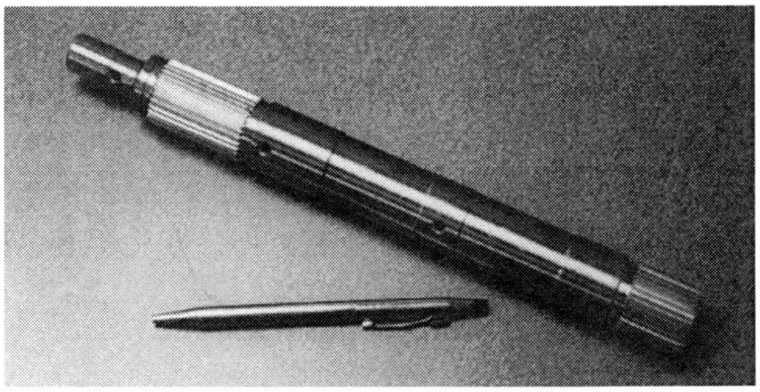

Figure 29: A Carbo-Austempered™ steel school bus transmission output shaft.

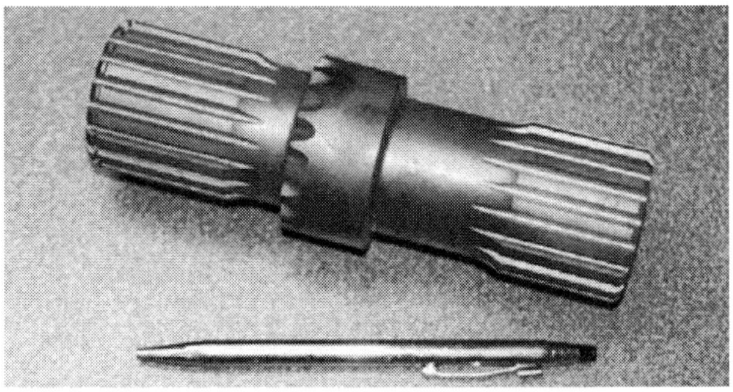

Figure 30: A Carbo-Austempered™ steel shaft coupler for mis-alligned shafts.

Figure 31: Carbo-Austempered™ steel side pinions can be found in many high torque differentials.

Figure 32: Starter Clutch shells in many light vehicles are austempered to withstand the high impact clattering engagement of the starter.

AUSTEMPERING- WHAT IT IS, AND WHAT IT ISN'T

Austempering is a high performance heat treatment but it is not a panacea. The application, as with all material/process combinations must fit.

ADI makes a quiet, low cost gear or shaft in its allowable loading range but it will not outperform carburized quenched and tempered alloyed, low carbon steel in bending or contact fatigue. Therefore, if a current product in carburized steel is failing in bending fatigue or pitting, ADI would not be a solution. However, if the contact and bending loads are in ADI's range, a considerable cost and noise advantage can be expected.

Below 40 HRC evidence would indicate that martensitic structures will outperform Bainitic structures. However, at hardnesses in excess of 40 HRC, Austempered medium carbon steels outperform through hardened martensitic components in impact strength and notched fatigue loading. The reduction in distortion as compared to a Q&T or carburized Q&T shaft often eliminates the need for post heat treatment straightening and elimintes losses due to cracking.

Carbo-Austempered™ steel will outperform HRC 60 carburized quenched and tempered steels in impact and bending fatigue, but, at 58 HRC, maximum hardness are limited to slightly lower contact loads. Therefore, applications where Carbo-Austempering™ apply are those applications where spike overloads in bending occur.

Thus, one should use Austempering, (as would be the case with other material/process combinations), as one option in their design "tool kit". The designer should work closely with the material provider and the heat treater to determine if Austempering would provide a benefit to his/her drive component application.

Conclusions

The Austempering process offers the designers of gears and power transmission components a viable, cost effective, high performance alternative to many conventional material/process combinations.

Austempering of irons and steels results in increased levels of fatigue strength, wear resistance and toughness. Benefits in the areas of noise reduction, manufacturability and wear resistance have also been documented.

Acknowledgments

The authors would like to thank the following individuals and/or companies for their assistance in preparing information for this manuscript:

ADI Treatments
AP Suzhou
AP Westshore
Applied Process, Inc.
Dana Holding Corp.
Demetri Golematis
Farrar Corp.
Ford Motor Co.
General Motors Allison Transmissions
General Motors Powertrain Div.
Getrag Gears of North America
John Keough
Meghan Haycock
Purolator
SEW Eurodrive
Svedala
Terry Lusk
Wells Manufacturing Durobar Div.

References

[1] ASTM A897/A897M-06, Standard Specification for Austempered Ductile Iron Castings, ASTM International, West Conshohocken, PA, www.astm.org.

[2] Ductile Iron Data for Design Engineers, Section IV Austempered Ductile Iron, revised 1998, Rio Tinto Iron & Titanium, Inc, Montreal, Quebec, www.ductile.org.

[3] Gladieux, S. and Hayrynen, K.., "A Preliminary Characterization of the Galling Wear Properties of Austempered Ductile Iron (ADI) and Carbo-Austempered™ Steel," www.appliedprocess.com.

[4] AGMA 939-A07, Austempered Ductile Iron for Gears, American Gear Manufacturers Association, Alexandria, VA, www.agma.org.

[5] "Austempered Ductile Iron Database," ASME Gear Research Institute, Final Report, June 1989.

[6] Brandenberg, K., Ravenscroft, J., Rimmer, A. and Hayrynen, K., "An ADI Crankshaft Designed for High Performance in TVR's Tuscan Speed Six Sports Car," Proceedings of SAE 2001 World Congress, March 2001, Paper # 2001-01-0408, www.sae.org.

[7] www.youtube.com/watch?v=DYZFZ7ug5YA

[8] Keough, W. R., "Equipment, Process and Properties of Modern Day Austempering," Proceedings of the International Heat Treating Conference: Equipment and Processes, April 1994.

[9] Tartaglia, J. M. and Hayrynen, K. L., "A Comparison of Fatigue Properties of Austempered vs Quenched & Tempered 4340 Steel," accepted for publication in JMEP.

Proceedings of the 26th ASM Heat Treating Society Conference
B.L. Ferguson, R. Jones, D.S. MacKenzie, and D. Weires, editors
https://doi.org/10.31399/asm.cp.ht2011p0108

Integral Preoxidation of Aerospace Gear Steel

Timothy De Hennis
The Boeing Company, Philadelphia, PA, USA
timothy.m.dehennis@boeing.com, 610.591.3956

Dale Weires
The Boeing Company, Philadelphia, PA, USA
dale.j.weires@boeing.com, 610.591.7993

Tyler Pounds
Northstar Aerospace (Chicago) Inc., Bedford Park, IL, USA
tpounds@nsaero.com, 708.728.2045

Abstract

A study of the use of integral preoxidation was conducted on aerospace gear steels. Some aerospace gear manufacturers make use of a preoxidation process that is separate from carburizing to produce parts with uniform case depths and surface hardness. The stand alone preoxidation process adds cost due to additional manufacturing steps, processing time, handling of parts, etc. Integral preoxidation may be a simpler surface activation method to facilitate uniform carburizing. Results and experimental data from the evaluation of X2M test coupons carburized with an integral preoxidation and carburization cycle are presented. The purpose of this document is to highlight pertinent information from the in-depth literature review regarding preoxidation of specific aerospace gears steels. A study of the use of integral preoxidation and carburizing processing of X-2M (BMS 7-223) Steel was conducted and the experimental data is documented and discussed herein.

Introduction

Aerospace grade gear steels could be described as extremely clean (VIM/VAR) high alloy steels, and typically have greater than 2% alloying elements by weight. In particular, X-2M Steel contains high amounts of chromium (~5%), molybdenum (~1.4%), tungsten (~1.3%), silicon (~0.9%), and vanadium (~.5%). These alloying elements are added to steel to improve specific properties such as wear resistance, hardness at high temperatures, resistance to pitting fatigue and resistance to scuffing/scoring, etc. The vanadium and tungsten as alloying elements improve the steel's ability to retain hardness at high temperatures [1]. However, certain alloying elements, such as chromium, can readily form a heterogeneous passive metal oxide barrier along the steel surface at room temperatures. This passive film will prohibit the uniform carburization of surfaces when case-hardening, leading to isolated areas of shallow case depths and low surface hardness. These conditions leave the manufacturer with un-salvageable parts.

Special preliminary treatments designed to remove passive films and facilitate uniform atmosphere carburizing have been developed for steels including Pyrowear 53, X-2M, and M50NiL. Prior to atmosphere carburizing, parts can be exposed to high temperatures in an oxidizing atmosphere where the passive low temperature oxide layer is replaced by a less passive, high temperature oxide layer. The high temperature oxide is a more porous iron-based oxide (Fe_2O_3/M_3O_4 and possibly FeO) compared to the low temperature oxide (Cr_2O_3 & SiO_2) [2]. Once established, the porous oxide present on the parts surface will allow for uniform carburization. This method is commonly known as a 'preoxidation' process.

Some aerospace gear steels make use of a preoxidation process that is separate from carburizing. A carburizing process was developed and patented by Boeing for X-2M Steel in the late 1970's that incorporated a separate preoxidation process with specific cleaning steps before and after the high temperature oxidation. These processing steps add cost due to additional manufacturing steps, processing time, handling of parts, etc. The parts are also sensitive to the additional cleaning steps. If all areas of the part are not properly abrasive blasted (using the correct media, dwell time, and pressure, etc.) the end result may be non-uniform or spotty carburization. Non-uniform carburization is typically not identified until later in the manufacturing process which increases the cost of the scrapped parts. An example of non-uniform carburization is shown in Figure 1, 2, and 3.

The preoxidation process can be performed as an integral part of the carburizing process. Integral preoxidation is a common processing technique that has been successfully used with Pyrowear 53; another aerospace gear steel. When X-2M steel was first put into use by Boeing, the metallurgical data on preoxidation processes in open literature was very limited and an integral preoxidation process was not developed. Integral preoxidation could be a simpler surface activation method to facilitate uniform carburizing of these types of X-2M and other similar gear steels.

Figure 1. Non-uniform carburization shown as light nital etch indications observed on three adjacent spur gear teeth (arrows).

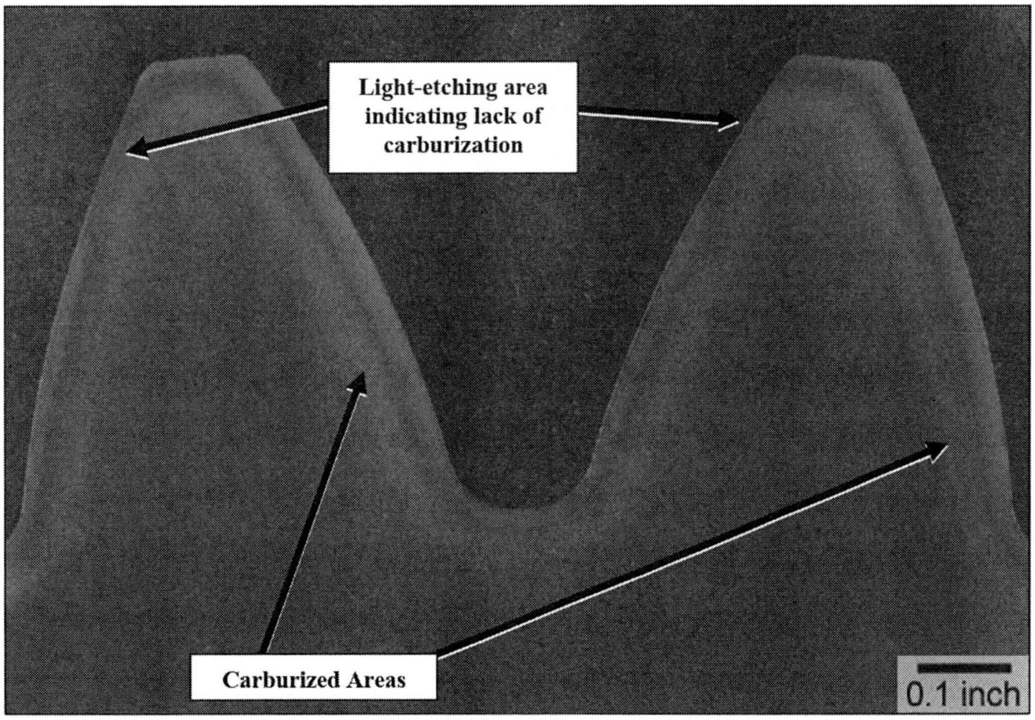

Figure 2. Macroscopic image of an etched cross-section prepared though the light-etching areas of two adjacent teeth shown in figure 1. This image reveals a visual indication of lack of carburization on the addendum of the gear tooth flanks with typical carburization in all other areas.

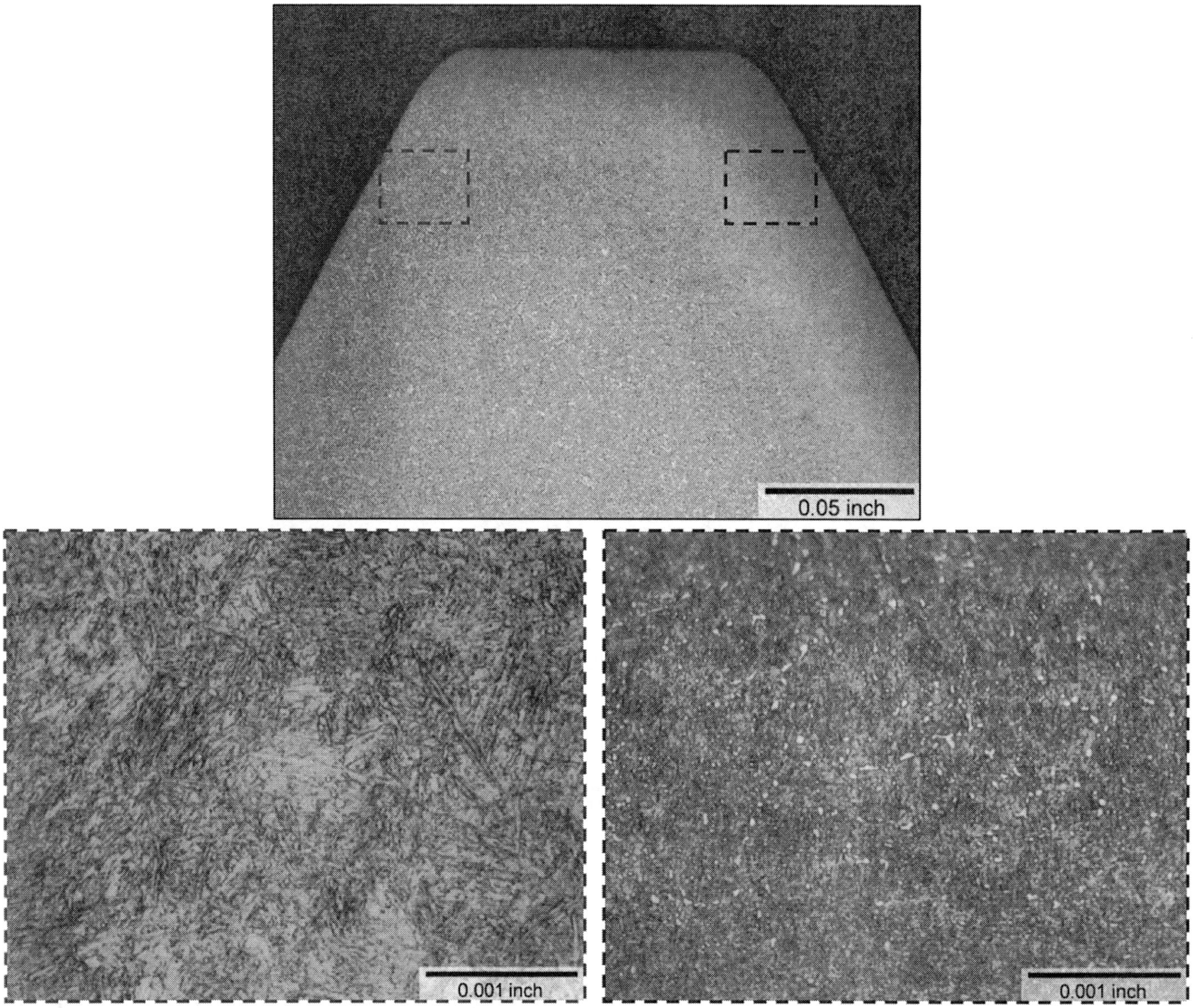

Figure 3. Micrographs of the etched metallographic cross-section through a light-etching indication revealing the microstructure in an area of non-uniform carburization (red box – bottom left) and an adjacent area considered as typical carburized and hardened X-2M steel (blue box – bottom right).

High Temperature Oxide Formation

Internal oxidation is known to occur during carburizing of steels and has been well documented by George Krauss [3], Geoffrey Parrish [4], and others. Most of the available literature concentrates on the detrimental affects of the oxides in the final product (i.e. change in fatigue resistance, reduction in hardenability, etc.). Although the use of internal oxidation of steels in preparation for carburizing is not very well documented, the principles are the same as described by Krauss and Parrish: Oxygen atoms are absorbed at the surface, diffuse inward along grain boundaries, into the bulk lattice, and then chemically combine with available substitutional elements that have high oxidation potential to form oxides [3]. Silicon, Manganese, and Chromium are elements that are very likely to oxide. X-2M contains relatively large amounts of these elements, see Table I.

Table I
Alloy composition for X2M Steel

	Carbon	Silicone	Manganese	Sulfur	Phosphorus	Tungsten	Chromium	Vanadium	Molybdenum
X-2 Modified BMS 7-223	0.13-0.16	0.80-1.00	0.20-0.40	.015 Max	.015 Max	1.20-1.50	4.75-5.25	0.40-0.50	1.30-1.50

Internal oxidation will occur during carburization and could have contributed to the spotty carburization identified in the early use of X-2M. By preoxidizing the parts prior to carburization, the substrate material is slightly depleted of the elements (chromium and silicon) that form passive oxide barriers to carburization at the near surface interface with the carburizing atmosphere since these elements form alternative oxides during high temperature oxidation. Due to this depletion, the preoxidized and cleaned surface has a chemistry that is less likely to reform the passive oxides such as Cr_2O_3 & SiO_2 allowing the carbon atoms to move more freely into the steel.

Integral Preoxidation of X2M Steel

Small scale carburization experiments were performed at Northstar Aerospace (Chicago) Inc. to study the effects of an integral preoxidation and carburization cycle on X-2M steel. Notched test samples were exposed to an integral preoxidation carburization process, using different preoxidation temperatures for each experiment. The preoxidation temperatures were 1500°F and 1800°F. Carburization was performed at 1700°F for 6 hours. This carburizing recipe was not a standard recipe for X-2M, but instead was selected to allow the authors to determine if the preoxidation process was effective in conditioning the surfaces to be carburized. The integral process was performed in a pit-type atmosphere carburizing furnace, and the test samples were placed in each of the three work zones (Top, Middle, and Bottom levels). After the carburizing process the test coupons were slow cooled to ambient temperature, subcritical annealed, hardened (austenitized and free quenched in oil), sub-zero treated, and double tempered. The carburization test coupons were then evaluated for surface hardness, effective case depth, microstructure, and carbon percentage profile. The results of these tests showed that the integral preoxidation carburization heat treat cycle at 1500°F was ineffective at producing a uniform case. The test samples that were integrally pre-oxidized at 1800°F produced promising results, and details are provided below.

The surface hardness of the samples pre-oxidized at 1800°F was measured via direct hardness using a Rockwell Instron Hardness Tester and HR30N scale. The results ranged from 60-65HRC after conversion using ASTM E 140, see Table II. Typical surface hardness for carburized X2M steel is 59-64HRC on finished parts (i.e. after grinding).

The effective case depth of the samples pre-oxidized at 1800°F was measured with Vickers microhardness. The effective case depth was defined as the perpendicular distance from the surface to a point where the microhardness indicates a hardness of HVN 513 (approximately 50HRC). The test samples were evaluated in three areas: the notch root, the notch flank, and the outer diameter. The resulting data was considered typical for carburizing per the test parameters (i.e time, temperature, and atmosphere). The results are shown in Table II.

Metallographic cross-sections were prepared through the notched test samples to evaluate the microstructures. Representative images of the case microstructure observed in the root of the samples are shown in Figure 4. The root areas of all test samples exhibited a typical microstructure.

Atomic emission spectrometry was used to generate a carbon percent profile for one of the samples pre-oxidized at 1800°F. Carbon percent levels were measured along the carburized surface and at specific depths within the case using a Leco Glow Discharge Spectrometer (GDS-500A). The resulting data is shown in Figure 5. A carbon profile of a conventionally carburized X-2M steel test sample is also shown in Figure 5 for comparison purposes.

Table II

Effective Case Depth and Hardness of Test Slugs Pre-oxidized at 1800°F

	Effective Case Depth[1] (inch)			Hardness (HRC)	
	Notch Root	Notch Flank	Outer Diameter	Surface[2]	Core[3]
Top Sample	0.040	0.046	0.048	63 - 65	41 - 41.5
Middle Sample	0.042	0.048	0.049	62 - 64	41.5
Bottom Sample	0.043	0.048	0.050	60 - 65	40.5 - 42

1-Measured with Vickers microhardness (500grams) and converted to HRC per ASTM E 140
2-Measured with HR30N scale per ASTM E384 and converted to HRC per ASTM E 140
3-Measured with HRC scale per ASTM E 18

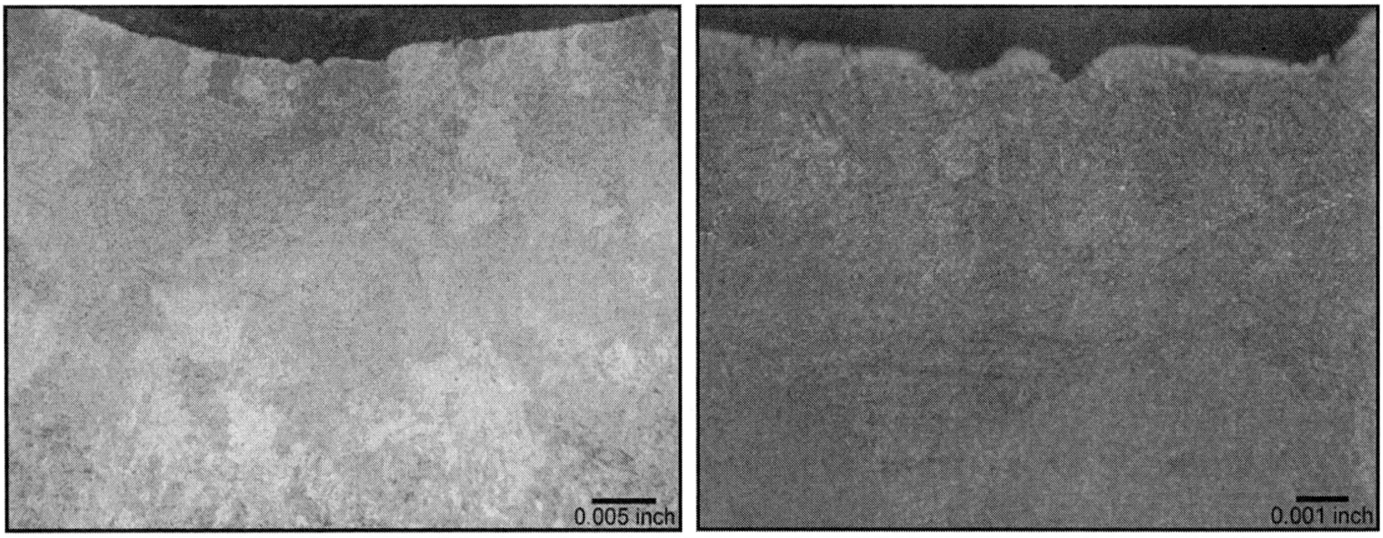

Figure 4. Representative micrographs of the carburized case in the roots of one of the bottom test samples from the 1800°F integral preoxidation. Images on the left were taken at 100x, and images on the right were taken at 400x.

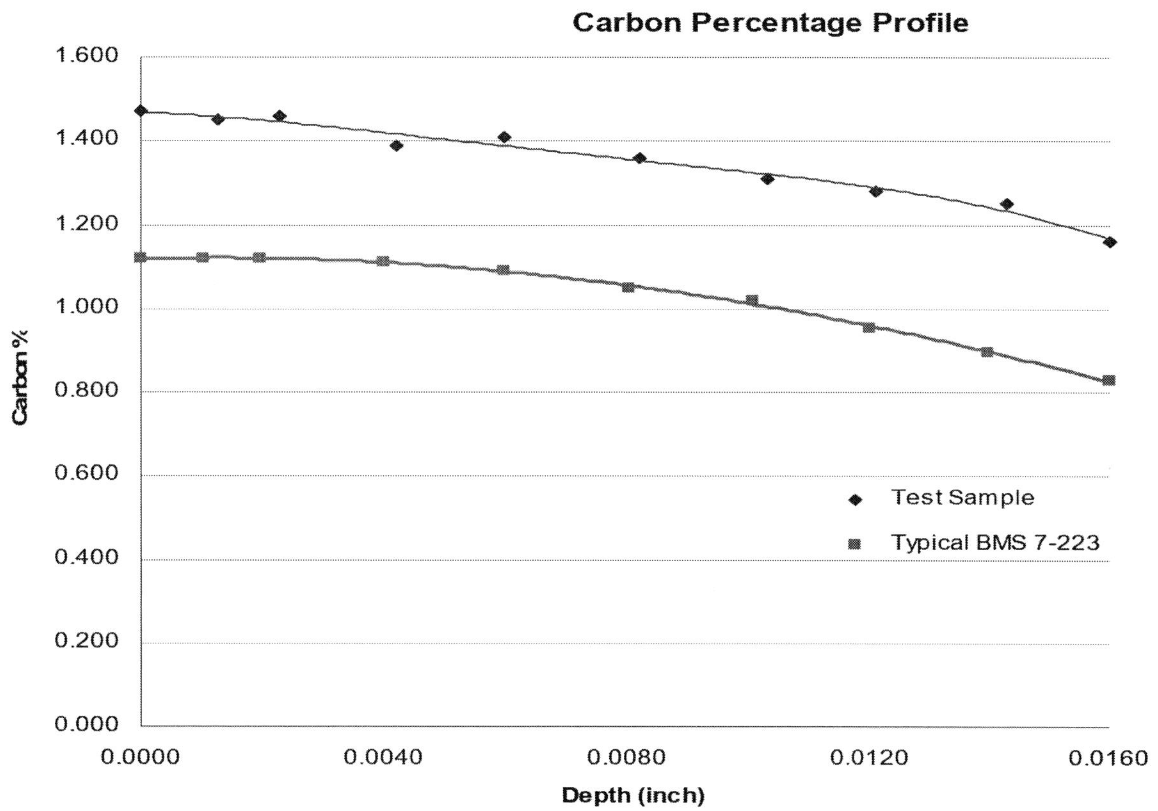

Figure 5. A carbon percent profile for one of the samples pre-oxidized at 1800°F using the high carbon potential test recipe and no subsequent diffusion step. A carbon profile of a conventionally carburized X-2M (BMS 7-223) steel test sample is also shown for comparison purposes.

Discussion

When the data obtained from the integrally preoxidized test specimen was compared with conventionally carburized X-2M some differences were noted. Surface hardness measurements were slightly higher than expected. The carbon profile showed a higher carbon percentage in the near surface of the case when compared to a sample that was processed through preoxidation separate from carburizing.

The slightly elevated surface hardness is most likely due to a higher than typical carbon percent near the surface of the carburized case. A typical carburizing cycle for X-2M steel consist of two stages which can be termed 'boost' and 'diffuse'. The first stage of high carbon potential, known as "boost", is followed by a second lower carbon potential stage, known as "diffuse", that allows better control of the microstructure. The carburization process used for the test samples evaluated herein did not receive a boost and diffuse process, only the boost step was performed as the purpose of the test was to determine if the integral preoxidation was functionally effective in facilitating uniform carburizing. The single stage carburization process with high carbon potential and no subsequent diffusion step is considered a contributing factor to this result.

Conclusions

Preoxidation prior to carburization is used in the processing of gears made from X-2M steel for consistent production of components with uniform case depths and high surface hardness. The preoxidation treatment introduces a high temperature oxide at the surface of the component, which promotes uniform carburization across the entire part surface. Integral preoxidation treatments are currently employed with the processing of several carburized aerospace gear steels, allowing manufacturers to eliminate additional manufacturing costs associated with stand alone preoxidation and carburization procedures.

Small scale tests were performed to determine the effectiveness of integral preoxidation and carburization of X-2M steel. Integral preoxidation of X-2M samples at 1500°F followed by carburization at 1700°F did not produce uniform carburization. Integral preoxidation of X2M samples at 1800°F followed by carburization at 1700 °F yielded samples with uniform carburization.

Future Work

Future work to be performed regarding integral preoxidation of aerospace gear steels includes the following:

- In-depth literature review of the thermodynamics and kinetics driving high temperature oxide formation in aerospace gear steels.
- Characterization of the metal-oxide formed during preoxidation.
- Continue development testing of integrally preoxidized and carburized test specimens using boost and diffuse stages to facilitate comparison with current manufacturing processes.
- Continue the experiments using integral preoxidation cycles will be performed using preoxidation temperature between 1600°F and 1800°F.
- Evaluate the effect of integral preoxidation on Copper plating integrity. Copper, typically used as a masking material during carburizing degrades at the preoxidation temperature used currently (1800°F). If a lower preoxidation temperature proves successful, copper plating may be used in the future for masking areas of a part made of X-2M steel that are to remain uncarburized. Current practice is to mask using machining stock in these areas, carburize, and remove the carburized layer by machining later in the gear making process.

Acknowledgments

The authors would like to acknowledge the following individuals for their contributions and assistance with this study: Sarah M. Byrnes, Julia C. Hess, Alexander G. Hollenbach, James Ireland, Thomas R. Kachelries, and Bounem Saysongkham.

References

[1] R. J. Cunningham and W. Lieberman, United States Patent 3,885,995, 'Process for Carburizing High Alloy Steels', May 27, 1975.

[2] M. L. Schmidt, 'Preoxidation of Pyrowear 53 Before Gas Carburizing: Theory and Its Effect on Pyrowear 53 Alloy', Journal of Heat Treating, Vol. 8, No. 1, 1990.

[3] G. Krauss, 'Steels: Heat Treatment and Processing Principles', ASM International, 6th printing, 2000

[4] G. Parrish, 'Carburizing: Microstructures and Properties', ASM International, December 1999

Proceedings of the 26th ASM Heat Treating Society Conference
B.L. Ferguson, R. Jones, D.S. MacKenzie, and D. Weires, editors
https://doi.org/10.31399/asm.cp.ht2011p0115

New Carbonitriding Processes

Prof. Franz T. Hoffmann, Matthias Steinbacher, PD Brigitte Clausen, Sebastian Bischoff, Dr. Heinrich Klümper-Westkamp and Prof. Hans-Werner Zoch
Stiftung Institut für Werkstofftechnik (IWT), Bremen
hoffmann@iwt-bremen.de, 0049 421 218 5330

Abstract

In the last centuries carbonitriding was mainly used to enhance the hardenability of unalloyed steels. IWT developed gas-carbonitriding and low-pressure-carbonitriding processes to increase fatigue behavior and quality compared to case hardening.
For example, modern gas-carbonitriding processes make it possible to extend materials´ strength, so that the limit of use of a given alloy can be expanded. The paper shows examples for the treatment of ball bearing and case hardening steels. The treatment results in microstructures, which are unusual, compared with conventional heat treated parts. They are characterized by high amounts of retained austenite and carbonitride precipitations. By a controlled process, which has been developed in cooperation with PROCESS-ELECTRONIC, it is possible to adjust surface carbon- and nitrogen content independently.
Low pressure carburized parts have the advantage that no internal oxidation occurs. So they have the potential of leading to a higher strength. Nowadays LP-carburizing is used in a wide range, whereas LP-carbonitriding processes are at a starting point. In this paper possibilities and limitations of this process are shown. So, inline controlling of LP-processes in a classical way is not possible, but simulation guided process control. The paper will give examples for LP-carbonitriding processes and the resulting microstructure.

Introduction

Carbonitriding is mainly used to enhance the hardenability of unalloyed steels. IWT developed carbonitriding processes with the focus on fatigue behavior and quality. This is done by using atmosphere compositions outside the usual range and controlled atmospheres. Today, when carbonitriding, constant ammonia additions, e.g. 2 to 5 Vol.-% are added to the carburizing atmosphere without controlling or sensoring the nitriding potential of the atmosphere [10].
The presented processes are controlled processes, where carbon and nitrogen potentials of the atmosphere are controlled. This allows to adjust surface-carbon and -nitrogen-contents to desired values independently and guarantees constant product quality.

Fundamentals

Nitrogen lowers the A1-Temperature in the system Fe-C and expands the phase field of austenite. This allows using lower carbonitriding and hardening temperatures compared to carburizing (Fig. 1).

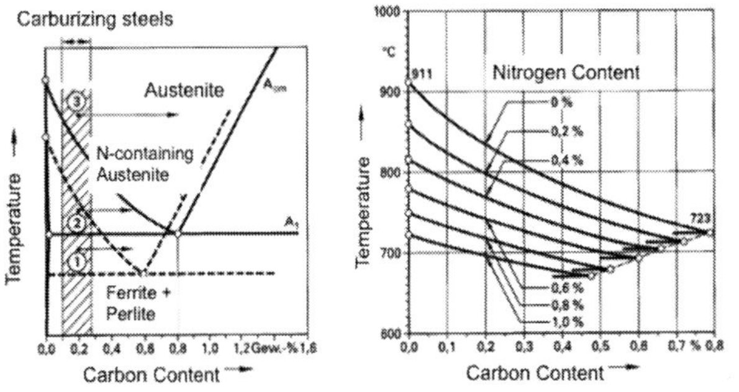

Fig. 1:Influence of Nitrogen on A1- and A3-temperature in the system Fe-C-N [1]

Besides this, by the addition of Ammonia to the carburizing system the carburizing reaction and therewith the carbon potential are influenced by lowering the CO_2-content of the atmosphere as main effect and additionally lowering the CO-content.

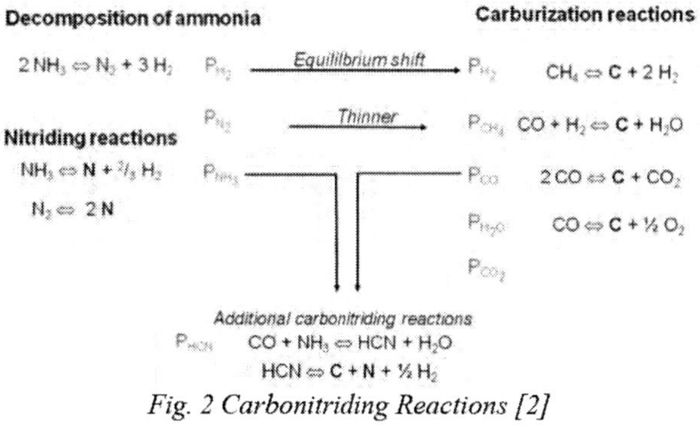

Fig. 2 Carbonitriding Reactions [2]

Fig. 2 shows the most important reactions of the nitriding process, the carburizing process, and the combined reactions. It can be seen that an additional combined reaction occurs, which has as well a nitriding part as a carburizing part via HCN. This interaction of the carburizing and nitriding reactions effects, that the ammonia addition influences the carbon potential (Fig. 3). In practice the interdependence of ammonia addition and carbon potential is experimentally determined for each plant to adjust the carbon potential, because these reactions take place not only at the surface of the batch but also at the inner surface of the furnace and at the charging rack. So, the same addition of ammonia to a carburizing atmosphere may result in different surface carbon contents.

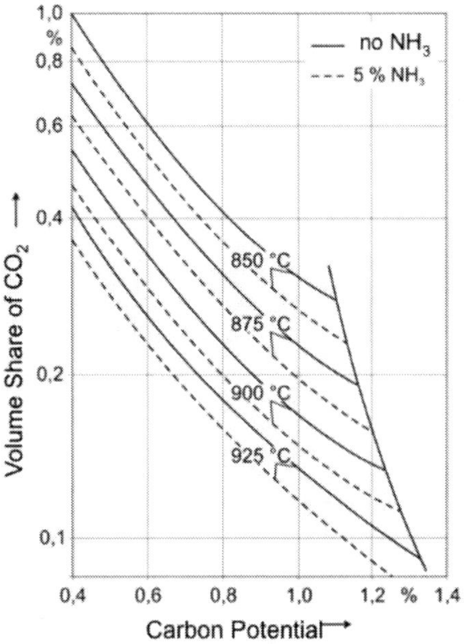

Fig 3: Influence of ammonia addition to the carburizing atmosphere on the carbon potential [3]

As shown before, Nitrogen expands the phase field of austenite to lower temperatures also. As the same addition of ammonia leads to higher Nitrogen contents in the steel, today carbonitriding is often done in a lower temperature range (Fig. 4). A second reason for using low process temperatures is, that the solubility of nitrogen in austenite decreases with increasing temperature. The disadvantage is the low diffusion velocity, so that these processes cannot be used to generate higher case hardening depths economically.

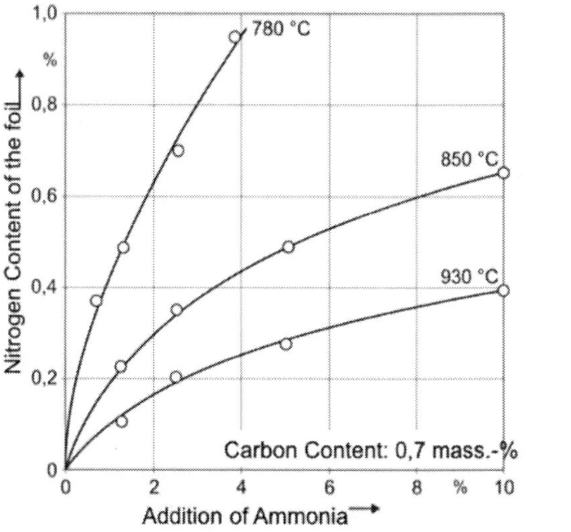

Fig 4: Influence of temperature and ammonia addition to the carburizing atmosphere on the nitrogen content [4]

Another limiting factor for choosing the process parameter is, that pores are often found, if certain amounts of the ammonia addition to the carburizing atmosphere will be excessed, as shown in fig. 5. This demand is the reason, that in practice the ammonia addition is limited to 2 – 5 Vol.-%.

Another important effect of carbonitriding is the increase of the hardenability by nitrogen. For this reason non or low alloyed steels are carbonitrided to increase the hardenability of the steel.

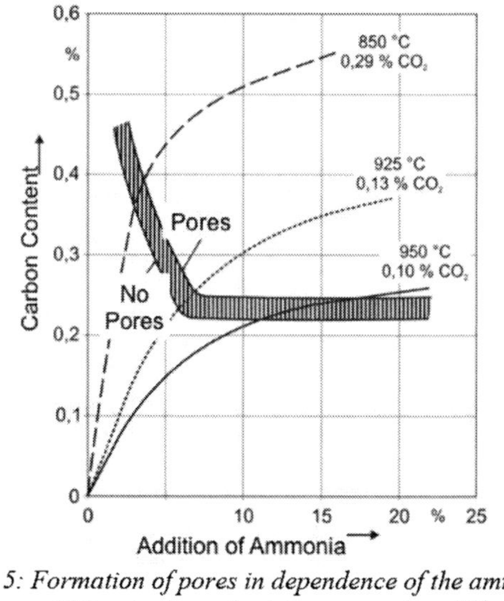

Fig. 5: Formation of pores in dependence of the ammonia addition to the carburizing atmosphere [5]

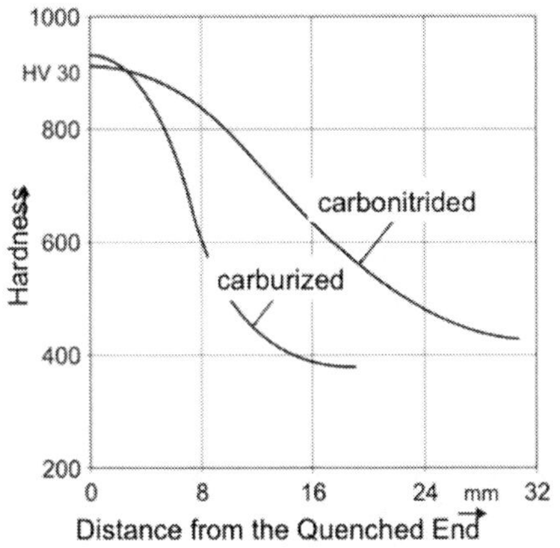

Fig. 6: Comparison of the hardenability after carburizing and after carbonitriding [5]

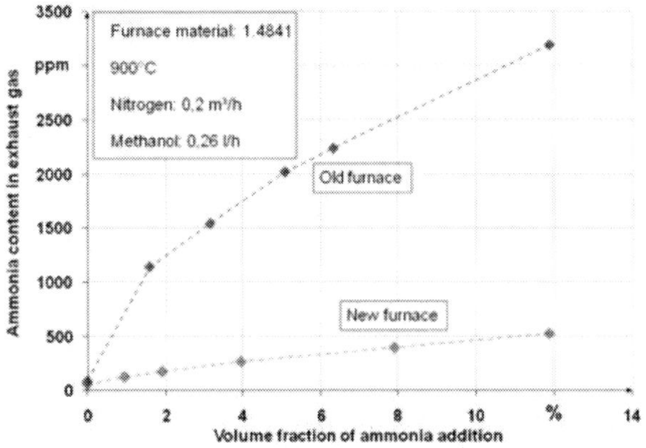

Fig. 7: Comparison of the amount of ammonia in the exhaust gas carbonitriding in an old and in and new muffle [2]

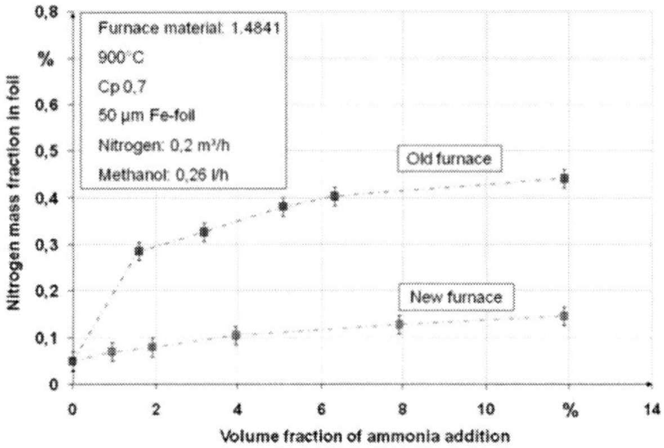

Fig. 8: Nitrogen content in foils depending of the state of the muffle material and the ammonia addition [2]

The aims of the processes described in the following had not been to increase the hardenability of non-alloyed steels, but increasing mechanical properties of carburizing steels as used for gears and of the ball bearing steel AISI 52100 (DIN-EN 100Cr6) to enhance life time. Further advantages can be reached using a controlled carbonitriding process.

Controlled Carbonitriding

Usually when carbonitriding fixed ammonia additions are used and the different carbonitriding behavior of the different furnaces are not taken into account. Fig. 7 shows the results of identical carbonitriding processes in 2 bell-type furnaces of the same size. The only difference is, that one furnace had an old muffle and the other a new one. Material of both muffles had been DIN EN 1. 4841. The surface of the new muffle is highly active; this results in high ammonia dissociation at the muffle, so that leaving only less ammonia to be detected in the exhaust gas. The material of the old muffle is saturated and less active due reactions during former carburizing and carbonitriding processes. So, dissociation takes mainly part at the surface of the batch. Identical processes- this means time-temperature profiles and time-atmosphere-composition of the gassing-profiles – lead to different compositions of the exhaust gas (fig.7) and therewith to different nitrogen contents in the surface of the heat treated parts (fig. 8).

By recent investigations of IWT Bremen together with process-electronic [8] a controlling system for carbonitriding systems has been developed. The system bases upon diffusion calculation. Data, necessary for the calculation are delivered by exhaust gas analysis. Additionally to the data used for carburizing ammonia is analyzed. Fig. 9 shows, that

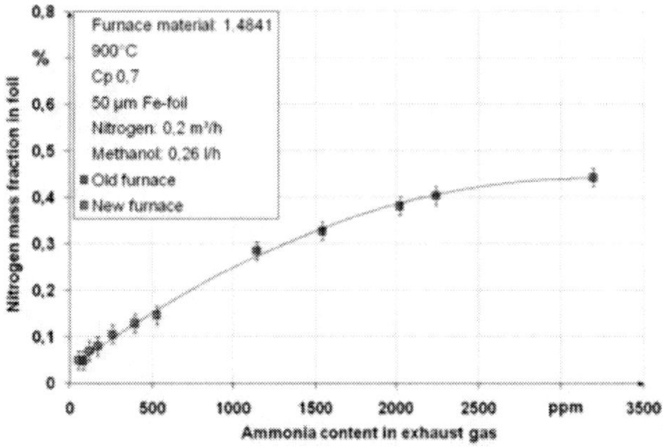

Fig. 9: Correlation between nitrogen content in a foil and the ammonia content of the exhaust gas [2]

it is possible by this method to control surface-carbon and -nitrogen content of the parts independently from the state of the furnace and the size of the batch. This method enables to adjust carbon- and nitrogen-content in the thermodynamically possible range independent from state of furnace, and size of the batch. Further on, the results showed, that it is possible to predict pore formation, based on thermodynamic calculations, as publishes by Slyke [6, 7]. The enormous scattering of the results, when treating with constant ammonia additions, can be minimized or dispelled. Besides this, it can be predicted, if pores will occur. Fig. 10 shows the limits for pore formation. In this figure experimental results were plotted, which showed no pores. Ammonia additions in the range, from which is said, that pores would occur, do not produce pores, when the nitriding potential is in the field of no formation of pores (fig. 11). Reason is, that there is no definite interdependence between ammonia content of the process gas and the nitrogen potential at the surface of he parts because of the side reactions at furnace parts and charging rack. Only controlled atmospheres would lead to reproducible results.

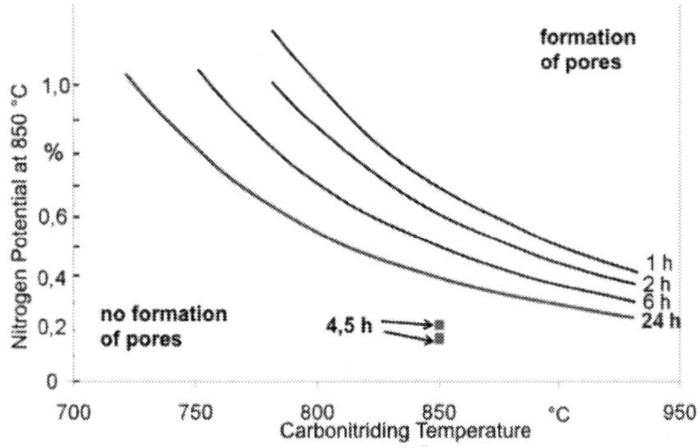

Fig. 10: Limits for pore formation in dependence of temperature and nitrogen potential [2], [6], [7]

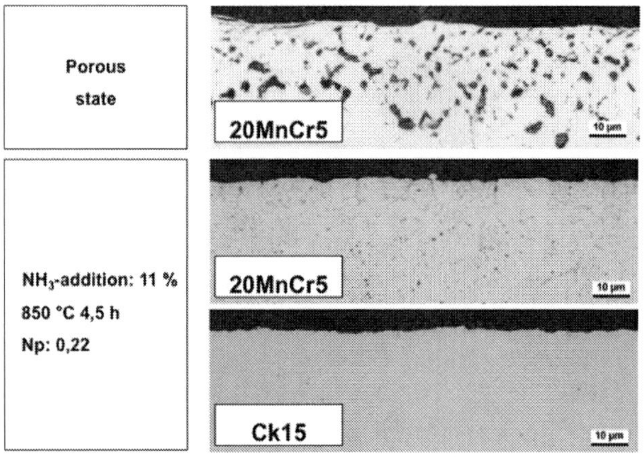

Fig. 11: Carbonitrided surface with and without pores

Carbonitriding of ball bearing steel AISI 52100

In future, besides enhancing hardenability increasing mechanical properties will be the new field for carbonitriding, if there is multiaxial load, e.g. combined Hertzian pressure and bending. Investigations [3] showed, that carbonitriding with the aim to produce very small fine dispersed carbides and carbonitrides (about 30 %) in a martensitic matrix with a content of retained austenite around 45 % an increase of lifetimes can be reached for dirty operating conditions. Fig. 12 gives an example. The samples had been heavily loaded in a testing machine. Normal life times under these test conditions had been 50 -75 h. The possible increase in lifetime is shown in fig. 14, where REF is the usually treated part and RA2 the special carbonitrided part. An increase of lifetime more than 2 times could be found.

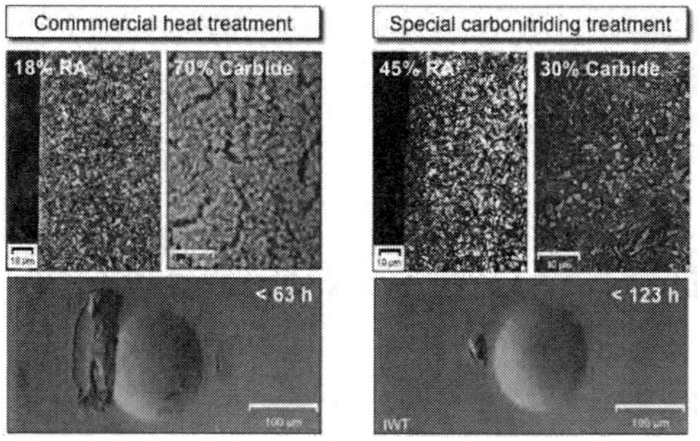

Fig. 12: Failure behavior of commercial treated and special carbonitrides ball bearing rings [9]

By carbonitriding AISI 50100, the outer surface is free of carbides and contains nitrides. As nitrogen increases the carbon potential, there is an "uphill" diffusion of carbon towards the core. So, below the nitride containing outer surface layer with low carbon content follows a carbide rich zone (fig. 13). The restriction of carbonitriding such parts is, that the case hardness depth is limited. So, it can be, that the grinding stock to be removed because of distortion may be too high, so that carbonitriding cannot be used.

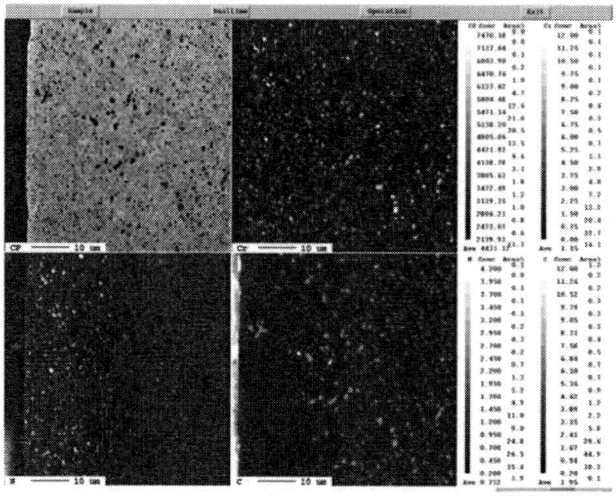

Fig. 13: Distribution of N, C, Cr in the surface region of special carbonitrided AISI 52100

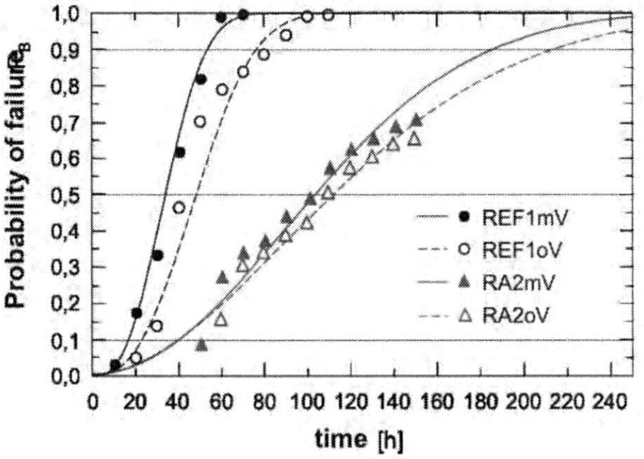

Fig. 14: Ball Bearing Rings with intendations, probability of failure. Ref: carburized, RA2: special carbonitrided [9]

Carbonitriding of Gears

Fig.15 shows a today unusual process scheme for carbonitriding gears. Depending on the combination of carbon- and nitrogen potentials different C- and N-profiles can be generated, as exemplarily shown in Fig. 16.

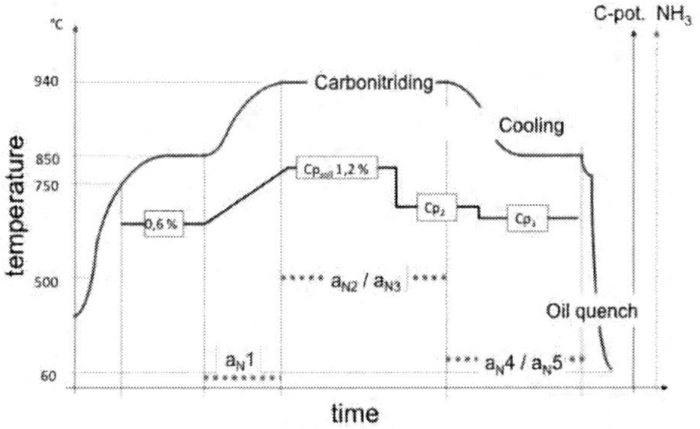

Fig. 15: Carbonitriding scheme for treating gears [10]

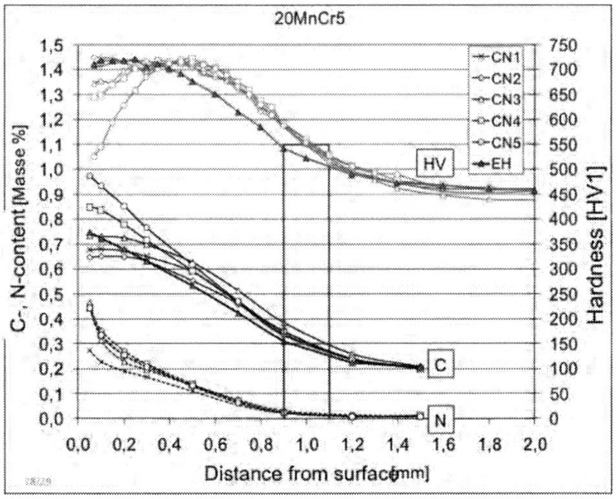

Fig. 16: C- and N-profiles depending on process parameters [10]

At these processes the interaction of nitrogen and carbon activities has to be taken into consideration. At high nitrogen activities carbon will diffuse towards core, because nitrogen increases the carbon-activity.

Treating gears by this method will have no positive effect at the root of the gears because of uniaxial stress conditions, but lifetime experiments showed advantages at multiaxial stress conditions as existent at the flanks. In some cases it may be a method to enhance fatigue life, especially when the fatigue strength of the material is near the limit.

Low Pressure Carbonitriding

Low pressure carburized parts have the advantage that no internal oxidation occurs. So they have the potential of leading to a higher strength. Nowadays LP-carburizing is used in a wide range, whereas LP-carbonitriding processes are at a starting point. One reason is, that existing vacuum furnaces have to be reconstructed for carbonitriding: Material and vacuum pumps have to be consistent against ammonia. The other reason is, that the nitrogen solubility depends on the pressure accordingly Sieverts Law, nitrogen solubility decreases with the root of the pressure. Fig. 17 shows this correlation by an example.

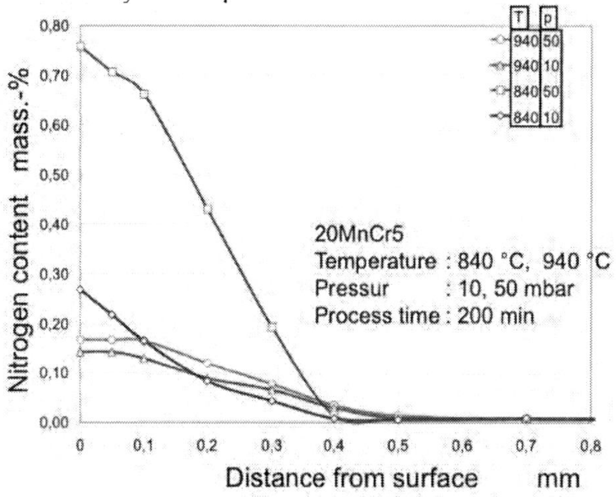

Fig.17: LP-carbonitriding: dependence of nitrogen intake on temperature and pressure [10]

So, the process scheme has to consider these relations (Fig. 18). Carbon boost steps are followed by ammonia boost, whereby a certain pressure has to be kept up. During carburizing steps at lower pressure nitrogen effusion can occur. During the last process cycles the parts are nitrided again.

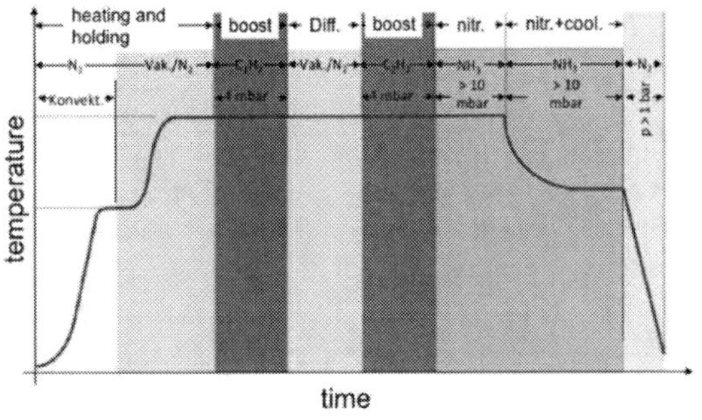

Fig. 18: LP-carbonitriding, process scheme [10]

By this method nitrogen contents up to 0.6 mass.-% can be reached (Fig. 19).

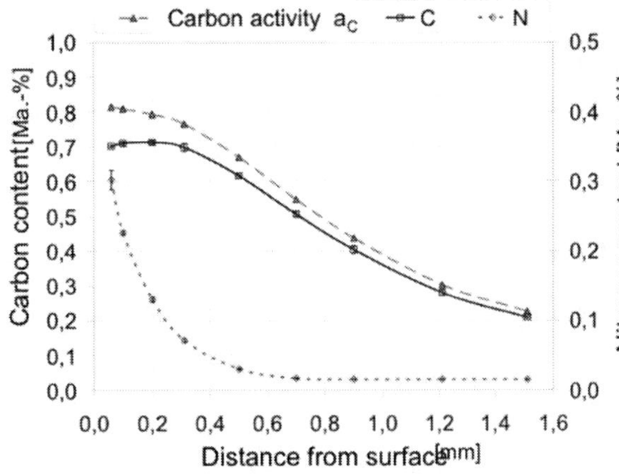

Fig. 19: [10]

Fig.20 shows the surface near region of a carbonitrided steel.

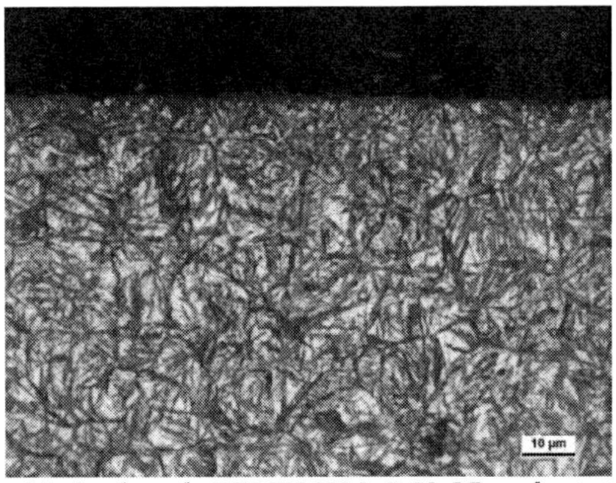

Fig. 20: Steel grade DIN EN 20MnCrS5, LP-carbonitrided at 940 °C, gas-quenched, tempered at 180 °C martensite, retained austenite, fine CN-precipitations

The surface is free of intergranular oxidation. In the stable retained austenite fine carbonitrides are precipitated.

Fig. 21: Steel grade DIN EN 20MnCrS5, LP-carbonitrided at 940 °C, gas-quenched, not tempered, Cr-depletion at the grain boundaries by formation of precipitations at the grain boundaries

If the process control is not optimal, formation of chromium rich precipitations at the grain boundaries take place combined with a chromium depletion at the grain boundaries.
Process control is carried out by process simulation before the treatment. Here the single steps of the process are calculated.

Conclusions

In the past, carbonitriding has been used to enhance hardenability of the parts, fatigue properties had been of minor interest. The process had been done by fixed gassing rates, not considering dissociation at furnace walls or charging racks. This led to results with high scatter bands and low process reliability. New carbonitriding processes raise process stability and reliability and enhance the field of carbonitriding:

- Controlled carbonitriding enables to set defined surface carbon- and nitrogen-contents. So, specified values can be reached safely, and desired properties are in small scatter band.
- Controlled carbonitriding balances different demands on ammonia depending on the state of material an furnace equipment.
- Special carbonitriding with high amounts of stable retained austenite and very small precipitations enhance increase fatigue properties, if parts are treated multiaxially. This could be shown by experiments at high loaded ball bearing under dirty conditions and at gears.

Acknowledgments

We thank the German "Bundesministerium für Forschung und Technologie" (BMFT) for the funding of the project 03X3522C
And we wish to thank the „German Bundesministerium für Wirtschaft und Technologie", (BMWi) for the funding of the project AViF A235 through „Forschungsvereinigung der Arbeitsgemeinschaft der Eisen und Metall verarbeitenden Industrie", (AViF) e.V.

References

[1] Läpple, V., Wärmebehandlung des Stahls, Verlag Europa Lehrmittel, 9. Auflage, 2006

[2] Bischoff, S., Klümper-Westkamp, H., Hoffmann, F., Zoch, H.-W., Development of a sensor system for gas carbonitriding – Part 1, HTM J. Heat Treatm. Mat. 65 (2010) 3, pp. 141-148

[3] AWT-FA5, AK4 (Herausgeber), Die Prozessregelung beim beim Gasaufkohlen und Einsatzhärten, ISBN 3-8169-1454-3,Expert-Verlag, Renningen-Malmsheim, 1997

[4] Chatterjee-Fischer, R.; Schaaber, O., Einige Betrachtungen zum Carbonitrieren im Gas Härterei-Techn. Mitt. 24 (1969) 2, S.121 – 124, 24 (1969) 4, S. 292 – 295, 26 (1971) 2, S. 108 -110

[5] Davies, R.; Smith, C. G, A practical study of the carbonitriding process, Heat Treatment of Metals 1(1978), S. 12 – 14

[6] Slyke, J, Carbonitriding, an investigation from the process point of view, Linköping University, Linköping, 1979, Dissertation

[7] Slycke, J., Ericsson, T., A study of reactions occurring during the carbonitriding process, J. Heat Treating 1981, Vol. 2 (No. 1), pp. 3-19

[8] Winter, K.-M., Independently controlled nitrogen and carbon potential – a new approach to carbonitriding processes, proceedings, heat treat 2011, Ohio, Cincinnati, Nov. 2011

[9] Günther, D., Partikelüberrollung, FVA Forschungsheft Nr.651, FVA, Frankfurt, 2001

[10] Hoffmann, F., Clausen, B., Steinbacher M., Carbonitrieren verspricht Verbesserung der Eigenschaften von Zahnrädern, MM Das IndustrieMagazin 30/2008, pp. 32-36.

[10] Steinbacher, M., Running investigations at IWT, Bremen

Proceedings of the 26th ASM Heat Treating Society Conference
B.L. Ferguson, R. Jones, D.S. MacKenzie, and D. Weires, editors
https://doi.org/10.31399/asm.cp.ht2011p0122

DETAILED SPECIFICATIONS FOR GLOBAL HEAT TREATMENT SOURCING AND MATERIALS

Jared Sponzilli
Navistar, Inc, Melrose Park, IL USA
Jared.sponzilli@navistar.com, 708-865-4018

John Sponzilli
Warrenville, IL USA
jsponzi@sbcglobal.net, 630-447-9144

Abstract

The very nature of global sourcing means that components must carry clear and detailed specifications for material, heat treatment, and test methods. Qualified global heat treat facilities can achieve good control of not only the common features such as surface and gradient hardness, but also of microstructure, core hardness, residual stress, and other critical metallurgical parameters. This paper will discuss a new concept for material specifications and more detailed heat treatment specifications for the global marketplace.

Introduction

The key word in the title of this paper is the word DETAILED. It's where the heart of all good specifications lies, and where many engineering drawings fall short. Drawing data such as dimensions, tolerances, bills of material, and assembly instructions must be correct. Engineers know just how critical these parameters are and they take pride in their accuracy. After all, if dimensions are incorrect, tolerances are too large, or the bill of materials is incomplete, parts cannot be made or simply won't work. But when it comes to materials and heat treatments, designers are often willing to copy old specs that date back years or even decades, or casually write new ones that are incomplete or vague.

Have you ever seen a drawing of a carburized part that called for something like this?

Carburize 57-63 HRC
Effective Case Depth 1.0-1.5

The heat treater might ask:

- How is surface hardness to be measured?
- Are oxide networks permitted?
- What surface microstructure is permitted? Is any transformation product allowed just below the oxide network?
- How much retained austenite is allowed?
- What core hardness is required?
- Does the spec allow carburize, slow cool, reheat and quench?
- Is grinding after carburizing permitted? Is it desired?

The list of questions can get pretty long and the answers have a lot to do with the kind of heat treat shop that will be capable of producing a part and how the part will ultimately perform. One can only imagine how even more confusion can be introduced when a specification is translated to another language to facilitate global sourcing.

Material and heat treat specifications are every bit as important as dimensioning and tolerancing when it comes to ensuring quality components. Vague or incomplete specifications may be adequate with a captive shop or for sole sourced components, but as soon as a part is sourced to another supplier, or the supplier resources the heat treatment of his parts, problems can arise. This paper will provide some guidelines for providing the supply base with comprehensive, clear and concise material and heat treatment specifications.

Material Specifications

When a supplier produces a part, he is required to meet the specifications for the print material and sometimes to verify it. What happens if a part is designed in Germany with a DIN spec and produced in the US? Can the US equivalent material be used?

Technically, the answer would be 'no', but in practice, using the US equivalent may be the only option.

There are many types of material specifications including national, regional, technical society, corporate, and generic, among others. Some of the better known standards organizations include ASTM, ISO, DIN, BS, and JIS. The European community has now standardized on Euronorms (EN), but some national and society specifications continue to be used.

In general, the more common a material is, the fewer differences exist between specification requirements under different specification systems. Hence, common materials will have more equivalents. For example, most specification systems have similar chemical requirements for common steels, such as 1018, 8620 and 4140 or for iron castings, such as gray iron and ductile iron. It is possible, however, that some specification systems for these materials include requirements for properties such as internal cleanliness and some do not. Some specifications may have requirements for minimum tensile and yield requirements and some do not. Even though the grades are similar, properties may vary by as much as 10% from one specification to another. A few examples are yield strength for as cast ductile iron, tensile strength for an as cast gray iron and tensile and yield strength for aluminum sheet. If certain properties are required and must be verified, then care must be taken when allowing different but equivalent material specifications for that application.

Sometimes it is possible to add special requirements to a material with little cost penalty if the requirements are within normal production practices and most producers can achieve them routinely. It is usually more feasible to do this if you have a stable supply base that is used to handling special restrictive requirements. The wider the global supply base is for a given component, the more difficult it will be to apply special requirements to component material specifications. For example, today most H-band steels are made to a restricted chemistry and hardenability, even though they may not be specified or sold that way. If tighter control of incoming material (to help with the control of heat treat results or distortion) is required, it is entirely possible to specify half band hardenability with little or no cost penalty (reference SAE J1868). Of course, this may require buying a large quantity of steel or sourcing the material with only one or two main suppliers. Another common but critical parameter that can be restricted is carbon content, which is important for heat treat response and repeatability as well as core hardness and strength. While many standard steels are specified with .07% carbon ranges, this can be narrowed to .05% with little cost or procurement impact. As a general rule, mid range restrictions are preferred over high side or low side restrictions. For steel with a specification range of .16-.23% carbon, a restriction of .17-.22% might be applied.

Equivalent Materials

One means of simplifying the material conversion issue for the global supply base is to establish a table of equivalent materials. Ideally, this is in place before the design stage so that differences between various material specifications for a given material can be taken into account during design. The designer would know that his part could be made from a material produced under one of several different global materials/specifications. Then he could design for the material, within the group of equivalent materials, with the lowest properties or the one with poorest formability, etc. The design's "safety factor" would have to accommodate different materials with slightly differing properties.

When Case Corporation and New Holland (CNH) merged in 1999, this is the approach that was taken. After several years of development, the material specs were successfully combined so that parts could be specified with a grade of material that has several global equivalents. Depending upon where the part would be made, the appropriate local specification could be used to make the part. Today, engineers at CNH understand that when they specify a material for a part designed in North America, the part may be produced in Europe, South America, or elsewhere with equivalent materials. Since the specific materials that can be used are clearly defined in the CNH specification, including properties, the engineer can take the alternate materials into account at the design stage. Note that only particular alternate materials may be used for a part as designated in the CNH specification. Some critical parts require separate, unique specifications, and some drawings must restrict material usage. After nearly 10 years, this method has improved greatly on previous approaches that sometimes resulted in incorrect materials being substituted by the global supply base.

Table 1 shows a few examples of materials that are equivalent to one another and can generally be used interchangeably.

Table 1

TYPE / SPECS	HR Carbon Steel – 1045	HR Low Alloy Carburizing Steel	Gray Iron – mid range strength
ASTM/SAE	A576,1045	A322, 8620 A304, 8620H	SAE J431, G10H18 or G11H18 (G3000)
Euronorm	EN10083-1, C45E	EN10084, 16MnCr5+H or 20NiCrMo2-2+H	EN1561 EN-GJL-200
DIN	DIN17200, Ck45	DIN17210, 16MnCr5 or 21NiCrMo2	DIN1691, GG20

There are many such tables of equivalency in existence today, but to our knowledge, none have been sanctioned as preferred on a global basis. This would be an excellent next step in the development of global specifications with clear equivalencies and minimum design properties.

Guidelines

When specifying a material, a few simple rules will help make sure that parts can be produced by as many suppliers and in as many regions as possible. These include:

- Specify the most common material possible
- Use national or regional specifications that have the widest global use, such as Euronorms, JIS, or ASTM; avoid company specifications, especially for common materials, because they are inherently unique and have limited availability
- While often beneficial and sometimes required, restrictions and special requirements will limit availability, so avoid restrictions or apply them sparingly
- If restrictions are applied to a material, attempt to make sure that the restrictions are within normal limits for current production capabilities for the material
- If possible, furnish a list of alternate materials / specifications that will also be acceptable for your parts; use caution and understand the differences (properties, test requirements, etc.) between your original specification and the alternates allowed
- Never place a material on a drawing without an accompanying specification reference
- Understand that some material conditions and product forms can be quite common in one country and virtually not available in most others. One such example is cold drawn steel bar, which is available in the US, but has limited availability and different properties elsewhere.

Heat Treatment Specifications

A heat treatment specification used to procure parts from a global supply base should be as detailed and explicit as possible. The purchaser should specify the type of heat treatment, metallurgical and mechanical property requirements, and methods of test, at a minimum. A poorly written specification can result in inconsistent part quality, time lost for repeated clarifications between the purchaser and the supplier, or in the worst case, unexpected failures.

Too often, design margins are just that - margins on the physical design of the part based on expected minimum mechanical properties, cross sectional area, and geometry. They generally do not account for metallurgical variations resulting from heat treatment that can arise through normal variation, poor quality control or vague heat treatment specifications. With modern heat treatment practices and improved process controls, it is possible, and indeed mandatory, to develop specifications that utilize these improvements and take full advantage of the supply base capabilities. A good heat treatment specification can improve confidence in the design and justify the use of lower design margins, which in turn can result in weight and cost savings.

Definitions

Any specification should incorporate a list of definitions explaining key terminology. This includes a definition for the heat treat process itself, as well as any terms that may be difficult to translate directly to other languages, such as decarburization, pitch line, runout, etc.

Process Details

Certain general requirements dealing with quality systems and equipment should be common to all heat treatment specifications.

Following are a few examples of general requirements:

- The supplier shall have a quality system in place that complies with or is equivalent to ISO/TS 16949 (or as appropriate for the industry).
- All processing equipment shall be capable of measuring and recording temperature, time, and atmosphere composition.
- All processing equipment shall be tested and certified to AMS 2750 or an equivalent internal or industry standard for pyrometry.

Usually, engineering specifications do not specify process details, preferring to leave them the supplier. For the most part, process parameters in a heat treat specification should be limited to those items that are unique to the process or material and are absolutely required to produce the desired result.

For example:

- Steel parts with an as quenched hardness of 45 HRC or greater shall be tempered within 30 minutes of quenching.
- Cryogenic treatment is required for carburized parts containing >2.0 wt% Ni

In rare cases, such as austempering which has highly developed thermal cycles, it may be necessary to specify even more details of the process such as time, temperature, heating and quenching conditions, etc.

Metallurgical Requirements

This is where the heart of the specification lies, the metallurgical features that need to have specific, well defined, and measurable results. Properly written metallurgical requirements are what allow specifications to have limited processing requirements. The supplier will have to be capable of controlling his process very closely, if they are to meet detailed metallurgical requirements.

The following tables provide some examples of detailed heat treat specification parameters for Quench and Tempering (Table 2), Carburizing (Table 3), and Induction Hardening (Table 4). Note that not all parts will require the level of detail shown in these tables. For example, a heat treat specification for a carburized wear part will not require the same level of microstructure control that a carburized gear might. This paper does not attempt to deal with these differences.

Table 2. Suggested elements for quench and tempering specification

Critical Feature	Requirement	Test Method	Comments
Surface Hardness	Should be specified on drawing based on part strength requirement. 5 point HRC range is typical for standard steels (e.g. 30-35 HRC).	ASTM E10; ASTM E18	• Indicate the correct scale for testing based on material, heat treatment, and part geometry. • Surface hardness is required over the entire part, but a specific test location may also be specified. • Measure after tempering.
Core Hardness	• Depending on material and part geometry, core hardness may differ from surface hardness. • Location and requirement should be specified on the drawing; 10 point HRC range is typical	ASTM E10; ASTM E18	• Indicate the correct scale for testing based on material, heat treatment, and part geometry. • Core hardness should be specified at a specific depth below the surface.
Surface Microstructure	• All surfaces should contain tempered martensite. • No decarburization or carbon enrichment that would change surface hardness or microhardness measured at 0.2 mm depth	light microscopy at 100X with appropriate etch ASTM E1077	• Decarburization may be acceptable on some less critical applications. • Critical fasteners may require more stringent control on decarburization.
Core Microstructure	• Parts shall exhibit a tempered martensitic microstructure to a depth within the hardenability limits of the specified steel. • Primary ferrite, which is evidence of incomplete austenitization, is not acceptable.	light microscopy at 100X with appropriate etch	Core microstructure, like hardness, is dependent on material and geometry.
Grain Size	ASTM No. 7 or finer.	ASTM E112 or equivalent	
Tempering	Required, unless otherwise stated.		Temper as required to achieve hardness

Table 3. Suggested elements for carburizing specifications

Critical Feature	Requirement	Test Method	Comments
Surface Hardness	58-62HRC	ASTM E18 or conversion from microhardness reading at 0.1 mm below the surface.	Direct readings preferred, superficial loads should be used for shallow cases.
Core Hardness	Application dependant: Fatigue critical component, 32-42 HRC Other, 25 HRC minimum	ASTM E18	Requires cross sectioning to measure
Effective Case Depth (ECD)	Application dependant: • See AGMA standards for gears • Appropriate values for other parts should be determined through testing and/or experience. • Maximum ECD should be ~30% higher than minimum ECD	Depth to 50 HRC per ASTM E384	50 HRC is the most common criteria, but others may be defined depending on materials and/or application.
Surface Carbon	Range depends on alloy content: • 0-3%: 0.75-1.00% C • >3%: 0.65-0.95% C	Optical emission spectrograph <0.15mm of surface; carbon bars and Leco analysis.	
Case Microstructure	Tempered high carbon martensite.	Light microscopy @ 500x.	• Evaluate at grinding depth, if appropriate • Reference micrographs can be helpful
Distance to first non martensitic transformation (DFB)	≥30% of minimum ECD		
Carbides	No blocky carbides, no continuous intergranular carbide network.		
Grain size	ASTM No 7 or finer	ASTM E112 for measurement, see ASTM E407 for etching methods	Measured in the case after carburizing.
Retained Austenite	Expected range between 10 and 30%	Visual comparison with standard micrographs at 1000x or via X-ray diffraction after chemical removal of 0.75mm	XRD is preferred for applications where dimensional stability in service is critical (in the μm range).
Intergranular oxide network (IGO)	Allowable limit should be based on ECD: • 0-0.75 mm ECD, ≤ 20 μm • 0.75-3.0 mm ECD, ≤ 25 μm • >3.0 mm ECD, ≤ 30 μm	light microscopy @ 500x	Reference micrographs can be helpful
Surface Non Martensitic Transformation Products (NMTP)	Limits based on application: • Fatigue critical components, none • Other, ≤ 20 μm	light microscopy @ 500x	Reference micrographs can be helpful

Table 4. Suggested elements for induction hardening specifications.

Critical Feature	Requirement	Test Method	Comments
Surface Hardness	• Should be specified on drawing based on part strength requirement. • 5 point HRC range is typical for standard steels (e.g. 50-55 HRC).	ASTM E10; ASTM E18	• Indicate the correct scale for testing based on material, heat treatment, and part geometry. • Measure at the ends and center of hardened zone. • Measurements should be made after tempering, when performed.
Effective Case Depth	Application dependant: • See AGMA standards for gears • Appropriate ranges for other parts should be determined through testing and/or experience	Depth to 40 HRC microhardness per ASTM E384 for case depths up to 3.0 mm; Rockwell traverse per ASTM E18 acceptable for deeper case depths	• Alternative criteria may be appropriate depending on material or application such as: ○ Depth to 45 HRC ○ Depth to (Surface Hardness – 7 HRC) • Full depth hardness is required underneath any features (keyways, splines) unless otherwise specified
Visual Case Depth	Measure on a perpendicular section through hardened zone and through runout areas	Macroetch per ASTM E340	Full depth hardness pattern required along the entire hardened zone
Runout Zones	Areas of pattern runout may be specified on the drawing, if needed; hardness should taper to base hardness within a distance along the surface of 1.5X effective case depth		Specify areas in the runout zone that do not permit hardening
Case Microstructure	Essentially tempered martensite; no primary ferrite to a depth equal to the effective case depth; irregular parts may require some relief on microstructure	Light microscopy @500X magnification; nital etch	
Grain Size in Hardened Zone	Grain size ASTM No.7 or Finer; corners and difficult to harden areas may require ASTM No. 5 or finer	ASTM E112	
Cracking	None permitted	Magnetic Particle Inspection, ASTM E709 or equivalent	

Supplier Qualification

A thorough and well thought out specification does not ensure the receipt of a quality heat treated component in and of itself; supplier selection is a critical part of the process. While suppliers are usually selected based on their ability to meet requirements for cost, lead time and capacity, technical capability can be overlooked. To help ensure that technical capability plays a key role in supplier selection, it is advisable to develop a set of guidelines describing what type of equipment, personnel, and quality systems are required of the supplier in order to meet the requirements of the specification. This can be conveyed in the form of a "Supplier Questionnaire", or can form the basis of a more formal, on site audit. Specific processes may warrant their own documents, but many of the key points are the same across many processes. As an example, some key points constituting a checklist for suppliers performing atmosphere heat treating can be found below.

Quality System. Various standards organization have developed quality system guidelines. Third party registrars then perform audits and certify that a company or specific location within a company meet those requirements. ISO/TS 16949 is common throughout the automotive industry, while AS9100 is common in the aerospace industry.

Personnel qualification and certification.

- Is there a degreed metallurgist or an equivalent individual who can evaluate and interpret microstructures?
- Have operators and inspectors received (documented) training or certification through ASM, the Metal Treatment Institute or other internal or external source?

Laboratory. If laboratory work is performed in house, does the lab have the following equipment?

- Metallograph for evaluating microstructures
- Microhardness tester (should be part of the internal gage calibration system)
- Suitable sectioning, mounting, grinding, and polishing equipment. For anything but the simplest analysis, samples for microstructural evaluation should be mounted prior to etching and examination. For examination of near-surface features like decarburization or inter granular oxidation, automatic polishing equipment is strongly recommended.
- If laboratory work is subcontracted, are the laboratories accredited to ISO 17025 *for the discipline they are being contracted to perform*?

Processing equipment.

- Is thermal processing equipment maintained per AMS 2750? If not, does a system exist to monitor the following:
- Furnace temperature uniformity
- Thermocouple accuracy
- Instrument accuracy
- Are furnace records maintained that allow traceability of part number, job number, work order, purchase order number, or other unique identifying information to the time, temperature, and atmosphere records?
- Is atmosphere controlling and monitoring equipment suitably calibrated and maintained? Oxygen probes should be burned out at some regular interval, 3 gas (IR) analyzers should have a PM schedule that includes cleaning all sample lines and filters. Is atmosphere control periodically checked via a secondary means such as shim analysis or hot wire testing? Are generators properly maintained?
- Is properly maintained hardness test equipment available to the operators?

Conclusion

The competitive nature of global manufacturing dictates the need for flexibility in sourcing of both materials and processing. Selecting common materials with many global equivalents will help insure that the correct materials are used. It also helps the designer to better account for differences in equivalent global materials at the design stage. A pre-established list of equivalents that are permitted further simplifies the global supplier's material selection.

Because the global nature of the supply base can reduce the amount of direct contact between the engineer and suppliers, the best tools for maintaining control of component quality are the specifications governing their manufacture. Providing clear, detailed, and relatively tight specifications will aid in the selection of a competent supplier and ultimately will ensure consistent part quality.

References

Restricted Hardenability Bands for Selected Alloy Steels, SAE J1868, February 2010, Published by SAE International, Warrendale, PA

Pyrometry, SAE AMS 2750, September 2009, Published by SAE International, Warrendale, PA

ASTM Standard E10, 2010, "Standard Test Method for Brinell Hardness of Metallic Materials" ASTM International, West Conshohocken, PA, 2010, DOI: 10.1520/E0010-10, www.astm.org.

ASTM Standard E18, 2010b, "Standard Test Methods for Rockwell Hardness of Metallic Materials" ASTM International, West Conshohocken, PA, 2010, DOI: 10.1520/E0018-08B, www.astm.org.

ASTM Standard E1077, 2001 (2005), "Standard Test Methods for Estimating the Depth of Decarburization of Steel Specimens" ASTM International, West Conshohocken, PA, 2010, DOI: 10.1520/E1077-01R05, www.astm.org.

ASTM Standard E112, 2010, "Standard Test Methods for Determining Average Grain Size" ASTM International, West Conshohocken, PA, 2010, DOI: 10.1520/E0112-10, www.astm.org.

ASTM Standard E384, 2010e2, "Standard Test Method for Knoop and Vickers Hardness of Materials" ASTM International, West Conshohocken, PA, 2010, DOI: 10.1520/E0384-10E02, www.astm.org.

ASTM Standard E407, 2007e1, "Standard Practice for Microetching Metals and Alloys" ASTM International, West Conshohocken, PA, 2010, DOI: 10.1520/E0407-07E01, www.astm.org.

ASTM Standard E340, 2000 (2006), "Standard Test Method for Macroetching Metals and Alloys" ASTM International, West Conshohocken, PA, 2010, DOI: 10.1520/E0340-00R06, www.astm.org.

ASTM Standard E709, 2008, "Standard Guide for Magnetic Particle Testing" ASTM International, West Conshohocken, PA, 2010, DOI: 10.1520/E0709-08, www.astm.org.

Proceedings of the 26th ASM Heat Treating Society Conference
B.L. Ferguson, R. Jones, D.S. MacKenzie, and D. Weires, editors
https://doi.org/10.31399/asm.cp.ht2011p0131

Quality Improvement in Heat Treatment Based on Necessary Information Exchange

Volker Ermert
Wolf Behälter- und Apparatebau GmbH & Co. KG, Wilnsdorf, Germany
vermert@rjwolf.de

Arnold Horsch
AHOTEC e.K., Remscheid, Germany
info@ahotec.biz

Dieter Klein, Thorsten Wuest
BIBA - Bremer Institut für Produktion und Logistik GmbH, Bremen, Germany
{kln},{wue}@biba.uni-bremen.de

Rainer Kohlmann
Werkstoffberater, Siegen, Germany

Ralph Mahlig
VHK Vakuum-Härterei Köllner GmbH, Schmerbach, Germany
rmahlig@vhkoellner.de

Britta Rentrop
Glüh- und Härtetechnik Unna GmbH & Co KG, Unna, Germany
britta.rentrop@ghtu.de

Abstract

Insufficient and missing information can lead to avoidable costs in heat treatment. The costs can arise on the one hand by failures and on the other hand by time consuming information acquisition. Therefore, it is a wish of the research group 25 "Quality assurance in heat treatment" of the AWT (Association for Heat Treatment and Materials Technology) (www.awt-online.org) to sensitize all contractors of a heat treatment order to exchange necessary information. Thus, this should be reached by first writing down necessary information and second presenting the reasons why the information exchange is so important. This will be done by the illustration of three examples in which the effect of the information on the product result is shown. To provide these explanations to the general public a concept and the first implementation of a knowledge system is introduced. To set up a documentation about this issue this publication is a continuation of [1].

Introduction

Among others successful heat treatment depends on the availability of information about the material, the previous as well as the following manufacturing processes. The basis to accomplish the expected results of heat treatment according to customer requirements, like hardness, is already determined to a large extend by the chemical composition in the steel plant. However, if information about special alloying elements is missing, it may lead unintentionally to failures during the heat treatment. Therefore, with necessary information available it can already be pre-checked, if the customer requirements can be achieved by the heat treatment. Information about a previous manufacturing process can be just as important. E.g. remainder of cooling lubricant on workpieces can lead to scrap after a special heat treatment. Information about the manufacturing processes after the heat treatment can be relevant if a finishing process follows. So if e.g. the information of a following chromium coating process is provided, the annealing process can be adjusted in a way that the performance characteristics correspond to the requirements in the specification. If the above mentioned necessary information is not given it is obvious that the costs increase. To get the needed information e.g. of the chemical composition you may have to ask for or telephone to somebody which is time-consuming. But you can also have costs during and after heat treatment. Think about surprisingly results (e.g. remainder of cooling lubricant) and failures in production (e.g. not consider chromium coating) or parts which will fail before reaching lifetime end. In these cases you have to find out reasons, prove if the heat treatment shop is responsible for the failures and maybe prove innocence by law. All these actions have to be done because of the missing necessary information. The heat treater should receive this information from the customer voluntarily in order to achieve a win-win situation with a satisfying result for all stakeholders. To point out the necessity of this information and to support the voluntary information sharing, the research group 25 "Quality assurance in heat treatment" has taken the task to supply the respective reasons for the necessary information. Based on the table of the IHT (Industrial Association of Hardness Engineering) respective "necessary information of the customers of the heat treatment shop" during placement of an order (http://www.haertetechnik.org) [2] the research group 25 has extended this table by additional influencing factors for a successful heat treatment. The term "influencing factor" is only another expression for necessary information in the case of a successful heat treatment. While [1] focuses mainly on the presentation and explanation of influencing factors of successful heat treatment, this paper focuses on examples for necessary information. Therefore, the influencing factors will be shown but not discussed in detail. Afterwards, three examples will be presented and discussed in detail. A main task is the publication and dissemination of these examples to the general public. Consequently, the examples will be collected in a knowledge system and will be prepared for public access. The structure and the design are presented next to the examples. A conclusion and an outlook on further activities in that case complete this paper.

Influencing Factors of Successful Heat Treatment

Fig. 1 shows an extraction of influencing factors of successful heat treatment [1].

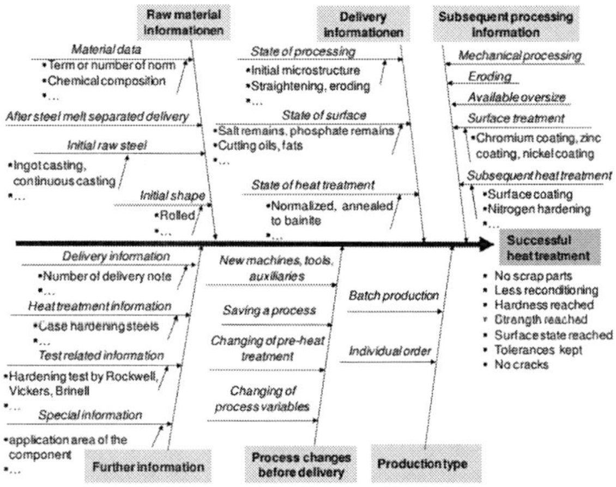

Figure 1: Factors of influence on successful heat treatment

The main influencing factors are given by necessary raw material information up to the production type. Beneath the main influencing factors are further more detailed sub factors like information of material data up to information of the initial shape in the case of raw material information. In the third level the sub factors were named more detailed like information about term or number of norm or information of chemical composition in the case of material data information. But as already mentioned, for a more detailed discussion of the influencing factors see [1]. In this paper three examples of the following influencing factors are given. First: Main influencing factor "Raw material information" - showing the reasons for necessary information in the field of material data. Second: Main influencing factor "Raw material information" and "Process changes before delivery"- showing the reasons in the field of material data and missing information of a changed pre-heat treatment. Third: Main influencing factor "Process changes before delivery" – showing the reasons in the field of "changing of pre-heat treatment". This collection of influencing factors in Figure 1 is the basis of the knowledge system which will be discussed later.

All examples will be presented in the same way. First the material with its background is introduced. Next we have a look at the given information at the time of the order and the customer's specification. After this follows a description of what is shown by the example.

First Example - Material data

This is an example for missing information about material data. The problem here is that only the material specification 51CrV4 (SAE 6150) of the heat-treatable spring steel (hardening and tempering) is given by the customer without any further information about the chemical composition and tempering conditions of the steel. By this example it will be presented how much different information and interpretations are possible about the chemical composition, the hardenability and the tempering behavior.

If the heat treater tries to find out more information about 51CrV4 (SAE 6150) he can find several data sheets e.g. in the internet (see Figure 2) [3], [4], [5].

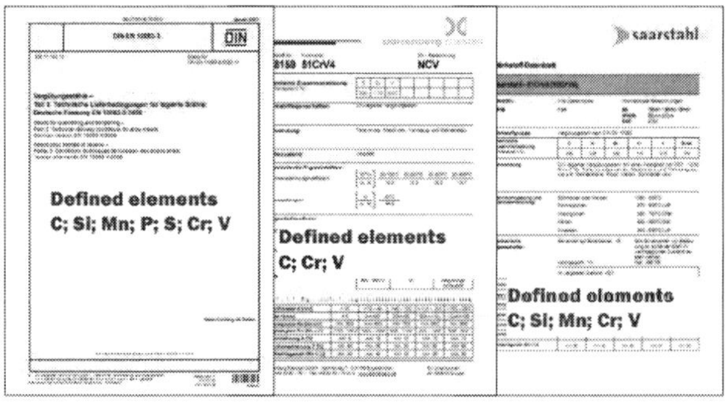

Figure 2: Extraction of material data sheets from internet

Figure 2 shows specifications of the alloying elements of 51CrV4 (SAE 6150) from three different data sheets. The mentioned number of defined elements varies from three to seven. In this case the DIN EN 10083-3 [3] specifies with seven the highest number of elements. Since it is known that the hardenability can be evaluated by knowing the alloying elements the information from these three data sheets is not enough to give the heat treater a sufficient answer about the hardenability. This becomes clear with a look at Figure 3. After [3] the scatter band of hardenability can be divided into three parts. The "HH-grade" is in the top of the specification area and the "HL-grade" at the bottom. For both the scatter band is closer than for the "H-grade" which will combine the other grades seen in the pictures of Figure 3. In the table of Figure 3 it can be read for a Jomminy distance of 1.5 mm that the difference is 8 between the highest and the lowest value.

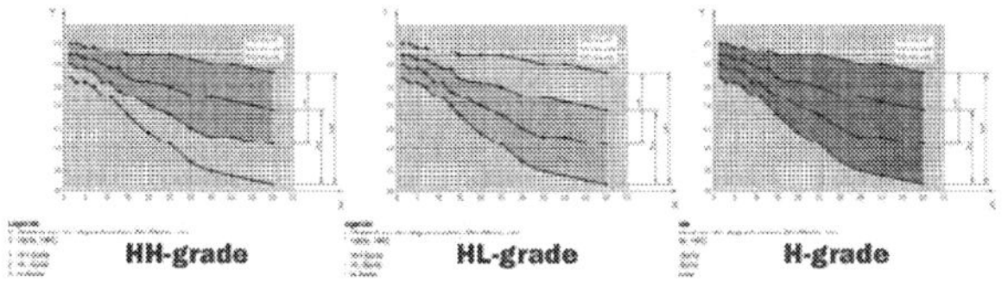

51CrV4/1.8159		1,5	3	5	7	9	11	13	15	20	25	...	mm
H max	HH max	65	65	64	64	63	63	63	62	62	62	...	HRC
	HH min	60	59	59	58	56	56	54	53	50	48	...	HRC
	HL max	62	62	61	61	60	59	59	57	56	55	...	HRC
H min	HL min	57	56	56	55	53	52	50	48	44	41	...	HRC
HH max-HL min	Δ	**8**	**9**	**8**	**9**	**10**	**11**	**13**	**14**	**18**	**21**	...	HRC

Figure 3: Hardenability scatter band of 51 CrV4 (SAE 6150) according to [3]

If a special hardness value is needed there are two ways to find it out. One way to get the necessary information is the certificate of the melt from the steel plant. The other way is to measure the hardness but this costs extra money.
Unfortunately, missing information about 51CrV4 (SAE 6150) can lead to another problem. The material specification 51CrV4 (SAE 6150) has two different material numbers, 1.8159 and 1.2241. This will lead to different tempering curves (see, Figure 4).

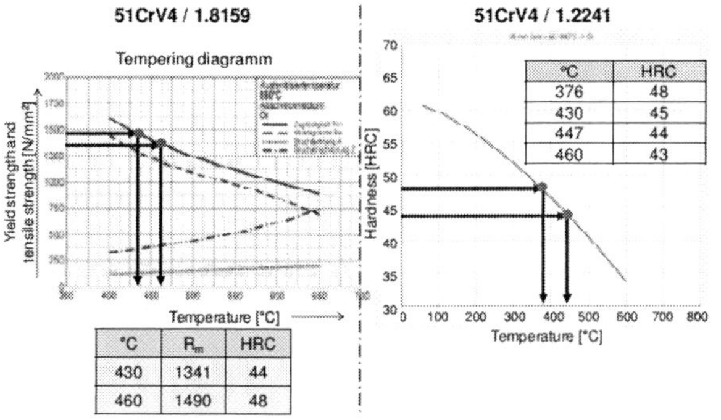

Figure 4: Different behavior for the same tempering temperature, left graphic from [4], right graphic from [6].

Out of the left diagram (1.8159) it can be seen that at 460°C a tensile strength of R_m = 1490 N/mm^2 is given (see also left table). This is equivalent to 48 HRC according to [7]. From the right table it can be seen that 48 HRC are reached at 376°C. The problem occurs if the values of the left table for 1.8159 are used while the 51CrV4 has the specification of the material number 1.2241. Taking the left table a temperature of 460°C is used to get 48 HRC. In the case of 1.2241 (see the right table) a temperature of 460°C will lead to a hardness of only 43 HRC. In both shown cases only further investigations can give an answer about the status of the material. A safe way to reach the customer-demanded hardness consists of three steps:

First step: Temper the workpiece at a low temperature. The workpiece is harder than the specification. Measure the hardness.

Second step: Temper again with a higher tempering temperature to understand the material behavior according to hardness. Measure hardness again and interpret the steep drop of hardness.

Third step: Temper again. Due to the knowledge from the first and the second tempering the specified hardness can be adjusted.

Nevertheless this procedure costs a lot of additional money.

Second Example - Material data and state of heat treatment

This example shows that if customer and supplier discuss a problem it can lead to positive effects. To illustrate this we will use the following scenario. The customer delivers parts out of C45 (SAE 1045 - material number 1.0503 according to [8]) in the status +U (untreated, cooled at air) machined for inductive surface hardening to > 550 HV. Now the heat treatment process ends in pretty much differing results: Sometimes the hardness is not reached. This might be because the carbon content is too low respectively the microstructure of +U has a very large secondary grain size. The customer discusses the problem with the heat treater. Out of this consultation results the suggestion to choose a different material and a different pre-heat treatment. The customer changes to Cf45 (SAE 1045 - material number 1.1193, unalloyed high-grade structural steel for flame and induction hardening) according to [9]. Cf45 usually has a constricted carbon content, defined content of Al = 0.015-0.03 percent by weight and N = 0.008-0.012 percent by weight. The state of delivery is now +MN (thermo mechanical normalized formed) which guarantees secondary fine grain. The question arises, why is the hardness now reached without problems. It will be shown that a defined Al and N content and finer grains lead to higher hardness.
At this point it should be stated that the volumes 1-4 of the atlas for the heat treatment of steels [10-13] are taken as basis of the explanation of the examples two and three. The explanation here is given by [10].
To explain the reason the two melts C45 and Cf45 have to be compared. Due to the fact that in the series of atlas [10-13] no comparison of these two melts is possible another material is used for explanation instead. To elaborate the example two different melts of the material 20MnCr5 (SAE 5120) as a reference to C45/Cf45 were considered which can be found in [10]. The comparison will be performed on the basis of time temperature austenization diagrams (TTA). After explaining the reason on the basis of 20MnCr5 the results and statements of the two melts from 20MnCr5 were reconverted to C45/Cf45. Starting point of the example is the difference in Al und N content of the two melts of the material 20MnCr5 given in Table 1. The Al and N content of melt 1 is very close to the defined content of Cf45. Therefore, melt 1 supplies later the explanation for Cf45. Melt 2 is not in the defined area and provides the explanation for C45. Melt 2 consists of a coarsed

secondary grain, because of the lower content of Al and N [10].

Chemical composition in percent by weight of 20MnCr5 (SAE 5120)											
	C	Si	Mn	P	S	Al	Cr	Cu	Mo	N	Ni
Nominal value min - max	0.17 0.22	0.15 0.40	1.10 1.40	0.035	0.035		1.00 1.30				
Melt 1	0.20	0.18	1.30	0.017	0.033	0.033	1.24	0.21	0.01	0.011	0.02
Melt 2	0.20	0.19	1.19	0.016	0.004	0.004	1.22	0.21	0.01	0.006	0.03

Table 1: Melts 1 + 2 of 20 MnCr5 (SAE 5120) from [10]

The defined state of Al and N leads to a finer grain size and always guarantee the specified conditions of the customer. In detail:

The following explanations refer to a heating velocity between 300°C/s und 1000°C/s. Melt 1 is fine grained and reaches the homogenous austenite earlier than melt 2. In Figure 5 a Vickers hardness of 460HV is reached for a temperature of 910 °C. Melt 2 has coarse grain and reaches 460 HV at a temperature of 970°C (see Figure 6). The same result (460 HV) can be reached 60°C lower by melt 1 which means less energy and costs. Comparing the Vickers hardness at the temperature of 910°C between the two melts the hardness of melt 2 is only 430 HV and therefore 30 HV lower than the hardness of melt 1.

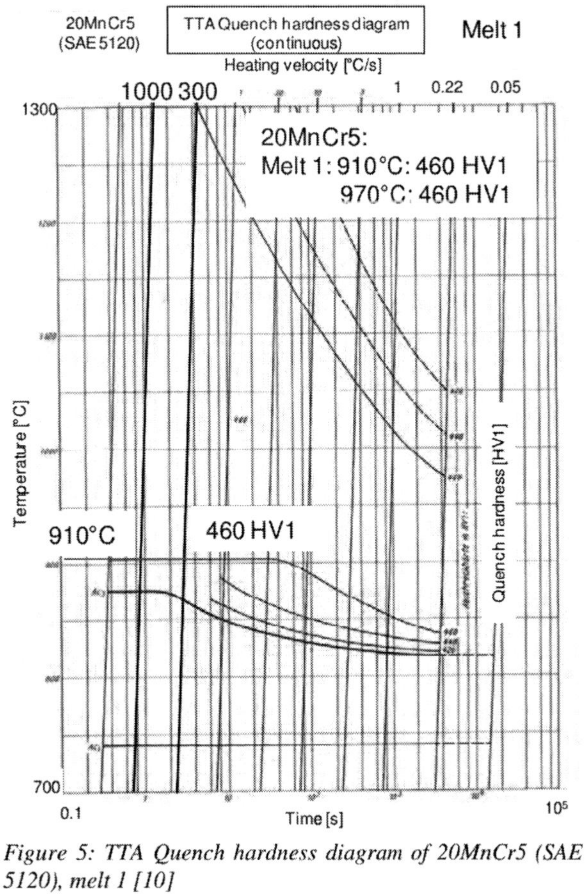

Figure 5: TTA Quench hardness diagram of 20MnCr5 (SAE 5120), melt 1 [10]

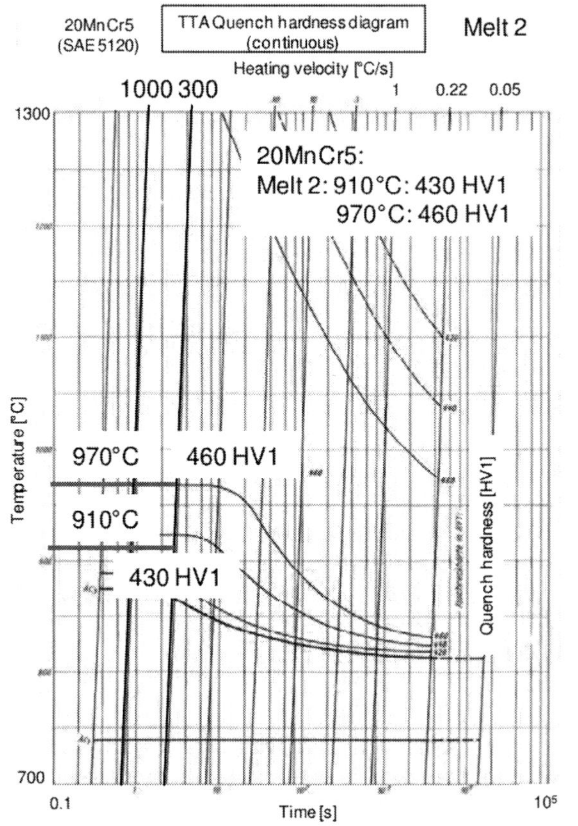

Figure 6: TTA Quench hardness diagram of 20MnCr5 (SAE 5120), melt 2 [10]

Conversion of results from 20MnCr5 melt1/melt2 to Cf45/C45

The example of 20MnCr5 shows, that the hardness can be increased by a defined adjustment of Al and N while the carbon content has the same values (see Table 1). Since this behavior is also true for the two melts Cf45/C45 only the hardness values have to be determined now. It should be said, that a special formula was used to calculate the conversion of the hardness values [14]. The calculation was done for a carbon content of 0.45 percent per weight. The conversion of melt 1 of 20MnCr5 leads to 600 HV at 910°C for Cf45 and reaches the specification of the customer with >550 HV. For melt 2 the conversion leads for the C45 at 910°C to a hardness of only 450 HV. This shows that the conversion is not linear. In the case of melt 2 it is considered that at 910 °C the melt is not already austenized, it will be formed less martensite which leads to less hardness. A hardness of 600 HV for melt 2 is reached only with a temperature of 1050 °C.

Results of this example: The hardness is sometimes reached and sometimes not. This depends on the not defined content on Al and N and the grain size, hence the coarse grain size. Cf45 guarantees these conditions whereas C45 has no defined areas and will reach the specification sometimes and sometimes not at 910 °C for the hardening procedure. A cooperative discussion between customer and heat treater can save time and money to everybody.

Third Example – Changing of pre-heat treatment

The last example is about the heat treatable steel 50CrMo4 (SAE 4150- material number 1.7228). The heat treater gets an order from the customer for a heat treatment of parts which he has done several times before. The customer sends parts of the material 50CrMo4 after DIN EN 10083-3 [3] in the condition pre-tempered, machined for surface hardening by heating through induction and quenching in a aqueous polymer solution. Because of the high effort of machining, the customer changes the parts to +AC treated (annealed on globular cementite), cold forged (near net shaped) to reduce the costs of machining dramatically. The customer does not give the information about the changed process to his heat treater.

Surprisingly for the heat treater, the demanded hardness of > 700 HV cannot be reached with the well-known repeated heat treating process. Why? The customer has changed his process chain and thereby the delivery state of the microstructure from "tempered" to "annealed on globular cementite" without communication to the heat treater. With the same heat treatment process parameters the demanded hardness cannot be reached. The reason can be found in [10] by comparing the two different states tempered (Figure 7) and annealed (Figure 8) which is close to +AC.

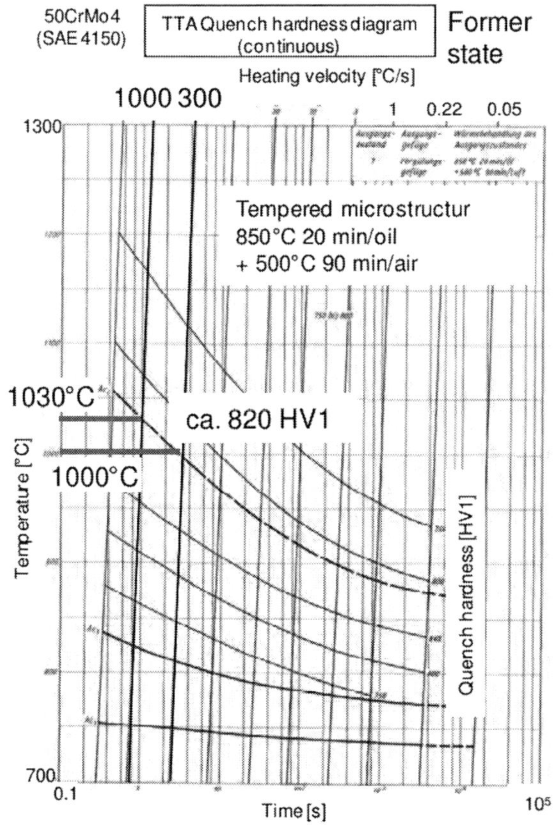

Figure 7: TTA Quench hardness diagram of 50CrMo4, Tempered microstructure [10]

The explanations will be carried out again for a heating velocity between 300 – 1000°C/s. The comparison starts at the dashed line "homogeneous austenite" by the tempered state (Figure 7) and the reached hardness in HV. It will be shown first (1) how much softer the new state (+AC) is and second (2) how much more energy e.g. temperature is needed to reach the same hardness as for the tempered state.

(1) Before the customer changed his process a hardness of around 820 HV could be reached (Figure 7). Now only 600 HV can be reached (Figure 8). As a result the customer parts now have a lesser hardness of 220 HV.

(2) To reach 820HV with the new material (+AC) a temperature of around 1260 °C is needed and therefore a higher temperature of 260°C to get the same hardness as for the tempered parts.

This illustration shows that a heat treatment which was correct for the previously used material (tempered) cannot be used to get the same results with the changed process of previous pre-heat treatment. The reason in this case is that the bigger carbides of the soft annealed condition need more time or a higher temperature to dissolve and to release free carbon for hardening. So the heating process for the tempered state is more successful than the heating process for the +AC state.

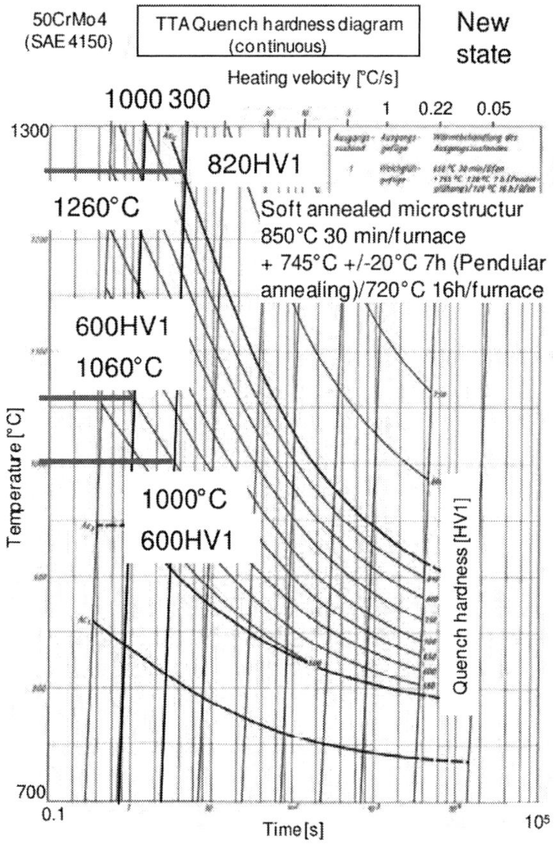

Figure 8: TTA Quench hardness diagram of 50CrMo4, soft annealed [10]

Knowledge System

One of the main tasks of the AWT research group 25 is to disseminate the reasons for necessary information exchange. A good way is to provide the knowledge online for public use. Due to this a first knowledge system which is up to now online but only accessible for members is shown in Figure 9. The structure of the knowledge system is taken from Figure 1. Example one is sorted to raw material information (material data), example 2 to raw material information (material data) and delivery information (state of heat treatment) and the third example is sorted by process changes (changing of pre-heat treatment).

AWT FA25 › Wissenssystem Waermebehandlungsauftrag

Wissenssystem Waermebehandlungsauftrag

Was this page helpful?

Knowledge system "Heat treatment order"

Haupteinflussgrößen/Main influencing factors	Erste Untereinflussgrößen / First sub factors
Informationen zum Ausgangsmaterial / Raw material information	Werkstoffdaten/Material data
	Nach Stahlschmelzen getrennte Anlieferung/After steel melt separated delivery
	Rohstahlbasis / Initial raw steel
	Ausgangsform / Initial shape
Informationen über Anlieferzustand / Delivery information	Bearbeitungszustand / State of processing
	Oberflächenzustand / State of surface
	Wärmebehandlungszustand / State of heat treatment
Informationen über Weiterbearbeitung / Subsequent processing information	Mechanische Bearbeitung / Machanical processing
	Erodierarbeiten / Eroding
	Vorhandenes Aufmaß / Available oversize
	Oberflächenbehandlung / Surface treatment
	Nachgeschaltete Wärmebehandlung / Subsequent heat treatment
Weitere Informationen / Further information	Lieferinformationen / Delivery information
	Wärmebehandlungsinformationen / Heat treatment information
	Prüfinformationen / Test related information
	Spezialinformationen / Special information
Prozessänderungen vor Anlieferung / Process changes before delivery	Neue Maschinen, Werkzeuge, Hilfsmittel / New machines, tools, auxiliaries
	Einsparen eines Prozessschritts / Saving a process
	Änderung einer Vorwärmebehandlung / Changing of prea-heat treatment
	Änderung von Prozessvariablen / Changing of process variables
Produktionstyp / Production type	Serienplanung/-fertigung / Batch production
	Einzelaufträge / Individual order

Figure 9: Prototype of the online knowledge system

By choosing a link in the column "First sub factors" a selection of information due to the order criteria will be shown and finally the information and the reasons can be seen. This knowledge system is based on an open source wiki-type. In this case useful functions for research information are already included in the system. It is possible to use the search function or to build up a tag cloud to provide information [15].

Conclusions and Outlook

The paper shows the necessity of information exchange for the heat treater. It also shows that voluntary information sharing can lead to a win-win situation for customer and heat treater. In a first step necessary information for a successful heat treatment was arranged and listed in a diagram (Figure 1). In the next step reasons why this information in Figure 1 might be important were demonstrated by three examples. Finally, the knowledge system to disseminate the information easily in an online wiki platform was presented. The next steps are to fill up the knowledge system with reasons of all influence factors of successful heat treatment.

Acknowledgments

The work on this paper was accomplished within the Collaborative Research Center SFB 570 "Distortion Engineering" at the University of Bremen. The authors acknowledge the financial support by the Deutsche Forschungsgemeinschaft. Details on the SFB 570 can be found at the internet site (www.sfb570.uni-bremen.de).

References

[1] Ermert, V. *et al*, "Necessary Information for Successful Heat Treatment," *Proc. of the European Conference on Heat Treatment 2011 "Quality in Heat Treatment"*, Wels, Austria, 2011, pp. 12-19, CD-Rom, ISBN: 978-3-9503135-0-5.

[2] N.N., „Angaben zum Wärmebehandlungsauftrag," http://www.haertetechnik.org/cms/File/tabelle%20angaben%20zum%20wb-auftrag.pdf, accessed 2011/07/21.

[3] N.N., Steels for quenching and tempering - Part 3: Technical delivery conditions for alloyed steels; German version EN 10083-3:2006, Beuth Verlag (Berlin, 2006).

[4] N.N., "Werkstoff-Datenblatt 1.8159 - 51CrV4", http://www.doerrenberg.de/, accessed 2011/07/21.

[5] N.N., „Werkstoff-Datenblatt 51CrV4," http://www.saarstahl.com, accessed 2011/07/21.

[6] N.N., „Dr. Sommer Datenbank Stahlwissen NaviMat 1.0," 2002.

[7] N.N., Metallische Werkstoffe - Umwertung von Härtewerten, EN ISO 18265:2003, Beuth Verlag (Berlin, 2003).

[8] N.N., Steels for quenching and tempering - Part 2: Technical delivery conditions for non alloyed steels; German version EN 10083-2:2006, Beuth Verlag (Berlin, 2006).

[9] N.N., Flame And Induction Hardening Steels - Quality Specifications, DIN 17212:1972-08, Beuth Verlag (Berlin, 1972).

[10] Orlich, J. *et al*, Atlas zur Wärmebehandlung der Stähle. Band 3: Zeit – Temperatur – Austenitisierung – Schaubilder, Verlag Stahleisen mbH (Düsseldorf. 1973).

[11] Wever, F. *et al*, Atlas zur Wärmebehandlung der Stähle. Band 1: Zeit – Temperatur – Austenitisierung – Schaubilder, Verlag Stahleisen mbH (Düsseldorf. 1961).

[12] Rose, A. *et al*, Atlas zur Wärmebehandlung der Stähle. Band 2, Verlag Stahleisen mbH (Düsseldorf. 1972).

[13] Orlich, J. *et al*, Atlas zur Wärmebehandlung der Stähle. Band 4: Zeit – Temperatur – Austenitisierung – Schaubilder. 2. Teil, Verlag Stahleisen mbH (Düsseldorf. 1976).

[14] Atkins, M.: Atlas of continuous cooling transformation diagrams for engineering steels. British Steel Corporation, 1977.
[15] Hassan-Montero, Y. *et al*, "Improving Tag-Clouds as Visual Information Retrieval Interfaces," *Proc 1st International Conference on Multidisciplinary Information Sciences and Technologies*, Mérida, Spain, October. 2006.

Proceedings of the 26th ASM Heat Treating Society Conference
B.L. Ferguson, R. Jones, D.S. MacKenzie, and D. Weires, editors
https://doi.org/10.31399/asm.cp.ht2011p0138

Cleaning for Heat Treating

D. Scott MacKenzie, PhD, FASM, Robert Johnston
Houghton International, Inc., Madison and Van Buren Aves., Valley Forge PA 19482
smackenzie@houghtonintl.com, rjohnston@houghtonintl.com

Abstract

Heat treating often requires changing the surface chemistry of the part from carburizing or Nitriding. In addition it is important to remove residual quench oil from parts after quenching. Discussion of various cleaner chemistries will be discussed, with application to heat treating. Case histories will be shown showing the effect of improper cleaning has on parts.

Introduction

The demands of heat treated goods are increasing. Not only do they have to meet property and straightness requirements, but they also have to meet increasingly stringent appearance specifications.

Cleaners are used in nearly all industrial plants where metal working operations are found. Almost all soils need to be removed to some extent. Thus, if a plant uses coolants, drawing compounds, rust preventatives or heat treating oils, then the plant will likely have a strong requirement for cleaning.

Cleaning is the removal of dirt and soil from a surface. A cleaner is a formulated chemical system or product, dispersed in water, which removes unwanted dirt, soil, grease and other similar matter from a surface.

Cleanliness is a relative term. The total process will dictate the degree of cleanliness. For example, a part is being cleaned during various states of manufacture to remove chips or machinery fluids. The cleaner will leave behind a residual rust protective film. Although there is residual film on the part, it is considered clean. In the case of electroplating, blackening or enameling, the part must be chemically clean. This is often referred to by the term "water break free." After cleaning, if a part is rinsed in clear water, the water should run down the part in a continuous unbroken film. This indicates a water break free part. If the film is interrupted, this indicates some soil is remaining on the surface. Cleanliness should be determined by the customer. During a cleaner trial, it is good policy to allow the customer to make the initial comments concerning cleanliness. A part which may appear to be dirty to an operator could be quite acceptable to the final customer.

Cleaning Process Considerations

To establish quality cleaning of parts, all process variables should be considered. This includes previous process and soils from previous processes. The subsequent process should also be considered, as the residues and soils from the previous operation may not be tolerable. The washer capability and capacity should be understood in terms of the types of soils it is capable of removing. The existing water quality should be considered, and if water treatment should be considered (excessive water hardness). The bio-stability of the cleaner should be reviewed to avoid the growth of bacteria and fungus, and to avoid the use of biocides and fungicides. Finally, the part orientation should be considered. The part orientation evaluation should include examination of nesting of parts, and the orientation of the spray nozzles with relationship to the parts.

The residues remaining after cleaning should also be considered. These residues should not cause smoking in tempering operations; not cause flash rusting; or discoloration of the parts.

A typical cleaning process schematic for heat treating is shown in Figure 1 - Process schematic for cleaning in a heat treating shop..

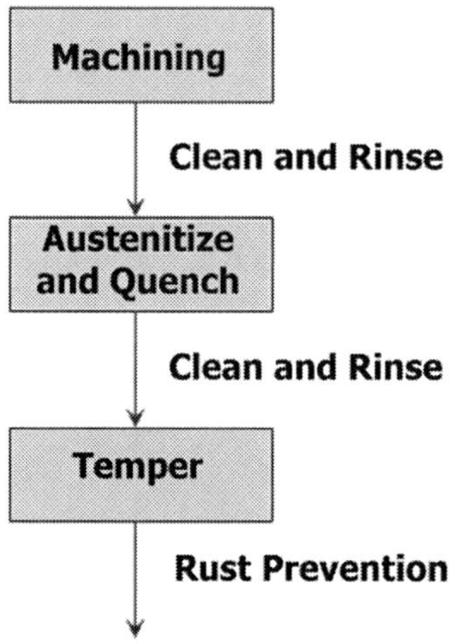

Figure 1 - Process schematic for cleaning in a heat treating shop.

Types of Soils

The soil is an important factor in determining the type of cleaner to be used as well as conditions for cleaning. Soils can be broadly classified into six groups:

Pigmented Drawing Compounds. The presence of these pigments makes cleaning difficult, especially after the heat generated by a drawing operation and the impacting of pigment into the metal. Mechanical cleaning methods such as ultrasonic, electrolytic, and spray cleaning aid in the removal of such compounds.

Un-pigmented Drawing Compounds, Oils and Greases. Hot alkaline cleaning normally removes such soils. If high viscosities are encountered , pre-cleaning in a solvent or solvent emulsion cleaner will dissolve some soil and lower the viscosity thus enabling the alkaline cleaner to perform more efficiently.

Machining Lubricants. Soluble oils, synthetic and emulsion cutting fluids are readily removable with synthetic and alkaline cleaners. The presence of sulfur and chlorine components may make the soil more difficult to remove with alkaline cleaners and solvents or emulsions are sometimes necessary. The presence of fatty acids in the cutting fluids promotes better cleaning through reaction with the alkaline cleaner.

Polishing and Buffing Compounds. A Buffing compounds in general are difficult to remove because of the heat generated during the buffing operation. Aging and over-buffing are to be avoided. Aging causes polymerization and over-buffing leaves excessive amounts of compounds on the work. Products containing high degrees of detergency and solvency are normally recommended.

Carbonaceous Soils. Dry finely divided carbon particles are some of the more difficult soils to remove. Direct agitation, such as high pressure spray impingement, sometimes removes such soils. An emulsion cleaner is often effective as is ultrasonic cleaning.

Rust Scale and Paint. Removal of these soils is considered a specialty area. High caustic alkaline cleaners may be used for removal of light rust, paint, lacquer, etc. Heavy scale must be removed with acids. Paint removal from aluminum and zinc metals must be accomplished with a paint stripper.

Besides the type of soil considered above, there are additional soil variables that should be considered to determine the proper method of cleaning. These soil variables that should also be considered are:

- **Physical form of the soil**. This includes the viscosity of the liquid and the ,elting point of any solids.
- **Fine particulate**. The presence of finely divided particles can make soil removal very difficult to remove.
- **Corrosion or Oxidation products in the soil**. Proper neutralization of these soils is import so that damage will not occur to the part.

Cleaners work by both a chemical action and a physical action. The chemical actions are:

- **Reaction**: Some soils are removed by reacting with the cleaner involved. For example, alkaline cleaners will react with fatty soils to form soaps. These soaps will aid the cleaning process.
- **Dissolving**: Other soils are water soluble and will be removed by being dissolved in the cleaner.
- **Emulsification**: An emulsifying cleaner works by dissolving oils and greases and forming a milky white emulsion.
- **Wetting**: A wetting agent or surfactant, a key ingredient in a cleaner solution, functions by lowering surface tension, thus allowing cleaner solution to penetrate between the soil and the surface.

The physical actions are:

- **Impingement** is involved in spray cleaning processes. The part is hit by a cleaning solution flowing through a series of spray nozzles at rather high pressures. This scrubbing action is critical to the removal of soil from the surface.
- **Agitation** is found where soak cleaning is used. A soak cleaner depends solely on detergency for removal of soil. Thus, agitation in the form of air or steam being bubbled through the solution is extremely helpful in soil removal.

In many cases, it is necessary to remove soils after one metalworking operation and prior to another. An example would be if a part was processed through a grinding process and was scheduled to go into Heat Treating. After grinding, the part would have grinding fluid and metal fines on the surface which would cause problems in the subsequent heat treat process. An in-process cleaner would be used.

A finish cleaning operation is a process where the part goes through a complete cycle such as painting, plating, enameling or anodizing, etc. Cleaning is of the utmost importance. The part must have a water break free surface prior to the application of a coating.

Rehabilitation cleaning is the cleaning of parts so they can be put back in service. Removal of rust, paint and grease are examples of the soils being removed. A typical application is the cleaning of gas and diesel engines.

Typical Cleaning Problems

Cleaning problems usually manifest themselves after the tempering operation. This is because any residues left on the parts. Typical problems are rinse stages being contaminated; poor cleaning due to nesting of parts; alkaline residues from poorly rinsed parts; rust and caustic burns; and excessive staining from abused quench oils.

Rinse Stage Contaminated. After a recent shutdown, parts from a furnace where showing evidence of a dried yellow solid material on the "cleaned" part (Figure 2). The parts are heat treated, quenched and then cleaned and rinsed. The cleaner used was silicated cleaner, and the machining is being done with a chlorinated oil that is made up on site from spend hydraulic fluid.

The dried yellow material was scraped out of the grooves and analyzed via KBr pellet-FTIR. It was identified as probable inorganic silicate, and iron oxide. Since the parts were rinsed after the cleaning operation, it was thought that the rinse station had an excessive concentration of silicated cleaner from the cleaning cycle.

Figure 2. - Parts cleaned and rinsed after heat treatment. Analysis shows that the yellow residue was due to inorganic silicate and iron oxide. Contaminated rinse station with high levels of silicate was thought to be the cause of the residue.

Poor cleaning due to nesting of parts. After the tempering operation, parts were found to be discolored (Figure 3). The discoloration matched the outlines of adjacent parts. The "cleaned" part was rinsed with ethyl ether, which was then evaporated. An oily residue remained. Fourier Transform Infrared Spectroscopy (FTIR) shows that it is an ester (possible vegetable oil) with an apparent high chlorine level.

Figure 3 - Parts that were heat treated and cleaned, and found to have high levels of esters and other sludge.

The "cleaned" part shows that the rinse stage must be highly contaminated with cleaner, and also with oil. This is drying on the parts, and at heat-treating temperatures it will fuse and leave chemical burn marks as shown in the third photograph. The solids from the quench oil were collected and were found to contain 7% iron, 2% nickel, 2% chromium, 1% sodium, 1% aluminum, 2% silicon, 3% sulfur, 10% calcium. This is similar to previously analyzed sludge samples, and confirms that parts are not clean, and are leaving swarf and "cooked" cleaner residue in the quenching oil. The oil portion was analyzed via FTIR and identified as a mixture of mineral oil and ester (estimated 10%). The ester could be due to hydraulic fluid contamination. FTIR analysis of quenching oil from the furnace shows a slightly elevated level of oxidation in the quench oil, but it was probably insignificant. This appears to be a cleaning related issue, rather than a quenching oil problem.

Alkaline Residue. A bolt manufacturer had a series of customer rejects for white residue in the threads. The white material (Figure 4) was analyzed via reflectance Fourier Transform Infrared Spectroscopy (FTIR) and identified as primarily carbonate residue. Investigation of the rinse tank found that high carbonate concentrations were present, which is drag-out from the cleaner operation. The customer was instructed to dump the rinse tank, and to check the alkaline levels in the rinse tank on a regular basis.

Figure 4 - Alkaline residues found in the threads of a fastener.

Caustic Burns. A customer manufacturing bolts had a series of rejections from his customer for rusted and oxidized parts (Figure 5). A package containing various sizes was submitted to the Houghton Cleaning Laboratory for evaluation. Each bolt has an even black oily finish (from the rust preventative), but there are round orange areas visible. In addition, round areas that were bluish in color were observed.

The incoming bar stock is coated with calcium stearate (to aid in cold heading and threading). Parts are cold headed with chlorinated lube, washed with caustic cleaner, then go directly to heat treating (no rinse), quenched in a cold quench oil, washed in an alkaline cleaner, blow-off (no rinse), tempered at up to 1100°F, and then dipped (hot) into an aqueous rust preventative which flashes off the water. The red spots are then observed.

Figure 5 - Caustic burns found on the body of a fastener caused by inadequate rinsing after cleaning prior to heat treatment.

After degreasing with ether in the laboratory to remove the rust preventative, the bolt was noticeably covered with round spots, and several red areas as described by the customer.

The spots are caused by a condition known as "caustic burn". Essentially, caustic cleaner and/or chlorinated material and/or calcium compounds are burned into the steel during the heat treating process, leaving spotty alkaline areas on the surface of the part. The steel is quenched in an oil based product, which does not remove the caustic material since it is burned onto the surface of the part from the high temperatures during austenitizing. The parts are then probably further impacted in the tempering furnace (although the damage is already present).

Recommendations to the customer were to thoroughly wash and rinse the part prior to heat treatment to eliminate any caustic residues that would cause caustic burns on the part.

Inadequate Rinsing. A customer was having rust issues on one of their induction hardeners, on one machine only (Figure 6). A polymer quenchant is used, and delivered to the machines from a 14,000 gallon central system. All machines use the same quenchant. The rust is always on the inside of the part. Parts are formed with zinc stearate, machined with synthetic coolant, and cleaned with alkaline cleaner and rinse prior to induction hardening.

The inside of the part was rinsed with a small amount of distilled water and Phenolphthalein. It turned a deep red color. The rusted area was rubbed onto a piece of ashless filter paper, and placed into an EDXRF spectrometer and analyzed for elemental content. In addition to the iron oxide, large amounts of sodium, potassium, and sulfur were detected. The inside of the part is coated with alkaline cleaner, probably based on potassium hydroxide and sodium sulfonate (or similar chemistry). This caused subsequent rusting. It was recommended to investigate the rinse stage on the discrepant induction hardener, and to determine if either rinse nozzles were clogged, or the rinse had excessive contamination. More frequent dumping of the rinse station was also recommended.

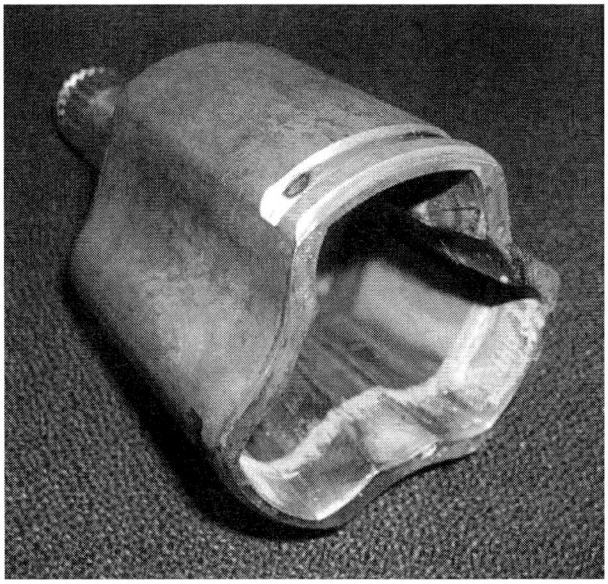

Figure 6 - Rusting on a single induction hardening machine that was caused by improper rinsing of the part after heat treatment.

Excessive Staining. A customer was experiencing severe black streaks on parts coming out of several furnaces, immediately after quenching, and prior to the quench washer (**Error! Reference source not found.**). As a test, parts were hand cleaned in organic solvents and inorganic cleaners prior to heat treatment. The quenched parts still had the streaking on their surface, even after through cleaning before heat treatment. The part in **Error! Reference source not found.** was collected immediately after quenching, and was hand cleaned with water and isopropanol (a total of 160 parts were cleaned in this manner). We were asked to analyze the black residue and determine corrective action.

Figure 7 - Excessive staining of parts observed immediately after quenching, prior to the quench washer.

The part was wiped thoroughly with a lab wipe and isopropanol, in order to remove any residual quenching oil. The part was then placed into an Energy Dispersive X-Ray Fluorescence Spectrometer (EDXRF) and the surface was analyzed for contaminants. Sulfur and calcium are present at significant levels. Following this, a piece of laboratory filter paper was dampened with isopropanol containing a small amount of acetic acid. This was vigorously rubbed on the black lacquered area and some of the black material was removed. The paper was analyzed in the same way (EDXRF). Calcium, sulfur, iron, manganese, and chromium are all prominent elements in the analysis.

The presence of calcium and sulfur in the stain (after removing residual quenching oil) shows that the material is baked-on quenching oil. What is also interesting is that the staining shows the progression of the vapor phase during quenching.

The mar-tempering quench oil was analyzed and found to be severely oxidized, with a excessively high Precipitation Number and Total Acid Number. Because of the extent of the oxidation, it was recommended to dump and recharge the system with fresh quench oil, and to institute proper maintenance and monitoring procedures.

Basic Factors Affecting Industrial Cleaning

With few exceptions, there are certain principles which apply to all types of cleaning. These are:

- Increased temperature usually improves cleaning.
- Agitation to move the soil (rather than the cleaner) improves cleaning. Agitation to move the cleaner is important when the layer of cleaner next to the surface has become heavily contaminated with soil or has cooled off.
- A minimum concentration of cleaner is needed for cleaning; above this level cleaning improves with increased concentration but each increment of cleaner has a lesser effect than the previous. A point of diminished returns is eventually encountered.
- Adequate time must be provided for detergency or reaction of cleaner with soil. Otherwise agitation or mechanical removal effects become more important.
- Rinsing away of soil and cleaner is necessary and must take into consideration factors such as: The amount that can be left behind without harm; how much cleaner residue can be tolerated; and Pressure rinse water or an agitated rinse is far more effective than a "still" rinse.
- Soil must be kept from redepositing on the work. This might take the form of cleaner components to suspend the soil, or design of the vessel holding the cleaner to provide room for soil to settle to the bottom (away from the cleaning area), or an overflow for oil to float away from the surface of the rinse.

Concentration. The concentration necessary for good cleaning is geared to the method of agitation. Thus, disregarding any foam problems, it is possible to get similar cleaning from the same alkaline cleaner at 8 to I2 oz./gal. by soaking, 1/2 to 1oz./gal. by spraying, 1/10 to 1/2 oz./gal. by steam cleaning, or 2 to 4 oz./gal. in a high pressure water spray at room temperature.

Time Required for Cleaning. This is determined by the conditions of temperature, soil, concentration, agitation, etc.. Keep in mind that it does take some time for the detergency process to take place. In a static bath, cleaning may take 5 to I5 minutes; in a high pressure spray of quality detergent, it may take a few seconds. It is expensive in equipment, labor and chemicals to reduce cleaning time, and there are situations in which the expense is not justified.

The time available for cleaning are very closely related to the economics of the cleaning operation not only from a direct production viewpoint, but also on the basis of time lost in re-cleaning, rejects or customer complaints when inadequate cleaning results from an unsuccessful balance of cleaning factors.

Temperature. The effect of temperature depends on the type of soil and cleaner. A first consideration is that enough temperature to melt a soil that can be melted, such as a fat, grease, or wax, makes a tremendous difference in the rate of cleaning. For most oils, increase of temperature reduces the viscosity of the oil, makes it more mobile and therefore more easily removed. There is a well established principle that the rate of chemical reaction is doubled for every 10°C or I8°F increase in temperature. If cleaning is by chemical reaction between a fatty oil and alkali, or if a paint or foodstuff is being decomposed by chemical action, or if a rust or scale remover is acting chemically, this same relationship occurs. Most industrial cleaning is carried out at temperatures of I40 - I80°F.

Agitation. Variation in agitation represents the greatest difference in the methods of cleaning that we have discussed so far. Equipment design should have the objective of getting the amount of agitation necessary to remove the soil as rapidly as is needed at the lowest cost. Agitation should be utilized with the idea of moving the soil. Circulation or stirring the cleaner is far less effective but is sometimes useful. High pressure spraying of cleaner is most effective. A great deal of attention has been paid to obtaining good agitation to improve cleaning. This may vary from the unsophisticated scrub brush to mild pressure sprays (5 - 40 psi), to high pressure sprays (300 - 1000 psi), to the current density electrocleaning, to high intensity ultrasonic cleaning, etc. Bubbling air through the cleaner or stirring with a mixer, or sloshing the work piece around in the cleaner are far less effective methods of agitation. Major limitations of agitation are: equipment expense, foaming, generation of irritating, fog, toxic vapors, or flammable gasses.

Water Hardness. Water hardness depends upon the amount of chemical salts dissolved in the water. These salts, in the presence of heat or alkalinity, may react to produce insoluble material which precipitates out of solution in the form of hard water scale. Unless this formation of hard water scale is prevented, a number of undesirable results are possible. The major undesirable result is the buildup of hard water scale. This scale may be precipitated on the heating coils of wash tanks, reducing the efficiency and wasting energy and money. Scale may build up in spray nozzles and on belt conveyors, impairing cleaning efficiency. The effectiveness of the spray is reduced and scale may block the solution from reaching and cleaning the objects on the belt. Scale may also build up in the rinse tanks in extremely hard water situations due to alkaline carry over of cleaner into the rinse tank. Hard water salts regularly react with fatty acid soaps to yield insoluble residues on parts being cleaned. The harder the water is, the greater the susceptibility to rusting of parts and the washer. Hard water is not conducive to good cleaning.

Control Procedures for Washers:
It is important that the washer system be properly monitored to achieve the desired cleaning. A disciplined approach to washer and cleaner maintenance is critical if quality cleaning is to be achieved.
Daily or shift (depending on use):

- Concentration determination by titatration
- Verification of system pressures and temperatures
- Verify cycle time
- Visual verification that the oil skimmer or overflow is working
- Check filtration equipment (screens or filters)

Rinse stations require the same tests. If an air blow off is used, proper operation should be verified to ensure that all nozzles are operating correctly.
Weekly control measures:

- Inspect exhaust equipment for proper functioning
- Visually inspect nozzle arrangement for misalignment or improper functioning (clogged nozzles, etc.)
- Interior spray guards for proper location and function.

This is in addition to the routine preventative maintenance done on the quenching system and quench oil.

Conclusions

In this article, typical heat treating cleaning issues were illustrated. The different types of cleaners and washers were described. Recommendations for the proper maintenance of cleaner solutions and rinse solutions were provided. If these simple precautions are followed, then clean parts will result, with fewer customer complaints and fewer rejections.

Acknowledgements

We would like to thank Houghton International for permitting the publication of this paper.

Proceedings of the 26th ASM Heat Treating Society Conference
B.L. Ferguson, R. Jones, D.S. MacKenzie, and D. Weires, editors
https://doi.org/10.31399/asm.cp.ht2011p0144

Bottleneck Oriented Load Planning in Heat Treatment – Optimizing the Production Flow Saves on Time and Resources -

Dipl.-Ing. (FH) Karl-Michael Winter,
PROCESS-ELECTRONIC GmbH, Heiningen,
Tel. 0 71 61 / 94 888 0 Email: km.winter@process-electronic.com

Abstract

Making money in heat treating is getting more and more demanding. There is obviously still the economical crisis that has caused the biggest impact on all of us, nevertheless knowing the true dynamic capacity of our - still partially turned down - equipment and being able to squeeze out even the smallest capability will optimize the cost to earnings ratio.

Everyone is using computer systems in order to plan their work-flow; however, often a disadvantage of such systems is their missing knowledge of what has to be done with the parts when it comes to breaking down the shop orders into real loads.

This article is giving an overview on the existing planning layers and what different methods are used and explains how a consequent planning around bottlenecks will result in an optimized throughput while at the same time trying to minimize the total number of equipment used – hence saving resources.

Introduction

Having a look on the various computer aided systems used in a modern company with a heat treating department we can group those systems according to their tasks on hand. Based on the number and the positioning within the company the result can be presented as a pyramid (Figure 1).

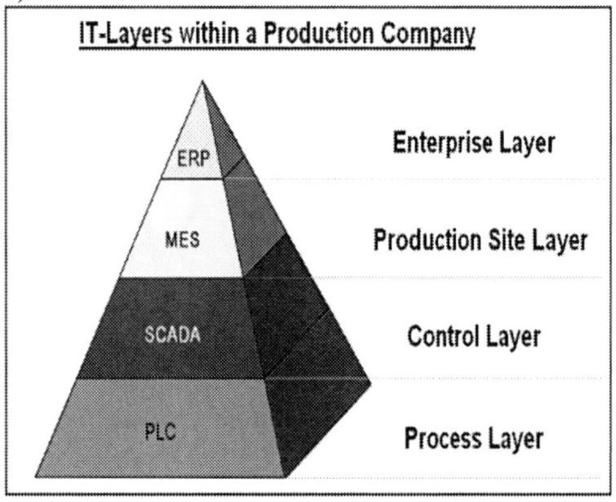

Figure 1: IT-Layers used in production companies

On the top we have the Enterprise Resource Planning (**ERP**) System handling all commercial activities in the company, like quoting, order processing, purchasing or invoicing and accounting.

In addition, as the name already suggests, this is the level where the planning of all resources takes place. In other words all existing equipment, the personnel and the money is planned in a manner to run all upcoming business processes as efficiently as possible.

Below the ERP is the production site control level where so-called Manufacturing Execution Systems (**MES**) are managing and (fine) planning the production processes. At this level all quality related data will be collected as well.

Below the MES is the control layer where Supervisory Control and Data Acquisition Systems (**SCADA**) finally execute and survey the production, interfacing with the control and process layer.

The boundaries between the various layers are fuzzy and, depending on the language used such a set of interconnected systems might also be described in a different manner.

On the top we will then find an inventory control system communicating with one or more production control systems. The production control systems are exchanging data with control centers that are sending the parameters and commands necessary to operate the tasks on hand to the connected PLC's and controllers.

In this setup the boundaries between the different layers are as well fuzzy. Depending on the complexity of the production chain respectively depending on the production depth and the number of equipment used layers might be omitted, in which case their tasks will be distributed between the neighboring layers.

The roles of the different levels in the production planning differ depending on the abstraction level, i.e. the distance to the executing machine. Therefore from the ERP system down to the control centers and PLC's a more and more detailed refinement is taking place.

A production planning on the enterprise level for that reason is not able to consider the real manufacturing situation at every single machine. The planning is based on assumptions where theoretical production capacities are balanced with calculated production times for the orders on hand. The goal is to reach an optimum in terms of usage of the machinery, through-put time and adherence to delivery dates. At the same all of this should require a minimum stock of material and parts in production in order to have as little as possible money locked up.

The resulting list of shop orders, roughly planned, will then be sent to the MES. The MES will now break down the shop orders into manageable production lots using its knowledge about the actual situation on shop floor derived by the machinery and shop order data acquisition and the operator schedules. These production lots will be scheduled onto the available equipment in such a way that the given production times will not be exceeded. The goal on this level is to come up with as short as possible throughput times while at the same time trying to optimize the machinery usage. To cut it short: whatever the ERP feeds into the MES has to go through the production in order to fulfill the set targets.

But the MES still is having a more abstract view onto the real production as it has to deal with a huge variety of machines and processes available in a company.

Therefore it might be necessary to do a further refinement of the planned shop order operations, especially if the equipment that is assigned to execute such an operation is not a single machine but a production cell consisting of several machines. Very often the MES has a view onto cells as being black boxes, not knowing about the machines within the cell. An operation in such a case will be something like „Heat Treating". The scheduling of the included sub-operation like „Cleaning", „Case Hardening" and „Tempering" onto the machines within the cell will then be done by the cells control centre. The scheduling might be done either manually or partially automated.

Typical Planning Methods

The different methods used in the production planning can be distinguished into the one that are using the push principle and the others using the pull principle.

By using the **push principle** new orders will be pushed into the actual planning and manufacturing situation and planned onto the available resources by running a continuous scheduling based on the given start and delivery dates. Using the production flows (work plans) attached to the parts to be produced the orders will be split into operations and every operation will be scheduled onto an according capacity. A capacity in this context is the sum of the machinery hours needed to execute a given operation, for example „gaseous nitrocarburizing". Having done that, a capacity leveling function is checking for possible collisions. Resource overloads can be cleared by postponing non-critical orders or by scheduling overtimes. The planning result is a detailed list of operations defining what material respectively what order will have to be started on what machine or group of machines at a given date and time and when this operation has to be finished.

By using the **pull principle** on the other hand a new order will be planned at the end of the production chain. Again the order first has to be split into operations and the last operation is used as the starting operation for the further planning. The next-previous operation will get the order to deliver the material in time. The scheduling is therefore done backwards, pulling the orders through the production. This is reducing the complexity and also the time needed for the planning as now each workplace will manufacture the parts by taking them out of a buffer in front of the machine into a buffer at the end of the machine that is treated as the buffer in front of the next workplace according to the production flow. Each workplace can easily launch a re-order at the workplace in front of itself when seeing a low level in its front buffer. The goal of this principle is to have as little as possible stock between the operations and provide optimum delivery reliability. It is best suited to deal with serial production and high quantities, keywords: KANBAN, Just-in-Time.

Highly sophisticated systems are using algorithms applying mathematical models onto the complex planning problems in order to optimize the planning result. More practical oriented systems are aiming for getting a local optimum in a short time by running a series of trials and errors. The latter systems are acting similar to a human being and therefore their results will better match the expectations of a human production planner.

The requirements for the planning result are partially contradicting. Ideally it should reach:

- Minimum throughput times
- High delivery reliability
- Minimum stock
- Maximum capacity usage
- Minimum costs

Reducing the throughput times will come to lower parts on stock but on the other hand the usage of the machinery might be low. In an order oriented manufacturing the focus will be more on the delivery reliability and short throughput times. But on the other hand we don't want to spend a lot of money on providing over capacities.

To solve this dilemma it is possible to pick a system that favors different requirements. We differentiate between inventory oriented planning and resource oriented planning.

An **inventory oriented** planning is trying to reach an optimum product flow at the end of the production chain by adjusting the stock of parts between the various operations. These buffers are exactly that big that the delivery reliability can be reached. The inventory oriented planning can be done load-oriented (push) or according to the KANBAN principle (pull).

Resource oriented planning systems try to avoid bottlenecks within the production chain. The Optimized Production Technology (OPT) approach optimizes the product flow through the operation having the smallest capacity. A bottleneck will always arise if the available capacity is smaller or maximum equal to the capacity needed for the orders to process. The bottleneck oriented disposition (EOD, German: Engpassorientierte Disposition) is focusing on material bottlenecks.

In short, OPT optimizes the in capacity bottlenecks by varying the production lots whereas EOD adjusts the send-ahead quantities, i.e. the number of parts that will be delivered to the next operation in one shot; the optimization of the product flow at the capacity bottleneck will be solved implicitly.

Fine Planning of Loads in Heat Treatment

But how is this related to heat treatment?

The planning activities within the heat treatment differ in type and complexity.

In a captive heat treatment treating parts out of a serial production there will typically be an ERP system that might or might not be connected to a MES system, depending on the size of the company. The planning has to adjust to a steady flow of parts in order to grant optimum delivery reliability.

In a commercial heat treatment we will find completely different conditions and requirements. Here, bigger companies are as well using the ERP-MES approach in their IT environment whereas smaller companies which do not have to network a lot of production sites will rather use a highly specialized inventory control system, connected to production management and control computers.

Accordingly, the resulting task list (operations) delivered to the machines respectively to the production management and control computers will vary (Figure 2). This might be based on shop orders with production lots optimized to the manufacturing of parts but also already planned loads given to an automated heat treatment cell. In addition, the quality of the planning result also depends on the in-build knowledge of the planning software about the special requirements in heat treating.

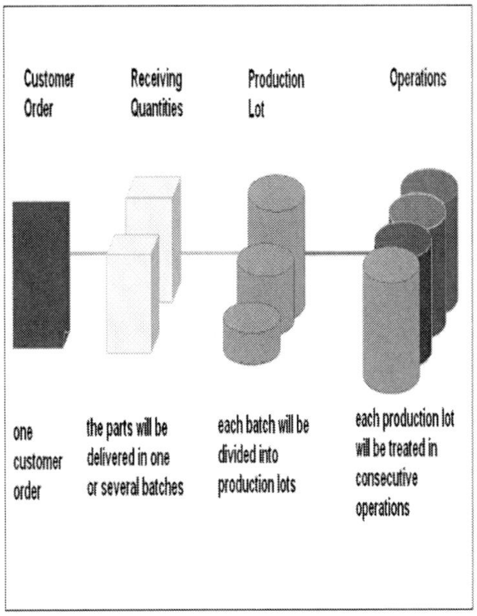

Figure 2: Breaking down a customer order into consecutive operations on a production batch

It is kind of simple to plan operations around machines and workplaces that operate with a direct reference to an order. In this case the workplace is only able to execute one order after another. If a shop order is registered on the machine all parts being produced by the machine will belong to this shop order and can be reported within the machinery and order data acquisition system to the planning system. This type of workplace does not have an internal storage.

This changes immediately if we are dealing with industrial furnaces. Conveyor belt furnaces and pusher furnaces are treating parts that might belong to many shop orders. The registered shop order will be at the front of the furnace whereas parts coming out of the furnace will belong to a shop order that has been de-registered several hours ago. It is getting even more complex if we are dealing with furnaces treating parts packed in baskets on trays. A load for a pusher or a batch furnace is able to hold parts of various shop orders (Figure 3). To make it even more demanding the size of the loads for the different machines in a heat treating shop can be different, resulting in the need of re-packing the parts between the operations.

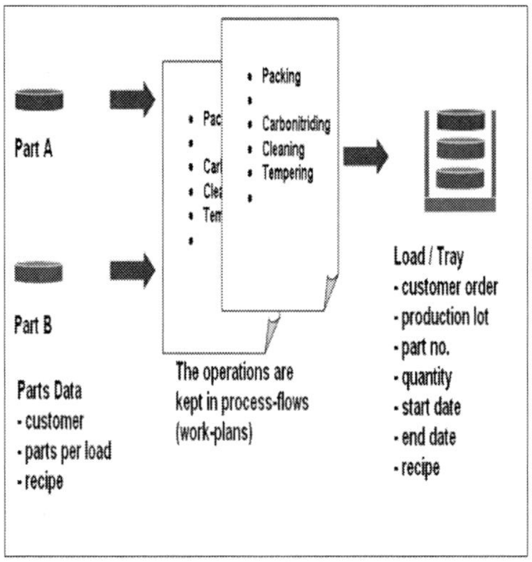

Figure 3: Assigning parts of different orders to one load

This is a situation that cannot be covered correctly by classical planning systems, this requires a human being. It needs years of experience in order to arrange heat treating parts in such a manner that it is possible to reach an as big as possible throughput while still all parts will be treated according to their specification.

Sequencing and Load Preparation

In the case that readily packed loads are entered into a fully automated heat treating cell the controlling computer system has to schedule the loads efficiently in order to give as short as possible through-put times.

An adequate method for this task will be to optimize the usage of the capacity bottlenecks.

According to the OPT philosophy:

- The usage of a non-bottleneck does not depend on it's own capacity but is given by other constraints restrictions within the system

- Activation and utilization of a resource are not synonymous
- An hour lost on a bottleneck is an hour lost of the total system
- An hour saved on a non-bottleneck is just a mirage
- Bottlenecks govern both and inventories

Imagine a cell consisting of several batch/iq furnaces, some drawers but only one washer, in this setup the washer will turn into a bottleneck the moment only loads with shallow cases have to be treated. The sequencing must assure that there will be no idle time on the washer. An additional constraint would be not to have too long waiting times for a load between case hardening, cleaning and tempering to avoid any hydrogen embrittlement.

If the cell consist of several pit type furnaces working into a common oil bath there will be a totally different requirement. In many installations moving the load into the quench has to be done manually or at least requires an operator to take care that there will be no fire due to a system failure. With the carburizing furnaces working 24x7 but the operators only available in two shifts during the working days the loads have to be planned in such a way that one after another will be ready for quenching without too much delay. If a load would have to stay within the furnace for a too long time the parts could end up outside the specification. And obviously, having the quench waiting during operator times will slow down the overall performance.

The planning tool cyclically simulates all loads within the cell and as well all loads that are planned to enter the cell and creates a projection of the found local optimum displaying the load and operation sequence with start and finish times. Planned outages like maintenance etc will also be part of the planning (Figure 4).

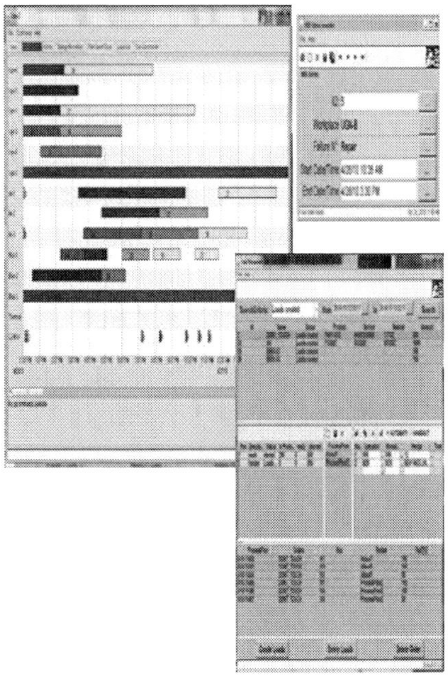

Figure 4: A GANTT-Diagram displays the result of the planning. Outages are taken into account.

If we now consider a department or even the whole heat treating shop as an automated cell - with a manual load transport - this planning approach can be applied with some extensions.

Chained machines or fully automated cells will be treated as single machines within the production planning and can be planned as either recursive cells within the department acting as the „main-cell" or separated into single machines within the department.

Similar to the big brother ERP the planning tool combines machines into groups according to their capabilities. This enables the control computer to sequence both, a series of operations – case hardening, cleaning, tempering – and to resolve a complex operation – case hardening complete – in such a way that even a possible back-up manufacturing can be considered automatically.

Assigning priorities to the machines enables to schedule loads onto the equipment until a set maximum capacity is reached before requesting the next machine of the same group for the further planning. Assigning priorities to shop orders enables an optimized sequencing able to handle even fast-track orders that have to be done immediately. In order not to have the lower prioritized orders moved to the very end of the planning list the assigned priority will change dynamically depending on the total delay time.

Another problem that has to be solved is the differing sizes of the production lots of the machines we have to plan; the tray of a central washer need not hold the same number of parts like the tray of a batch/iq. In addition we might have to re-pack the trays if two shop orders share one operation but the next operation needs to be done on different machines or requires different process parameters.

Data Interface to the Superior Planning System

When applying this type of fine planning the control computer also has to provide the machinery and order data and has to report all information like produced quantity, scrap quantity, outages, register and de-register time stamps to the superior computer system in the expected format. In this way the data can be used in the same way as usual and easily integrate into the planning tool on enterprise level.

Conclusions

Putting the focus onto bottlenecks while planning the complex material flows within the heat treatment, offers an elegant tool to optimize the throughput but and at the same time the machinery usage. If this task is performed on the specialized control layer the result will much better adopt to the actual manufacturing situation on shop floor as it would be possible with the high level planning software on the ERP or MES with their more abstract view.

Proceedings of the 26th ASM Heat Treating Society Conference
B.L. Ferguson, R. Jones, D.S. MacKenzie, and D. Weires, editors
https://doi.org/10.31399/asm.cp.ht2011p0148

Adoption of Automation and Process Control in a Job Shop

Tom Benoit, President
The Flame Treating and Engineering Company
West Hartford, CT USA
tbenoit@flametreating.com, 860 953-3519 F 860 953-1723

Introduction: Flame Treating was founded in 1940 by three Union Carbide employees to provide flame hardening services to the machine tool and capital equipment industry in the Connecticut River Valley. The market generally ran from Bridgeport to Providence and from New Haven to Springfield, VT with occasional opportunities in Worcester and the North Shore of Massachusetts. In 1956 the company purchased the first of it's induction machines and expanded the services to induction hardening.

Early Process Controls: At this time the control of the flame hardening process was provided by the trained eye of the operator. At the point of his interpretation of the correct temperature the manipulation of equipment of the end of the heat cycle occurred manually. While this was satisfactory in many applications it was far from without it's short comings. The induction process provided some improvement. The operator determined the cycle of the first piece in the same visual manner. However, after inspection of the first piece he would set a timer to insure the repeatability of the cycle time.

New Applications: Using the induction heating phenomena many new opportunities presented themselves for processing. Small gears, sintered alloy components, screw machine products and stampings became potential new opportunities for localized heat treating. These smaller components lent themselves readily to induction heating. And initially the trained operator and the process timer provided a level of control that was capable of producing the customers requirements.

As our customer base grew we began to see new materials for us to process. Initially 1045, 4140 and the 1100 alloys were the most common. But we were soon presented with stainless and air hardening materials that were more critical in temperature requirements. When approaching 1900F the ability of the trained operator to make an accurate determination of the temperature decreases. The straw that broke the camels back so to say was the annealing of the shank of a cutter shaft made of D-2. Lots of lost time and useless experiments working without temperature sensing.

Optical Infrared Pyrometry: Knowing we needed an accurate way of observing and controlling temperatures we began the investigation of optical pyrometers. Our efforts lead us the Ircon. We purchased one instrument with the sole intention of using it as a set-up tool to insure the correct processing temperature is being reached. We now had real confidence in knowing the temperature of the processed parts. We were also able to show and train operators in observing temperatures of processed parts. To date we had made no efforts to marry temperature control with any type of automation.

The Sensor Ring: The drive to marry temperature sensing and automation came when a potential customer proposed that we assist them in processing sample sensor rings for the first ABS breaking system for GM. A round part, sintered and coined required hardening on a bearing diameter and the internal spline teeth. Initial indications were that 20,000 would be required weekly
The first task was the coil design. Experiments lead us the a three turn

coil that heated both the ID and the OD simultaneously. Parts were processed in a single spindle lift and rotate device and immersion quenched in polymer to product the proper hardening. After processing several sample lots the customer asked us to process lots of 1000 until they got their equipment running. We felt confident that our ability would overshadow theirs and began to develop automated tooling.

Tooling Concepts: We designed a 4 position index table with lift and rotate capability at the heating and quenching stations. Using a small PLC with only 6 inputs and 4 outputs we merged the optical pyrometer with the index table to control the process. We now controlled the hardening temperature of each of the sensor rings processed. The controller indexed the table, raised the spindle, initiated the heating, ended the heating cycle, dropped the spindle and indexed the table to start the cycle again.

Problems: The basic concept was good and success was enjoyed. But over time we realized we needed additional control and sensing. We occasionally found incomplete indexing, incorrect lifting of the spindle to the required space for heating, lack of proper dropping of the spindle before index to the quench and mixed parts in the shipments.

Solutions: The small Omron controller was updated and expanded. A signal was sent from the table to tell the PLC that proper index had occurred. Prox switches were installed to indicate that the spindle was in the correct position for heating to occur and for index to begin. Additionally visual sensors were used to allow the PLC to know that the wrong part was in the spindle and that the processing stopped prior to the heating cycle. If the wrong item was processed the coil arched and production stopped until the tooling could be replaced. Additionally controls were added to the quenching phase of the process and now parts could be quenched in the heat position of indexed to the spray quench station. With these modifications to the equipment many other items were processed. By changing spindles and coils a wide range of gears, sprockets, shafts and copper connectors can be processed efficiently thru the equipment with complete confidence in the processing parameters.

The Next Evolution: Shortly thereafter we were presented with an opportunity to process shafts for a family of steering columns. Four parts were presented for discussion. The expected volume would be 164,000 monthly for nearly 5 years. Here we merged our index table tooling concepts with a PLC controlled scanner. In this application the two PLC's communicated and controlled index and scanning functions. Additionally we added temperature recording at the shutoff point of the that the cycle was complete and accurate.

Horizontal Scanning: Shortly thereafter a customer presented us with the opportunity of induction hardening a shaft for a fuel control. In this instance we used what we had learned about PLC's, sensing and automation to develop a horizontal scanner. Here we controlled location, heat on, heat off, index and part sensor location to insure meeting the customer requirements.

Flame Hardening Enhancements: Along this journey with the induction requirements we began discussing with Ircon the possibility of using the pyrometer to measure the surface temperature of flame hardening operations. It was explained to us that this is easy. The instrument will see the heated component but not the burning gasses.

A customer presented us with the opportunity to flame harden the leading edge of industrial components for electrical generation equipment.

Their specification called for heating the edge of these items for a length of approximately 5" using a hand torch. We told them that we would use mechanical equipment we possessed for some samples but would be developing an automated system using a PLC, optical pyrometer machine tool to provide accurate and repeatable heating. We did and it worked excellent and now automatically heat components of stainless steel, 52100 and 4150 along liner edges.

When another customer presented a similar application we held a meeting with his retired metallurgist. We met with them and wrote a specification. Then we visited the operation hardening a set of vee rails. The operator lit the torch and walked away. The metallurgist commented that someone needed to control the operation. He was encouraged to watch the temperature output and the machine screw. At the set point the tooling began to move and the temperature was maintained within about 10F. He commented to the general manager to "get out of their way. These people are out of the dark ages.

Result: By controlling and monitoring thru the PLC we are able to;

- Confirm the table index
- Sense the correct item is presented for processing
- Confirm that the part is in the correct position
- Control the process temperature
- Sense the proper location of the spindles for motion to occur
- Delay the quench to the proper time
- Quenching can occur at the heating station or at a subsequent location
- Flame hardened parts can be processed using the same techniques of control and automation.

Proceedings of the 26th ASM Heat Treating Society Conference
B.L. Ferguson, R. Jones, D.S. MacKenzie, and D. Weires, editors
https://doi.org/10.31399/asm.cp.ht2011p0151

Cost-Effective Technology for Induction Contour Hardening of Bevel, Hypoid and Pinion Gears

Semyon Brayman*, Anatloy Kuznetsov*, Sergey Nikitin*, Bob Binoniemi*, Valery Rudnev**

* *ERS Engineering Corp., 6346 Orchard Lake Rd # 103, West Bloomfield, MI 48322, SBrayman@ersengine.com*
** *Inductoheat Inc., 32251 N. Avis Dr., Madison Heights, MI 48071, rudnev@inductoheat.com*

Introduction

Presentation discusses proven and cost-effective induction technology that allows contour hardening of bevel, hypoid and pinion gears as well as gear-like complex-geometry components utilizing inexpensive steels. This unique technology was specifically developed to replace carburizing process. Variety of case studies will be discussed in this presentation demonstrating chemical composition of steels, achieved hardness patterns, and microstructure of hardened area, transition zone and core of the workpiece. Uniqueness of this technology is not only related to its capability to induction contour hardening of complex-geometry parts, but also allowing to obtain fine-grained martensitic structures (typical grain size range 8-11) with substantial compressive residual surface stresses (up to 600 Mpa/85ksi) that dramatically improve mechanical properties of induction hardened components.

Conventional approach to gear hardening

The traditional gear heat treatment is gas carburizing in batch or continuous furnaces for an extended time and then oil quenched. The gears are typically low temperature furnace tempered for a minimum one hour at temperature. Majority of gearing is made of low carbon alloy steels such as 8620, 8822, 4320 and 4817. After carburization, carbon contents of the case region are typically within the range of 0.7-1.0%. Surface hardness range is 58-63HRC. Post heat treat processing may include grinding and shot peening to improve fatigue strength.

Over the years, gear manufacturers have gained knowledge about improving quality of heat-treated parts using lean and environmentally friendly technologies; such as induction hardening that can be easily automated and incorporated into manufacturing cell. Gear performance characteristics (including load condition and operating environment) dictate the required surface hardness, core hardness, hardness profile/pattern, residual stress distribution, grade of steel and its prior microstructure [1-3].

Medium- and fine-sized teeth gears are typically induction hardened using encircling inductors (spin hardening). A major goal of induction gear hardening is to provide a fine-grain martensitic layer on specific areas of the part. The remainder of the part is unaffected by the induction process. Hardness, wear resistance, contact fatigue and impact strength increase allowing to improve load-carrying capacity to handle higher speeds and torques.

Another goal of induction gear hardening is to have individual gear traceability and formation of considerable compressive residual stresses at the surface and within the case depth. Compressive stresses help inhibit crack development and resist tensile bending fatigue. External and internal spur gears, worm gears, racks and sprockets are among the parts that are routinely induction hardened.

Gear operating conditions, the required hardness levels and cost are important factors to consider when selecting materials for induction-hardened gears. Induction hardening is generally limited to certain types of steels. Unlike carburizing, during which carbon is infused into the part, induction hardening requires higher-carbon steels to begin with. Typical induction hardening steels include AISI 4140, 4340, 1045, 4150, 1552 and 5150. In some cases, high carbon content steels (i.e., AISI 1065, 1080, 52100) are used.

Depending on the application, tooth case hardness after hardening and tempering is typically in the 54 to 62 HRC range. Core hardness is primarily depends upon steel chemical composition, and steel condition prior to induction hardening. For example, for quenched and tempered prior microstructures, core hardness is usually in the 28 – 42 HRC range.

There was a believe that not all gears and pinions were well suited for a conventional induction hardening. Hypoid and bevel gears, spiral bevel automotive pinions and noncircular gears used to be rarely induction hardened and typically carburized. This situation has been changed. As an example, Figure 1 shows three examples of inductively case hardened gear-like components.

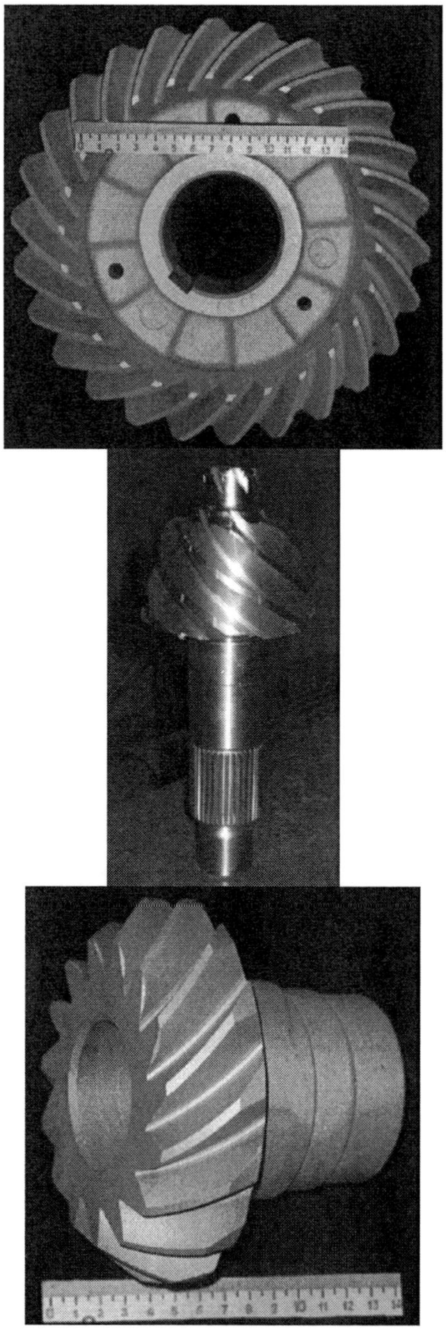

Figure 1. Complex-shaped gear-like components that used to be difficult to be hardened by conventional induction hardening. In the past, all of these parts used to be carburized, not any more

Limitations of conventional approach to gear hardening

Hardened case depth obtained upon induction hardening is directly related to an eddy current penetration depth that, in turn, affects the depth of the austenized surface layer. Non-uniformity and inconsistency of austenized surface layer would lead to correspondent deviations of hardened case depths. Two major factors that restricted using conventional induction hardening techniques for hardening complex-shaped gears and gear-like components:

1. Poor electro-magnetic coupling between the encircling-type coil and the tooth root compared with the tooth tip in combination with a complexity of gear geometry made it difficult to uniformly induce energy in the root, flank and tip areas of the teeth. Power surplus in the tip of the tooth and gear edge regions compared with that in the root were often taking place when medium and, in particular, with RF frequencies were used. This leads to a tendency to overheat edges, sharp corners and areas, which have better electromagnetic coupling with the inductor, negatively affecting the hardness patterns.

2. Presence of a significantly greater thermal "heat sink" located below the gear root and neighboring regions compared to the tooth tip added to a difficulty of obtaining a contour-like hardness pattern in particular in cases of bevel, hypoid, and pinion gears.

Novel development in induction gear hardening. TSH steels.

Michigan based, ERS Engineering Corporation has worked with its European partners to develop unique steel metallurgy and process that allow to replace carburizing with induction hardening for a wide range of complex-shaped components, including parts previously thought to be impossible to induction harden successfully. Inductoheat has been chosen by ERS to be their exclusive manufacturer of induction heat-treating systems for TSH Technology in the North American market.

TSH Technology is a synergistic combination of low hardenability (LH) advanced steels and special induction hardening techniques. TSH steels were successfully used in Europe for more than decade. However, recent breakthrough in further excelling of TSH steels, substantially widen the range of their applications.

TSH steels are low hardenability low-alloy carbon steels are characterized by limited controllability and reduced grain growth during heating into the hardening temperature range. They can be substituted for more expensive standard steels typically used for conventional induction hardening or carburizing grades. TSH steels have significantly less alloying elements such as manganese, molybdenum, chromium and nickel, making them less expensive than majority of conventional low alloy steels. Their chemical composition is somewhere between micro-alloy steels and plain carbon steels, providing fine-grain martensite with extremely high compressive stresses at the tooth surface.

Different applications may call to slightly different steel chemical compositions and process recipes. As an example, Table 1 shows two of the most typical compositions of the TSH steels (60LH and 80LH) vs. some carburized steel grades (8620H and 4320H).

Table 1. A comparison of typical chemical composition (%) of two low hardenability TSH steels vs. carburized grades.

	60LH	80LH	8620H	4320H
C	0.55-0.63	0.77-0.83	0.17-0.23	0.17-0.23
Si	0.10-0.30	0.10-0.25	0.15-0.35	0.15-0.35
Mn	0.10-0.20	0.10-0.20	0.60-0.95	0.40-0.70
Cr	< 0.15	< 0.25	0.35-0.65	0.35-0.65
Ni	< 0.25	< 0.25	0.35-0.75	1.55-2.0
Cu	< 0.30	< 0.30	-	-
S	< 0.04	< 0.04	< 0.04	< 0.04
P	< 0.04	< 0.04	< 0.035	< 0.035

With TSH technology components are usually through heated or partial heated (depth of heating needs to be 2-3x deeper than required harden depth) and then are rapidly quenched. The hardened depth is mainly controlled by the steel's chemical composition. Even though, components made from TSH steels are often heated through, their limited hardenability allows obtaining well-defined and crisp hardness case depth with well-controlled hardness pattern having minimum case hardness deviations even when hardening complex-shaped parts (Fig.2).

In the past, it was practically impossible to induction harden parts shown on Figures 1 and 2, due to a combination of shape complexity and gear sizes. Now it is possible to get those beautiful uninterrupted hardness patterns by using a simple operation: through heating those parts using low frequency inverters and quenching.

Noticed, that the hollow shaft (Figure 2,left) was induction hardened on OD, ID and teeth

region using a single operation producing a un-interrupted case hardness pattern.

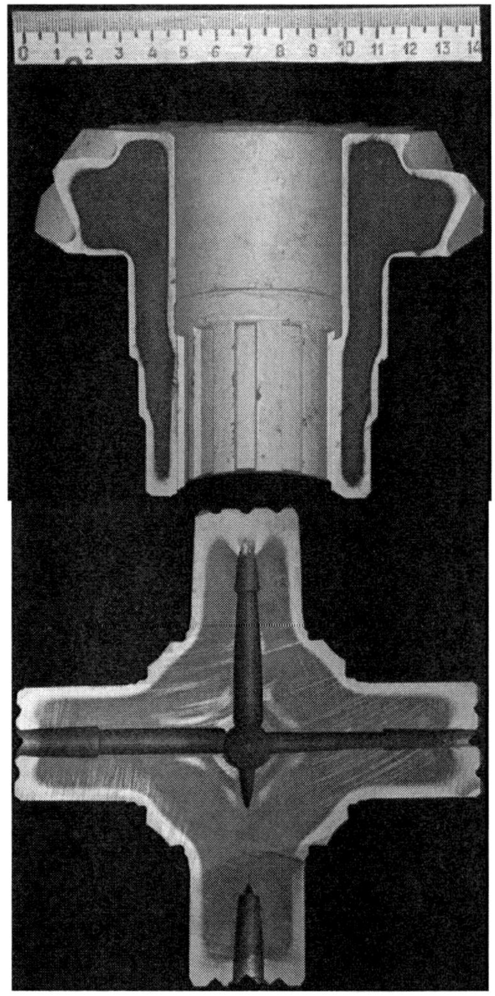

Figure 2. With TSH Technology, complex-shaped components are induction hardened using low or medium frequencies and rapidly quenched, resulting in contour-like induction hardening patterns. Notice an uninterrupted hardness patterns on complex-shaped geometry parts. (Courtesy of ERS Engineering Corp.)

Table 2 shows a chemical composition of the spiral bevel side gear that is shown on Figure 2, left. (Study was conducted by STORK® Climax Research Services, Inc.)

Element	%	Method
C	0.79	ASTM E1019-08
S	0.015	ASTM E1019-08
P	< 0.005	ISO-14707:2000E (Glow Discharge)
Si	0.08	ISO-14707:2000E (Glow Discharge)
Mn	0.06	ISO-14707:2000E (Glow Discharge)
Cr	0.17	ISO-14707:2000E (Glow Discharge)
Ni	0.06	ISO-14707:2000E (Glow Discharge)
Mo	< 0.02	ISO-14707:2000E (Glow Discharge)
V	< 0.008	ISO-14707:2000E (Glow Discharge)
Al	0.0089	ISO-14707:2000E (Glow Discharge)
Cu	0.04	ISO-14707:2000E (Glow Discharge)
Ti	0.085	ISO-14707:2000E (Glow Discharge)
Nb	< 0.008	ISO-14707:2000E (Glow Discharge)

There are several important factors associated with TSH Technology:

- TSH steels are relatively inexpensive, incorporating significantly smaller amounts (3-8 times less) of alloying elements such as manganese, molybdenum, chromium, nickel, etc. compared to traditionally used low-alloy steels.
- TSH steels require using low and medium induction hardening frequency (1 to 15kHz) than conventional induction contour hardening processes. This helps to reduce equipment capital cost (including the reduction of power supply cost).
- Formation of high magnitude of the desirable compressive surface residual stresses (up to 600 Mpa/85ksi)
- The hardened depth is primarily controlled by the steel's chemical composition and initial microstructure. Heat-treating process is repeatable and robust.
- Formation of fine-grain and in some cases ultra-fine grain martensitic structures within the case depth. This allows improving both: toughness and strength of induction hardened components at the same time. Core structure is a mixture of bainite and pearlite resulting in a good combination of core ductility and strength.
- An ability of TSH technology to utilize lower heating frequencies compared to conventional induction gear hardening processes dramatically reduces probability of overheating part's edges, tips and sharp corners that could occur due to electromagnetic end effect and proximity effect when RF heating frequencies are used.
- TSH technology provides improved repeatability and reduced overall process sensitivity. It is far less sensitive than conventional induction hardening to imperfections of part's positioning during heating (for example, part's wobbling due to a bearing wear, when component rotates during the heat cycle).

In order to obtain such a contour-like hardness pattern using encircling inductor and conventional induction hardening technique and taking into consideration the size of the helical gear shown on Figure 3, it would require applying medium frequency and large amount of heating power for very short time to form austenitic layer at gear surface.

Figure 3. Helical gear induction hardened according to TSH Technology (Courtesy of ERS Engineering Corp.)

In contrast to conventional induction hardening process, required kW of power supply can be dramatically reduced with TSH Technology, since it does not demand formation of contour-line austenized layer prior to quenching and does not require having short heat time. This permits using lower power inverters.

Figure 4 shows fine-grain martensitic structure of the case depth region of the helical gear shown on Figure 3.

Table 3 shows a chemical composition of inductively hardened helical gear shown on Figure 3. (Study was conducted by STORK® Climax Research Services, Inc.)

Element	%	Method
C	0.77	ASTM E1019-08
S	0.014	ASTM E1019-08
P	< 0.005	ISO-14707:2000E (Glow Discharge)
Si	0.08	ISO-14707:2000E (Glow Discharge)
Mn	0.06	ISO-14707:2000E (Glow Discharge)
Cr	0.17	ISO-14707:2000E (Glow Discharge)
Ni	0.06	ISO-14707:2000E (Glow Discharge)
Mo	< 0.02	ISO-14707:2000E (Glow Discharge)
V	< 0.008	ISO-14707:2000E (Glow Discharge)
Al	0.0088	ISO-14707:2000E (Glow Discharge)
Cu	0.04	ISO-14707:2000E (Glow Discharge)
Ti	0.09	ISO-14707:2000E (Glow Discharge)
Nb	< 0.008	ISO-14707:2000E (Glow Discharge)

Figure 4. Fine-grain martensitic structure (400X) of helical gear shown on Figure 3

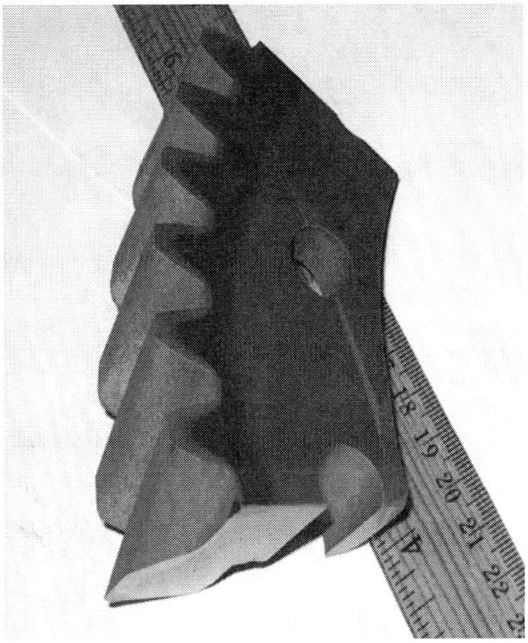

Figure 5. Induction harden pattern (zoomed) of large helical gear shown on Figure 3

Conclusion

The TSH technology is extremely promising technology that was developed to replace thermo-chemical heat treating processes; such as carburizing, carbonitriding, nitriding; or furnace direct hardening currently widely used. Its benefits are particularly noticeable when induction case hardening medium and large size complex-shaped components.

Applicability of the TSH Technology is not only limited to gears. Bushings, transmission shafts, king pins, ball joints, driveshaft and differential crosses, bearings, coil and leaf springs are also good candidates for TSH Technology. Components heat treated utilizing TSH technology are stronger and more durable than some made of conventionally heat-treated standard steels.

TSH technology leads to the obvious cost savings. Case hardened components that utilize TSH Technology have an improved cost structure and superior properties over traditional induction and thermo-chemical heat-treating processes. It is lean and green process.

References

1. V.Rudnev, D.Loveless, et.al., Handbook of Induction Heating, Marcel Dekker, NY, 2003.
2. K.Shepelyakovskii, Induction Surface Hardening of Parts, Mashinostroenie, Moscow, 1972.
3. ERS Engineering Inter-Company Report, 2010.

Proceedings of the 26th ASM Heat Treating Society Conference
B.L. Ferguson, R. Jones, D.S. MacKenzie, and D. Weires, editors
https://doi.org/10.31399/asm.cp.ht2011p0158

Unique Computer Modeling Approaches for Simulation of Induction Heating and Heat Treating Processes

Dr. Valery Rudnev, FASM

Inductoheat Inc., 32251 N. Avis Dr., Madison Heights, MI 48071

e-mail: rudnev@inductoheat.com Tel. 248-629-5055,

Introduction

In the last decade, when discussing subjects related to a computer modeling of induction heating, the word *"usefulness"* has been replaced by the word *"necessity"*. Modern computer simulation is capable of effectively simulating electromagnetic and thermal phenomena for many processes, including those that involve electromagnetic induction. Combination of a sophisticated engineering background with advanced process simulation software provides to induction heating professionals the unique ability to analyze, in few hours, complex technological processes. It offers the ability to predict how different, interrelated and non-linear factors may impact the transitional and final thermal conditions and what must be accomplished to improve process effectiveness, determine the most appropriate process recipes and serves as a comfort factor when designing new systems.

In 2007, ASM International has begun an ambitious undertaking to compile all-new, comprehensive resources on modeling as it applies to a computer simulation of different metal processing technologies. Carefully selected world-recognized experts from leading universities, national research laboratories and industrial corporations from 13 countries were chosen to submit materials. As a result, a brand new two-volume set (Fig. 1) was published as a part of ASM Handbook series. The first part, Volume 22A, *Fundamentals of Modeling for Metals Processing*, appeared in 2009. The second part, Volume 22B, *Metals Process Simulation*, was published in 2010.

This two-volume set covers a wide range of subjects including heating, heat treating, casting, solidification, forming, forging, rolling, coating, joining, machining, powder metallurgy, integration modeling and simulation in design of equipment and many others.

Among other useful information, Volume 22,B consists of two articles that are solely devoted to a computer modeling of induction thermal technologies: *Simulation of Induction Heating Prior to Hot Working and Coating* (pages 475 – 500) and *Simulation of Induction Heat Treating* (pages 501 – 546) [1,2]. It was my privilege to be chosen to exclusively cover intricacies of computer modeling of induction heating and induction heat-treating processes for this Handbook and to author both articles.

Figure 1. New two-volume set of ASM Handbooks on computer modeling and metal process simulation.

How It Has Been Done In The Past

An estimation of the process parameters based on a single formula and "rules of thumb", as well as using the analytical methods and equivalent circuit coil design methods were popular in the 1960's through 1990's. Though those techniques were easy to employ, they were very subjective with inherent major restrictions limiting their use for quick estimation of only ballpark parameters of induction heating systems [1-4]. There was always a danger in obtaining erroneous and inadequate results with such overly simplified estimation techniques

Recent advancements in manufacturing high-performance computers, increased complexity of induction heating applications, further increasing demands to manufacture higher quality parts in combination with a necessity of improving cost-effectiveness of development stage by shortening the learning

curve and reducing development time have significantly restricted the usefulness of those simplified formulas.
Rather than using estimation techniques with many restrictions and disputable accuracy, modern induction heating specialists turned to highly effective numerical simulation methods such as finite-differences, finite-elements, edge elements, boundary elements, mutual impedance method and others. Each of those numerical simulation techniques has certain *pros* and *cons* and has been used alone or in a combination with others [2,4].

For each application or family of similar applications, certain numerical methods or software are preferred. It is obviously quite difficult if not impossible to find a single medicine that will be equally effective for both constipation and diarrhea. The medicine will most likely stop diarrhea or it will help constipation but not both. The same reasoning holds true in searching for an efficient and universal computational tool. There is not a single universal computational method that optimally fits all cases and is optimum for modeling all induction heating applications.

In recent years finite element method became a dominant numerical simulation tool for a variety of engineering applications. Though finite element method (FEA) is very effective modeling technique, it cannot be considered as an ultimate computational tool for all induction heating applications. In some cases a combination of different numerical methods is more effective, in others FEA is a preferred choice.
Any numerical computer modeling techniques require a network mesh of the modeling area. That network includes induction coil(s), the heated workpiece, fixtures and other electrically conductive bodies located in the close proximity to induction coil. Specifics of mesh generation affects accuracy of simulation, required time for pre- and post-processing, as well as actual time to run numerical simulations. As an example, Figure 2 shows network meshes for three the most popular numerical computer modeling techniques.

Even a cursory look at a comparison of network meshes reveals that depending upon specifics of a particular induction application; a particular type of mesh generation technique could be preferable and might be a decisive factor in a selection of numerical simulation techniques that would suit the best to modeling of a particular application [2,4]. For example, with BEM, unknowns are only located on the boundaries or interfaces. Therefore, it requires discretization only of the boundaries of the electrically conductive components of induction system (Fig.2, right). This substantially simplifies one of the most time-consuming parts of numerical model preparation compared to FDM and FEA. References [2,4] discuss PROS and CONS of different computer modeling methods and their applicability to simulate induction heating processes.

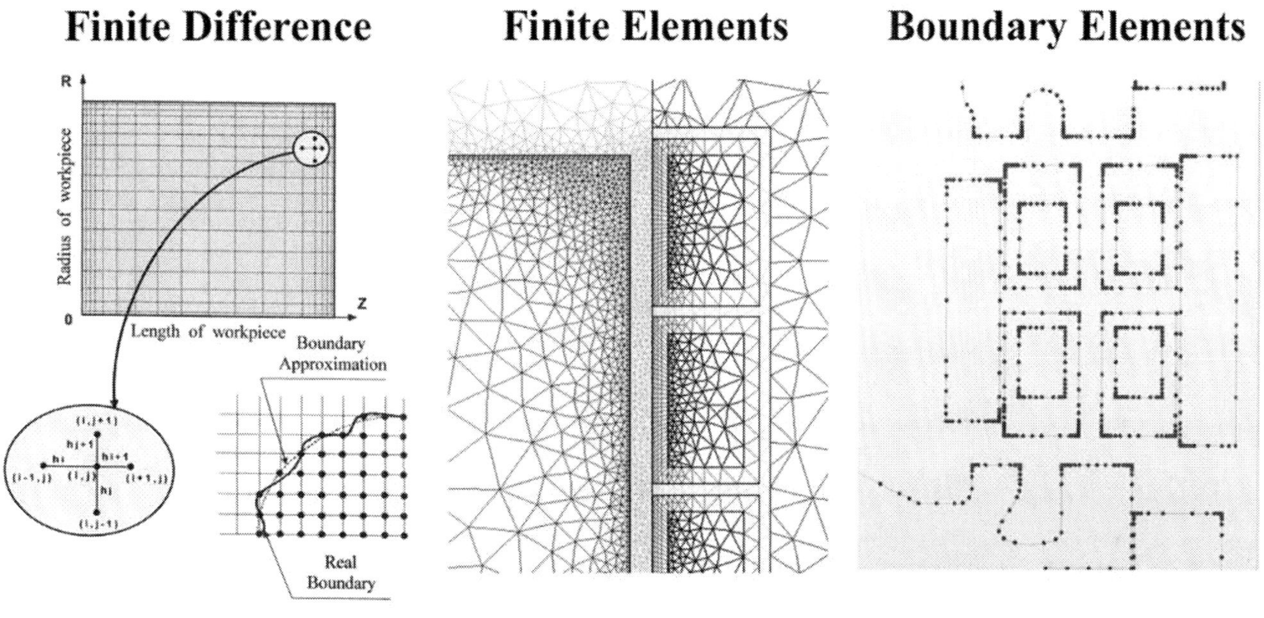

Figure 2. A comparison of network meshes for three the most popular numerical techniques used in computer modeling of induction heating: finite difference (FDM), finite element (FEA), and boundary element (BEM) [2].

Limitations of Generalized All-Purpose Commercial Programs

Majority of commercial codes used for computer modeling of induction heating processes are all-purpose programs that were originally developed for modeling processes taking place in electrical machines, motors, circuit breakers, transformers, magnetic recording systems and were later adapted to induction heating needs. Regardless of well-recognized impressive capabilities of modern commercial software, many of generalized programs experience difficulties in taking into consideration certain features of a particular induction heating application. This includes

- The presence of thermal refractory and necessity to take into consideration thermal radiation view factors.
- Heated workpiece can simultaneously move, rotate or oscillate in respect to induction coil.
- Some operations combine heating and quenching stages.
- Existence of non-uniform *initial* temperature distributions.
- Presence of end plates, guides, fixtures, liners, etc.

Therefore, it is important to keep in mind, that some critical feature(s) of a particular induction heating application could be limiting factor for all-purpose software and creating considerable challenges for generalized commercial codes forcing an analyst to make not well-defined assumptions that might dramatically affect an accuracy and usefulness of simulations.

Our experience shows that there is not a single universal computational method that optimally fits all induction heating applications. Taking into consideration specifics and subtleties of a wide variety of induction heating applications, it is preferable to have number of subject-oriented and highly effective software rather than searching for a single universal code. For family of similar problems, certain numerical methods or software are preferred.

Scientists and engineers of Inductoheat Inc. utilize a variety of generalized commercial software as well as number of proprietary codes in their professional activity allowing selecting the most appropriate modeling technique and optimizing a process of computer simulation.

Case Study #1: Induction Scan Hardening

Numerical computer simulations allow manufacturers of induction equipment to determine comprehensive details of the process that would be difficult, if not impossible, to determine experimentally. As an example, Fig.3 shows the results of computer modeling the sequential dynamics of induction scan hardening a hollow shaft using a two-turn MIQ (machined integral quench) inductor with an "L"-shaped flux concentrator ring (frequency = 9 kHz).

At the beginning (Fig.3, A and B), a 2.6 second power dwell is applied to properly heat the shaft fillet area. During this stage, an inductor is energized but does not move and quenching is not applied. Upon completing the dwell stage, the shaft fillet is sufficiently preheated and scanning begins. Scan rate and coil power are varied during scanning to allow proper accommodation of changes in shaft geometry. Computer modeling reveals several important process subtleties [3]:

- During scanning, appreciable heating of the shaft begins at a distance a good deal above the top copper turn creating a preheating effect. Factors responsible for preheating are heat flow in the axial direction due to thermal conduction and propagation of the external magnetic field generating heat sources outside of induction coil.
- Presence of an external magnetic field outside the induction coil is also responsible for the post-heating of shaft areas located immediately below the bottom turn, and, in some cases, even in regions where the subsequent quenchant impinges the shaft surface. With insufficient quenching, the latter can dramatically reduce quenching severity and potentially create conditions for crossing the nose of the CCT (continuous cooling transformation) curve, resulting in the formation of mixed structures with the presence of upper transformation products (e.g., bainitic/pearlitic structures or "ghost" networking). Such microstructures are notorious for scattering and lower hardness readings.
- Electromagnetic proximity effect and coil end effect both cause the hot spots appearing on shoulder near a shaft diameter change (regardless of the coil power reduction there). During scanning, the magnetic field preferably couples to the shoulders leading to a power density surplus at those locations. The presence of hot spots produced by power surplus necessitates having prolonged cooling to remove excessive heat, ensuring martensitic formation and obtaining sufficient hardness. At the same time, a heat deficit could occur in the undercut region and transition area near the shaft's smaller diameter.

"Comet- Tail" Effect:

It is imperative to take into consideration a comet-tail effect when developing a scan hardening process recipe. Figure 4 (left & right) shows magnified temperature patterns of intermediate process stages (Fig.3, E & F respectfully). The comet-tail effect manifests itself as a heat accumulation in shaft subsurface regions below the scan inductor, being pronounced in the areas of a diameter change. Upon quenching, the temperature of shaft surface can be cooled sufficiently below the M_s temperature. At the same time, the heat accumulated in the shaft subsurface might be sufficient for tempering back of as-quenched surface regions and could potentially result in appearance of soft spots within the case depth. Sufficient quench-out is essential to prevent this undesirable phenomenon.

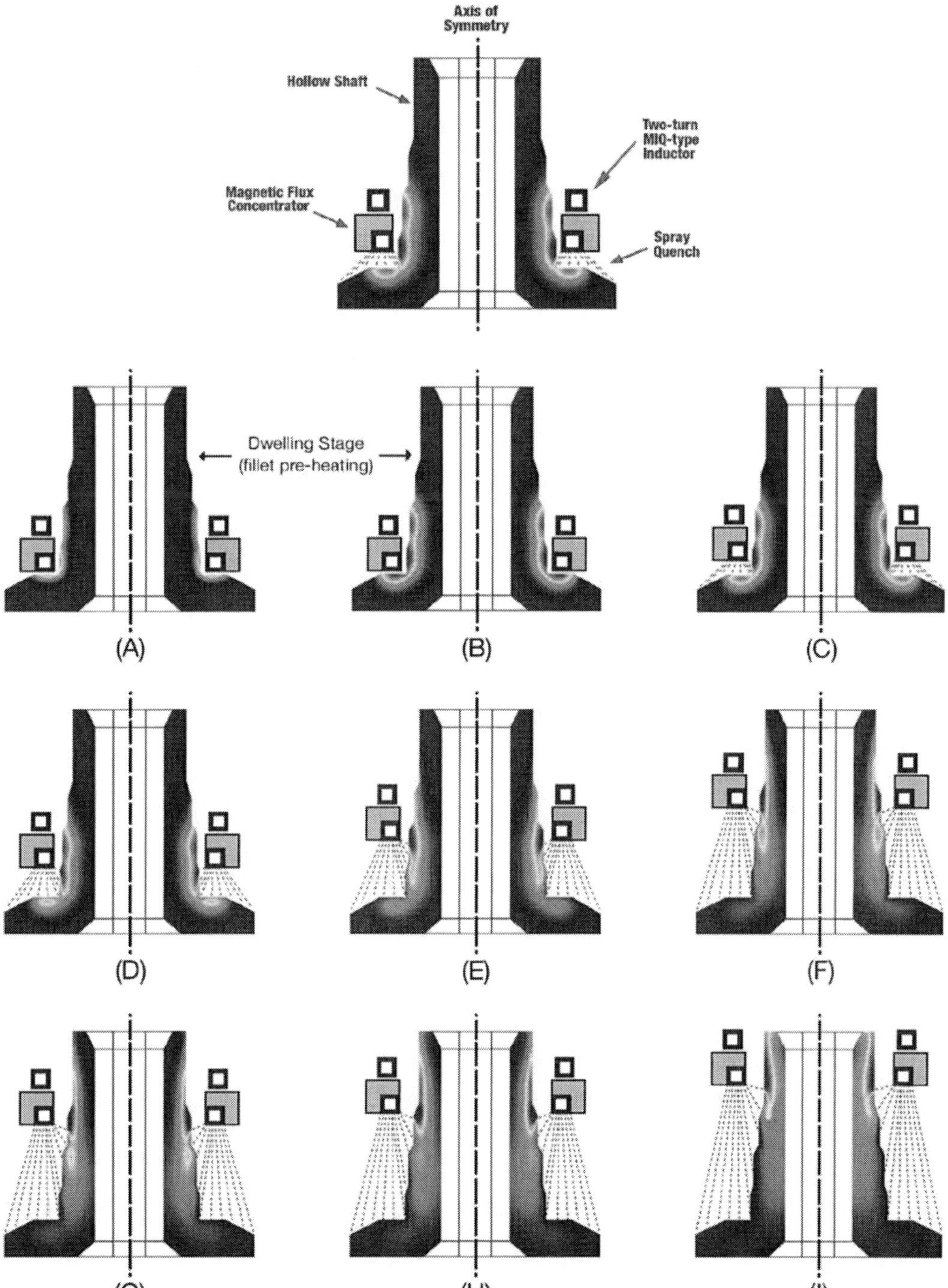

Figure 3. Computer simulation of the sequential dynamics of induction scan hardening a hollow shaft using a two-turn machined integral quench (MIQ) inductor with an L-shaped magnetic flux concentrator ring (frequency = 9kHz) [2].

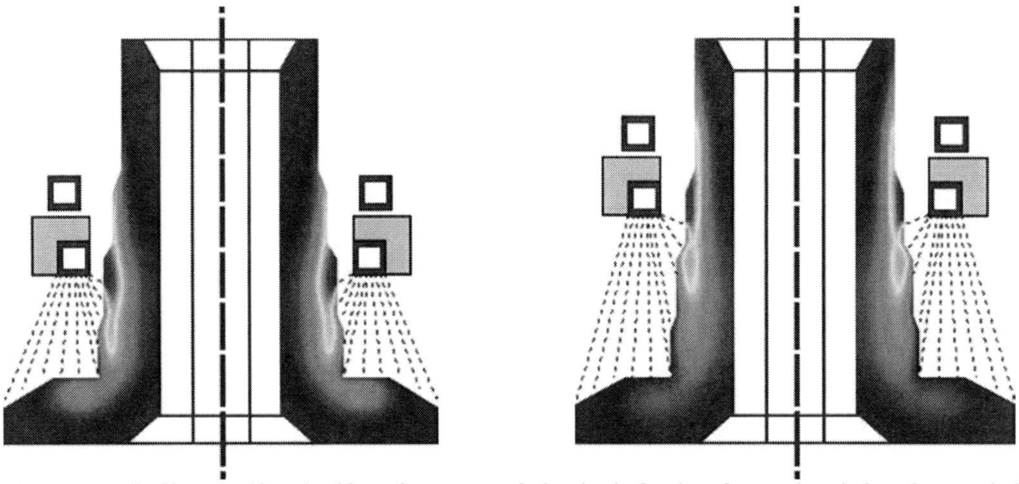

Figure 4. The comet-tail effect manifests itself as a heat accumulation in shaft subsurface regions below the scan inductor [2].

Unfortunately, a limitation of the great majority of commercially available all-purpose software is that they are not capable of taking into consideration a comet-tail effect when trying to model induction scan hardening. In addition, some software cannot properly handle pre- and post-heating effects appeared due to the external magnetic field propagation and axial heat flow due to thermal conduction. Make sure that applied software is free of these restrictions and properly models all-important physical phenomena.

When designing inductors and developing optimal process recipes, it is imperative to properly model not only heating but the spray-quenching stage as well. Otherwise, crucial aspects of the process might be missed, having a negative impact on modeling accuracy and its usefulness.

Case study # 2: A Common Incorrect Assumption

Some practitioners incorrectly assume that with induction billet/rod/bar heating prior to hot and warm forming, the coldest temperature is always located at the core of the workpiece and the maximum temperature is always located at its surface. It is also often assumed that overheating does not occur if surface temperature measured by a pyrometer or thermocouple does not exceed the maximum permissible level. Besides that, process control systems that predict (simulate) rise of an average (mean) temperature and temperatures of surface and core of the billet often assumed to be sufficient to guarantee proper heating providing a comfortable factor for some practitioners. Example of such typically predicted "surface-to-core" temperature profile is shown on Figure 5. However, it is imperative to recognize that under certain but very realistic conditions, the presence of heat losses from billet's surface may shift the temperature maximum further away from the surface marking its location somewhere beneath the billet's surface.

Study shows [5,6] that positioning and magnitude of the subsurface temperature surplus, potential steel overheating and even burning is a complex function of four major factors: *frequency*, *refractory*, *final heating temperature*, and *power distribution along the heating line*.

Lower *frequencies* increase the current penetration depth resulting in more "in-depth" heating and leading to a faster temperature raise at billet's core. This shortens the induction line, but on another hand, under some conditions, it can also increase the subsurface overheating by making it more pronounced and shifting the location of the maximum temperature further away from the surface.

The use of an appreciably thick *refractory* with improved thermal insulation properties does just the opposite, reducing subsurface over-heating and shifting billet's maximum temperature towards its surface.

Increase of the *forging temperatures* lead to an effect similar to an effect of lowering frequency in respect to a location of maximum temperature and severity of the subsurface heat surplus

An effect of *power distribution along the heating line* on billet's radial temperature distribution is more complex and it is seldom discussed in the literature. In most publications devoted to progressive induction heating of billets, it is strongly suggested to have a graded (profiled) power distribution along induction line by putting more power into the coils at the beginning of the line. Putting more power up-front might sound as a universal "rule of thumb" since it forces more energy into the billet at the front of the heating line, allowing more time to soak into the core and shortening the length of the line. This approach typically utilizes a single inverter that powers several coils with graded number of copper turns or/and series/parallel coil circuit connections.

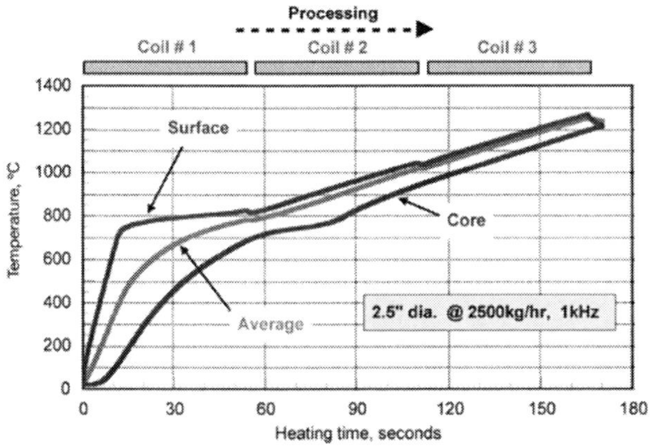

Figure 5. Conventional "surface-to-core" time-temperature profiles [1,5,6]

The problem with this approach, however is that the power distribution along the heating line in some installations cannot be easily modified if the production rate, kind of metal or billet size changes. For example, if the production rate is reduced, a subsurface overheating typically worsens with a conventional induction design potentially negatively affecting the billet's subsurface microstructure. It is also very common to find an appearance of billet-sticking problems to be more pronounced with graded power distribution along induction line when the system runs at a rate slower than the nominal for which it was designed. Since the system puts more energy into the billet in the beginning of the heating line, too much energy soaks down into billet's subsurface area in cases when the line runs slow. The presence of surface heat losses can reverse a traditionally expected radial temperature profile leading to the subsurface temperature being greater than at its surface.

In many cases, subsurface temperature might be hot enough to cause the billets to fuse together. The effect of subsurface overheating is particularly pronounced when heating smaller size billets at a lower rate using an induction line designed for heating larger billets at a nominal rate. As an example, Figures 6 shows "surface-to-core" profiles when heating 2"(50.8mm) diameter billets (Fig.6, left) at a slower rate utilizing a conventional induction heating line designed for processing 2.5"(63.5mm) billets (Fig.6, right) at a nominal rate (see Fig.5). Note that in both cases the billet's surface temperature that would be recorded by pyrometer is the same. Further reduction in billet's diameters could worsen a severity of subsurface overheating [5,6].

Besides a potential danger of a premature die wear on hammers and presses, as well as other issues related to altering a quality of forged parts improperly heated billets can raised some safety concerns.

Practice shows that when heating large billets at nominal rates, more power should be shifted towards the beginning of induction line. At slower rates however, when heating smaller than nominal size billets it is desirable to re-distribute power by its shifting towards the end of the induction line.

Taking into consideration that pyrometers can only reliably measure billet's surface temperature, there is always a danger to "miss" an appearance of the sub-surface overheating. Therefore, precise temperature prediction is imperative in order to avoid localized over-heating that is related to "hot shortness" problems due to residuals and possible variations in the chemistry of given steel.

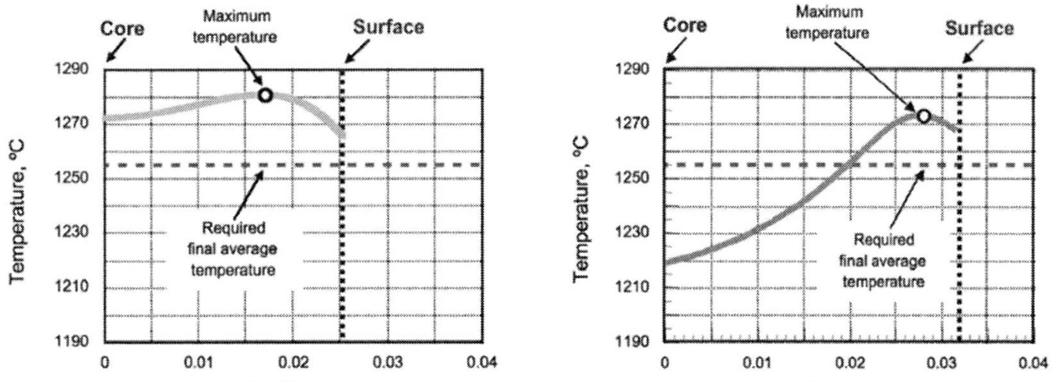

Figure 6. Final "surface-to-core" temperature profiles when heating 2"(50.8mm) dia. billets at a slower rate (left) utilizing a conventional induction line designed for processing 2.5"(63.5mm) billets at a nominal rate (right) [1,5,6].

Case Study # 3: FLUXMANAGER™ Technology for Pipe End Heating

Stress relieving of steel tube/pipe ends is typically done prior to machining of the thread. To stress relieve using induction heating, the tube/pipe end is placed in a multi-turn induction coil and heated for the specified time and specified process parameters including final part temperature, required heated pipe end length, allowable radial and longitudinal temperature deviations, and production rate.

Some applications call for a sharp longitudinal heat transition zone while others require certain transient temperature profiles. Pipe diameters typically range from 0.375 to 20 in. (approx. 10 to 510 mm) with wall thicknesses between 0.150 and 1.25 in. (3.8 and 32 mm). The length of heated tube/pipe end ranges from 2 to 18 in. (50 to 460 mm) depending upon application specifics. Achieving axial and radial temperature uniformity is imperative for a quality product

Figure 7. Thick-wall oil country tubular goods

In the past, there have been a number of successful induction tube/pipe end heating systems – utilizing frequencies in the range of 1-30 kHz – supplied to industry. Recent trends to increase pipe wall thicknesses for oil country pipes (Fig.7) in combination with tighter requirements for heat uniformity [7], however, have outlined several drawbacks of using higher frequencies versus line frequency when heating thick-wall magnetic steel pipes to temperatures suitable for stress relieving. This includes

- When heating thick-wall pipes the "skin" effect (even at line frequency) is very pronounced and the ratio of "wall thickness-to-eddy current penetration depth" is quite large [4]. Higher frequencies have a tendency to further increase this ratio. For example, this ratio can easily exceed 15 when heating thick-wall pipes utilizing a frequency of 1kHz. With 10kHz this ratio can exceed 40. Therefore, there will be a danger of localized overheating the outside surface of thick-wall magnetic pipes while trying to provide required temperature at pipe's inside diameter. This could result in an undesirable heterogeneous stress-relieving structure.
- Higher frequencies are noticeably more sensitive to pipe positioning inside the induction coil, meaning that even slight variation in coil overhang and coil-to-pipe proximity due to a non-symmetrical positioning could lead to an appriciable temperature variation at the pipe-end area. This lowers process repeatability and negatively affects controllability typically resulting in the appearance of the "hot" and "cold" spots.
- High frequencies require using solid-state inverters that, in some cases, could appreciably increase the capital cost of the machinery.

Computer simulation is an ideal tool to determine the appropriate coil design and heat recipe for this application. Figure 7 shows an example of FEA computer modeling of induction heating a 12 in. (305 mm) long end portion of a 14.625 in. diameter (372 mm) carbon steel pipe with a wall thickness of 0.625 in. (16 mm) from ambient temperature to 600°C (1110°F). The required temperature uniformity of ±20°C (±36°F) at any point within the required end region of the pipe is achieved using Inductoheat's FluxManager™ technology (frequency = 60 Hz). The induction coil comprises five groups of turns (8 + 4 + 4 + 4 + 12turns). Lamination shunts placed outside the induction coil enhance heat efficiency and repeatability.

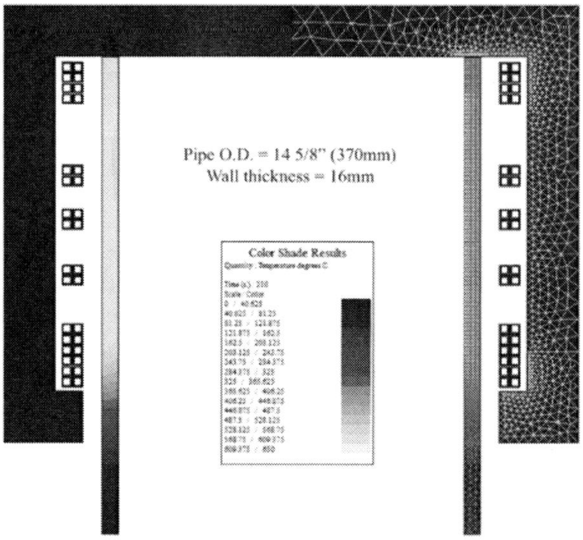

Figure 8. An example of FEA computer modeling of induction heating a 12 in. (305 mm) long end portion of a 14.625 in. diameter (372 mm) carbon steel pipe with a wall thickness of 0.625 in. (16 mm) from ambient temperature to 600°C (1110°F). FEA mesh is indicated on the right side [8].

Crucial Tips Executives Must Know Regarding Computer Modeling of Induction Heating

Tip # 1. It is important to remember that any computational analysis can at best produce only results that are derived from the correctly defined theoretical model, governing equations and boundary conditions. Therefore, before you hire somebody to do computer simulations, make sure that analyst(s) has clear understanding of the process specifics, as well as an appropriate education *in the area* where you are seeking help. For example, when you are flying an airplane, you expect from a pilot to have an appropriate training. When you need medical assistance, you expect from a doctor to have a proper education and appropriate medical degree. Same principle should be applied when you are choosing a company or analyst(s) to do computer simulation for you. Otherwise, you might get pretty pictures but erroneous and technically inadequate results and numerous excuses why modeled results do not match practical data. Therefore, simple question what is the basic education of computer modeling analyst that you are about to hire to do simulations for you might be important step in avoiding unpleasant surprises and wasting your time and money.

Tip # 2. Make sure that physical properties of heated materials are properly defined. Well-know but rude saying "garbage in - garbage out" clearly indicates a necessity of having accurate physical properties of heated material. Experience shows that poorly defined material properties are responsible for appriciable amount of simulation errors. Only reliable sources should be used when adapting physical properties.

Tip # 3. In order to minimize the risk, it makes sense to deal with companies that offer, so-called, one-stop service (Figure 9, right). In this case one company would be responsible for all stages of R&D (including computer modeling), design, fabrication, testing, start-up and aftermarket support, rather than dealing with number of companies with not well defined overall process responsibilities (Figure 9, left) in cases, if something goes wrong.

Tip # 4. It is important to recognize that the use of modern numerical software (including finite elements, boundary elements, finite difference, edge elements, etc.) does not guarantee obtaining correct simulation results. It must be used in conjunction with experience in numerical computations, proper education of analyst(s) and his/her engineering background. This is especially so because even in modern commercial software, regardless of the amount of testing and verification, a computer modeling program may never have all of its possible errors detected. The analyst must consequently be on guard against various kinds of possible errors. The more powerful the software, it is more complex with potentially having the greater probability of errors. Be aware that computer generated attractive pictures might be misleading if were obtained by amateur. Common sense, engineering "gut feeling" and advance education in area of modeling are always the analyst's helpful assistants.

Conclusions

In fast-pace world economy an ability of induction heating manufacturers to minimize time between a customer request for quotation and quotation through efficient computer modeling is critical for company's success. In opposite to academia, a rapidity of industry does not often have a luxury to wait several days in order to obtain the results of modeling but demand having reliable results of computer simulation within a few of hours.

Measures should be taken in making sure that properly educated analysts use proper simulation software to perform a computer modeling. This is the reason why some induction heating manufacturers (for example, Inductoheat Inc.) besides using commercially available generalized software also developed various problem-oriented highly effective proprietary programs that allow to effectively simulating particular group of induction heating and heat treating applications taking into consideration all important process features.

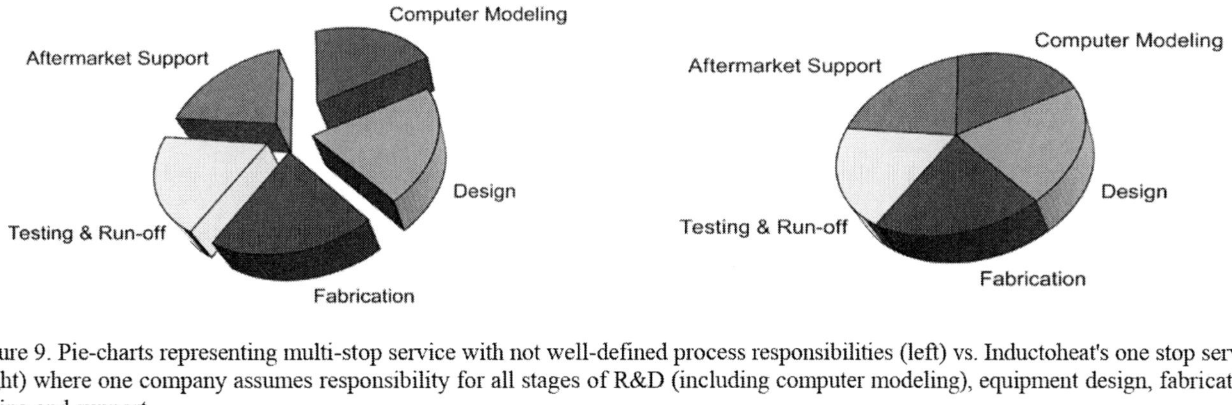

Figure 9. Pie-charts representing multi-stop service with not well-defined process responsibilities (left) vs. Inductoheat's one stop service (right) where one company assumes responsibility for all stages of R&D (including computer modeling), equipment design, fabrication, testing and support.

References

[1] Rudnev,V., Simulation of Induction Heating Prior to Hot Working and Coating, *ASM Handbook series*, vol. 22B: Metals Process Simulation, editors D.U.Furrer and S.L.Semiatin, ASM Int., 2010, p.475 – 500.

[2] Rudnev,V., Simulation of Induction Heat Treating, *ASM Handbook series*, vol. 22B Metals Process Simulation, editors D.U.Furrer and S.L.Semiatin, ASM Int., 2010, p.501-546.

[3] Rudnev,V., Computer Modeling of Induction Heating: Things to be Aware of, Things to Avoid, *Industrial Heating*, May, 2011.

[4] Rudnev, V., Loveless, D., et. al., *Handbook of Induction Heating*, Marcel Dekker, NY, 2003.

[5] Rudnev,V., Brown.D., Van Tyne, Ch., Clarke,K., Intricacies for the successful induction heating of steels in modern forge shops, *Proc. of 19th Int. Forging Congress*, Chicago, 2008, p.71-82.

[6] Brown,D., Rudnev,V., Lin,J., Nakagawa,T., *Superior induction heating technologies for modern forge shops*, Proceedings of Japanese Forging Association (JFA), July, 2009.

[7] Ross,P., Rudnev,V., Gallik,R., Elliott,G., Innovative Induction Heating of Oil Country Tubular Goods, Industrial Heating, May, 2008, pp.67-72.

[8] Rudnev,V. Tips for computer modeling induction heating processes, Part 1, *Forge*, July, 2011, pp.21-23.

Proceedings of the 26th ASM Heat Treating Society Conference
B.L. Ferguson, R. Jones, D.S. MacKenzie, and D. Weires, editors
https://doi.org/10.31399/asm.cp.ht2011p0167

Data acquisition for numerical modelling of induction surface hardening – process specific considerations

Maximilian Schwenk, Jürgen Hoffmeister, Volker Schulze
Karlsruher Institute of Technology (KIT), Institute for Applied Materials (IAM-WK), Karlsruhe, Germany
maximilian.schwenk@kit.edu, juergen.hoffmeister@kit.edu, volker.schulze@kit.edu

Abstract

Induction surface hardening is a widely used manufacturing process in order to improve the mechanical properties of components. However, better process understanding as well as process development requires numerical modeling. The modeling itself depends on the input data regarding the process and the material behavior. Data acquisition is a rather difficult task due to very short process times, as seen in contour hardening of gears. The paper will give an overview over critical aspects regarding input data and their acquisitions as well as a comparison between experiments and simulation in order give a better understanding for improving the modeling.

Introduction

Modern mass production of mechanical components in the automotive industry faces an ever increasing pressure of high quality products at the lowest cost possible. A very critical process step is the heat treatment close to the end of the manufacturing chain. No other process step influences the final properties of the component in terms of wear resistance, load bearing capacity, distortion and residual stress distribution as much. Due to several reasons such as its ease of integration in a manufacturing chain, reproducibility and low cause of distortion, induction surface hardening can be found in modern production chains [1].

Most often, the process development necessary in order to integrate this technology in manufacturing chains is based on "trial and error", leading to rather high development costs as well as long development times. Increased efforts in switching from case hardening to induction surface hardening have proven to be cumbersome because not only the process, but also the material changes. Especially in the automotive industry, this leads to long development times because internal benchmarks in terms of mechanical properties need to be met.

The reduction of development time can be achieved by the use of numerical models. However, the numerical modeling of this process still falls short in comparison to general heat treatment simulations. One of the main reasons is the difficulty of data acquisition, since material and process parameters are very difficult to determine, especially when dealing with high power density applications such as in contour hardening of gears, where heating times can range from 0.1-1 second. General models describing induction heating and hardening have been presented numerously, addressing the main features of this process [2,3].

The first part of this paper will give an overview over the material and process parameters to consider as well as possible solutions to extract them from experiments. The second part deals with a direct comparison of induction surface hardened specimens made of quenched and tempered AISI 4140 and their corresponding simulation in order to point out specific aspects to be considered in order to further improve numerical models.

Induction Hardening – Process Complexity

Induction surface hardening is a very complex process because several physical fields interact with each other as shown in Figure 1. The coil's electromagnetic field induces eddy currents in the workpiece which lead to ohmic losses. Hysteresis losses in ferromagnetic materials can usually be neglected because their contribution is rather small [4]. The distribution of the so-called Joule losses depends on the electromagnetic properties such as the magnetic permeability μ and electric conductivity σ as well as the frequency f. This relation leads to the skin depth δ defined as follows:

$$\delta = \sqrt{\frac{1}{\pi f \mu \sigma}} \qquad (1)$$

Both, the magnetic permeability μ and the electric conductivity σ are temperature dependent, whereas the permeability μ also depends on the magnetic field strength H. Considering these material properties to be dependent on the phase fraction would be possible too, but not considered worthwhile. Therefore the electromagnetic field can be decoupled from the mechanical/metallurgical field.

The temperature distribution within the workpiece can be determined by the heat conduction equation [5]:

$$\underbrace{\frac{\partial}{\partial x_i}\left[\lambda_i(T,f)\frac{\partial T}{\partial x_i}\right]}_{1}+\underbrace{\frac{\sigma\varpi^2|\vec{A}|^2}{2}}_{2}+\underbrace{\sum_k \Delta H_k \frac{df_k}{dT}}_{3}+\underbrace{\sigma_{ij}\,\dot{\varepsilon}_{ij}^{pl}}_{4}=$$

$$\underbrace{\rho c_p(T,f)\frac{\partial T}{\partial t}}_{5} \qquad (2)$$

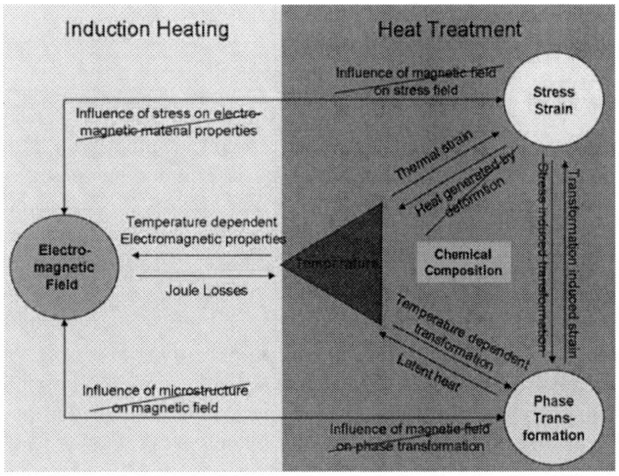

Figure 1: Physical interactions in induction heating [6]

The first term on the left hand side describes the heat conduction, which is reflected by the thermal conductivity λ. The remaining terms on the left hand side describe the heat sources/sinks, where ϖ is the angular frequency, A the magnetic vector potential, ΔH_k the enthalpy of transformation, f_k the phase fraction of phase k, σ_{ij} the stresses and $\dot{\varepsilon}_{ij}$ the plastic strain rates. The second term expresses the heat source due to the induced eddy currents, the third term the latent heat and the fourth term the heat dissipation due to mechanical deformation. Usually the last can be neglected, since the plastic strains occurring are rather small [7]. The term on the right hand side (5) represents the temperature evolution with time. Here, the density ρ and the specific heat c_p influence the temperature field. The thermal properties depend on the phase fraction and temperature as indicated in equation 2.

The spatial and temporal temperature evolution leads to corresponding phase transformations and causes thermal gradients. Phase transformation kinetics is primarily influenced by the chemical composition and the initial microstructure [8]. During the transformation itself, so-called transformation strains occur due to the difference in density between the old and new phase. Besides the thermal strain ε_{ij}^{th} caused by temperature gradients and the transformation strain ε_{ij}^{tr}, also elastic strains ε_{ij}^{el}, plastic strains ε_{ij}^{pl} as well as strains due to transformation plasticity ε_{ij}^{tr} occur [5]. In case of the elastic strain, the Young's modulus E and Poisson's ratio ν depend on phase fraction and temperature. For the plastic strains, the flow stress σ_f must be given for each phase, depending on temperature and equivalent plastic strain ε^{pl}. The strain due to transformation plasticity depends on the transformed phase fraction as well as its rate and can be expressed by the transformation plasticity constant K, which is material dependent [9].

Furthermore, it is possible to consider the influence of stresses on the phase transformation as described in [10,11]. As mentioned before, the interaction between the metallurgical field and the electromagnetic field could be considered as well. The following sections shall give an overview of how to extract specific material data required to model induction surface hardening.

Process Parameters

Process parameters are a crucial step towards good simulation results. However, it must be clear that general induction heating causes some major problems in extracting them as well as matching the numerical model to the actual process. Considering contour hardening of gears, a single shot hardening process is applied, which means that the gear is rotated at rather high speed, while being enclosed by the induction coil. For this reason, induction hardening of such components is power controlled and not temperature controlled. During the heating cycle, the generator will induce a constant power into the workpiece by increasing the current in order to compensate the power drop caused by the loss of ferromagnetism at the Curie temperature.

In most high power applications concerned with contour surface hardening, the current measurement directly at the coil will not be possible. Therefore, the current measurement with a Ragowski coil has to be conducted at the outlet of the frequency converter. Depending on the data recording equipment and heating time, the whole heating cycle might be recorded. In our case, only a regular scope TPS2000B from Tektronix was available and therefore only a part of the heating cycle could be recorded. In order to use the information for the modeling, the generator control was switched from power to current control. Further on, this might allow a better understanding since no correlations between heating parameters and temperature evolution are possible so far.

The measurements were conducted separately for several current settings for medium and high frequency of 12 and 325 kHz, respectively. The dual frequency generator was provided by Eldec Schwenk Induction GmbH with 200 kW medium frequency (MF) and 250 kW high frequency (HF). The current measurements were related to the defined input current displayed at the beginning of the heating cycle. The rms value or effective current was determined by integrating over the measured data. Figure 2 shows the relation between the defined input current and the effective current. The linear correlation shows that the evaluation of the limited data recorded gives good results.

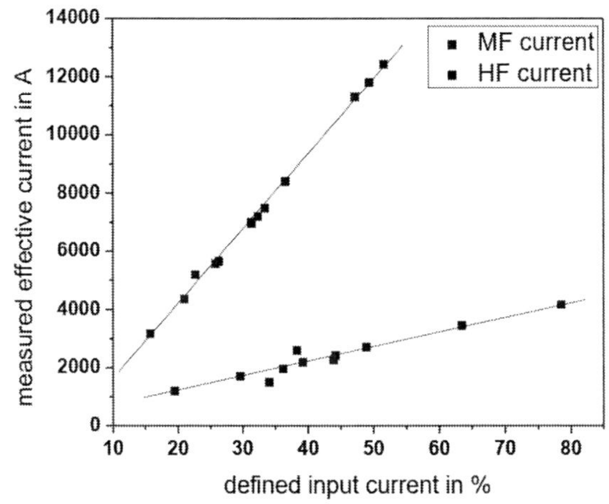

Figure 2: Correlation between measured current and defined input current

Besides the current input, the frequency and heating time need to be determined as input data for numerical models. The frequency will change during the heating cycle due to the change of the oscillating circuit caused by changing material properties of the workpiece and hence changing inductance of the resonant circuit. However, the frequency change will not have any major influence on the heating itself since the change will be within the same order of magnitude [12]. Therefore, the frequency displayed by the generator may be used. More important will be the heating time, which can be approximated in two ways. The first possibility would be the use of the current measurement. Once the heating cycle is started, the scope will start recording as shown in Figure 3. Around 100 ms, full power is supplied to the induction coil.

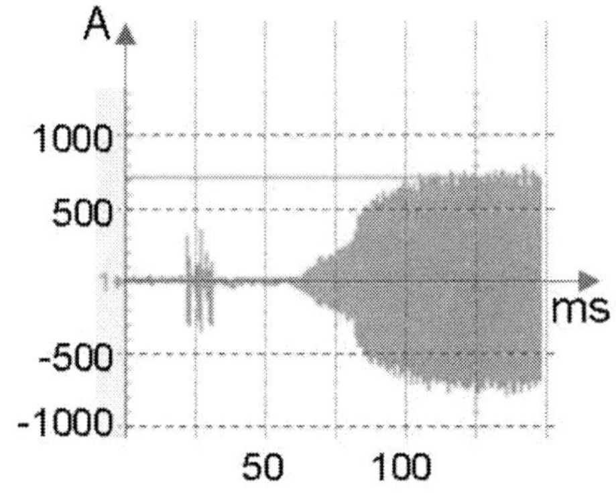

Figure 3: Estimation of actual heating time using scope measurements of current (MF measurement 15.8%)

Another possibility is the use of temperature measurements, which are needed for validation purposes. For the experiments conducted in this study, cylindrical specimens of quenched and tempered AISI 4140 (570°C/3h) were used. Figure 4 shows the dimensions of the specimen as well as the locations of the thermocouples for the MF and HF. A fifth thermocouple was spot welded to the surface of the specimen. In order to conduct the temperature measurements, the cylindrical specimens were not rotated, disregarding the inhomogeneous heating towards the leads of the inductor.

From the temperature measurement at the surface it is possible to estimate the actual heating time. This might lead to a small error at the beginning of the simulation, since the transient part of the current rise will not be reflected correctly as seen in Figure 3, if a constant current density is applied.

The time needed to move the quenching ring into its quenching position can be calculated from the travel distance and the travel speed or from the surface temperature measurement, which will give the exact time when the quenching fluid actually wets the surface, seen by a rapid decrease of the measured temperature. During the delay time needed to move the workpiece into its quenching position, the temperature drop at the surface is mainly dominated by heat conduction into the workpiece.

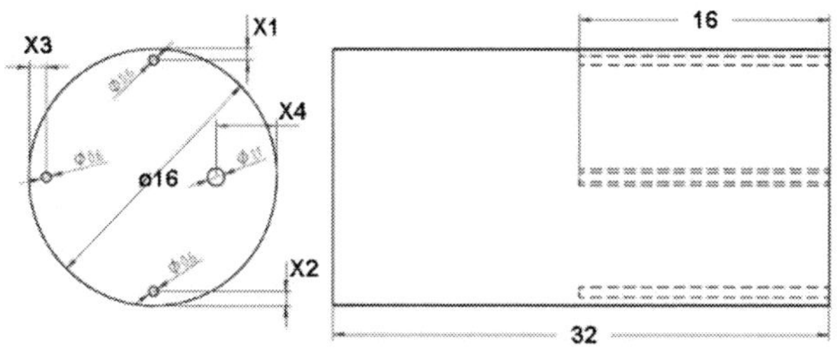

	Location of Thermocouple [mm]			
	X1	X2	X3	X4
MF	0.7	0.9	1.1	4.0
HF	0.7	1.2	1.7	4.0

Figure 4: Thermocouple positioning in cylindrical specimen

For the quenching part of the simulation, the heat transfer coefficient (HTC) as thermal boundary is needed. There are standard methods to determine the quenching intensity of quenching fluids [13]. In the work of [14], the influence of quenching parameters such as temperature, polymer concentration and flow rate were investigated in terms of quenching degradation during the use in induction hardening. The ivf smartquench® used complies with ISO standards, which means that the diameter of the test probe is relatively small. For this reason a direct use in determining the quenching intensity of quenching systems used may only be practical if the component has approximately the same diameter in order to reflect the correct wetting kinetics. In our simulation, the HTC for the quenchant with a polymer concentration of 8% at a flow rate of 20 l/min and a temperature around 28°C was used.

In [15] the basic principle of the finite difference scheme to solve the heat conduction problem using the inverse method for cylindrical specimens is given. Having developed such a tool, allows the use of arbitrary diameters, as long as the assumption of the heat conduction in radial direction is fulfilled. Very important to notice is the sensibility of such methods, because the results dependent on the discretization as well as time steps chosen, showing the main difficulty when using inverse methods for the determination of the HTC.

Another approach is the use of commercially available inverse method as offered by Deform-HT®, for example. In case of complex components with locally differing HTC's, such methods are the only possibility to retrieve input data for the simulation. However, it must be clear that the HTC calculated has no physical meaning which means that an interpretation on how to influence the HTC for heat treatment design purpose will not be possible. In [16], a comparison of different inverse heat transfer methods is made, showing the problem mentioned when using commercial inverse method tools.

Material Data

Numerical models depend on good material data in order to describe the material behavior during the specific process. Besides the thermal and mechanical properties usually needed in heat treatment simulations, the electromagnetic/thermal coupling requires also the electric conductivity and relative permeability. The electric conductivity can be determined by measuring the voltage drop over a predefined distance applying a constant current. Since the conductivity depends on the length and cross section of the specimen used, the increase due to thermal expansion has to be considered.

Far more complicated is the extraction of the relative permeability as discussed in [17]. The magnetization curves used in [1] are retrieved from a French PhD thesis, which were measured using a Fluxmeter installed at European energy efficiency research center (ECLEER) of Électricité de France (EDF) at Moret sur Loing. Another possibility is the use of a so-called vibrating sample magnetometer (VSM), which is usually used for thin film characterization. The device measures the magnetic moment due to the applied magnetic field. Commercial devices reach several Tesla and temperatures up to 1273 K.

The magnetization curves at different temperatures can be used to describe the nonlinear behavior of the material. In [18] the concept of the magnetic co-energy is explained. A similar approach is used in the finite element software Flux 2D/3D [19]. Figure 5 gives the temperature dependent parameters of the so-called Frohlich equation. The parameters were extracted by combining the magnetization curves given in [20,21] and measurements conducted using a VSM.

Thermal material data will not be discussed at this point because there is vast amount of data available. The thermal conductivity can be determined by the so called laser flash method. In order to determine the thermal conductivity, the density as well as specific heat needs to be known. The specific heat can be determined with the differential scanning calometry (DSC). In [22] a comparison of thermo physical property measurements of austenitic steel from different research institutions is given, discussing the good agreement of measurements within the group of participants. In the present simulation model, thermo physical data from [23] were used.

In order to model the mechanical material behavior in terms of distortion and residual stress distribution, the flow stress curves of each phase present needs to be determined for the temperatures at which this phase may be stable along the thermal history of the heat treatment. Further, it is important to ensure that only the desired phase is present during the experiment. Depending on the initial microstructure, different strategies need to be applied.

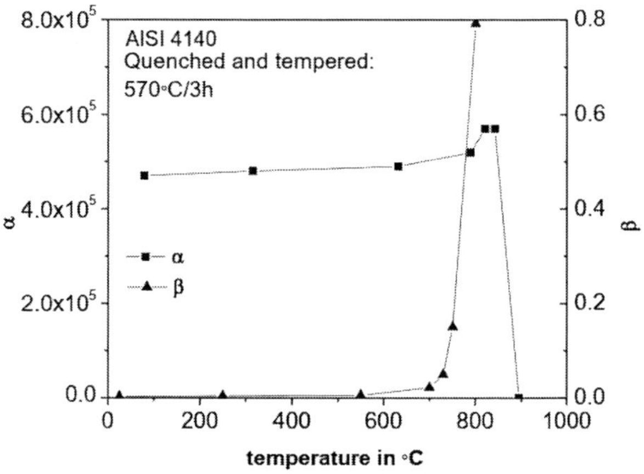

Figure 5: Material parameters α and β for nonlinear magnetic behavior versus temperature

In general, the initial microstructure is regarded as one phase, not distinguishing between ferrite and pearlite. In case of a normalized microstructure, standard tensile test can be used up to temperatures below the A_{c1}-temperature, neglecting thermally activated recovery processes. In case of a quenched and tempered microstructure, standard tensile tests can only be conducted up to temperatures below the tempering temperature. Above it, thermo mechanical simulators such as a Gleeble need to be used in order to minimize the effect of tempering effects on the mechanical properties.

The phases martensite, bainite and austenite need to be determined using a thermo mechanical simulator because high heating and cooling rates are necessary in order to prevent any unwanted transformations (cooling) or tempering (heating). For bainite, the sample needs to be cooled down from austenization temperature to the test temperature (assuming that 100% bainite will form at isothermal holding) followed by the tensile test after isothermal holding has ensured a complete transformation into bainite.

For martensite, the specimens need to be austenized and quenched in order to ensure 100% of martensite. For temperatures below the M_f-temperature, the tensile test can be conducted right after the cooling. Since no pure martensite exists at temperature above the M_f-temperature, the specimen must be cooled below the test temperature and reheated, which may lead to minor tempering effects. The temperature range between M_s and M_f temperature upon cooling may be used for extrapolation purposes.

The most challenging phase is austenite, since the specimen needs to be cooled down before any transformation can take place. This will strongly depend on the alloying elements of the steel. The time available for the tensile test can be taken from the TTT diagram of the corresponding steel. In the case of AISI 4140, the ferrite/pearlite nose in the TTT diagram initiates after approximately 10 seconds. This means that the testing at this temperature must be conducted within this time window. Hence, special thermo mechanical simulators are being used for such experiments. Besides the standard Gleeble simulator, [25] presents a special testing device which consists of a servo hydraulic testing machine, vacuum chamber, clip gauges to measure longitudinal and transversal strain as well as conductive and inductive heating units. For the conductive heating of the test specimen, a welding generator as in a Gleeble is used. The gauge length, the inner and outer diameters of the test specimen are almost identical to the ones used in a Gleeble [26].

In general, heat treatment will cause cyclic loading of the material due to the heating and cooling cycle [27]. For this reason, the so-called Bauschinger effect should be considered in order to use a so-called linear kinematic hardening model. In a more recent publication, the Bauschinger effect of supercooled SAE 52100 was investigated, using a Gleeble 3500® [28]. The conclusions drawn suggest the need for more sophisticated hardening models in order to describe the complex material behavior during heat treatment.

A further material parameter needed for the simulation of heat treatments is the so-called transformation plasticity coefficient K. During phase transformations, the softer phase will deform plastically, even though the stresses do not reach the yield strength of this phase. Greenwood-Johnson and Magee described this phenomenon and numerical simulations have shown that its consideration is important [9,29,30]. In [31] the experimental determination of the transformation plasticity coefficient is discussed thoroughly. Since transformation plasticity is related to the transformation rate, the experiments conducted to determine K should reflect the thermal history of the process. In the case of induction surface hardening this would mean that high heating and cooling rates are necessary. Therefore, only a Gleeble or similar test devices as mentioned can be used.

A recent publication discusses the consideration of transformation plasticity during heating, since most work has been focused on the transformations during quenching [32]. Even though the heating rates were relative low compared to induction surface hardening, the authors came to the conclusion that its consideration is important in terms of distortion prediction. This insight matches our point of view regarding the consideration during the heating step, since high temperature gradients lead to considerable plastic deformation under compression which affects the residual stress state at the end of the hardening process. This is discussed in more detail in [33].

Another aspect worth noting would be the anisotropic strain caused by banded microstructures [34-36]. The effect of the anisotropy in banded microstructures due to segregations can be determined by dilatometric studies, measuring in longitudinal and transversal direction [35,36].

From the aspects to consider in modeling specific processes, it becomes clear that extensive experimental work is needed. Further, the correct evaluation and implementation in numerical models requires a lot of work. For these reasons, numerical modeling expertise has to be built up and maintained in order to justify investments in determining the material parameters mentioned.

Phase Transformation Kinetics

The correct description of the phase transformation kinetics is of great importance due to the dependency of the material properties as well as the latent heat and strains. A common way to determine the kinetic parameters for the model used to describe the phase transformations taking place is the dilatometry, especially when dealing with high heating rates. There are basically two different heating sources possible, namely conductive heating or induction heating. The main problem concerning the induction heating system is the possibility of inhomogeneous temperature distribution in axial direction. Using hollow specimens can minimize possible temperature gradients in radial direction. In the paper by [37], the problem mentioned can be seen, as shown in Figure 6. A closer look reveals that heating rates of only 30K/s show an increase of strain rate before the phase transformation takes place. This may cause problems during evaluation, since the austenite fraction is related to the measured strain.

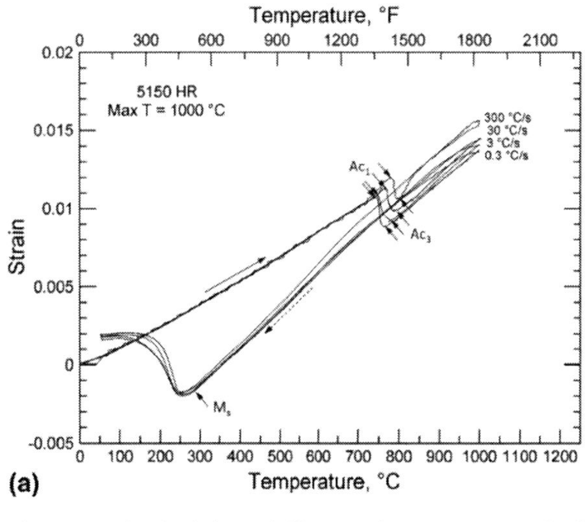

Figure 6: Inductively heated dilatometric measurements [37]

Own dilatometric measurements revealed that the temperature gradient in axial direction for heating rates between 100 – 1000 K/s can reach up to 200 K. Figure 7 shows the temperature measurements along the dilatometric specimen for an average heating rate of 100 K/s. The temperature measurements show that the temperature at the edges is lower compared to the center. This means that during the heating cycle, the average temperature lies below the control temperature and therefore the strain is underestimated. Once the Curie temperature is reached, more energy input occurs towards the edges, leading to equalization. This will cause the strain to increase, leading to the strain increase just before the phase transformation. A possible solution would be the measurement of the transversal strain in the middle of the probe, preventing edge affects from influencing the strain measurement.

This problem can be avoided by using conductive heating. In [38] high heating rates up to 500K/s were used in order to describe the transformation kinetics for practical applications. In this study, a DSI Gleeble 1500 was used. Due to the conductive heating, the problem pointed out above does not occur. Figure 8 shows experimental results of dilatational tests conducted with the test device described earlier. Constant heating rates of up to 10000 K/s were used in order to describe the phase transformation kinetics of laser hardening [39].

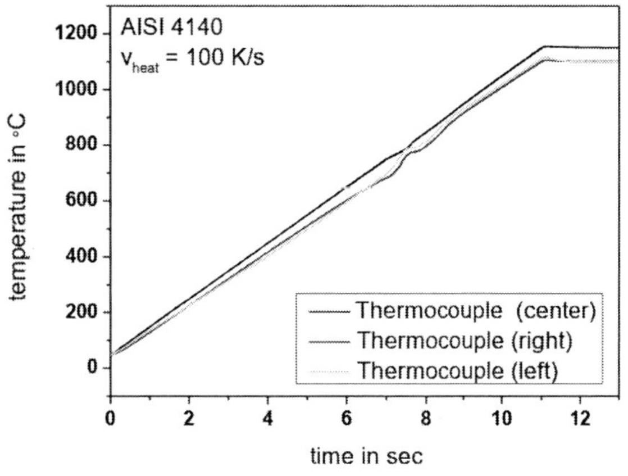

Figure 7: Measured temperature gradient for an average heating rate of 100 K/s

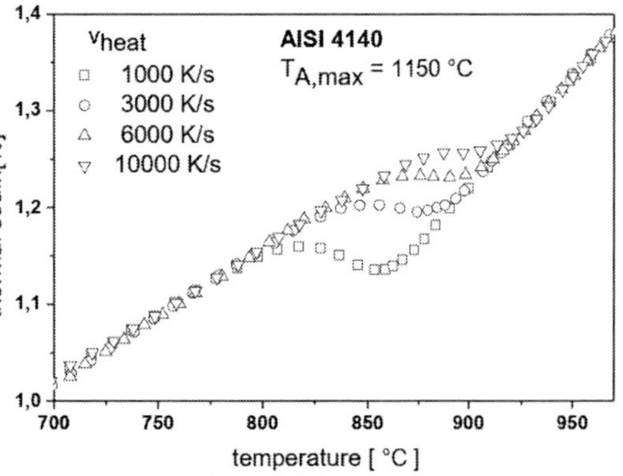

Figure 8: Conductively heated dilatometric measurements [39]

Since the heating rates used in this study cover one extreme in terms of induction heating rates, further experiments are needed. A first approach using a dilatometer with induction heating led to the aforementioned discovery of strain increase before the transformation. Certainly, the dilatometric specimens are much more convenient in comparison to the ones used in a Gleeble or the special test device.
The evaluation of the strain data from dilatometric studies are described in several papers [8,26,38,40]. Depending on the initial microstructure, the phase transformation may need to be described by two consecutive transformations, namely pearlite to austenite followed by ferrite to austenite (normalized). In the case of a quenched and tempered microstructure, no differentiation may be needed and a simple lever rule may be used to evaluate the austenite phase fraction. Figure 8 shows dilatometric measurements of quenched and tempered AISI 4140. From the measurements it can be seen that no distinction can be made.

Comparison of Experiment and Simulation

This section shall give a comparison of experiment and simulation in order to point out aforementioned aspects to consider which have not been conducted as of now. The 2-D simulations were conducted with the commercial finite element code MSC.Marc2010® and ABAQUS/Standard®. The finite element model of the coil and workpiece are shown in Figure 8. The thermal boundary conditions during the heating were applied as in [1]. For the electromagnetic/thermal coupling, MSC.Marc2010® was used, considering the nonlinearity of the permeability using the approach implemented as in [41]. For the thermal part, the specific heat was modeled as described in [27], also calculating the phase fractions. The thermo-mechanical-metallurgical part of the simulation was run in ABAQUS/Standard®, using the thermal history from the induction heating simulation. The phase transformation model is based on the work by Miokovic [39]. For the mechanical model, only the initial microstructure up to 500°C has been characterized in terms of flow curves, whereas for higher temperatures ideal-elastic-plastic data from [42,24] was used. The same source was used for bainite data. Martensite and austenite data was taken from [24]. Transformation plasticity is considered as described in [9].

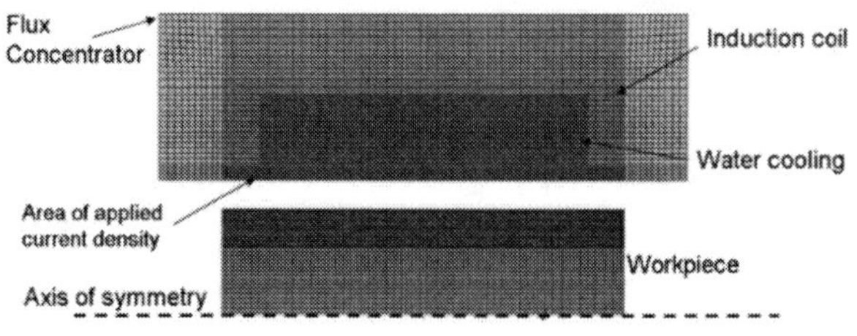

Figure 9: Finite element model

Figure 9 shows the temperature history for an experiment and simulation, applying a current of 7439 A with a frequency of 12 kHz for approximately 0.9318 sec, followed by quenching.

The quenching was modeled using the HTC data published in [14]. The energy conversion factor in the induction heating simulation was set to match the measured peak temperature within +/- 10 K. The Figure shows that the temperature measurement lags behind the simulated temperature due to the inertia of the thermocouple. The HTC used from literature fits fairly well. Upon wetting with quenching fluid around t = 2 sec, the temperature signal is distorted due to the hazardous geometry of the thermocouple, which influences the wetting kinetics The temperature drop of the temperature inside around t = 7.5 sec can be attributed to evaporating quenching fluid.

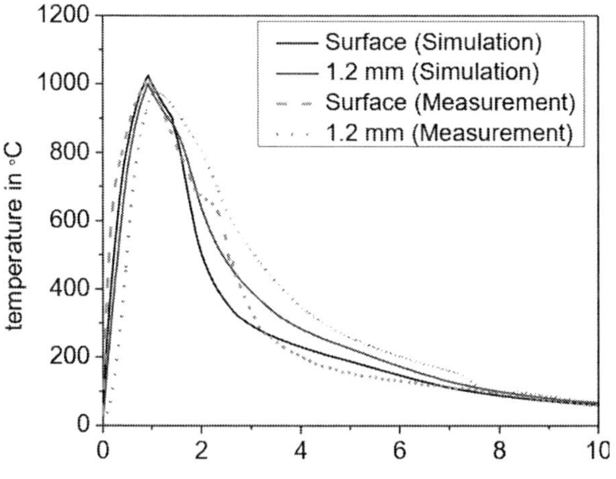

Figure 10: Comparison of experimental and simulated temperature evolution at selected locations

Results of the hardness profile are shown in Figure 11. The hardness calculation is based on the assumption that the martensite hardness for equilibrium carbon concentration of 0.42 % equals 805 HV. The variation of hardness arises from the banded microstructure present. The simulation results in a slight overestimation of the effective hardness depth (EHD), defined as the depth where the hardness value reaches 550 HV. The predicted transition zone is overestimated by 105 μm.

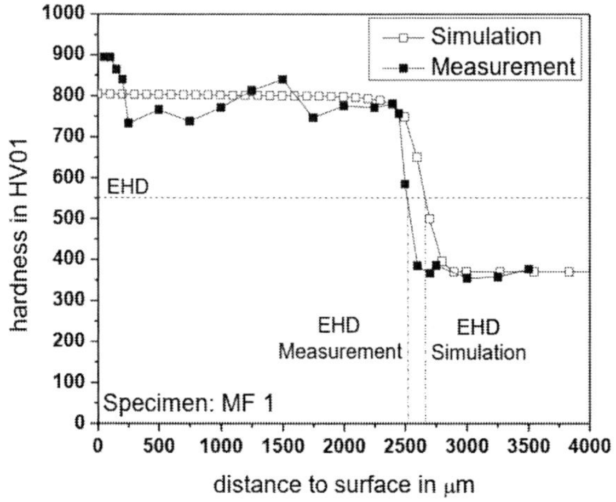

Figure 11: Hardness depth and EDH in comparison

In order to get a more global perspective of the hardening zone, Figure 12 shows a direct comparison of the hardness zone and the martensite distribution calculated. Although they match very good, minor differences can be seen towards the edges of the hardness zone. The curvature in the simulation is smaller in comparison. The micro section as well as the simulation show nicely that insufficient heating of the edges will occur due to the medium frequency. However, applying a homogeneous distribution of the current within the thickness of the skin depth might not reflect appropriately the actual current distribution because further experiments have shown a widening of the hardness zone with increased current whereas the simulation shows a less profound effect.

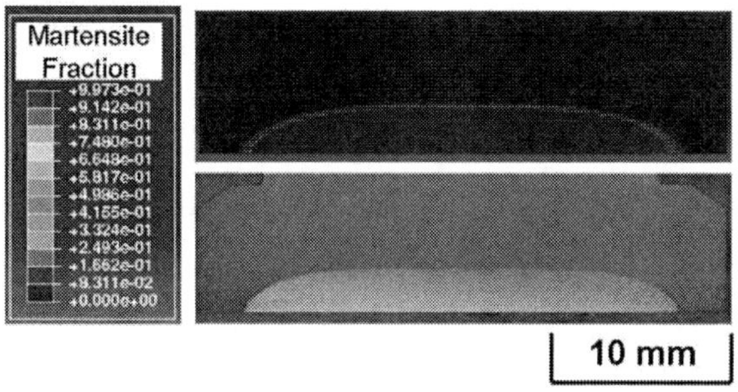

Figure 12: Comparison of micro section

Figure 13 depicts the superficial residual stress difference between the measurement and simulation ($\Delta\sigma$) versus the maximum compressive stress calculated during the heating cycle at the surface ($\sigma_{D,max}$). Although the model overestimates the residual stress by up to 50%, it can be seen that the stress difference tends to increase with increasing compressive stress, which is an indication that the transformation plasticity during heating needs to be considered. During the heating cycle, plastic deformation occurs under compression due to the high temperature gradients. The higher the deformation will be, the lower the resulting residual stress. Considering the transformation plasticity during heating would lead to an increase in plastic deformation and therefore lower residual stresses.

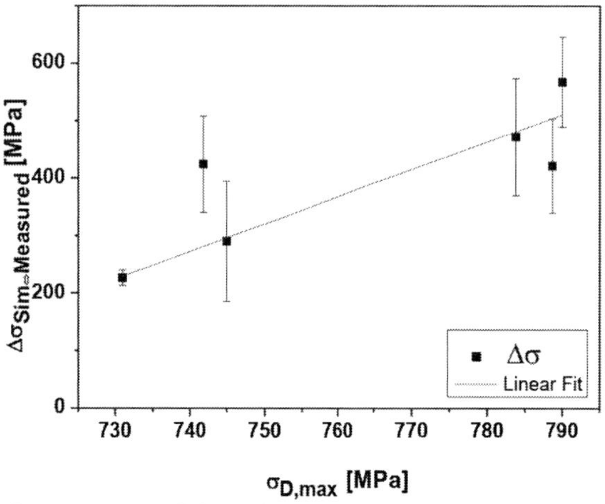

Figure 13: Superficial residual stress difference over maximum compressive stress at surface during heating

Conclusions

The paper presented gave an overview regarding the input data for the numerical simulation of induction hardening as well as possibilities to retrieve them through experiments. A thorough description of the phenomenon occurring in induction surface hardening requires an extensive database. The numerical result presented shows that the experiment can be predicted fairly well in terms of the hardness zone. However, better input data or rather further inclusion of specific aspects is needed in order to allow a more general prediction. Determining the transformation plasticity during heating as well as better mechanical material data are needed in order to further improve the presented model.

Acknowledgments

The authors would like to acknowledge the financial support by the German Research Foundation (DFG), which has funded this work in the framework of the Graduate School 1483. Further thanks go to the working group of L.-E. Lindgren, hosting me at the Luleå University of Technology and the KHYS of the KIT for financial support during this time. We would also like to thank Dr. Roth from Leibniz Institute for Solid State and Material Research for conducting magnetic measurements and Dr. Krause from Eldec Schwenk Induction GmbH for providing the induction hardening equipment.

References

1. Coupard D, Palin-luc T, Bristiel P, Ji V, Dumas C. Residual stresses in surface induction hardening of steels: Comparison between experiment and simulation. Materials Science and Engineering: A. 2008 Juli 25;487(1-2):328–39.

2. Bay F, Labbe V, Favennec Y, Chenot J. A numerical model for induction heating processes coupling electromagnetism and thermomechanics. International Journal for Numerical Methods in Engineering. 2003 Okt 14;58(6):839–67.

3. Hömberg D. A mathematical model for induction hardening including mechanical effects. Nonlinear Analysis: Real World Applications. 2004 Feb;5(1):55–90.

4. Rudnev V. Handbook of induction heating. New York: Marcel Dekker; 2003.

5. Rohde J, Jeppsson A. Literature review of heat treatment simulations with respect to phase transformation, residual stresses and distortion. Scandinavian Journal of Metallurgy. 2000 Apr 1;29(2):47–62.

6. Grum J. Induction Hardening. In: Handbook of Residual Stress and Deformation. ASM International; 2002. p. 220–47.

7. Denis S. Considering Stress-Phase Transformation Interactions in the Calculation of Heat Treatment Residual Stresses. Le Journal de Physique IV. 1996;06(C1):16.

8. Surm H, Kessler O, Hoffmann F, Zoch H-W. Modelling of austenitising with non-constant heating rate in hypereutectoid steels. International Journal of Microstructure and Materials Properties. 2008;3(1):35–48.

9. Besserdich G, Scholte B, Mueller H, Macherauch E. Consequences of transformation plasticity on the development of residual-stresses and distortion during martensitic hardening of SAE-4140 steel cylinders. STEEL RESEARCH. 1994 Jan;65(1):41–6.

10. Denis S, Sjöström S, Simon A. Coupled temperature, stress, phase transformation calculation. MTA. 1987 Juli;18(7):1203–12.

11. Inoue T, Arimoto K. Development and implementation of cae system "hearts" for heat treatment simulation based on metallo-thermo-mechanics. J. of Materi Eng and Perform. 1997 Feb;6(1):51–60.

12. Stutz DE, Semiatin SL. Induction Heat Treatment of steel. Metals Park, Ohio: American Society for Metals; 1986.

13. Liscic B. Heat Transfer Control During Quenching. MATERIALS AND MANUFACTURING PROCESSES. 2009;24(7-8):879–86.

14. Troell E, Kristoffersen H, Loevgren M, Strand N-E. Influence on quenchant performance during induction hardening. Heat Processing. (4):329–34.

15. Archambault P, Denis S, Azim A. Inverse resolution of the heat-transfer equation with internal heat source: Application to the quenching of steels with phase transformations. J. of Materi Eng and Perform. 1997 Apr;6(2):240–6.

16. Sugianto A, Narazaki M, Kogawara M, Shirayori A. A comparative study on determination method of heat transfer coefficient using inverse heat transfer and iterative modification. Journal of Materials Processing Technology. 2009 Juni 1;209(10):4627–32.

17. Kawaguchi H, Enokizono M, Todaka T. Thermal and magnetic field analysis of induction heating problems. Journal of Materials Processing Technology. 2005 Apr 10;161(1-2):193–8.

18. Labridis D, Dokopoulos P. Calculation of eddy current losses in nonlinear ferromagnetic materials. Magnetics, IEEE Transactions on. 1989;25(3):2665–9.

19. Cedrat. Volume 2 - Physical description. In: Flux 10 User's guide. Cedrat; 2009.

20. Barrere E. Contribution à la modélisation des contraintes résiduelles lors de la trempe superficielle au défilé après un chauffage par induction, PhD thesis. ENSAM CER de Bordeaux; 1992.

21. Melander M. Theoretical and experimental study of stationary and progressive induction hardening. J. Heat Treating. 1985 Dez;4(2):145–66.

22. Rudtsch S. Intercomparison of Thermophysical Property Measurements on an Austenitic Stainless Steel. International Journal of Thermophysics. 2005;26(3):855–67.

23. Lakhkar RS, Shin YC, Krane MJM. Predictive modeling of multi-track laser hardening of AISI 4140 steel. Materials Science and Engineering: A. 2008 Mai 15;480(1-2):209–17.

24. Weise A. Entwicklung von Gefüge und Eigenspannungen bei der thermomechanischen Behandlung des Stahls 42CrMo4. Chemnitz: FLUX; 1998.

25. Obergfell K, Schulze V, Vohringer O. Layout of a new thermo-mechanical test device - Experiments in the short-time range. MATERIALPRUFUNG. 2002;44(4):139–43.

26. Miokovic T, Schwarzer J, Schulze V, Vohringer O, Lohe D. Description of short time phase transformations during the heating of steels based on high-rate experimental data. JOURNAL DE PHYSIQUE IV. 2004 Dez;120:591–8.

27. Simsir C, Gür CH. 3D FEM simulation of steel quenching and investigation of the effect of asymmetric geometry on residual stress distribution. Journal of Materials Processing Technology. 2008 Okt 16;207(1-3):211–21.

28. Simsir C, Dalgiç M, Lübben T, Irretier A, Wolff M, Zoch HW. The Bauschinger effect in the supercooled austenite of SAE 52100 steel. Acta Materialia. 2010 Aug;58(13):4478–91.

29. Greenwood GW, Johnson RH. The Deformation of Metals Under Small Stresses During Phase Transformations. Proceedings of the Royal Society of London. Series A. Mathematical and Physical Sciences. 1965 Jan 19;283(1394):403–22.

30. Magee CL. Transformation kinetics, microplasticity and ageing of martensite in Fe-3l-Ni. Carnegie Mellon University, Pittsburh: 1966.

31. Taleb L, Cavallo N, Waeckel F. Experimental analysis of transformation plasticity. International Journal of Plasticity. 2001 Jan;17(1):1–20.

32. Hunkel M, Dalgic M, Hoffmann F. Plasticity of the Low Alloy Steel SAE 5120 during Heating and Austenitizing. Berg Huettenmaenn Monatsh. 2010 März;155(3):125–8.

33. Markegaard L, Kristoffersen H. Residual stress after surface hardening - an explanation of how residual stresses are created. Proceedings of IFHTSE 2009. 2009;:359–66.

34. García Navas V, Gonzalo O, Quintana I, Pirling T. Residual stresses and structural changes generated at different steps of the manufacturing of gears: Effect of banded structures. Materials Science and Engineering: A. 2011 Juni 15;528(15):5146–57.

35. Hunkel M. Anisotropic transformation strain and transformation plasticity: two corresponding effects. MATERIALWISSENSCHAFT UND WERKSTOFFTECHNIK. 2009 Juni;40(5-6):466–72.

36. Jaramillo RA, Lusk MT, Mataya MC. Dimensional anisotropy during phase transformations in a chemically banded 5140 steel. Part I: experimental investigation. Acta Materialia. 2004 Feb 23;52(4):851–8.

37. Clarke KD, Van Tyne CJ, Vigil CJ, Hackenberg RE. Induction Hardening 5150 Steel: Effects of Initial Microstructure and Heating Rate. J. of Materi Eng and Perform. 2011 Jan;20(2):161–8.

38. Dykhuizen RC, Robino CV, Knorovsky GA. A method for extracting phase change kinetics from dilatation for multistep transformations: Austenitization of a low carbon steel. Metall and Materi Trans B. 1999 Feb;30(1):107–17.

39. Miokovic T, Schulze V, Vöhringer O, Löhe D. Prediction of phase transformations during laser surface hardening of AISI 4140 including the effects of inhomogeneous austenite formation. Materials Science and Engineering: A. 2006 Nov 5;435-436:547–55.

40. Lee S-J, Clarke KD, Tyne CJ. An On-Heating Dilation Conversional Model for Austenite Formation in Hypoeutectoid Steels. Metall and Mat Trans A. 2010 Juni;41(9):2224–35.

41. Hansson S, Fisk M. Simulations and measurements of combined induction heating and extrusion processes. Finite Elements in Analysis and Design. 2010 Okt;46(10):905–15.

42. Mioković T. Analyse des Umwandlungsverhaltens bei ein- und mehrfacher Kurzzeithärtung bzw. Laserstrahlhärtung des Stahls 42CrMo4. Aachen: Shaker; 2005.

Proceedings of the 26th ASM Heat Treating Society Conference
B.L. Ferguson, R. Jones, D.S. MacKenzie, and D. Weires, editors
https://doi.org/10.31399/asm.cp.ht2011p0177

Recent Inventions and Innovations in Induction Hardening of Gears and Gear-like Components

Dr. Valery Rudnev, FASM

Inductoheat Inc., 32251 N. Avis Dr., Madison Heights, MI 48071

e-mail: rudnev@inductoheat.com Tel. 248-629-5055,

Introduction

In recent years, gear manufacturers have obtained comprehensive understanding about how technology can be used to produce quality gears. The application of this knowledge has resulted in gears that are quieter, lighter, and lower cost, and have an increased load-carrying capacity to handle higher speeds and torques while generating a minimum amount of heat, dramatically reducing distortion and requiring minimum or no post grinding.

In contrast to carburizing and nitriding, induction hardening does not need heating the whole gear. With induction, heating can be localized to only those areas in which metallurgical changes are required forming a fine-grain martensitic layer and significant residual stresses on specific areas of the part. For example, the flanks, roots, and tips of gear teeth can be selectively hardened [1-4].

Gear-performance characteristics (including load condition and operating environment) dictate the required surface hardness, core hardness, hardness profile, residual stress distribution, grade of steel, and the prior microstructure of the steel. References 1,3-4 provide a comprehensive review of favorable and unfavorable prior microstructures (structure of the "green" part) of gears and their effect on selection of the induction hardening process parameters.

The first step in designing an induction gear heat treatment machine is to specify the required surface hardness and hardness profile. Insufficient hardness as well as an interrupted ("broken") hardness profile at tooth contact areas could have a dramatic negative impact on gear life due to poor load carrying capacity, premature wear, tooth bending fatigue, rolling contact fatigue, pitting, spalling and can even result in some plastic deformation of the teeth [1].

Coil Designs and Heating Modes

Depending upon the required hardness pattern and tooth geometry, gears are induction hardened by encircling the whole gear with a coil (so-called "spin hardening of gears") or, for larger gears, heating them "tooth-by-tooth". Tooth-by-tooth concept comprises two different techniques: "tip-by-tip" or "gap-by-gap" hardening [1-4].

"Tip-by-tip" and "gap-by-gap" hardening techniques

"Tip-by-tip" and "gap-by-gap" techniques can apply a single-shot heating (tip-by-tip) mode or scanning mode (gap-by-gap). Scanning rates can be quite high, reaching 15" per minute and even higher. Both "tip-by-tip" and "gap-by-gap" techniques are typically not very suitable for small and fine pitch gears (modules smaller than 6) [1].

Inductor geometry depends upon the shape of the teeth and the required hardness pattern. According to the "tip-by-tip" hardening, an inductor encircles a body of single tooth. At present time, this technique is used very seldom.

"Gap-by-Gap" hardening is much more popular technique compared too "tip-by-tip" concept. This is the reason why "tooth-by-tooth" hardening is often associated with the term "gap-by-gap hardening. "Gap-by-Gap" hardening requires the inductor to be symmetrically located between two flanks of adjacent teeth (Figure 1).

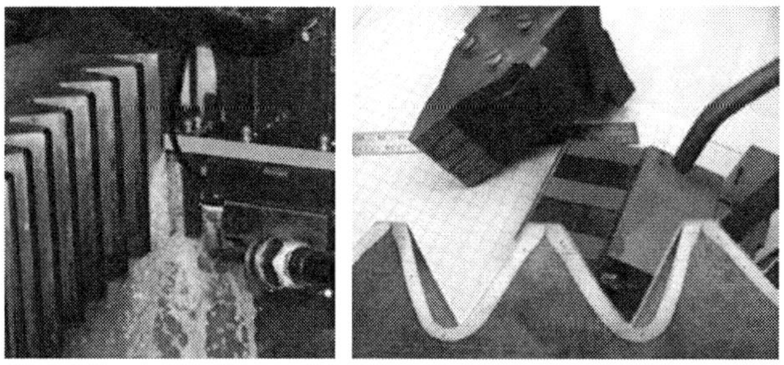

Figure 1. "Tooth-by-Tooth" ("Gap-by-Gap") induction scan hardening of large gears. (Courtesy to Inductoheat Inc.)

Typically, the inductor-scanning mode is applied for gap-by-gap hardening gears. Two scanning techniques used include one where the inductor is stationary and the gear is moveable, and the other where the gear is stationary and the inductor is moveable [1-4].

Inductors can be designed to heat only the root and/or flank of the tooth, leaving the tip and tooth core soft, tough and ductile (Figure 1). Though this is one of the oldest hardening techniques, recent innovations continue improving quality of gears heat-treated using this method.

- Pattern uniformity is quite sensitive to coil positioning and its symmetrical location in a gap between two teeth [1]. Non-symmetrical coil positioning results in a non-uniform hardness pattern. For example, an increase in the "inductor-to-gear" air gap on one side will result in a reduction of hardness and shallower case depth there. Shallow case depth can diminish the bending fatigue strength of the gear. Excessive wear of the working (contacting) surfaces of the gear tooth can also occur. Therefore, precise CNC inductor fabrication techniques, inductor rigidity, and superior alignment techniques are required. Special locators (probes) or electronic tracing systems are often used to ensure proper inductor positioning in the tooth space. Thermal expansion of metal during heating should also be taken into consideration when determining the proper inductor-to-tooth air gap. After gear loading and initial inductor positioning, the process runs automatically based on an application recipe. Figure 2 shows examples of induction tooth-by-tooth hardening machines.

Figure 2. "Tooth-by-tooth" ("gap-by-gap") induction scan-hardening machines. (Courtesy to Inductoheat Inc.)

- When developing tooth-by-tooth gear hardening, particular attention should be paid to electromagnetic end/edge effects and the ability to provide the required pattern in the gear end areas, as well as along the tooth perimeter. Upon scanning gear tooth, the temperature is distributed within gear roots and flanks quite uniformly. At the same time, since the eddy current makes a return path through the flank and, particularly through the tooth tip, proper care should be taken to prevent overheating the tooth tip regions, in particular at the beginning and at the end of the scan hardening. Improved system design helps to maintain required hardness uniformity (Figure 1, right). Specifics of gear geometry demand a particular process control algorithm. In the past, the process control recipe was limited to an available variation of power and scan rate vs. inductor position. Recent innovations bring unique ability of Inductoheat's novel inverters to independently control **both** power and frequency during scanning operation, which optimizes a control of electromagnetic and thermal conditions at initial, intermittent and final scanning stages. As an example, Figure 3 shows the revolutionary STATIPOWER® IFP™ (Independent Frequency & Power control) inverter. The ability to independently change the frequency and power of an induction system represents the long-awaiting dream of commercial induction heat treaters, since such type of setup would provide the greatest process flexibility. STATIPOWER® IFP™ is a revolutionary, IGBT type power supply specifically designed for hardening and tempering using a single inductor in a single station allowing independently adjustable frequency via CNC-program in a 5-40kHz frequency range and power in the range of 10-360kW. This concept substantially expands heat treat equipment capabilities for processing parts by programming power and/or frequency changes on the fly, maximizing heating efficiency and temperature uniformity while heating a complex shaped parts including gear-like components.

Figure 3. STATIPOWER® IFP™ is a revolutionary, IGBT-type power supply. It is only inverter of its kind to permit stepless variation of output power (10-360kW) and frequency (5-40kHz) via CNC-program during induction heating. (Courtesy to Inductoheat Inc.)

"Gear Spin Hardening (Encircling inductors)

Spin hardening is the most popular approach for induction hardening gears having fine- and medium-size teeth. Gears are rotated during heating to ensure an even distribution of energy. Single-turn or multi-turn inductors that encircle the whole gear can be used [1]. When applying encircling coils, it is possible to obtain substantially different hardness patterns by varying process parameters. Figure 4 illustrates a diversity of induction hardening patterns that were obtained with variations in frequency, heat time, and coil power [1].

Figure 4. Variety of induction hardening patters obtained using variations in frequency, heat time, and coil power. (Courtesy to Inductoheat Inc.)

As a rule, when it is necessary to harden only the tooth tips, a higher frequency and high power density should be applied; to harden the tooth roots, use a lower frequency. A high power density in combination with the relatively short heat time generally results in a shallow pattern, while a low power density and extended heat time produces a deep pattern with wide transition zones.

Quite often, to prevent problems such as pitting, spalling, tooth fatigue, and endurance and impact limitations, it is required to harden the contour of the gear, or to have gear-contour hardening (Figure 5). This often also maximizes beneficial compressive stresses within the case depth and dramatically minimizes distortion of as-hardened gears keeping it under 80-100 microns (0.003" – 0.004").

Figure 5. Contour hardened gears. (Courtesy to Inductoheat)

Many times, obtaining a true contour hardened pattern can be a difficult task due to the difference in current density (heat source) distribution and heat transfer conditions within a gear tooth. Two main factors complicate the task of obtaining the contour hardness profile [1,3-5].

The first factor is that with encircling-type coils, the root area does not have as good electromagnetic coupling with the inductor compared with the coupling at the gear tip. Therefore, it is more difficult to induce energy into the gear root. Secondly, there is a significant heat sink located under the gear root.

Simultaneous Dual Frequency Gear Hardening

Some induction practitioners have heard about simultaneous dual frequency gear hardening, which utilizes two appreciably different frequencies working on the same coil at the same time. Low frequency helps to austenitize the roots of the teeth and high frequency helps to austenitize the teeth flanks and tips.

However, it is not advantageous have two different frequencies working simultaneously all the time. Many times, depending upon the gear geometry, it is preferable applying lower frequency at the beginning of heating cycle and after achieving a desirable root heating, the higher frequency can complement initially applied lower frequency completing a job by working together.

For example, Figure 6 shows a sketch of the circuitry of Inductoheat's gear hardening system. Figure 7 shows waveforms of coil current and coil voltage when two appreciably different frequencies are applied at same time to a single inductor. This machine was designed specifically for induction hardening of internal wide-face, gear-like component having a minor gear diameter of 6.9 in. (176 mm) and a major gear diameter of 7.3 in. (186 mm) using a single-shot heating mode.

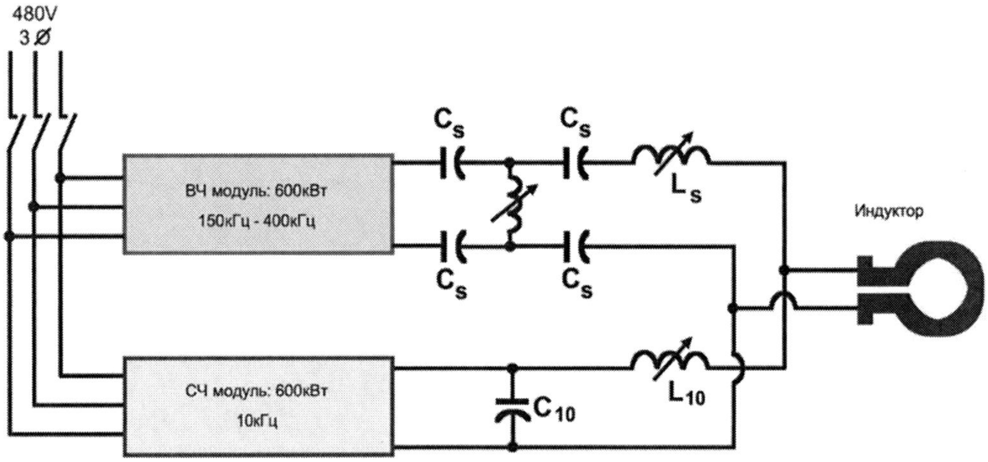

Figure 6. Sketch of the circuitry of Inductoheat's dual frequency gear hardening system. (Courtesy to Inductoheat Inc.)

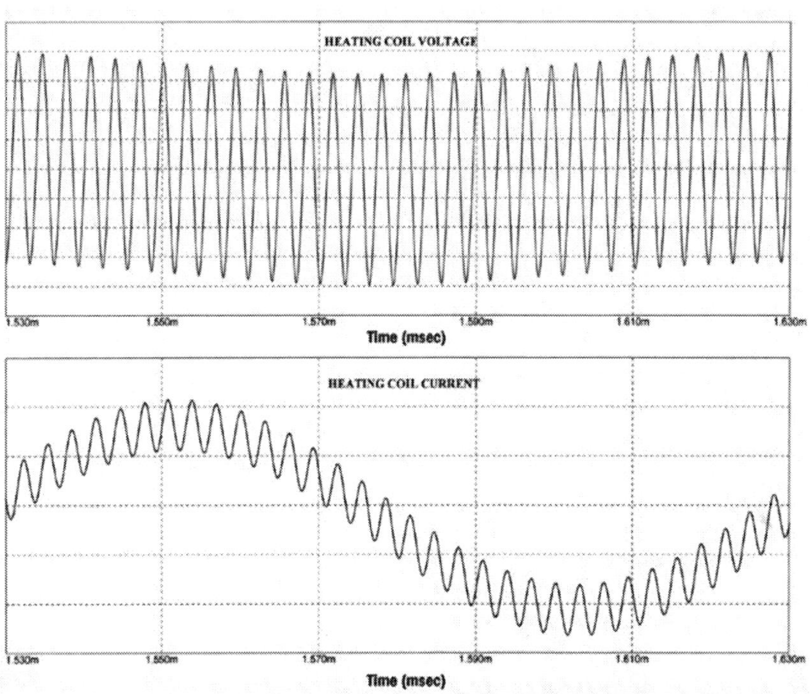

Figure 7. Waveforms of coil current and coil voltage of a single-coil simultaneous dual frequency gear hardening system. (Courtesy to Inductoheat Inc.)

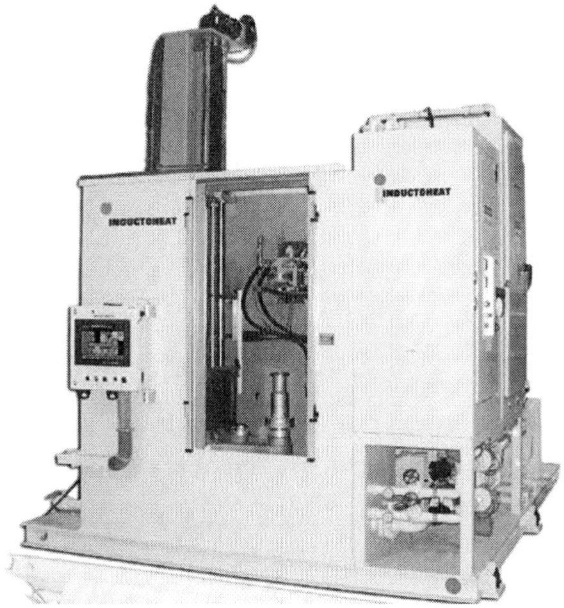

Figure 8. Inductoheat's single-coil dual frequency system comprises medium frequency (10kHz) and high-frequency (120 to 400kHz) modules working simultaneously or in any sequence desirable to optimize properties of the heat-treated gears. Total power exceeds 1,200kW. (Courtesy to Inductoheat Inc.)

The total power exceeds 1,200 kW comprising a medium frequency (10 kHz) and high-frequency (120 to 400 kHz) modules working not just simultaneously, but in ***any*** sequence desirable to optimize properties of the gears heat-treated using this unique technology. Total heat time was minimized to about 1.5 second. At the beginning of heating cycle a medium frequency is applied for 0.8 sec. providing required root heating. Remainder of the heat cycle, two frequencies was working together complementing each other. As expected, smaller gears will require less power.

The Inductoheat's two-frequency induction gear hardening system (Figure 8) also has some "auto-match" items to simplify tuning. It is rugged and will be used for high-volume single-shot hardening of several powertrain components, dramatically minimizing distortion of heat-treated parts and keeping it at level of about 80 microns (0.003") and providing a superior hardness pattern with favorable distribution of residual stresses.

Conclusions

Being environmentally friendly, green and lean technology induction heat-treating of gears and gear-like components becoming an increasingly popular choice. Recently developed inverters and process know-how, further expands an ability of this technology.

References

[1] Rudnev, V., Loveless, D., et. al., *Handbook of Induction Heating*, Marcel Dekker, NY, 2003.

[2] Rudnev,V., "Spin Hardening of Gears Revisited", *Heat Treating Progress,* ASM Int., March/April, 2004, p.17-20

[3] V. Rudnev, Induction hardening of gears and critical components, Part 1., *Gear Technology*, p 58-63, Sept./Oct. 2008

[4] Rudnev,V., Induction hardening of gears and critical components, Part 2., *Gear Technology*, p 47-53, Nov./Dec. 2008

[5] Rudnev,V., Single-coil dual frequency induction hardening of gears, *Heat Treating Progress*, ASM International, October, 2009, p.9-11

[6] Rudnev,V., "Systematic Analysis of Induction Coil Failures, Part 8: Gap-by-Gap Gear Hardening Coils", *Heat Treating Progress*, Nov./Dec., ASM, 2006, p.19-20

Proceedings of the 26th ASM Heat Treating Society Conference
B.L. Ferguson, R. Jones, D.S. MacKenzie, and D. Weires, editors
https://doi.org/10.31399/asm.cp.ht2011p0182

Stress and Distortion Evolution During Induction Case Hardening of Tube

Dr. Valentin Nemkov, Mr. Robert Goldstein, Mr. John Jackowski
Fluxtrol, Inc., Auburn Hills, MI, USA
vsnemkov@fluxtrol.com, +1 248 393 2000

Dr. Lynn Ferguson, Dr. Zhichao Li
Deformation Control Technology, Inc., Cleveland, OH, USA
lynn.ferguson@deformationcontrol.com

Abstract

Simulation of stresses during heat treating relates usually to furnace heating. Induction heating provides very different evolution of temperature in the part and therefore different stresses. This may be positive for service properties or negative, reducing component strength or even causing cracks. A method of coupled simulation between electromagnetic, thermal, structural, stress and deformation phenomena during induction tube hardening is described. Commercial software package ELTA is used to calculate the power density distribution in the load resulting from the induction heating process. The program DANTE is used to predict temperature distribution, phase transformations, stress state and deformation during heating and quenching. Analysis of stress and deformation evolution was made on a simple case of induction hardening of external (1st case) and internal (2nd case) surfaces of a thick-walled tubular body.

Introduction

Stress formation during induction hardening is more complicated than after furnace heating because of selective heating with high temperature gradients. At the same time it gives new possibilities for the optimization of stress distribution during surface treatment.

Stresses and distortions during induction treatment have been under study from the very beginning of development and industrial application of this method (middle of 1930's) [1, 2]. Multiple studies have been made since that time [3-10]. It has been found that induction surface hardening gives compressive stresses in the martensite layer, which are beneficial for the component service increasing bending strength of the part and its wear resistance [11]. However in some cases tensile stresses may develop that reduce service properties of the component and even cause local micro or macro cracks. Strong stresses also lead to the dimensional changes and distortions of the component. Typically these changes are undesirable but in some cases they may be beneficial. An example of a positive role is to provide structure or size correction from induction heating, such as shrink fitting or induction straightening of large assemblies or parts [12].

Traditional descriptions of residual stress distribution after surface induction hardening of a steel bar may be illustrated by Fig. 1 [13]. It shows all three components of residual stress: radial, axial and hoop, as well as hardness distribution. In this typical case, radial stresses σr are much smaller than axial σa and hoop σt stresses. Hoop stresses are typically the strongest and they are mainly considered in analysis. In this case they have a maximum value of about 700 MPa (compressive) on the surface, drop to zero at the distance approximately corresponding to the case depth, change sign, reach a value of approximately 300 MPa (tensile) at half the radius and then drop to zero at the bar center. Tensile radial stresses may cause circular internal cracks, separating the hardened layer from the part core.

Stress evolution during heat treatment is a very complicated process and there have been no analytical methods to predict it accurately. Only experimental methods have been used for analysis. Experimental methods for determining stresses [14] include dissection, x-ray, neutron, or ultrasound, all of which are labor consuming and expensive. Also, they can be made only after the end of processing and important factors and effects that can occur during the process are being missed.

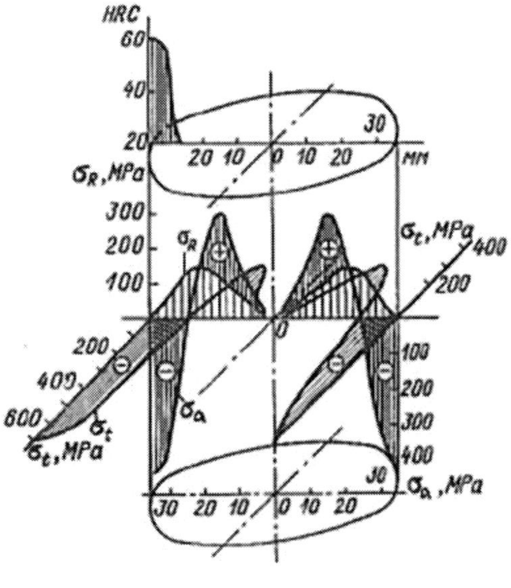

Figure 1: Residual stresses after hardening of steel 1045; cylinder OD is 65 mm; σt - hoop stresses, σr - radial stresses, σa - axial stresses; case depth is app. 6 mm

Development of internal stresses and distortions resulting from induction processing has a few root causes. In the block diagram of induction heat treatment process (Fig. 2), stress and distortion blocks are at the end of a long chain of mutually coupled phenomena.

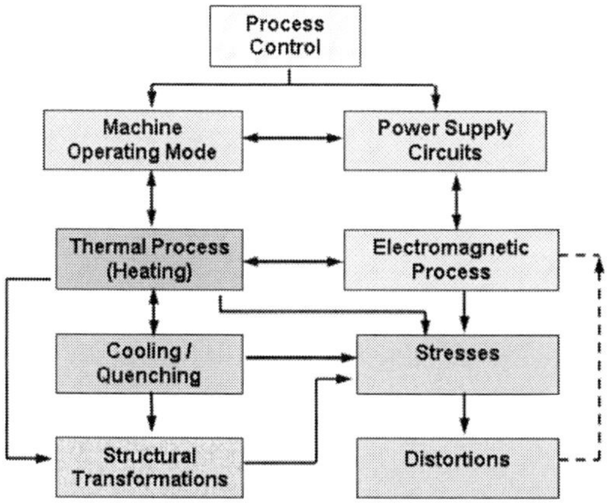

Figure 2: Chain of mutually coupled phenomena in the process of induction heat treatment

There are three types of stress sources – electromagnetic, thermal and structural. Stresses appear from the very beginning of heating due to thermal expansion of the surface layer and electrodynamic forces created by electromagnetic field on the part surface. Stresses and distortions due to electromagnetic forces are usually small compared to thermal and structural, and may be neglected except in the case of heating flat surfaces or long parts with a thin profile.

Computer simulation has been used for electromagnetic and thermal analysis for almost 40 years and has become a standard procedure for induction system design and optimization in recent years. There are multiple packages for induction system simulation. Fluxtrol, Inc. uses several programs from the relatively simple Elta to Flux 2D/3D Finite-Element package. In combination with structural analysis, using such program as Metal 7, computer simulation offers a less expensive and less time consuming method for tool and process development. Good accuracy of simulation and completeness of obtained information allows us to term this development procedure as "Virtual Prototyping of Induction Heat Treating"[15].

Simulation of stresses and deformation is an emerging technology, especially when dealing with induction, which started to be used for this purposes approximately 20 years ago [16-21]. Besides the variety and complexity of simulation algorithms, stress simulation requires large databases of thermal, metallurgical and mechanical properties of materials over the entire range of temperatures used during processing. In the present study the induction heating software Elta was used for electromagnetic and thermal simulation and the Dante package was used for thermal, structural, stress and distortion simulation.

Short description of simulation programs

Dante is finite element based software, which may be used for a complete analysis of a variety of heat treatment processes such as gear hardening after carburizing and furnace heating, quenching after furnace heating, etc. Dante has no electromagnetic block; therefore the combination with other programs is required for simulation of the processes that contain induction heating. The Dante software includes subroutines that mathematically describe the mechanical, thermal and metallurgical phase transformations that occur during heating and quenching of steel components so that phase fractions, hardness, dimensional change and residual stress can be predicted during the entire thermal treatment. Dante includes a database of steel alloys that cover the common carburizing and through-hardening grades of steel. It also includes data for surface heat transfer coefficients for more commonly used quenchants and quenching methods. References [22, 23] contain more detailed information on Dante.

Elta [24] is a program based on 1D Finite-Difference electro-thermal simulation of induction heating systems with analytical accounts for finite lengths of the part and induction coil. Elta has a database for electromagnetic and thermal properties of different materials and temperature-dependent heat transfer coefficients for a variety of quenchants.

Selection of case study

To minimize the number of variables and have the possibility to make a more or less general analysis of stress and deformation evolution during heat treatment, it was decided to select a case of OD and ID induction single-shot hardening of a thick walled tube. A 4140 steel tube with an outer diameter of 28 cm, inner diameter of 16 cm and a length of 16 cm was heated uniformly by external (case 1) and internal (case 2) inductors. The heating and quenching processes were designed using Elta. The goal was to obtain a case depth of 6 -7 mm with a maximum surface temperature of 1010 °C and a consistent heating time of 18 seconds. Under these conditions it was found that the hardness depth, corresponding to a temperature of approximately 800 °C, was 6 mm for ID heating and 6.8 mm for OD heating. A constant induction coil power regime was chosen. Process schedules are presented in Table 1.

Table 1: Induction hardening process schedule

Step	Time, sec	Frequency, Hz	Power, kW	
			OD	ID
Heating	18.0	2000	580	352
Hold	3.6	-	-	-
Spray Quench*	100.0	-	-	-
Air Cooling	200.0	-	-	-

*12% PAG polymer solution

Induction Hardening of Tube OD

The outer diameter surface heating example was examined first. At present time it is not possible to transfer results directly from Elta simulation into the Dante program. The power distribution predicted by Elta was used in the Dante model for temperature, stress, displacement and hardness calculations. Volumetric power density curves for selected moments of heating time for OD heating are presented in Fig. 3.

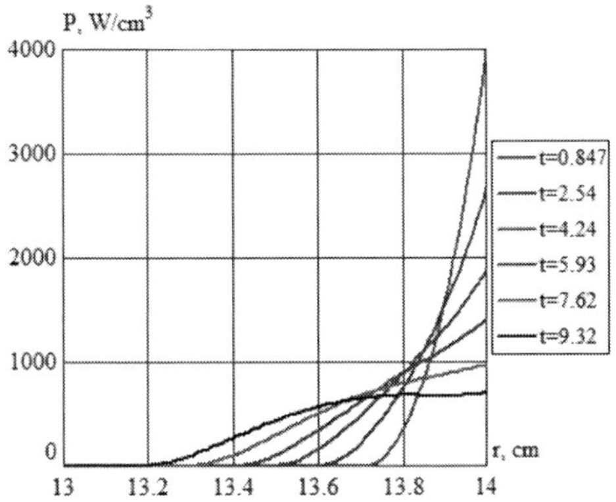

Figure 3: Power density distribution during OD heating

The heat transfer coefficient versus temperature used in simulation for spray quenching with a 12% PAG solution is displayed in Fig. 4.

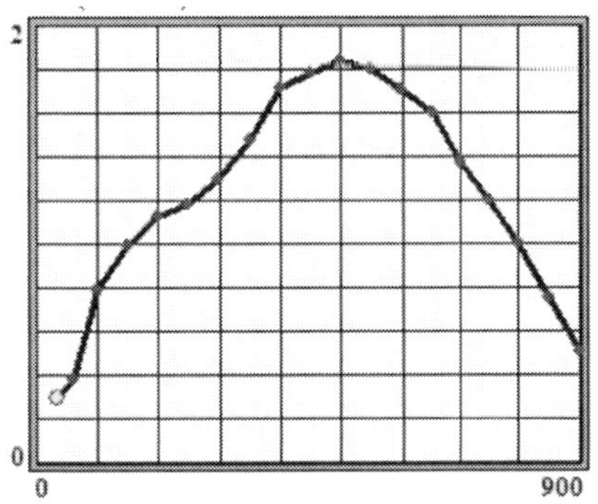

Figure 4: Heat transfer coefficient (W/cm²K) as function of temperature (°C) for selected spray quench

A history plot of temperature is shown in Fig. 5 for specified radii. Temperature fields calculated by Dante and Elta (Fig. 5) were very similar, which confirms accuracy of thermal blocks of both programs.

After 18 seconds of induction heating, the temperature at 1 mm below the OD surface has risen to over 1000 °C, while it is below 100 °C at a distance of 30 mm. During the hold period of 3.6 seconds, the surface temperature dropped to 825 °C due to intensive heat soaking into the part body. At a depth of 8 mm, temperature remains approximately the same (620 °C) and at depths greater than 15 mm, temperature continued to rise as heat diffuses from the hot OD toward the cool ID of the tube. Spray quenching causes an immediate and rapid drop in the near OD surface temperatures, while the bore temperature continues to rise slowly before final cooling. This temperature behavior, typical for induction surface hardening, in combination with structural transformations creates a complicated evolution of stress distribution during processing.

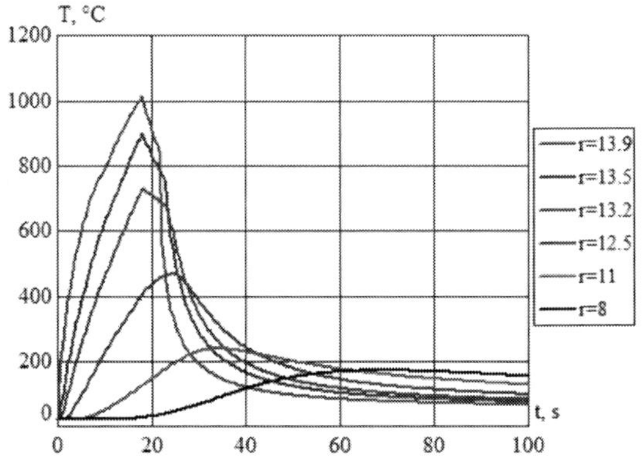

Figure 5: Temperature distribution during OD treatment for various radii (r)

The temperature distribution map at the end of the heat cycle is displayed in Fig. 6, along with the corresponding hoop stress distribution. A layer of compressive stress appears at a depth just below the region of formed austenite. As can be seen, the dimensional distortion due to thermal stresses is significant.

Fig. 7 shows almost instant austenite formation at the 1 mm and 5 mm depths during heating, and this austenite converts to nearly 100% martensite during spray quenching of the OD.

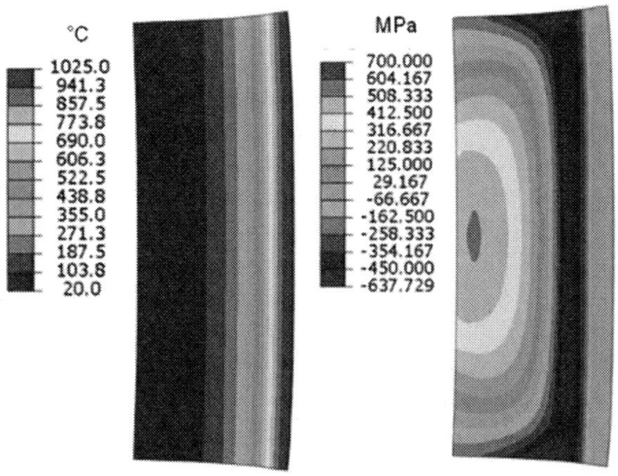

Figure 6: Temperature (left), hoop stress distribution (right) and part distortion at the end of heating for OD treatment. (Dimensional change is magnified 10X.)

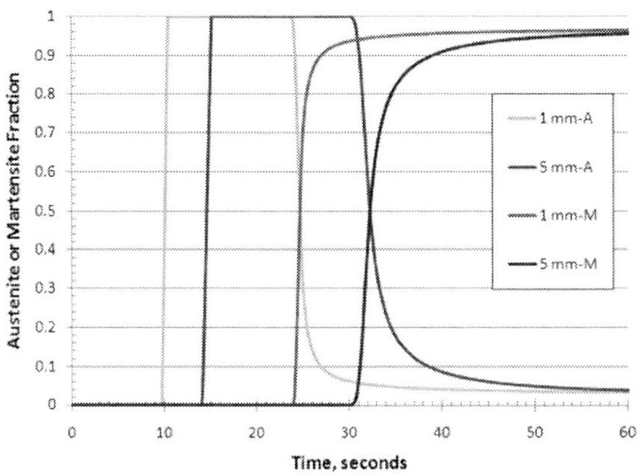

Figure 7: Austenite (A) and martensite (M) formation at depths 1 and 5 mm for OD treatment

The final stress distribution after quenching and cooling is displayed in Fig. 8. A layer of compressive stresses is present roughly to the depth of the case. In spite of uniform heating and quenching along the tube length, both hoop and axial stresses are distributed along the tube non-uniformly. Compressive hoop stresses are largest in the central zone of the tube and much lower near the outer top and bottom edges. The compressive layer is immediately backed by a layer of significant tensile stresses at depths from 7 to 15 mm. The remainder of the body maintains a slight state of compression. Stress reduction near the ends of the part may be termed "Stress End Effect of the Part". Distribution of axial stresses is similar to hoop stresses with greater end effect zone length and larger compressive stresses in the central zone compared to hoop stress values (700 MPa and 450 MPa correspondingly). It is worthy to mention that the part distortion is much lower than after the end of heating.

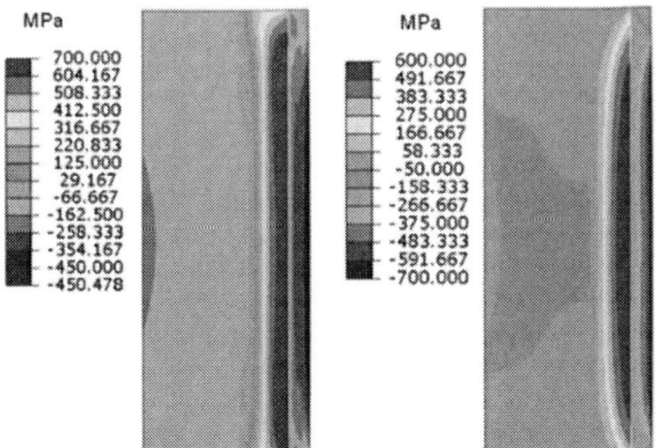

Figure 8: Hoop (left) and axial (right) stress distribution at the end of cooling for OD treatment. (Dimensional change is magnified 10X.)

It is very important to analyze stress and distortion evolution during processing. Fig. 9 shows the hoop stress evolution. Initially the tube OD is subjected to compressive hoop stress as the OD heats and its thermal expansion is restricted by the cold body of the tube. Considering the depth of 1 mm, notice how the stress starts to become less compressive after about 2 seconds. At about 10 seconds, evolution changes to an abrupt rise into the hoop tensile zone, followed by an equally abrupt drop back into the slightly compressive zone. The hoop stress' decline in compression after 2 seconds of heating is due to thermal expansion of the underlying material as it is being heated. At the 10 second point during heating, the abrupt rise in hoop tension is due to austenite formation (see Fig. 7) in first the outer layer (depth lower than and equal to 1 mm), and the accompanying density increase associated with austenite formation. Once austenite starts to form at depths greater than 1 mm, there is the resultant drop in hoop tension in the outer layer, so the 1 mm depth remains under a slight compressive hoop stress for the remainder of the heating stage.

At the 21.6 second mark of the process, the surface is sprayed and the 1 mm depth location experiences hoop tension due to thermal contraction of the austenite. Soon after, upon martensite formation (Fig. 7) and associated volumetric expansion of material, tension in the surface layer is relieved and hoop compression is experienced. From about 22 seconds to 34 seconds, the hoop stresses in outer layer are influenced by two factors: martensite forming at depths greater than 1 mm (leading to compression stress reduction) and overall thermal contraction of the tube (causing increase of compression stresses). As a result hoop stresses in outer layer become almost zero. After 34 seconds martensite formation is almost completed and the final thermal contraction of the tube (not in the range of the graph of Fig. 9) takes place. Hoop stresses in the hardened layer become more compressive with a final value of -430 MPa at a depth of 1 mm after complete part cooling.

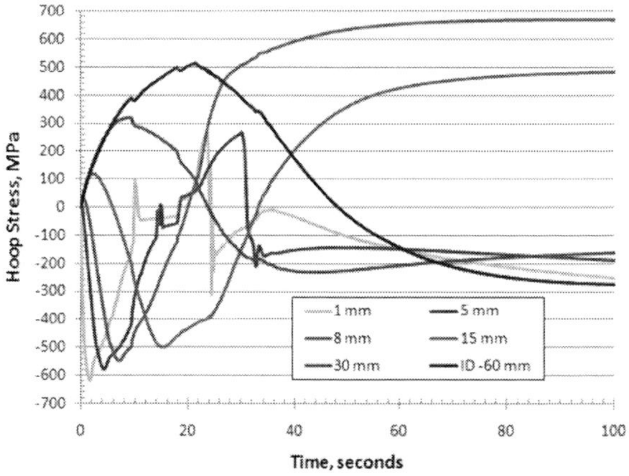

Figure 9: Hoop stress evolution at end of OD treatment at selected depths from OD Surface, at central height

Fig. 10 shows the predictions for all three components of stress in the central plane of the part after complete cooling, i.e. residual stresses. Radial stresses are very small as expected and their influence on the part properties may be neglected. The distribution of hoop stresses corresponds in general to the traditional opinion (Fig. 1). Hoop stresses are compressive in the hardened layer and become tensile beyond the case, reaching values up to 600 MPa at depths 8-16 mm. In the rest of the wall thickness there are moderate compressive hoop stresses around 100 MPa, which increase to 170 MPa on the ID surface. The axial stress distribution is similar to hoop stress behavior. It is interesting that on the ID surface, axial stress is stronger than hoop stress.

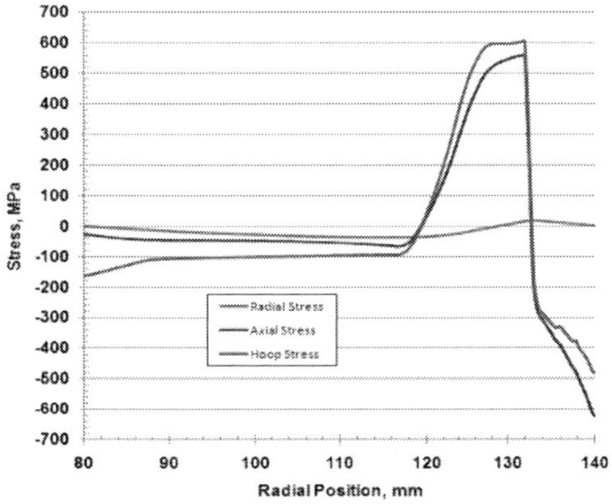

Figure 10: Distribution of hoop, axial and radial stresses in central plane of the part at the end of OD hardening

Evolution of the part distortion is also complicated and should be analyzed separately. The general character of deformation may be seen on Fig.s 6 and 8. After relatively large part deformation at the end of the heating stage, its shape returns nearly to initial dimensions with small distortion. Radial displacement evolution at different radii in the central plane is shown in Fig. 11. The increase of the external radius reaches 0.55 mm at the end of heating and then goes down to a value of +0.003 mm at the end of processing (outside the range of the figure). Much larger permanent displacements occur near the tube ends.

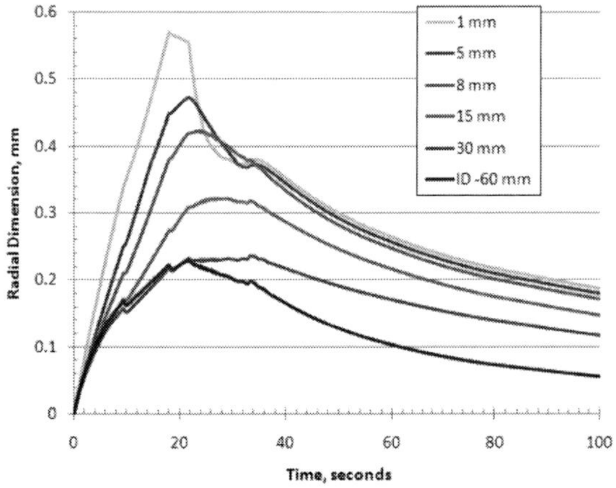

Figure 11: Radial displacement at selected depths in the central plane

Final radial and axial displacements at characteristic points of the part cross-section are given in Table 2. Axial position of the central plane is taken as zero. For OD heating, it is interesting that all radial displacements are negative except for an OD zone in the part center. Axial displacement on the OD is positive (part elongation) while on the ID it is slightly negative (part contraction). It means that there is a bending moment over the length of the part from the OD to the ID.

Table 2: Displacements of certain points on the cross-section of the tube at the end of the process cycle

Model	Location	Radial Displacement mm	Axial Displacement mm
Induction Heating of Tube OD	ID, corners	-0.040	-0.002
	ID, center	-0.050	0.0
	OD, corners	-0.061	+0.113
	OD, center	+0.003	0.00
Induction Heating of Tube ID	ID, corners	-0.065	+0.072
	ID, center	-0.146	0.00
	OD, corners	-0.061	+0.002
	OD, center	-0.058	0.00

Induction Hardening of Tube ID

Power densities at selected instances as functions of distance from the tube ID surface were taken from Elta (Fig. 11) and inserted into DANTE to predict residual stress and distortion. The distributions are very similar to those for OD heating with slightly higher power density values due to concave geometry causing the magnetic field and heat divergence in radius.

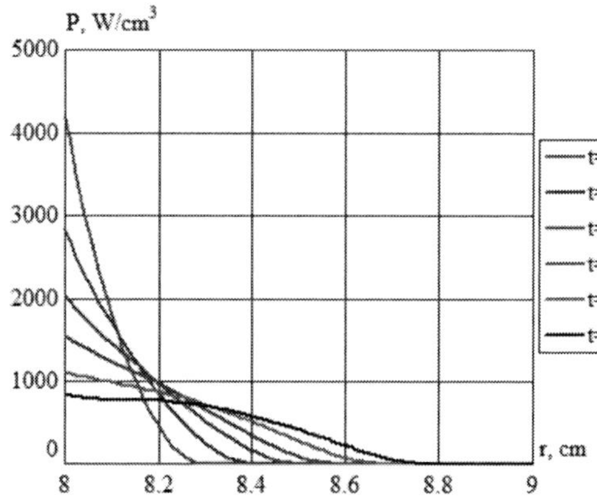

Figure 12: Power density distribution during ID heating

The temperature distribution map at the end of heating is displayed on Fig. 13. Included with these results is the radial displacement map, which shows the greatest radius expansion within .35 mm at the top and bottom, and approximately the same contraction in the central ID zone. The ID layer experiences elongation causing external bending of the part cross-section.

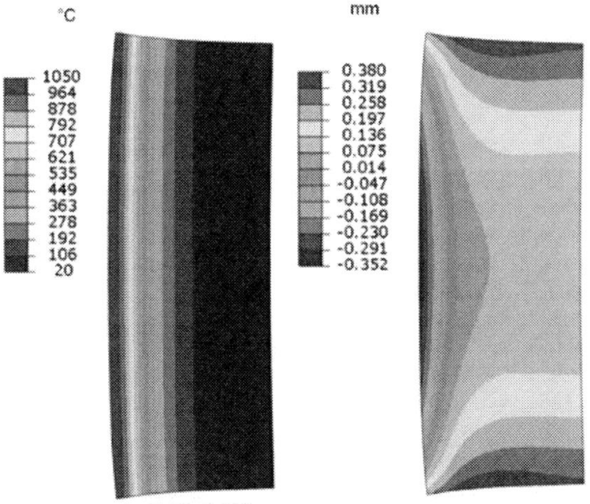

Figure 13: Temperature (left) and radial displacement (right) distribution at the end of heating cycle (18 sec) for ID treatment. (Dimensional change is magnified 10X.)

The final stress distribution after quenching and cooling is displayed in Fig. 14. It is surprising that hoop stresses of the inner surface of the central zone of the part are almost zero while great compressive stresses occur at the top and bottom edges of the tube. This is in opposition to the effects seen in the surface compression layer for the OD heating case. Once again there is a layer with strong tensile stresses just beyond the case and slight compression in the remainder of the body.

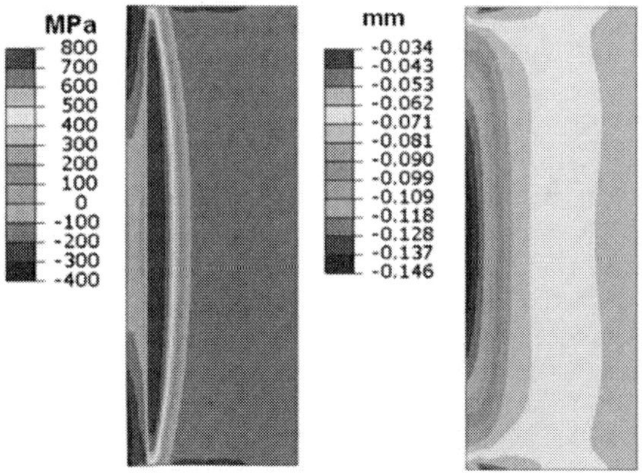

Figure 14: Hoop stress and radial displacement distribution at the end of cooling for ID treatment. (Dimensional change is magnified 10X.)

For tube ID hardening, both the bore and the OD are predicted to shrink in comparison to the starting dimensions. The radial shrinkage of the bore is predicted to be -0.146 mm and for the outer wall it is -0.058 mm. While the entire tube is predicted to shrink radially, the most significant shrinkage is at the tube ID center, and this is approximately two times the radial shrinkage at the tube ends. This is different than the radial dimensional changes for OD tube hardening (see Table 2). Similarly, the hoop stress evolution is considerably different for ID hardening than for OD hardening, comparing Fig. 15 to Fig. 9. The final hoop stress at 1 mm depth within the bore is only -84 MPa, while the final OD hoop stress is -98 MPa (out of range of figure).

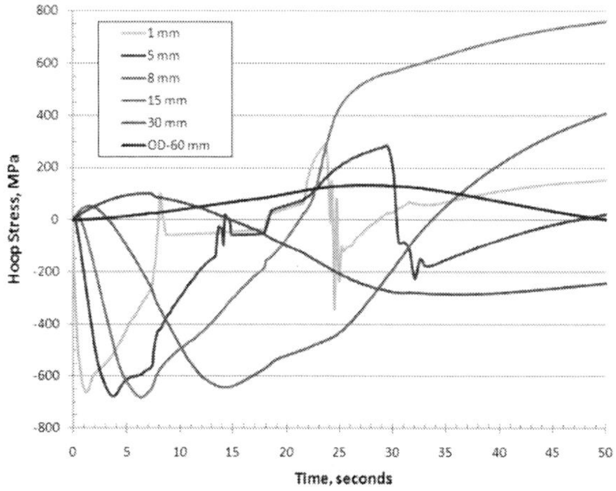

Figure 15: Hoop stress evolution at end of ID treatment at selected depths from ID Surface, at central height

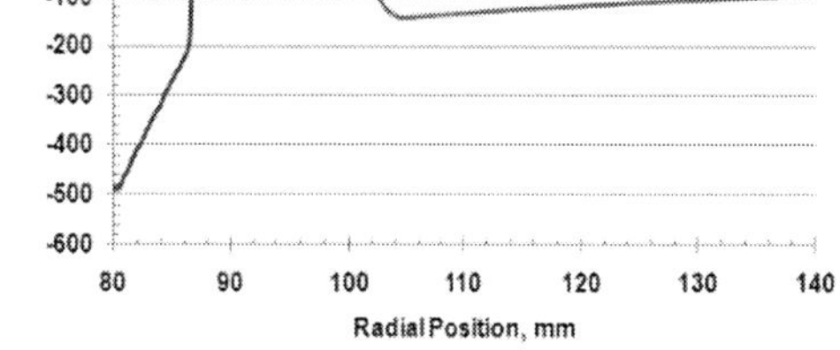

Figure 16: Distribution of hoop, axial and radial stresses in central plane of the part at the end of ID hardening

This study demonstrates how much valuable information about mutually coupled electromagnetic, thermal, structural and mechanical processes may be obtained using computer simulation.

Conclusions

Internal stresses due to temperature variation and structural transformation during induction processing may be positive (shrink-fitting, bending strength improvement) or detrimental/negative (deformations, cracks, service properties reduction). In order to achieve the desired final properties of a material, the phenomena that occurs during the cycle needs to be understood and predictable. Residual stresses and part deformation after induction surface hardening depend upon many factors: part geometry, alloy chemistry and initial microstructure of the material, case depth, heating and quenching parameters. It is very difficult or even impossible to predict the interaction of all these factors by empirical methods alone. Computer simulation is a powerful tool for analysis of stresses and deformation evolution during the entire processing cycle.

This study of a thick-wall tube induction treatment using the Elta and Dante programs showed that the final stress distribution is different for ID and OD surface hardening. Hoop stress distribution along the part length is not uniform with stress concentration near the tube ends for ID hardening and in the central zone for OD hardening. Dimensional distortion of the part may be significant during the treatment process even if the final deformation is small.

Stress evolution shows a reversal behavior with a possibility of strong tensile stresses during the martensite formation, which can lead to quenching cracks. This effect is stronger for ID hardening than for OD hardening.

Computer simulation can predict stress and deformation evolution and provides a means for finding scenarios for control and optimization of heat treating and following surface finishing processes with an account for residual stresses. International networking is important for better understanding of evolution of stresses and distortions during induction heat treatment and improvement of tools for their prediction.

References

[1] V. Vologdin, Surface Hardening by High Frequency Currents, Metallurg, no.12, 1937

[2] Kontorovich and L. Livshits, Internal residual stresses induced in steel by high-frequency surface hardening, Metallurg, No. 8, 30, 1940 (in Russian)

[3] J. Grum, Overview of residual stresses after induction surface hardening. International Journal of Materials and Product Technology, Volume 29, no.1-4, 2007

[4] M. Lozinski, Industrial Applications of Induction Heating, Pergamon Press, Oxford, 1969

[5] G. Golovin, Residual Stresses and Deformation during High-Frequency Surface Hardening, Mashghiz, Moscow, 1962 (in Russian)

[6] S. Jahanian. Thermoelastoplastic and Residual Stress Analysis during Induction Hardening of Steel, Journal of Materials Eng. and Performance, V. 4, Issue 6, 1995

[7] W. Dowling, et al., "Development of a Carburizing and Quenching Simulation Tool: Program Overview," 2nd Int. Conference on Quenching and Control of Distortion, eds, G. Totten, et al., ASM Int., 1996

[8] C. Anderson, et al., "Development of a Carburizing and Quenching Simulation Tool: Numerical Simulations of Rings and Gears," 2nd Int. Conference on Quenching and Control of Distortion, eds, G. Totten, et al., ASM Int., 1996.

[9] B. Liscic, Hardenability, in Steel Heat Treatment Handbook, edited by G. Totten and M. Howes, Marcel Dekker, NY, 1997

[10] J. Grum, Induction Hardening, in Handbook of Residual Stress and Deformation of Steel, edited by G. Totten, M. Howes, T. Inoue, ASMI, Materials Park, OH, pp 220-247, 2002

[11] Kudryavtsev, Internal Stresses as Resource of Strength in Machinary, M., Mashghiz, 1951 (in Russian)

[12] N. McPherson, Induction heat straightening – A distortion rework reduction tool for thin plates, Welding and Cutting, 7, no.23, 2008

[13] G. Golovin, N. Zimin, Technology of Heat Treatment of Metals Using Induction Heating, L. Mashinostroeniye, issue 3, 1979 (in Russian)

[14] Handbook of Measurement of Residual Stresses, Society for Experimental Mechanics, Jian Lu (Ed.): Fairmont Press, 1996

[15] R. Goldstein, V. Nemkov, J. Jackowski, Virtual Prototyping of Induction Heat Treating, Proceedings of the 25th ASM Heat Treating Society Conference, September 14–17, 2009

[16] D. Bammann, et al., "Development of a Carburizing and Quenching Simulation Tool: A Material Model for Carburizing Steels Undergoing Phase Transformations," 2nd Int. Conference on Quenching and Control of Distortion, eds, G. Totten, et al., ASM Int., 1996

[17] W. Dowling, et al., "Development of a Carburizing and Quenching Simulation Tool: Program Overview," 2nd International Conference on Quenching and Control of Distortion, eds, G. Totten, et al., ASM Int., 1996.

[18] H. Kristoffersen, P. Vomacka, Influence of process parameters for induction hardening on residual stresses, Materials & Design, Volume 22, Issue 8, December 2001

[19] P. Pacheco, M. Savi, A. Camara, Analysis of Residual Stresses Generated by Progressive Induction Hardening of Steel Cylinders, Journal of Stain Analysis, vol. 36, no. 5, 2001

[20] L. Hellenthal, C. Groth, Simulation of Residual Stresses in an Induction Hardened Roll, 23rd CADFEM Users' Meeting 2005, International Congress on FEM Technology, 2005

[21] C. Gur, J. Pan, Handbook of Thermal Process Modeling of Steel, CRC Press, 2009

[22] B.L Ferguson, A. Freborg, G. Petrus, M. Callabresi. Predicting the Heat-Treat Response of a Carburized Helical Gear, Gear Technology, December 2002

[23] Z. Li, A. Freborg, B. Ferguson, Application of modeling to heat treat processes. Heat Treating Processes, May/June 2008

[24] Elta (www.nsgsoft.com)

Proceedings of the 26th ASM Heat Treating Society Conference
B.L. Ferguson, R. Jones, D.S. MacKenzie, and D. Weires, editors
https://doi.org/10.31399/asm.cp.ht2011p0189

Dissolution of Second Phase Particles in 319-Type Aluminum Alloy

Leo J. Colley
Department of Materials Engineering, University of British Columbia, Vancouver, BC, Canada
ljcolley@interchange.ubc.ca

Mary A. Wells
Department of Mechanical and Mechatronics Engineering, University of Waterloo, Waterloo, ON, Canada
mawells@uwaterloo.ca

Robert MacKay
Nemak Engineering Centre, Windsor, ON, Canada
Robert.Mackay@Nemak.com

Wojciech Kasprzak
CANMET Materials Technology Laboratory, Hamilton, ON, Canada
Wojciech.Kasprzak@NRCan-RNCan.gc.ca

Abstract

The dissolution of second phase particles in a 319-type (Al-Si-Cu-Mg) aluminum casting alloy has been quantified by image analysis of metallographic specimens as well as an electron microprobe technique. The initial phase content of the as-cast material, and the change in volume fraction of each phase following solution treatment for various times at 480°C and 505°C, was determined by analysis of particles observed by backscattered electron microscopy. Furthermore, the change in dendritic alloy content during solution treatment was measured using electron microprobe analysis in order to estimate the relative volume fraction of second phase particles dissolved. Finally, a non-isothermal dissolution model was used to predict the dissolution behaviour during solution treatment and comparisons are made between the model predictions and experimental measurements.

Introduction

Aluminum casting alloys based on the Al-Si eutectic system are widely used in automotive components as a result of their high castability, good mechanical properties and low density. These alloys typically contain magnesium and/or copper as alloying elements, and cast components are usually heat treated to improve their mechanical properties. The most common heat treatment applied to these alloys is the T6 temper, which consists of solution treatment at high temperature to dissolve any copper- and magnesium-rich intermetallic phase particles, followed by quenching and ageing at an intermediate temperature to promote a fine dispersion of strengthening precipitates in the alloy matrix. The dissolution of intermetallic phases during solution treatment is a critical stage in this process because it controls the amount of each alloying element taken into solution and used for subsequent precipitation strengthening.

A number of investigations have been carried out over the years to determine the as-cast phase content of Al-Si casting alloys containing copper and magnesium, and the behaviour of these particles during solution treatment, and these are reviewed in a recent paper by Sjolander and Seifeddine [1]. Typically the as-cast structure consists of α-Al dendrites surrounded by an Al-Si eutectic phase that also contains coarse $CuAl_2$, Mg_2Si, Q-$Al_5Mg_8Si_6Cu_2$ and Fe-rich intermetallic particles. The $CuAl_2$ phase is usually present in two different morphologies; blocky, and eutectic, and the proportion of each has been shown to depend on several factors including cooling rate, level of modification and Cu and Mg content [2-5].

Solution treatment temperatures for Al-Si-Mg-Cu alloys are typically limited to about 500°C in order to avoid localized melting of low melting point $CuAl_2$ and Q-AlMgSiCu intermetallic particles [6,7]. This imposes a significant limitation in the ability to rapidly dissolve the intermetallics and longer solution treatment times impact the cost of production and reduce productivity. Researchers have investigated this problem and found that $CuAl_2$ and Mg_2Si dissolve relatively quickly, whereas dissolution of Q-$Al_5Mg_8Si_6Cu_2$ occurs more slowly [8], while the dissolution kinetics of the blocky $CuAl_2$ particle morphology are slower than those of the eutectic morphology [5,9]. Long solution treatments may also be problematic in other ways, as researchers have reported the formation of Al_7FeCu after long times at soak temperatures, which removes copper from solid solution and reduces the strengthening potential of the alloy [10,11]. In high Mg content alloys, the Fe-rich intermetallics may also transform to π- $Al_8Mg_3FeSi_6$ reducing the solute magnesium content.

Sokolowski et al [2] performed an experimental study to reduce the solution treatment time of 319-alloy components. They developed a two stage solution treatment procedure in which an initial lower temperature step allowed the dissolution of low melting intermetallic phases, and a second higher temperature step results in faster homogenization and dissolution rates. Chaudhry and Apelian [12] used fluidized bed heat treatments to study the effect of rapid heating on dissolution in Al-Si-Mg and Al-Si-Cu-Mg casting alloys. They found the high heating rates increase the dissolutuion rate and proposed an optimal treatment for the 319 alloy of 45 minutes soaking at 493°C using the fluidized bed. Others have shown that the response of the alloy during aging is improved if a fluidized bed is used for solution treatment [13].

Some researchers have used experimental techniques to analyze the evolution of dissolution processes during solution treatment. Lasa et al [14,15] have compared metallographic analysis of solution treated specimens with the results of calorimetry tests for two hypereutectic Al-Si-Cu-Mg alloys, and found that the heat flow characteristics of the alloy can be used to monitor the progress of dissolution, ideally when thermodynamic modeling is also used to provide information regarding the stability of the various intermetallic phases present in the alloy. Han et al [9] used electron microprobe and x-ray spectroscopy techniques to study the dissolution of $CuAl_2$ during solution treatment in 319-type alloys containing different magnesium additions. The maximum dendritic copper content was measured after 8 hours of soaking which was the longest time investigated at which point full dissolution of Cu-rich phases had not been achieved.

Although several models have been developed to predict dissolution in aluminum casting alloys [11,16,17], relatively little work has been targeted at the Al-Si-Cu-Mg alloy system. The benefit of developing models to predict dissolution during solution treatment include an improvement in our understanding of alloy behaviour during multi-stage heat treatment, which in turn can enable process optimization from the standpoint of the material and result in lower variability in final component properties. This is particularly important when casting large components with complex solidification patterns and non-uniform as-cast microstructures that respond to solution treatment locally at different rates [18]. In the work presented here, the dissolution behavior of an industrial 319-type alloy has been investigated by several experimental techniques, and a model has been developed to predict the evolution of the $CuAl_2$ particles as they dissolve. The model predictions and experimental results have been compared to assess the suitability of the model to make predictions for industrial solution treatment processes.

Experimental Methodology

Start Material

The alloy was a 319-type alloy, supplied in the form of industrially cast ingots by Nemak Engineering Centre, Windsor, Ontario. These were remelted and cast into a wedge mold at CANMET Materials Technology Laboratory, Hamilton, Ontario using an apparatus and procedure outlined in detail elsewhere [19]. No strontium or grain refiner additions were made for this research. The overall alloy composition, measured by an optical emission spectroscopy technique at CANMET is shown in Table 1. A quarter-section schematic of the copper wedge mould and resulting casting geometry is shown schematically in Figure 1. Metallographic specimens for investigation of the as-cast microstructure were taken vertically from the bottom of the chill at distances of 12mm, 25mm, 50mm and 75mm as shown in the diagram.

Table 1: Alloy Chemistry. One line separates the final line of the table and continuation of the narrative.

Element	Al	Si	Cu	Mg	Fe	Mn
Amount (wt%)	Bal.	8.3	2.8	0.5	0.45	0.34

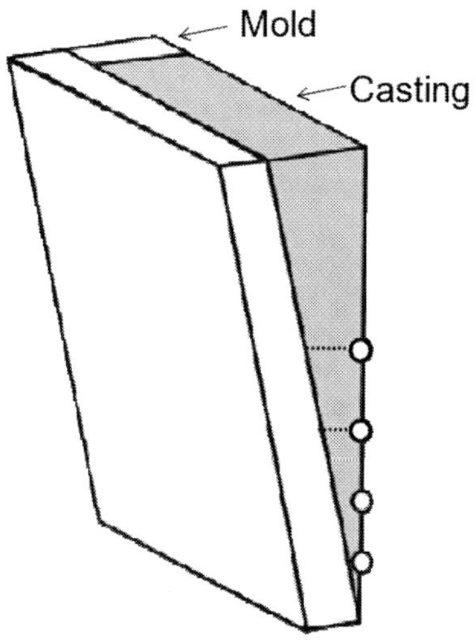

Figure 1: Schematic of the wedge mould and cast ingot showing the dimensions. Locations where metallogrpahic investigation of the as-cast structure are shown (O), and the region of the casting where heat treatment specimens were taken is shown by the shaded area.

Heat Treatment Experiments

Specimens for the heat treatment investigation were taken from a section of the casting located 50mm to 75mm vertically from the bottom of the chill as given by the shaded region in Figure 1. Specimen dimensions were 25mm (l) x25mm (w) x 4mm (h). Heat treatments were done using a Carbolite furnace at soak temperatures of 480°C and 505°C for various times up to 24 hours duration. In order to simulate typical industrial heating conditions, a 70 minute heating ramp was applied from room temperature to the soak temperature. Following heat treatment, the specimens were removed from the furnace and immediately quenched in water at room temperature.

Metallography and Image Analysis

All metallographic examinations were carried out on vertical sections which expose the microstructure perpendicular to the solidification direction. The specimens were mounted using a Buehler Epoxicure acrylic resin, then ground and polished by hand to a 0.06μm finish using a colloidal silica suspension.

Photomicrographs of the microstructures were taken using a Nikon EPIPHOT 300 series inverted metallurgical microscope equipped with a digital camera. Post-processing of the micrographs was carried out using Clemex Professional Imaging and Adobe Photoshop 7.0 software. Several microstructural characteristics were quantified by analysing the 2-D micrographs. In the as-cast alloys, the secondary dendrite arm spacing of the primary α-Al dendrites and the average size of the as-cast intermetallic particles were measured using the Imagetool software package.

As-cast and heat treated specimens were examined using a Hitachi S-3000 electron microscope in backscatter electron (BSE) mode. The imaging parameters used included an accelerating voltage of 20keV, working distance of 15mm. Backscatter electron micrographs were taken at random across the polished surface of the specimens and the volume fraction of each intermetallic phase present in the as-cast and solution treated specimens was quantified using a point count method. A minimum of 1600 points were used to estimate these values.

Microprobe

Electron probe microanalyses of the alloy microstructure were done using a fully automated CAMECA SX-50 electron-probe microanalyser, operating in wavelength-dispersion mode at the Electron Microbeam/X-ray Diffraction Facility in the Department of Earth and Ocean Sciences at the University of British Columbia. Quantitative analyses were made for Al Si, Mg, Cu and Fe using standards of known composition and the following operation conditions were used; excitation voltage, 15kV; beam current, 20nA; peak count time, 60s (10s for Al); background count-time, 30s (5s for Al); spot diameter, 1μm.

The interaction volume was estimated to have an approximate diameter and depth of 2.5μm and 2μm, respectively. Consequently, measurements were made across the width of the dendrite arms at spacings of 4μm, as well as, selected measurements in the eutectic regions and at the location of large (i.e. greater than 1μm) second phase particles. These results allow the spatial alloying element content across the dendrite to be estimated so as to determine the evolution of the dissolution and homogenisation processes. Between four and six dendrites were examined for each experimental condition.

Model Development

The dissolution of $CuAl_2$ particles has been modeled under the assumption that it is controlled by the diffusion of magnesium atoms in the α-Al matrix. The model is based on the steady-state dissolution of a single spherical particle of radius, r, in an infinite matrix, following the approach by Whelan [20] and applied to Mg_2Si dissolution in an A356 alloy previously [21].

$$\frac{dr}{dt} = -\left(\frac{C_o - C_i}{C_p - C_o}\right)\frac{D}{r}$$

The compositional terms C_i, C_o, and C_p are shown schematically in Figure 2. In the model formulation the microprobe measured copper content at the centre of the as-cast dendrite is substituted for the solute concentration at the infinite point, C_i. Furthermore, the temperature dependences of C_0 and D have been described using linear and Arrhenius relationships respectively. The dissolving particle is assumed to have $CuAl_2$ stoichiometry, and the initial particle size has been estimated using a mass balance approach in which the amount of copper contained in the particle is equated to the difference between the mass of copper measured in the as-cast and fully solution treated dendrite arms. The calculated mass of copper is then converted into the radius of the $CuAl_2$ particle assuming it is spherical and fully dense, with a density of 4.34kg/m^3 [22]. For simplicity it is assumed that no silicon or Fe-rich particles exist in the system.

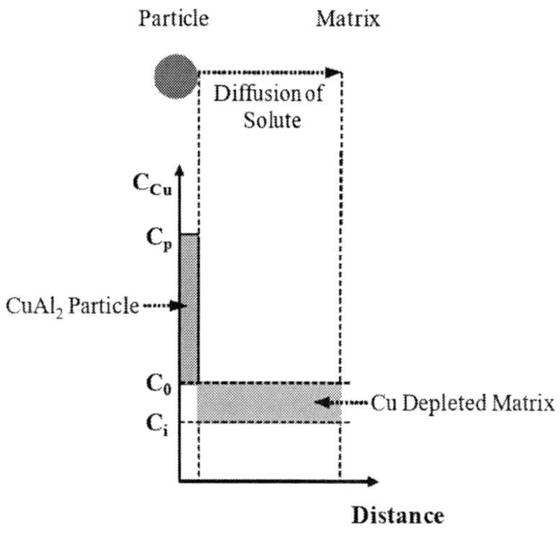

Figure 2: Schematic diagram showing the solute conentration surrounding a dissolving second phase particle.

The model calculates the rate of change in particle radius with time and hence the evolution of particle size can be determined. The relative volume fraction of the particle, f_r, during dissolution can be calculated using a similar approach to Nolfi et al. [23].

$$f_r = \frac{f}{f_0} = \frac{r^3}{r_0^3}$$

where the instantaneous particle radius, r, is predicted at each time step.

Results and Discussion

The results of the as-cast microstructure investigation are presented first, followed by the images and results of the point count analysis that indicate the changes in intermetallic content during solution treatment. The microprobe data is then presented, followed by estimates of the mean Cu and Mg content for the as-cast and solution treated conditions studied. Finally, model predictions for the dissolution of $CuAl_2$ are presented and compared with the results of the image analysis and microprobe investigations.

As Cast Microstructure

The optical micrographs show that the as-cast alloy has an unmodified microstructure consisting of the α-Al matrix, Al-Si eutectic and several intermetallic phases including $CuAl_2$, Q-phase and α-AlFeSi. Small Mg_2Si particles were also observed at the interfaces of other phases. The as-cast alloy is shown in Figure 3 and the phases observed are identified. Backscatter electron micrographs can be used to clearly indicate the differences in the various Cu-rich intermetallics. The blocky and eutectic morphologies of $CuAl_2$ are shown in Figure 4. The Q-phase is also observed to be closely associated with eutectic $CuAl_2$ particles. The script morphology of the α-AlFeSi particles is also clearly seen.

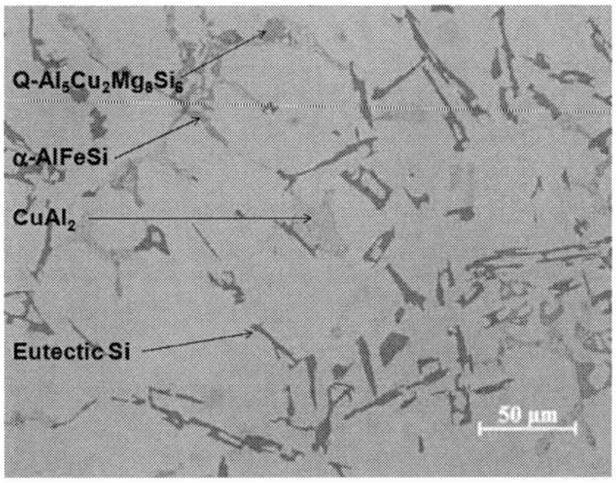

Figure 3: The as-cast microstructure of the 319-type aluminum alloy. The eutectic Si particles and intermetallic phases $CuAl_2$, $Q\text{-}Al_5Mg_8Si_6Cu_2$ and a-AlFeSi are identified using arrows. The Mg_2Si particles consist of small black particles at the interfaces of other second phase particles.

Microstructure Evolution

The dissolution of second phase particles in the 319-type alloy during solution treatment can be seen using backscatter electron micrographs. Figure 5 shows the change in phase content between the as-cast alloy and after solution treatment at 480°C for 150 minutes and 1440 minutes. The amount of $CuAl_2$ is seen to decrease significantly. The α-AlFeSi intermetallic is not affected under these conditions.

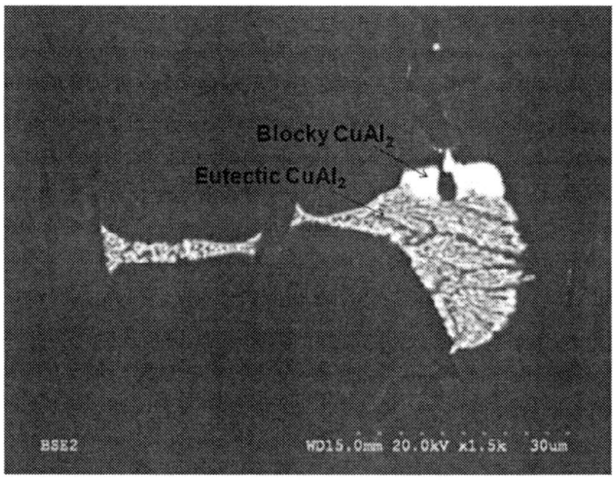

Figure 4: Backscatter elctron micrograph of the as-cast 319 alloy showing the blocky and eutectic $CuAl_2$ morphologies.

The results of the point-count analysis to estimate the volume fraction of each intermetallic are presented in Table 2. The volume fraction of $CuAl_2$ reported does not differentiate between blocky and eutectic morphologies. Estimates for the volume fraction of Mg_2Si are omitted as these particles are difficult to see in an aluminum matrix using the backscatter electron technique. The results presented in Table 2 confirm the observations of the backscatter electron micrographs shown in Figure 5. The amount of $CuAl_2$ decreases during solution treatment to near zero after 24 hours soaking at both temperatures investigated. In contrast, the volume fraction of the Q-phase only decreases at 505°C and remains fairly constant throughout solution treatment at 480°C. The Fe-rich intermetallic phases remain unaffected by solution treatment. Figure 6 shows a plot of the changes in volume fraction for each intermetallic phase.

These results suggest that 24 hours at 505°C is required for near-complete dissolution of the $CuAl_2$ and Q-phase particles, although most $CuAl_2$ is dissolved after 24 hours at temperatures below this. These results confirm other researchers results [2,9] that dissolution is not complete after 8 hours at temperatures below 500°C due to incomplete $CuAl_2$ dissolution but also because of the limited amount of total dissolution undergone by the Q-phase at these temperatures. Indeed, the two-step solution treatment procedure proposed by Sokolowski et al [2] may be beneficial due to dissolution of the Q-phase during the high second stage temperature. The differing dissolution behaviour of the two phases across this temperature range may explain some of the complexities of the heat treatment behavior of this alloy, including the level of strength achieved in the peak aged condition following different solution treatments. Increased copper and magnesium solute contents would be expected to arise from the slowly dissolving Q-phase, both of which contribute towards the later formation of strengthening precipitates.

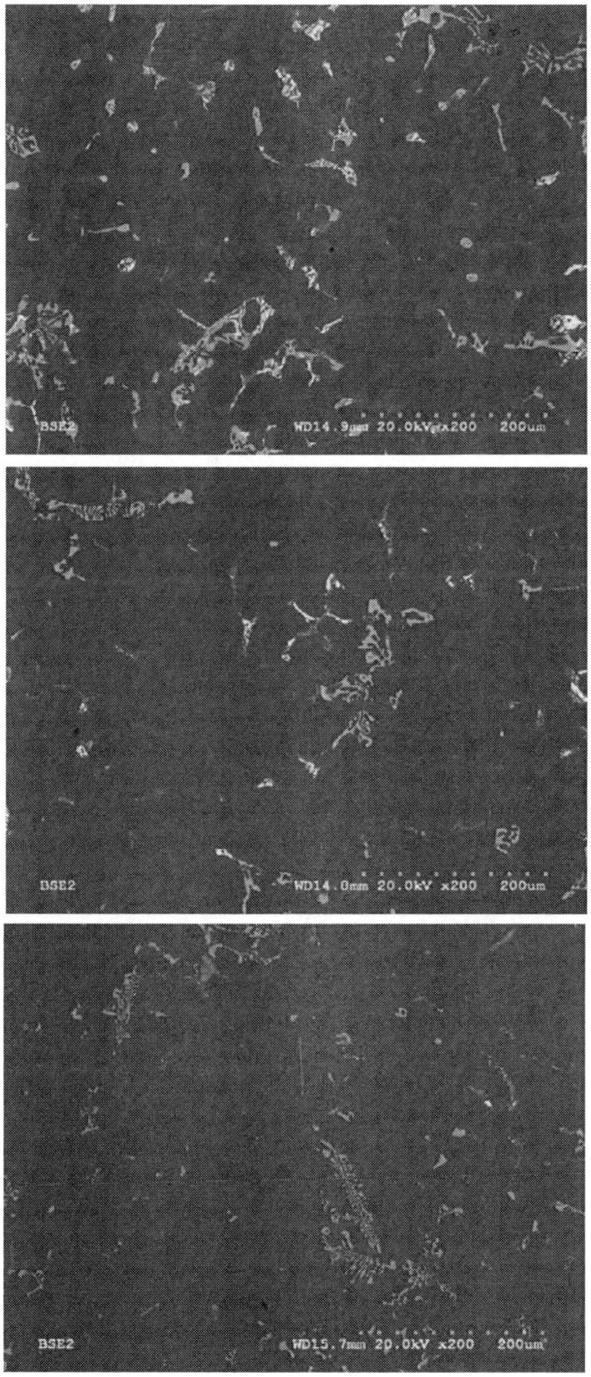

Figure 5: Backscatted electron micrograpahs for the 319 alloy: (a) in the as-cast condition and (b) following solution treatment for 150 minutes at 480°C, (c) for 1440 minutes.

Table 2: Estimated volume fraction for intermetallic phases n the 319-type alloy in the as-cast condition and following solution treatment using the point-count method.

Sol. Treat Temp. (°C)	-	480			505		
Sol. Treat Time (min)	AC	15	150	1440	15	150	1440
Vol Frac. $CuAl_2$	0.039	0.020	0.015	0.004	0.018	0.004	0.002
Vol. Frac. Q-AlCuMgSi	0.008	0.006	0.009	0.008	0.005	0.003	0.002
Vol. Frac. α-AlFeSi	0.030	0.030	0.036	0.024	0.032	0.026	0.027

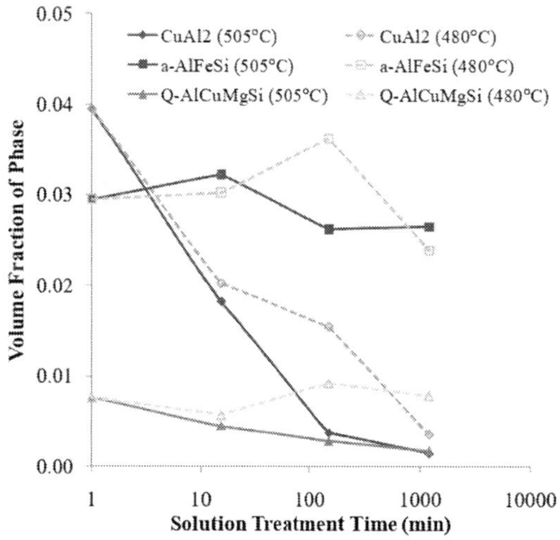

Figure 6: change in volume fraction of $CuAl_2$, α-AlFeSi and Q-phase in 319-type alloy during solution treatment at 480°C and 505°C.

Microprobe Results

Microprobe plots showing the change in distribution of copper and magnesium in the α-Al dendrites during solution treatment at 480°C and 505°C are shown in Figures 7-10. Microsegregation of copper and magnesium atoms is seen in the as-cast dendrites. The copper and magnesium content at the centre of the as-cast dendrite were approximately 0.6wt% and 0.12wt% respectively. Alloy content was significantly higher closer to the dendrite edge in both cases. The volume-averaged dendritic alloy concentration in the as-cast condition was calculated to be 1.05wt%Cu and 0.17wt%Mg assuming the dendrites are cylindrical. These values increase during solution treatment as second phase particles dissolved and the figures show increasing alloy content and decreasing microsegregation as homogenization occurs.

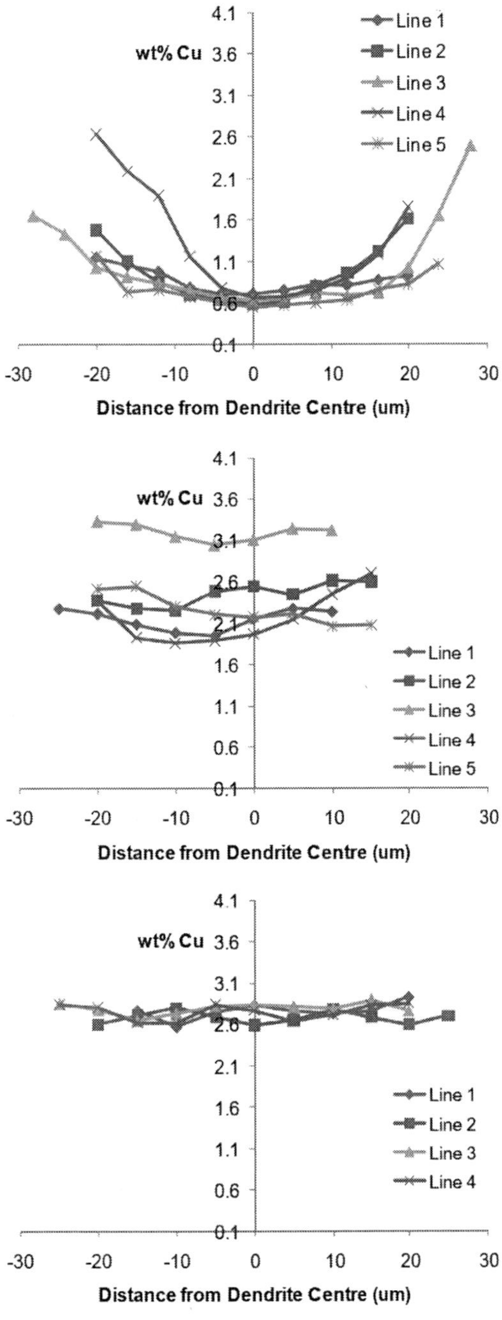

480

Figure 7: Electron microprobe data for the distribution of copper concentration across dendrites (a) in the as-cast condition, and after solution treatment at 480°C (b) for 30 minutes and (c) 1440 minutes.

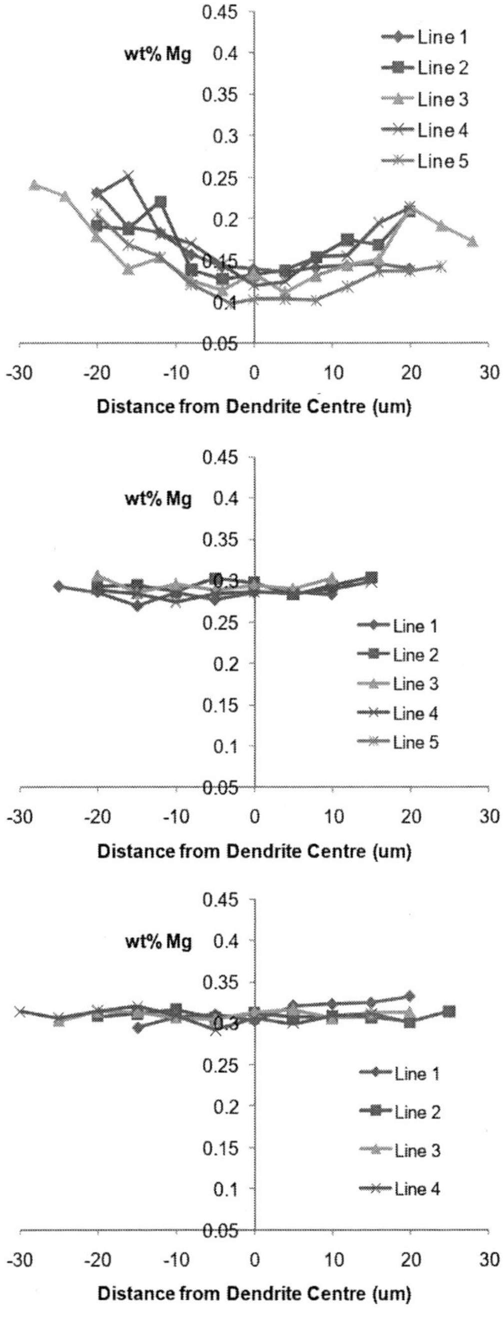

Figure 8: Electron microprobe data for the distribution of magnesium concentration across dendrites (a) in the as-cast condition, and after solution treatment at 480°C (b) for 30 minutes and (c) 1440 minutes.

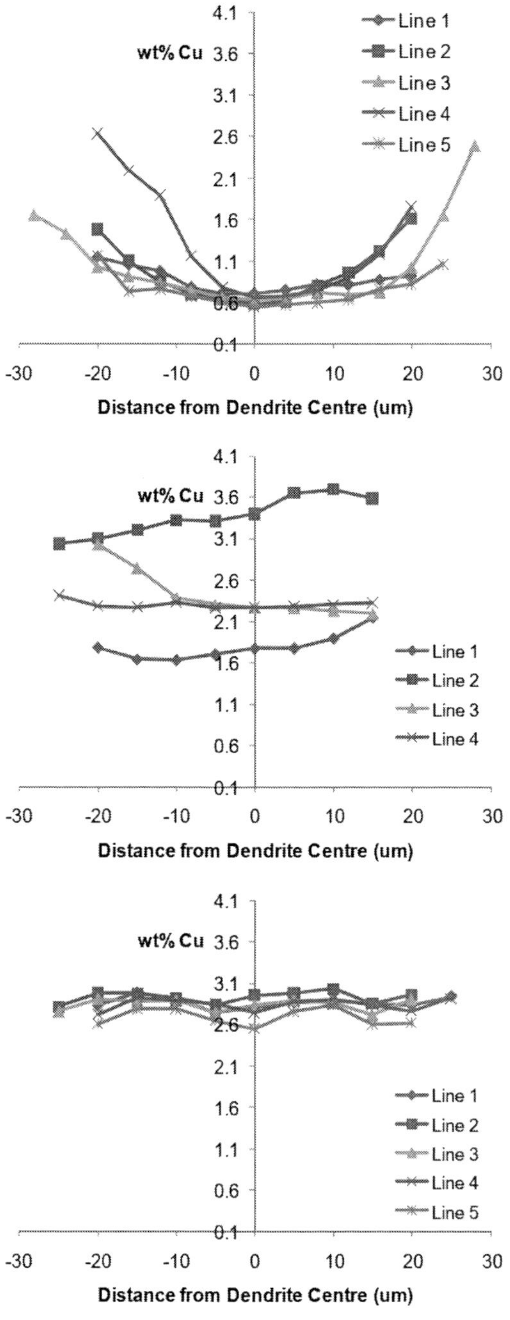

Figure 9: Electron microprobe data for the distribution of copper concentration across dendrites (a) in the as-cast condition, and after solution treatment at 505°C (b) for 30 minutes and (c) 1440 minutes.

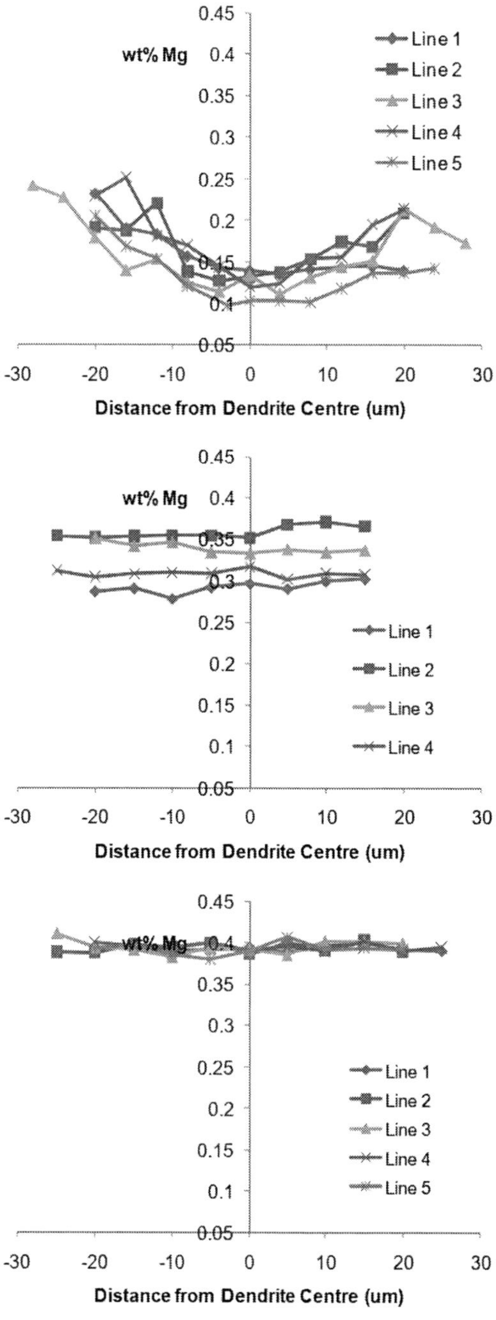

Figure 10: Electron microprobe data for the distribution of magnesium concentration across dendrites (a) in the as-cast condition, and after solution treatment at 505°C (b) for 30 minutes and (c) 1440 minutes.

The variation in alloy content between dendrites in the same specimen increases initially from the as-cast condition, and then decreases again. This is attributed to the non-uniform distribution of second phase particles and resulting enrichment of dendrites close to dissolving particles. This is an important observation in terms of the downstream processing of partially solution treated components in that some areas of the components will have significantly higher solute content than others, and therefore, not only would the values of the peak strength be higher in the solute enriched areas, but the kinetics of precipitation would be faster and resulting in more rapid arrival at the peak and overaged conditions compared to the less solute-rich areas.

The volume averaged copper and magnesium content of the dendrites studied in the microprobe examination have been calculated and plotted in Figure 11. Both similarities and differences in the evolution of the two alloying elements can be seen. In all cases, the solute content of the dendrite increases, however, the difference between the evolution of copper and magnesium at the two temperatures is noticeable in that the magnesium content increases much more at 505°C than at 480°C relative to the increases in copper content. This effect may be related to the dissolution of the Q-phase which contains a higher mass of magnesium than copper. It also suggests that the increased strength measured in this alloy following solution treatment at higher temperature is related to the formation of a larger amount of Mg-rich precipitates during aging, rather than copper-rich precipitates.

In comparison with the volume fraction calculations based on the backscatter electron micrographs, it is clear that the Mg and Cu content increases as the $CuAl_2$ and Q-phase dissolve. The increase in Mg content at 480°C measured by the microprobe is assumed to be related to the dissolution of the Mg_2Si particles observed on the interfaces of other intermetallics, and the increase of Cu at this temperature appears to be directly related to the dissolution of $CuAl_2$. At 505°C, dissolution of the Q-phase also occurs, resulting in an increased amount of Cu and Mg released into the dendrite.

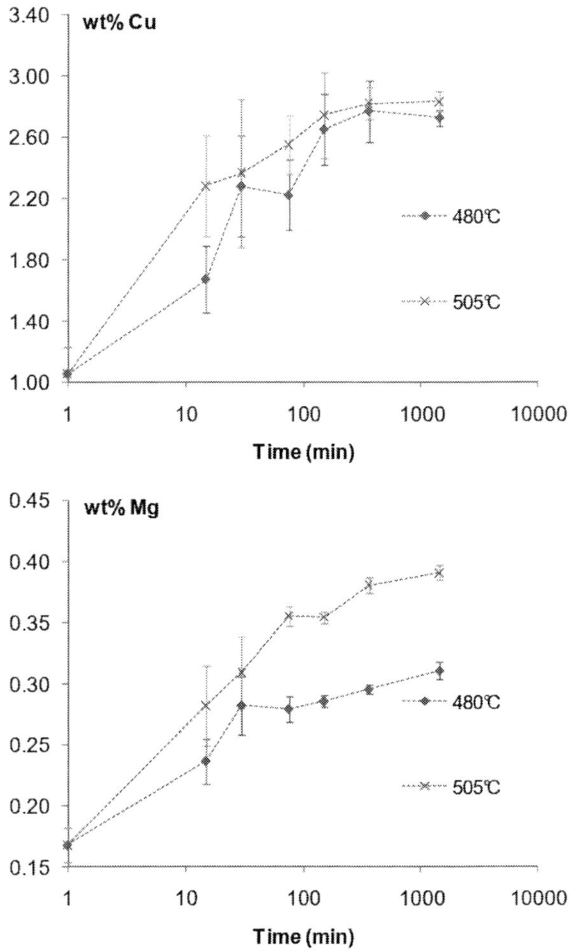

Figure 11: Volume averaged alloy content in dendrites examined by the microprobe: a) copper, b) magnesium.

Model Results

Model predictions for dissolution of $CuAl_2$ at soak temperatures of 480°C and 505°C following a 70 minute constant heating ramp are shown in Figure 12. The accuracy of the model predictions can be assessed by comparing them against estimates of the fraction dissolved from microprobe data of specimens in the partially solution treated condition. The average dendritic copper content is calculated from the microprobe data assuming the measured values from the microprobe are taken from a section that passes through the centre of a cylindrical dendrite, thereby capturing the minimum copper content at the dendrite core. The fraction dissolved is then calculated from the relative increase in copper content of the dendrite between the minimum and maximum cases (i.e. the as-cast and fully solution treated specimens) and assuming that the relative fraction increase is equal to the volume fraction of particle dissolved.

The model predictions fit the experimental data very well, although they slightly under-predict the rate of dissolution at the beginning of dissolution and over-predict the fraction dissolved at longer times, leading to an overall under-prediction of the total dissolution time. The under-prediction at short dissolution times is likely due to inaccuracies in the thermal dependence of the kinetic constants used for the model, as some dissolution would be expected during heating. Alternatively, the experimental heating rate the specimens were exposed to was not linear as has been applied to the model. In contrast, the over prediction at long solution treatment times is likely to be related to the assumption of a constant copper content at the infinite point (i.e. the dendrite centre), which in reality will increase significantly during solution treatment as the dissolved solute diffuses towards the low-solute regions. This effect causes a reduction in driving force for further dissolution, which also explains the difference in the overall dissolution time between the model and experimental results.

The results of the image analysis investigation have been converted to relative volume fraction based on the initial volume fraction measured in the as-cast material. This data has been compared with the model predictions and microprobe estimates in Figure 12. There is reasonably good agreement between the three techniques, suggesting that the microprobe approach used in this work to calculate the dendritic alloying element composition can be used to obtain good approximations of the relative fraction of second phase particles dissolved. The comparison also indicates that the modeling approach provides good estimates of the progress of dissolution. This is an important step towards the development of models that aim to improve our understanding of the complex metallurgical phenomena occurring during solution treatment of these alloys, with the long-term aim of increasing efficiency by finding optimum heat treatment conditions.

Conclusions

The following conclusions can be made based on the work presented here.

The as-cast microstructure of the 319 alloy studied here contained $CuAl_2$, Q-$Al_5Mg_8Si_6Cu_2$, Mg_2Si and a-AlFeSi as the main second phase particles.

CuAl2, Q-$Al_5Mg_8Si_6Cu_2$ and Mg2Si were found to dissolve during solution treatment. Mg_2Si and $CuAl_2$ dissolved more easily than the Q-phase. The Fe-rich intermetallic phase was unaffected by solution treatment.

Microprobe tests were able to provide information regarding the solute content of the dendrite throughout solution treatment. The copper and magnesium levels were observed to increase with solution treatment, and the increase in magnesium appears to be more sensitive to temperature than the increase in copper.

The point count method accurately described the change in second phase content during solution treatment. The higher sensitivity of the magnesium to temperature was related to the dissolution of the Q-$Al_5Mg_8Si_6Cu_2$.
The dissolution of $CuAl_2$ was modeled with reasonable accuracy using a steady-state dissolution model with input parameters derived from the microstructure characteristics and experimentally measured solute content of the alloy dendrite.

Dissolution models based on microstructural mass balance calculations as presented may potentially be used to determine the optimum solution treatment conditions for efficient heat treatment processing.

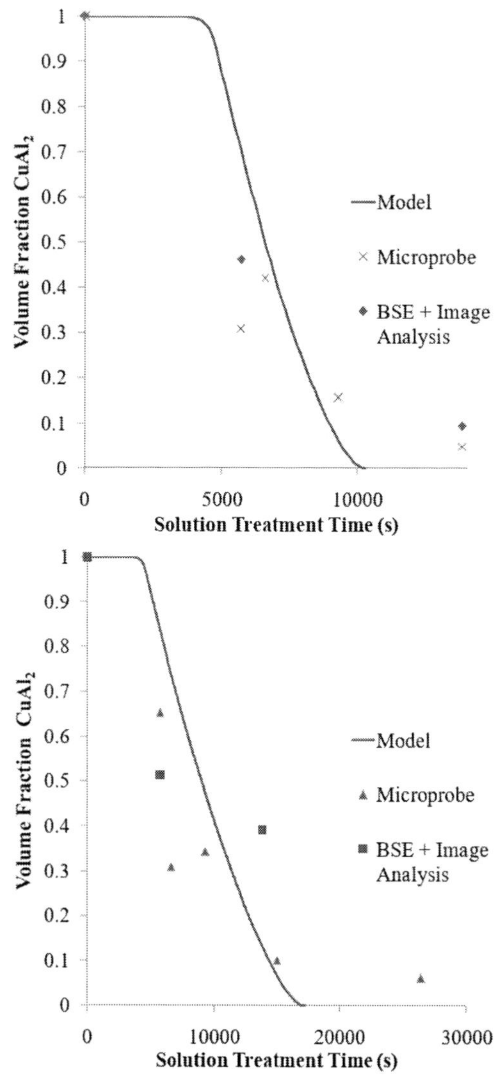

Figure 12: Model predictions for dissolution of $CuAl_2$ compared to calculated values from microprobe testing and image analysis of BSE micrographs: a) 480°C, b) 505°C.

Acknowledgments

The authors wish to thank Dr. M. Raudsepp and Ms. E. Czech at the X-Ray microanalysis facility in the Department of Earth and Ocean Sciences at the University of British Columbia for assistance with the microprobe testing. This research was supported by AUTO21, a member of the Network of Centres of Excellence of Canada program.

References

[1] Sjolander, E. and Seifeddine, S., "The heat treatment of Al-Si-Cu-Mg casting alloys", *Journal of Materials Processing Technology*, Vol. 210, (2010), pp. 1249-1259.

[2] Sokolowski, J.H., et al, "Removal of copper phase segregation and the subsequent improvement in mechanical properties of cast 319 aluminum alloys by a two stage solution treatment", *Journal of Materials Processing Technology*, Vol. 53 (1995), pp. 385-392.

[3] Samuel, F.H., Samuel, A.M. and Doty, H.W., "Factors controlling the type and morphology of Cu-containing phases in 319 Al alloy", *AFS Trans.*, Vol. 104, (1996), pp. 893-901.

[4] Samuel, A.M., et al, "Microstructural interpretation of thermal analysis of commercial 319 Al alloy with Mg and Sr additions", *AFS Trans.*, Vol.105 (1997), pp. 951-962.

[5] Li, Z., et al, "Effect of alloying elements on the segregation and dissolution of $CuAl_2$ phase in Al-Si-Cu 319 alloys", *Journal of Materials Science*, Vol. 38, (2003), pp. 1203-1218.

[6] Samuel, A.M., Gauthier, J., and Samuel, F.H., "Microstructural aspects of the dissolution and melting of CuAL2 phase in Al-Si alloys during solution heat treatment", *Metallurgical and Materials Transactions A*, Vol. 27A, (1996), pp. 1785-1798.

[7] Samuel, F.H., "Incipient melting of $Al_5Mg_8Si_6Cu_2$ and Al_2Cu intermetallics in unmodified and Sr-modified Al-Si-Cu-Mg alloys during solution heat treatment", *Journal of Materials Science*, Vol. 33, (1998), pp. 2283-2297.

[8] Moustafa, M.A., Samuel, F.H., and Doty, H.W., "Effect of solution heat treatment and additives on the microstructure of Al-Si automotive alloys", *J Mater. Sci.*, Vol. 38 (2003), pp. 4507-4522.

[9] Han, Y.M., et a., "Effect of solution heat treatment type on the dissolution of copper phases Al-Si-Cu-Mg type alloys", *AFS Trans.*, Vol. 116, (2008), pp. 79-90.

[10] Cerri, E., et al, "Effects of thermal treatments on microstructure and mechanical properties in a thixocast 319 aluminum alloy", *Mat. Sci. Eng. A*, Vol. 284, (2000), pp. 254-260.

[11] Dons, A.L., "The Alustruc homogenization model for industrial aluminum alloys", *J. Light Metals*, Vol. 1, (2001), pp. 133-149.

[12] Chaudhry, S.K., and Apelian, D., "Fluidised bed heat treatment of cast Al-Si-Cu-Mg alloys", *Met. Mat. Trans. A*, Vol. 37A, (2006) pp. 2295-2311.

[13] Ragab, Kh, A., et al, "Influence of fluidised sand bed heat treatment on the performance of Al-Si cast alloys", *Materials and Design*, Vol. 32, (2011), pp. 1177-1193.

[14] Lasa, L., and Rodriguez-Ibabe, J.M., "Characterisation of the dissolution of the Al_2Cu phase in two Al-Si-Cu-Mg casting alloys using calorimetry", *Materials Characterisation*, Vol. 48 (2002), pp. 371-378.

[15] Lasa, L., and Rodriguez-Ibabe, J.M., "Evolution of the main intermetallic phases in Al-Si-Cu-Mg casting alloys during solution treatment", *Journal of Materials Science*, Vol. 39 (2004), pp. 1343-1355.

[16] Rometsch, P.A., et al, "Modelling dissolution of Mg_2Si and homogenisation in Al-Si-Mg alloys", *Int. J. Cast Metals*, Vol. 12, (1999), pp. 1-8.

[17] Rometsch, P.A., et al, "Mass balance characterisation of Al-7Si-Mg alloy microstructures as a function of solution treatment time", *Int. J. Cast Metals*, Vol. 14, (2001) pp. 59-69.

[18] MacKay, R, and Sokolowski, J.H., "Comparison between wedge test castings and component engine block casting properties", *Int. J. Metalcasting.*, Vol. 4, (2010) pp 33-50.

[19] He, Y., et al, "Numerical simulation and experimental study of the solidification of a wedge-shaped AZ31 Mg alloy casting", *Canadian Met. Quarterly*, Vol. 48, (2009), pp. 145-155.

[20] Whelan, M.J., "On the kinetics of precipitate dissolution" *Met. Sci. J.*, Vol. 3, (1969), pp. 95-97.

[21] Colley, L.J., PhD Thesis, (2011), University of British Columbia.

[22] Mondolfo, L.F., "Aluminum Alloys: Structure & Properties", (1976), Butterworths, London.

[23] Nolfi F.V., Shewmon, P.G., And Foster, J.S., "Dissolution and growth kinetics of spherical precipitates" *Trans. AIME*, Vol. 245, (1969), pp. 1427-1433.

Proceedings of the 26th ASM Heat Treating Society Conference
B.L. Ferguson, R. Jones, D.S. MacKenzie, and D. Weires, editors
https://doi.org/10.31399/asm.cp.ht2011p0199

Properties of Semi-Finished Products and Welded Joints of Aluminum Alloy V92Zr after Prolonged Low-Temperature Heating

Dr. Alexey Sverdlin, Dr. Matthew Panhans
Milwaukee School of Engineering, Milwaukee, WI, USA

Abstract

The high-strength aluminum alloy V92Zr system Al-Zn-Mg is self-quenched alloy. The major alloying elements are 4.2 wt % Mg, 3.2 wt % Zn, 0.6 wt % Mn, and 0.15 wt % Zr. The most acceptable filler wires to weld this alloy are V92W, alloys AMg6, AMg4Zr and No.11 (Al-Zn-Mg). This alloy can be used in the aircraft production.

Prolonged heating at 50-70°C can lead to substantial structural changes in precipitation hardening of aluminum alloys due to the transition from zone to phase aging [1, 2]. According to [1-6], zone aging of Al-Zn-Mg alloys, particularly the weld seams, with repeated heating at 50-70°C substantially increases the strength and lowers the elongation, reduction in cross-sectional area, toughness, resistance to stress corrosion, and increases susceptibility to cracking. It was shown in [3, 6] that heating at temperatures even below the phase aging temperature changes the properties considerably.

This article deals with the effect of prolonged low-temperature heating on the mechanical properties, sensitivity to cracks in impact bending, and corrosion resistance of semi-finished products and weldments of aluminum alloys V92Zr after solution treatment and aging at the room and elevated temperatures.

Semi-finished products of aluminum alloy V92Zr.

The effect of heating at 70°C on the mechanical properties was investigated on sheets 5 mm thick, extrusions (220 mm x 12 mm), drop forgings (3000 x 350 x 45 mm), and forgings (2000 x 600 x100 mm). The results are shown in Fig. 1. Heating at 70°C after aging at 20°C (Fig. 1, I) leads to an increase of the tensile and (particularly) yield strengths, with a decrease of percent elongation, reduction in area, and fracture toughness. For example, the mechanical properties of the extruded strip in the transverse direction after heating at 70°C for 1000 hours are as follows: the ultimate strength increases by 50 MPa and the yield strength by 110 MPa; the elongation, reduction in area, and fracture toughness decrease by 10.0%, 5.0% and 5.0 J respectively. Evidently, [2] this is due to precipitation of large number of large number of fined particles of metastable phase in the grain boundaries, forming a continuous network. This change in the grain boundaries results in an increase of the strength, characteristics and reduction of plasticity, which are more notable for plates and extruded strip and less notable in the case of drop forgings and forgings made on the hydraulic press.

The properties of the step-aged material (60°C for 24 h +200°C for 1.5 h – "T65") after heating at 70°C remain almost unchanged (Figure 1, II), evidently due t the formation of a fairly stable structure as the result of such aging. After prolonged heating, the strength characteristics of the material aged at 20°C are higher than for the step-aged material.

The crack sensitivity in impact tests after heating at 70°C was investigated on plates 5 mm thick (Figure 2). Heating at 70°C raises the crack sensitivity of the material. The work of failure for the material with a crack after heating for 3000 h decreases by a factor of 2.4 (from 11.5 to 4.5 J) for the material with a crack after heating for 3000 hours decreased by a factor of 1.4 (from 6.5 to 4.7 J) for the step-aged material. Despite the substantial difference in the original work of fracture after aging at 20 for 4.5 months (11.5 J) and at 60°C for 24 h +200°C for 1.5 h (6.5 J), the work of fracture is almost identical (4.5 - 4.7 J) after 3000 h at 70°C.

The change in the mechanical properties as the result of corrosion was investigated on plates 2.5 mm thick (Table 1), while the stress corrosion resistance was investigated on sheet samples 5 mm thick, drop forgings, press forgings, and extruded sections (Table 2).

Table 1. Change in the mechanical properties of semi-finished products of V92Zr alloy

Aging conditions	Heating at 70°C, *h*	Before corrosion tests		Properties after corrosion test			
		TS, MPa	δ, %	TS, MPa	δ, %	$\Delta\sigma/\sigma$, %	$\Delta\delta/\delta$, %
20°C 9 months	-	460	15.8	420	9.2	8.4	41.7
	500	480	15.6	440	6.9	9.2	56.0
	1000	500	17.4	480	12.4	5.0	28.8
60°C for 24 h +200°C for 1.5 h	-	440	11.1	430	9.4	1.5	15.0
	500	440	10.9	440	11.2	10.5	0
	1000	440	10.7	440	11.0	0	0

Table 2. Corrosion resistance of semi-finished products after low-temperature heating at 70°C

Products	Location of specimens	Aging conditions	Holding time at 70°C	Number of specimens tested		Time to failure, days
Sheet 5 mm thick	Transverse	20°C 3 months	3,000	5	0	> 120
		60°C for 24 h +200°C for 1.5 h	3,000	5	0	> 120
Extruded sections	Longitudinal	20°C 3 months	1,000	5	0	> 120
		60°C for 24 h +200°C for 1.5 h	1,000	5	0	> 120
	Transverse	20°C 3 months	1,000	5	0	> 120
		60°C for 24 h +200°C for 1.5 h	1,000	5	0	> 120
Drop forging	Longitudinal	20°C 3 months	1,000	5	0	> 120
		60°C for 24 h +200°C for 1.5 h	1,000	5	0	> 120
	Transverse	20°C 3 months	1,000	5	3	36, 54, 70, > 120
		60°C for 24 h +200°C for 1.5 h	1,000	5	0	
	Vertical	20°C 3 months	1,000	5	3	30, 42, 53, > 120
		60°C for 24 h +200°C for 1.5 h	1,000	5	0	
Forging	Longitudinal	20°C 3 months	1,000	5	0	> 120
		60°C for 24 h +200°C for 1.5 h	1,000	5	0	> 120
		60°C for 24 h +200°C for 1.5 h	3,000	5	0	> 120
	Transverse	20°C 3 months	1,000	5	3	53, 71, 109
		60°C for 24 h +200°C for 1.5 h	1,000	5	2	85, 90
	Vertical	20°C 3 months	1,000	5	4	27, 35, 64, 98
		60°C for 24 h +200°C for 1.5 h	1,000	5	3	48, 64, 70

Heating at 70°C for 1000 h substantially reduces the decline of the mechanical properties due to corrosion (Table 1). No decline was noted for step-aged samples, while for samples aged at 20°C for 9 months the decline amounted to 5.0% in tensile strength and 28.8% in elongation.

According to the date in [2] heating at 70°C increase the susceptibility to stress corrosion cracking of Al-Zn-Mg alloys aged at room temperature but may not affect material subjected to phase aging. Stress corrosion tests for 4 months showed that heating at 70°C does not reduce the corrosion resistance of sheets or extruded sections after aging at 20 for 3 months or aging at 60°C for 24 h +200°C for 1.5 h (Table 2), but it does reduce the resistance of transverse samples from drop forgings and forgings aged at 20°C for 3 months. The average "life" of step-aged samples was higher.

Welded joints of aluminum alloy V92Zr

The study has been carried out on the welded joints after the prolonged heating at 70°C for 1000 and 3000 hours. The specimens from the extrusions, forgings, and sheets 2.5 and 5 mm thick were cut off and welded using an automatic GMAW (Gas Metal Arc Welding) process with various fillers (V92w, # 11, AMg6, and AMg4Zr [4]. Chemical composition of these fillers is shown in Table 3. Before heating, the specimens were aged at 20°C and at 60°C for 24 h +200°C for 1.5 h. It can be seen from the data in Tables 1 and 2 that prolonged heating has different effects on the properties of welded joints aged at room temperature and after artificial aging (60°C for 24 h +200°C for 1.5 h).

In case of aging at 20°C before and after welding, heating at 70°C for 1000 h raises the ultimate tensile strength of the welded joints, particularly those welded with filler wires V92w and # 11 (by 50-70 MPa), and to a lesser extent (by 20-30 MPa) those welded with AMg4Zr and AMg6. After artificial stepped aging, however, heating leads to a slight increase of the strength (15 MPa), evidently due to the coalescence of metastable phase and the formation of a fairly stable structure after this treatment. Failure of the samples occurs mainly in the zone of fusion; only in the case of aging at 60°C for 24 h +200°C for 1.5 h after welding does failure occur more often at a distance of 15-20 mm from the center of the weld.

Table 3. Chemical composition of the base metal and different filler wires for welding alloy V92Zr.

Material	**Chemical composition, %**									
	Mg	**Zn**	**Mn**	**Zr**	**Cr**	**Ti**	**Be**	**Cu**	**Fe**	**Si**
Forging,	4.4	3.39	0.75	0.13	-	-	-	0.015	0.15	0.14
Filler Metal:										
V92Sv	3.38	4.45	0.67	0.22	-	0.08	0.0021	0.050	0.27	0.18
No.11	4.12	2.20	0.40	0.11	0.015	0.10	0.0001	0.025	0.10	0.18
AMg6	6.24	0.08	0.58	-	-	0.07	0.0006	0.060	0.31	0.18
AMg4	3.8	2.15								

Table 4. Mechanical properties of weldments after different regimes of aging (Filler – V92w)

Aging		**TS, MPa**	**Place of failure**	**α, degrees**	**Impact, J**	
Before welding	**After welding**				**Seam**	**Fusion zone**
20°C 5 days	20°C 1 month	399	Fusion zone	32	9.4	7.7
	20°C 1 month + 70°C 1000 h	468		22	6.5	4.2
60°C 24 h + 200°C 1.5 h	20°C 1 month	391		29	9.1	6.9
	20°C 1 month + 70°C 1000 h	419		21	5.6	3.9
20°C 5 days	60°C 24 h + 200°C 1.5 h	419	15-18 mm from the center of seam	26	6.4	4.8
	60°C 24 h + 200°C 1.5 h + 70°C 1000 h	428	Fusion zone	28	4.5	3.6
60°C 24 h + 200°C 1.5 h	60°C 24 h + 200°C 1.5 h	417	Fusion zone	21	5.8	4.3
	60°C 24 h + 200°C 1.5 h + 70°C 1000 h	430		18	4.5	3.4

The change in the bending angle and the impact toughness match the change in tensile strength. The bending angle decreases (by 10-25°) as the result of heating after aging at 20°C, but only slightly (1-4°) after stepped aging. The impact toughness in the center of the weld and in the fusion zone also decreases after aging at 20°C but remains somewhat higher than after stepped aging. The bending angle and the impact toughness of joints welded with filler wire # 11, AMg6, and AMg4Zr are the same and 50-100% higher than for joints welded with V92w.

The susceptibility to cracking during impact bending was determined on welded joints in plates 2.5 mm thick welded with V92w (Table 3). Heating at 70°C raises the susceptibility to cracking in the weld and fusion zones as well as in the base metal. The effect of heating is more notable after aging at 20°C. Thus, the work of fracture of specimens with initial cracks in the fusion zone after aging at 20°C for 1 month decreases from 6.1 to 4.1 MPa/cm^2, while after stepped aging it decreases from 3.8 to 3.2 MPa/cm^2 (the specimens were aged at 20°C for 3 months before welding in both cases).

Table 5. Mechanical properties of weldments after different regimes of aging (filler – # 11)

Aging		TS, MPa	Place of failure	α, degrees	Impact, J	
Before welding	After welding				Seam	Fusion zone
20°C 5 days	20°C 1 month	390	Fusion zone	65	19.7	12.0
	20°C 1 month + 70°C 1000 h	436		48	11.5	5.8
60°C 24 h + 200°C 1.5 h	20°C 1 month	378		57	-	-
	20°C 1 month + 70°C 1000 h	410		45	-	-
20°C 5 days	60°C 24 h + 200°C 1.5 h	411	15-20 mm from the center of seam	44	1.1	0.66
	60°C 24 h + 200°C 1.5 h + 70°C 1000 h	426	Fusion zone	41	8.7	4.4
60°C 24 h + 200°C 1.5 h	60°C 24 h + 200°C 1.5 h	418	12-13 mm from the center of seam	40	-	-
	60°C 24 h + 200°C 1.5 h + 70°C 1000 h	430	Fusion zone	38	-	-

Table 6. Mechanical properties of weldments after different regimes of aging (filler – AMg6)

Aging		TS, MPa	Place of failure	α, degrees	Impact, J	
Before welding	After welding				Seam	Fusion zone
20°C 5 days	20°C 1 month	376	Fusion zone	52	1.70	1.20
	20°C 1 month + 70°C 1000 h	390		45	-	-
20°C 5 days	60°C 24 h + 200°C 1.5 h	392	10-15 mm from center of seam	43	1.3	0.79
	60°C 24 h + 200°C 1.5 h + 70°C 1000 h	400	Fusion zone	40	11.0	6.0

Table 7. Mechanical properties of weldments after different regimes of aging (filler – AMg4Zr)

Aging		TS, MPa	Place of failure	α, degrees	Impact, J	
Before welding	After welding				Seam	Fusion zone
20°C 5 days	20°C 1 month	371	Fusion zone	66	20.0	11.0
	20°C 1 month + 70°C 1000 h	388		52	15.2	8.0
20°C 5 days	60°C 24 h + 200°C 1.5 h	390		49	16.0	8.0
	60°C 24 h + 200°C 1.5 h + 70°C 1000 h	400		45	14.0	6.9

Table 8. Mechanical properties of weldments (extrusions) after different regimes of aging (filler – V92Zr)

Aging		TS, MPa	Place of failure	α, degrees	Impact, J	
Before welding	After welding				Seam	Fusion zone
20°C 5 days	20°C 1 month	426	Fusion zone	41	12.2	7.3
	20°C 1 month + 70°C 1000 h	469		30	8.7	4.4
20°C 5 days	60°C 24 h + 200°C 1.5 h	460	10-15 mm from center of seam	28	9.3	6.5
	60°C 24 h + 200°C 1.5 h + 70°C 1000 h	464		25	8.4	4.7

Table 9. Mechanical properties of weldments (sheets 5.0 mm thick) after different regimes of aging (filler – # 11)

Aging: Before welding	Aging: After welding	TS, MPa	Place of failure	α, degrees	Impact, J: Seam	Impact, J: Fusion zone
20°C 5 days	20°C 1 month	375	Fusion zone	90	21.4	13.5
	20°C 1 month + 70°C 1000 h	408		66	15.5	8.7
20°C 5 days	60°C 24 h + 200°C 1.5 h	460		67	-	-
	60°C 24 h + 200°C 1.5 h + 70°C 1000 h	464		61	-	-

Table 10. Mechanical properties of weldments (sheets 2.5 mm thick) after different regimes of aging (filler – V92Zr)

Aging: Before welding	Aging: After welding	TS, MPa	Place of failure	α, degrees
20°C 5 days	20°C 1 month	390	Fusion zone	70
	20°C 1 month + 70°C 1000 h	453		54
60°C 24 h + 200°C 1.5 h	20°C 1 month	385		72
	20°C 1 month + 70°C 1000 h	418		52
20°C 5 days	60°C 24 h + 200°C 1.5 h	395	15-20 mm from center of seam	50
	60°C 24 h + 200°C 1.5 h + 70°C 1000 h	403	Fusion zone	48
60°C 24 h + 200°C 1.5 h	60°C 24 h + 200°C 1.5 h	394	15-20 mm from center of seam	40
	60°C 24 h + 200°C 1.5 h + 70°C 1000 h	402	Fusion zone	37

Table 11. Impact properties of weldments (plates 2.5 mm thick) after different regimes of aging (Filler – V92Zr)

Aging		Impact properties with an initiated crack, J					
		Without heating			70°C 1000 h		
Before welding	After welding	Center of the weld	Fusion zone	Base metal	Center of the weld	Fusion zone	Base metal
20°C 3 month	20°C 1 month	8.3	6.1	12.0	5.6	4.1	7.1
	60°C 24 h + 200°C 1.5 h	5.2	3.8	7.2	4.8	3.2	6.7
60°C 24 h + 200°C 1.5 h	20°C 1 month	7.5	5.2	8.5	5.2	3.7	6.4
	60°C 24 h + 200°C 1.5 h	4.7	3.5	6.8	4.3	3.0	6.2

The general corrosion resistance of the welded samples was investigated in sheets 2.5 mm thick welded with the automatic TIG process and the filler wire V92Zr (Table 12). Heating at 70°C for 3000 h reduces only slightly the loss of the strength resulting from corrosion.

The resistance to stress corrosion was determined on samples of sheets 2.5 mm thick and extrusions welded with the automatic TIG process and the filler wire V92Zr, No. 11, and AMg6 (Table 13) Heating at 70°C for 1000 h impairs the corrosion resistance of the welded joints more after aging at 20°C for 1 month than after stepped aging.

The corrosion resistance of the welded joints made with filler wire No.11 was higher than for the joints welded with V92Zr.

High-temperature aging, inducing more complete decomposition of the solid solution, nay improve the stress corrosion resistance after heating. The investigation confirmed the positive effect of high-temperature tempering at 200°C for 2-4 h and 210°C for 2 h on the resistance to stress corrosion. No failures occurred in tests for 4 months.

The susceptibility to stress corrosion was lowest for joints welded with the filler wire No.11 and higher (about 40 days to failure) for the joints welded with AMg6. This is evidently due to the large chemical heterogeneity of the transition zone in the joints welded with AMg6 and also the formation of a continuous network of β-phase at elevated temperatures (200-210°C).

Table 12. Corrosion properties of weldments (aluminum alloy V92Zr)

Condition of the metal		Tensile strength, MPa		Losses after corrosion	
Before welding	**After welding**	**Before corrosion**	**After corrosion**	**Δ TS, MPa**	**Δ TS, %**
20°C 1 year	20°C 1 month	388	356	32	8.3
	20°C 1 month + 70°C 500 h	431	339	9.2	21.4
	20°C 1 month + 70°C 1000 h	453	419	3.4	7.6
	20°C 1 month + 70°C 3000 h	459	444	1.5	3.2
	60°C 24 h + 200°C 1.5 h	397	322	7.5	18.8
	60°C 24 h + 200°C 1.5 h + 70°C 500 h	401	395	1.05	2.5
	60°C 24 h + 200°C 1.5 h + 70°C 1000 h	408	382	2.6	6.3
	60°C 24 h + 200°C 1.5 h + 70°C 3000 h	423	414	0.9	2.3
60°C 24 h + 200°C 1.5 h	20°C 1 month	383	250	13.3	35.1
	20°C 1 month + 70°C 500 h	402	287	11.5	28.4
	20°C 1 month + 70°C 1000 h	418	417	0.1	0.8
	20°C 1 month + 70°C 3000 h	445	313	13.2	29.7

Table 13. Corrosion resistance of weldments (aluminum alloy V92Zr)

Filler	Base metal	Condition of material		Samples failed	Time of failure, days
		Before welding	**After welding**		
V92Zr	Sheet, 5mm thick	20°C 1 year	20°C 1 month + 70°C 1000 h	2	55, 74, > 120
			60°C 24 h + 200°C 1.5 h + 70°C 1000 h	2	57, 64, > 120
		60°C 24 h + 200°C 1.5 h	20°C 1 month + 70°C 1000 h	-	> 120
			60°C 24 h + 200°C 1.5 h + 70°C 1000 h	-	> 120
# 11		20°C 1 year	20°C 1 month + 70°C 1000 h	1	60, > 120
			60°C 24 h + 200°C 1.5 h + 70°C 1000 h	1	83, > 120
		60°C 24 h + 200°C 1.5 h	20°C 1 month + 70°C 1000 h	-	> 120
V92Zr	Extrusion	20°C 10 days	20°C 10 days + 70°C 1000 h	1	34, > 120
			60°C 24 h + 200°C 1.5 h + 70°C 1000 h	3	39, 51, 57, > 120
		60°C 24 h + 200°C 1.5 h	20°C 1 month + 70°C 1000 h	-	> 120
			60°C 24 h + 200°C 1.5 h + 70°C 1000 h	-	> 120

References

[1] Fridlyander, I. N., *et al*, Aluminum Alloys, 4, Metallurgia, Moscow, 1966.

[2] Sverdlin A., E. Kuznetsova, et al., *MITOM*, 8, 1966.

[3] Fridlyander,I I.N., *et al*, Aluminum Alloys, 5, Metallurgia, Moscow, 1968.

[4] Sverdlinm A., Pospelov K., Aluminum Alloys, 5, Metallurgia, Moscow, 1968.

[5] Fridlyander I. N., Aluminum and Its Alloys, *Znanie*, 1965.

[6] Fridlyander I.N., Gerchikova N., Zaitseva N., *MITOM*, 8, 1966.

[7] Sverdlin A., Properties of Aluminum Alloy V92Zr, Unpublished Research, #1489N, 1986.

[8] Sverdlin A., Fusion Welding of Aluminum Alloy V92Zr, Unpublished Research, #2K07D, 1987.

Proceedings of the 26th ASM Heat Treating Society Conference
B.L. Ferguson, R. Jones, D.S. MacKenzie, and D. Weires, editors
https://doi.org/10.31399/asm.cp.ht2011p0205

Heat Treatment Development for Rapidly Solidified Heat Resistant Cast Al-Si Alloy

W. Kasprzak
CANMET Materials Technology Laboratory, Hamilton, Ontario, Canada
wkasprza@nrcan.gc.ca

D. L. Chen
Ryerson University, Toronto, Ontario, Canada
dchen@ryerson.ca

E. Thibodeau
McGill University, Montréal, Quebec, Canada

Keywords: Al cast alloy, heat treatment, solidification rate, thermal analysis, dilatometer analysis, dimensional stability.

Abstract

Existing heat treatment standards do not properly define tempers for thin walled castings that solidified with high solidification rates. Recently emerged casting processes such as vacuum HPDC (High Pressure Die Casting) should not require long solution treatment times due to the fine microstructures caused by rapid solidification rates. The heat treatment studies involving rapidly solidified samples with Secondary Dendrite Arm Spacing (SDAS) between 35-10μm were conducted for solution times between 30 minutes and 9 hours and were carried out for various aging parameters. The metallurgical analysis revealed that an increase in microstructure refinement could enable reduction of solution time up to 50% without compromising the alloy's mechanical properties. The highest strength of 330 and 300MPa (47.9 and 43.5ksi) for UTS and YS respectively was achieved for T6 peak aging conditions. Additionally, thermal analysis and dilatometer results are presented to assess phase's dissolution during solution treatment, aging kinetics as well as dimensional stability analysis.

1. Introduction

Growing concerns over vehicle emissions and the price of energy has forced the automotive industry to take renewed interest in ways of improving fuel economy. One way of increasing fuel efficiency is reducing the overall mass of the vehicle. This could be achieved by employing lightweight alloys for structural and powertrain applications as well as by innovations in efficient manufacturing (such as short cycle heat treatment) for improved strength and consequent weight reduction [1,2].

Al-Si (3xx.x series) alloys are extensively used in the automotive (i.e., 356, 319, etc.) and aerospace industries (i.e., A357, C355, etc.) due to their good castability, mechanical properties, and corrosion resistance but can not maintain strength at temperatures exceeding 150°C (302°F). For this reason new alloys are being developed that could withstand higher operating temperatures (up to 250°C (482°F)) observed in contemporary powertrain components [1,3-7].

Heat treatment is an important part of the casting manufacturing process, thus, cost has to be kept low to gain economical advantages. Up to date, the existing heat treatment standards do not precisely define tempers for thin walled castings that solidified with high solidification rates [8]. The R&D efforts were undertaken to reduce heat treatment process duration but have not led to formalized implementation of revised heat treatment standards [9-15].

Precise relations between solidification rate, its effect on microstructure refinement, and its consequent impact on heat treatment effectiveness must be defined for any alloy of interest. The presented work attempts to optimize heat treatment processing for newly developed alloy solidified with various solidification rates. The wedge shaped mold castings were used to produce microstructures having Secondary Dendrite Arm Spacing (SDAS) that is typically observed in the cast components manufactured using High Pressure Die Casting (HPDC) or Low Pressure Permanent Mold (LPPM) technologies [13-14].

2. Experimental Procedures

The prototype Al-7%Si-1%Cu-0.5%Mg casting alloy with additions of transition metals such as Zr, V and Ti was used in this study [1] (Table 1). This alloy was cast in a water-cooled copper mold [16] that resulted in the microstructure variation having SDAS between 35-10μm measured along the mold height.

The solution treatment experiments were performed at 510°C (950°F) for times between 30 minutes and 9 hours. Next, test samples were subjected to the identical aging conditions at 150°C (302°F) for 100 hours (T6) and 250°C (482°F) for 16 hours (T7). Selection of optimum T6 and T7 tempers were done by carrying out artificial aging experiments at 100, 150, 200, 250, 275, 300, 330, 360, and 400°C (212, 302, 392, 482, 527, 572, 626, 680, 752°F) with variable times ranging from a few minutes (as-quenched) to 128 hours [1].

Table 1. Alloy chemical composition and thermal analysis results.

Alloying elements (wt %)				
Si	Cu	Fe	Mg	Zr-Ti-V
7.0	1.0	0.12	0.47	0.2
Selected thermal analysis results during alloy melting				
T_{BAM},(°C)	T_{ECuM},(°C)	$T_{MAl\text{-}Si}$,(°C)	T_{EAM},(°C)	
507	552	589	643	

Note: T_{BAM} (#1a) – Beginning of alloy melting, T_{ECuM} (#2a) – End of Cu based phase melting, $T_{MAl\text{-}Si}$ (#3) – Melting of Al-Si eutectic, T_{EAM} (#4) – End of Alloy Melting.

Hardness measurements were done using a Rockwell tester in the B scale on the test samples in as-cast and heat treated conditions. Twenty five (25) measurements were done and the average value with corresponding standard deviation was calculated and plotted.

Tensile testing was carried out for ASTM flat bars extracted from wedge castings using the computerized United Tensile Testing machine at a strain rate of 1×10^{-3} s^{-1} with a minimum of two samples at each level of SDAS.

The BAHR DIL 805 quench dilatometer was used to evaluate the alloy's dimensional stability using the sample's relative

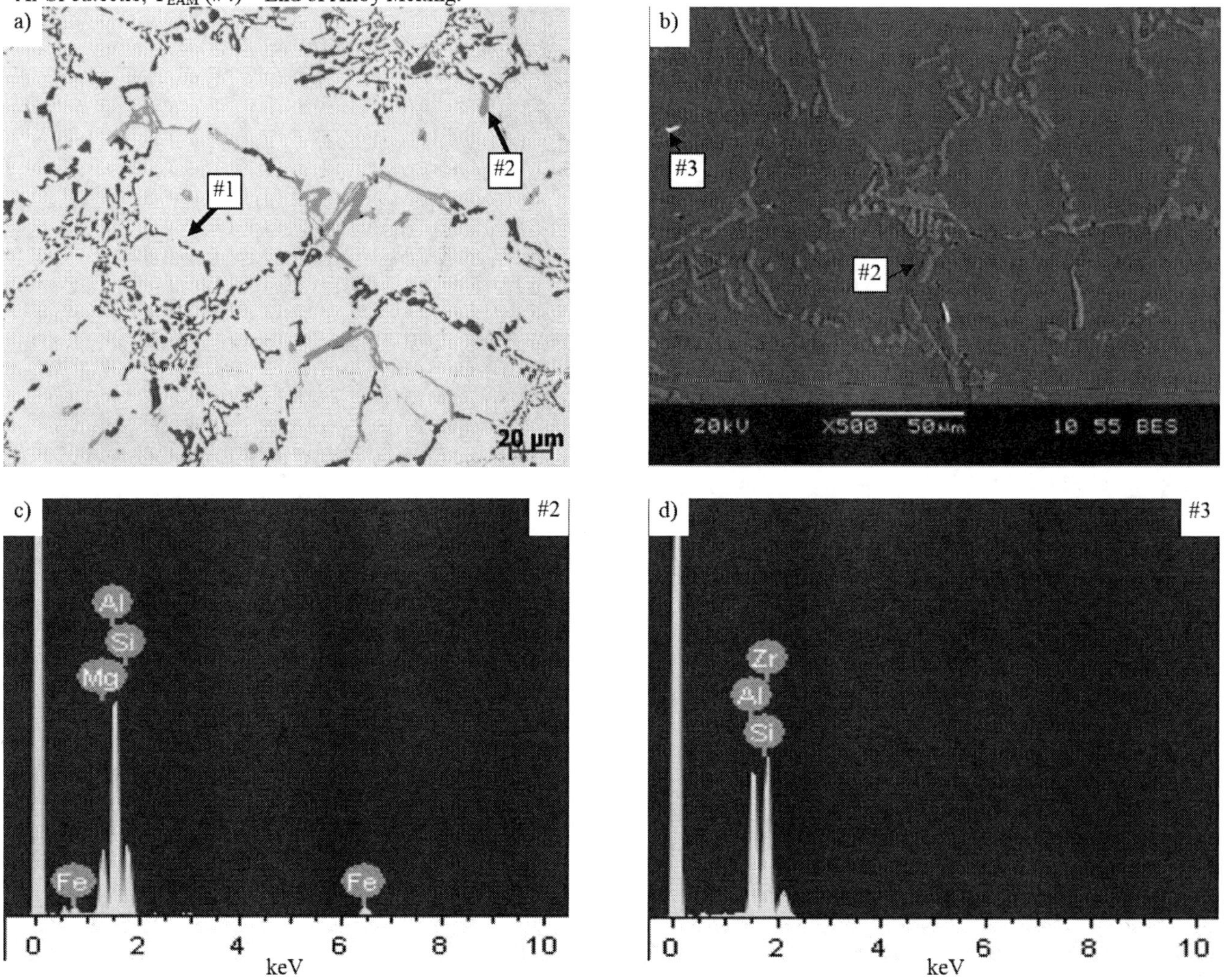

Fig. 1. Microstructures and X-ray microanalysis results for the alloy in the as-cast condition (SDAS=25μm): a) Optical micrographs – note the Al-Si eutectic (#1) and Al-Si-Mg-Cu phases (#2), b) SEM/BSE image, c-d) EDX spectra taken from Al-Si-Mg-Fe based phases (#2) as well as Al-Si-Zr based phases (#3).

length change and its derivative curves (dL/dt) as a function of temperature. Test samples having OD=5mm, L=10mm were extracted from tensile test bars and UTS was correlated with aging transformation kinetics remaining after various heat treatment operations.

Test samples were prepared using standard metallographic procedures as described elsewhere [13]. The alloy microstructure was examined using a light optical microscope (LOM) and scanning electron microscope (SEM) in the backscatter electron (BSE) mode, coupled with energy-dispersive X-ray spectroscopy (EDS) analysis.

3. Experimental Results

3.1. Alloy Characteristics in the As-Cast Condition

Microstructure analysis in the as-cast condition predominantly showed α-Al dendrites, unmodified Al-Si eutectic (#1) and Al-Si-Mg-Fe phases (#2) located within interdendritic regions as well as micro size Al-Zr-V and Al-Si-Zr-V-Ti particles (#3), both primary (nucleated during solidification) (Fig. 1) and secondary (precipitated during heat treatment processing).

Rapid solidification in a water-cooled mold resulted in significant microstructure refinement, having SDAS vary between 35-10μm, and consequent thermal modification of structural constituents (Fig. 2). Such microstructure variation was suitable for heat treatment optimization experiments required to determine the effect of solidification rate (expressed as SDAS) on heat treatment parameters.

Thermal analysis during the melting cycle of the test samples in the as-cast condition were carried out to: i) establish maximum allowable solution treatment temperature and, ii) to evaluate the effectiveness of solution treatment duration. It was revealed that incipient melting temperature, corresponding to the melting of the Al-Si-Mg-Cu phases, began at approximately 507°C (945°F) and was completed at 552°C (1026°F) (Fig. 3, curve as-cast). For this reason, the solution treatment temperature was selected as 510°C (950°F).

Hardness measurements in the as-cast condition showed a linear relation with solidification rate, indicating hardness increase with reducing SDAS size (Fig. 4). The increase in the as-cast hardness indicated that analyzed solidification rates were high enough to cause some solid solution saturation (strengthening effect) (Fig.4).

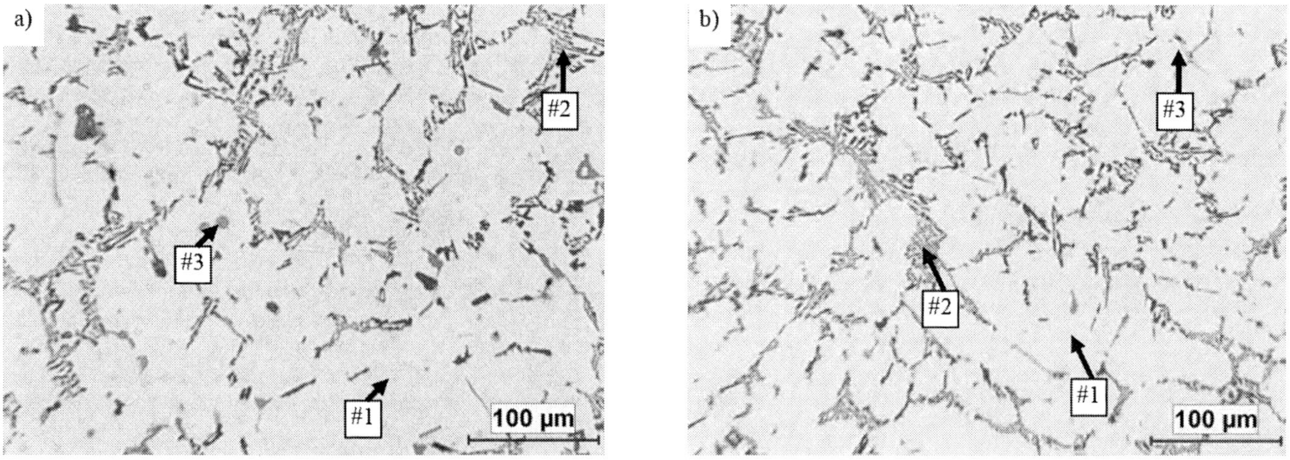

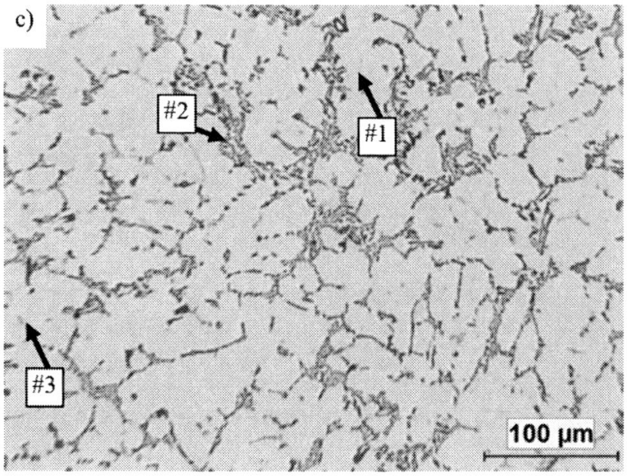

Fig. 2. Microstructure refinement levels obtained for wedge castings and used for the subsequent heat treatment optimization studies. The SDAS values and as-cast hardness are as follows:
a) 35μm / 21.3±5.2 HRB
b) 25μm / 27.7±4.2 HRB
c) 12μm / 36.0±1.8 HRB

Note: #1: α-Al dendrites, #2: Al-Si eutectic, #3: Al-Si-Mg-Cu phases.

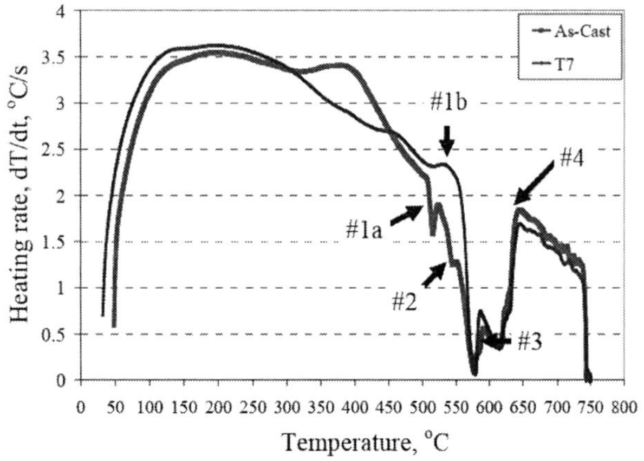

Fig. 3. Thermal analysis melting curves for the alloy in the following conditions: a) as-cast, b) T7. Note that alloy in the T7 condition had incipient melting temperature 43°C higher (550°C (1022°F)) than in the as-cast condition (507°C (945°F)). Note that the transformation temperatures are indicated by arrows (see Table 1 for details).

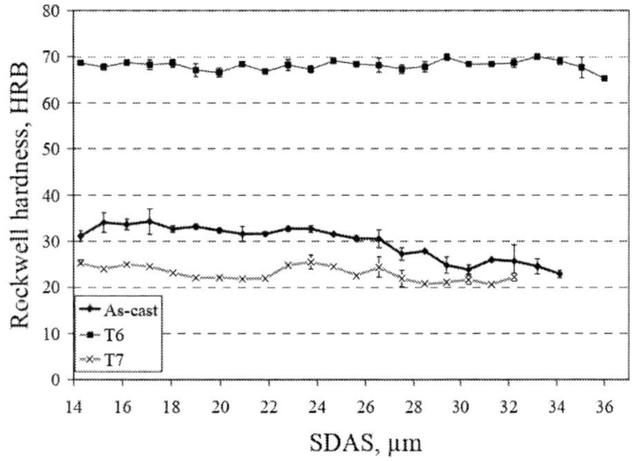

Fig. 4. The relation between hardness and SDAS for alloys in the as-cast, T6 and T7 conditions. Note increasing as-cast hardness with reducing SDAS size (higher solidification rates) and almost constant hardness value over analyzed SDAS range for T6 and T7 conditions. Note that it is evident that T6 and T7 correspond to peak-aged and over-aged conditions accordingly.

3.2. Alloy Characteristics in the Heat Treated Condition

Thermal analysis of the test samples in the solution treated, quenched and over-aged condition revealed that alloy solidus temperature (beginning of melting) was shifted up from 507 (#1a) to 550°C (945-1022°F) (#1b) (Fig. 3, curve T7). This indicates that 9 hours of solution treatment at 510°C (950°F) was sufficient for dissolution of Al-Si-Mg-Cu. The exact determination of the alloy melting temperature should minimize the risk of incipient melting and improve the dissolution kinetics of the Al-Si-Mg-Cu phases. In turn, if properly implemented it should lead to consequent process reduction time.

Metallurgical analysis of the test samples subjected to T6 temper showed approximately 40 points hardness increase from 30 to 69 HRB which remained constant for the analyzed SDAS range. For the T7 condition hardness dropped to 25 HRB which is below that of the as-cast value (i.e., 30) and was constant for the analyzed SDAS range. Hardness changes reflected the peak-aged (T6) and over-aged (T7) heat treatment conditions as optimized within this study.

For the test samples in the T6 condition it was found that an ultimate tensile strength (UTS) of 330MPa (47.9ksi) and a yield strength (YS) of 300MPa (43.5ksi) was achieved for solution treatment time of 9 hours at 510°C (950°F) and SDAS of 18µm (Fig. 5a). It was observed that reducing the SDAS (increasing a solidification rate) from 35 to 18µm resulted in an approximate increase of up to 20% in T6 tensile strength i.e., from 220-310 to 270-330MPa (32-45 to 39.2-48ksi) for solution treatment time of 30 minutes and 9 hours respectively. Increasing a solution treatment time from 30 minutes to 9 hours resulted in up to 20% increase in T6 tensile strength i.e., from 250-270 to 310-330MPa (36.3-39.2 to 45-47.9ksi) for SDAS of 35 and 18µm respectively.

Reducing SDAS from 35 to 18µm had approximately 10% effect on T6 yield strength (YS) development for 30 minutes of solution treatment at 510°C (950°F) i.e., from 240 to 270MPa (34.8 to 39.2ksi). In contrast increasing solution treatment time from 30 minutes to 9 hours at 510°C (950°F) resulted in an insignificant effect on T6 yield strength i.e., from 290 to 300MPa (42 to 43.5ksi) (Fig. 5b). Similar trends were observed for the test samples in T7 condition.

In the T6 condition the maximum elongation of 1.8% was observed that increased to 2.6% in the T7 condition for alloys having SDAS of 18µm and solutionized for 9 hours at 510°C (950°F) (Fig. 5c). In general elongation values were low since alloy processing was optimized for maximum strength. Further improvement could be accomplished by combined thermal and chemical modification of Al-Si eutectic structure and aging optimization.

It was evident that the solidification rate had a more pronounced effect on the strength development than the increase of solution treatment time, particularly for UTS. It was observed that increasing a solution treatment time beyond 4 hours at 510°C (950°F) is not economically justified since it did not bring a significant increase in hardness, or strength. Similar observations were made in previous optimization studies carried out for hypereutectic Al-Si alloys having SDAS of ~ 20µm [13,14].

3.3. Dimensional Stability Assessment

For the majority of cast components the selection of heat treatment parameters led to compromise between maximum achievable strength vs. dimensionally stable material that does not experience any transformation during in-service operation. Dilatometer analysis of the T6 and T7 test samples showed that alloys which were solution treated and artificially aged at

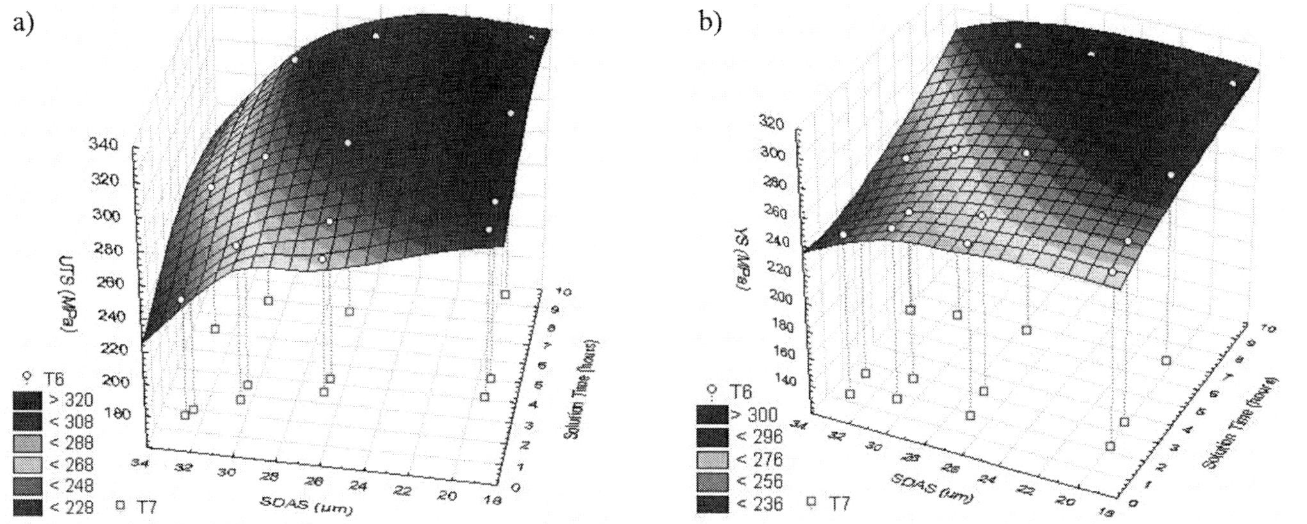

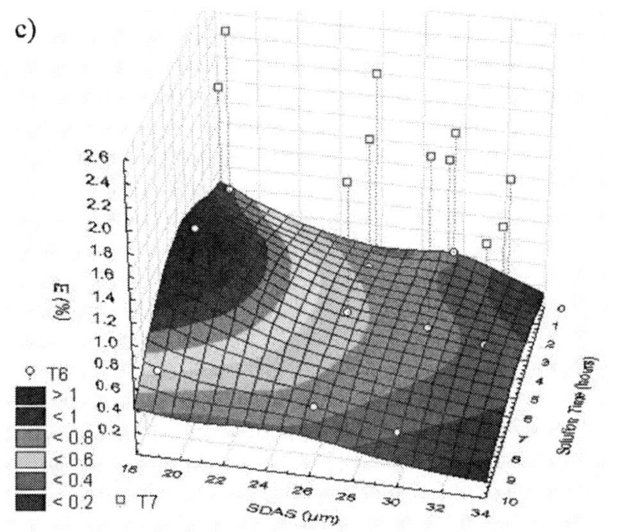

Fig. 5. The effect of solution treatment time on alloy mechanical properties in the T6 and T7 conditions obtained for SDAS between 18 to 32µm: a) UTS, b) YS, c) Elongation. Note increase in alloy strength for smaller SDAS. Solution treatment time longer than 4 hours does not bring significant gains in the strength improvement.

250°C (482°F) for 16 hours (T7) having UTS of 210MPa (30.5ksi) are the most dimensionally stable because they did not show any solid state transformation during subsequent heating after completion of the heat treatment operation (Fig. 6). Alloys aged at 150°C (302°F) for 100 hours had the highest strength but are less dimensionally stable as manifested by strong transformations observed on the derivative curve (Fig. 6). From a physical metallurgy point of view aging transformation must be completed and no further precipitation can occur during the in-service operation. The presented approach allows the selection of heat treatment parameters yielding dimensionally stable microstructure and corresponding optimum strength.

3.4. Comments on Heat Treatment Development

The presented heat treatment studies demonstrated the possibility to flexibly control alloy mechanical properties as a function of solidification rate and solution treatment duration allowing the selection of optimum parameters. Approximately up to 50% process time reduction for alloy with fine microstructures (i.e., SDAS-18µm) as compared to alloy having coarse microstructure (i.e., SDAS-35µm) can be achieved since solutionizing for more than 4 hours brings only a minor effect on UTS and YS strength development. As far as hardness is concerned, up to 95% process time reduction could be achieved since increasing time beyond 30 minutes of solution treatment does not bring hardness increase within SDAS of 35-10µm. Reduction of the heat treatment process duration was achieved by i) selecting a solution treatment temperature in the vicinity of alloy melting (i.e., 507°C (945°F)) but avoiding incipient melting and ii) by minimizing the diffusion distances and the time of the subsequent solution treatment due to reduced interdendritic spaces for samples having small SDAS size.

Rising energy costs should provide a positive stimulus to revise the existing heat treatment standards - particularly for cast components that solidified under high solidification rates

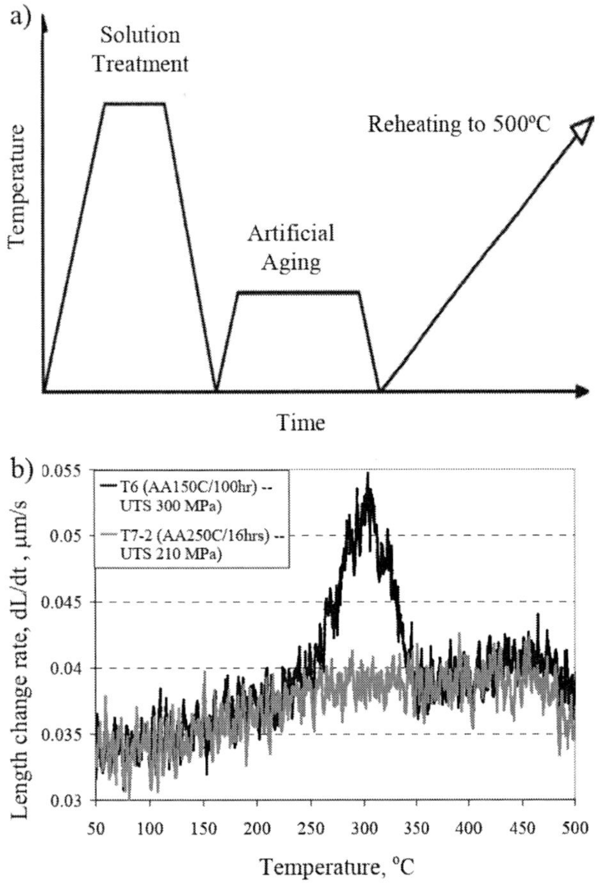

Fig. 6. Dimensional analysis assessment for the alloy in the T6 and T7 conditions.
a) Schematic of alloy thermal cycle. Note after completion of heat treatment the alloy was tested for dimensional stability assessment using dilatometer during reheating to 500°C.
b) Length change rate (dL/dt) vs. temperature curve.
Note that T7 condition renders the most dimensional stable condition where no transformation peak is observed during reheating cycle. T7 yields UTS of approx. 210 MPa (30.5ksi) while T6 of 300 MPa (43.5ksi).

- paying careful attention to and taking great advantage of the thermal as-cast microstructure with its potential to leverage the metallurgical response of the component to subsequent heat treatment processing [13,15].

4. Conclusions

Based on the heat treatment results for the Al-7%Si-1%Cu-0.5%Mg alloy with additions of Zr, V, Ti the following conclusions were drawn:

1. Rapid solidification experiments in a water-cooled wedge mold resulted in microstructure refinement suitable to evaluate the effect of solidification rate on heat treatment effectiveness.
2. Alloy incipient melting temperature was determined to be at ~507°C (945°F) and was used to select the solution treatment temperature at 510°C (950°F).
3. Hardness measurements in the as-cast condition showed a linear increase with the reduction of SDAS from 35 to 10µm. For T6 condition, the hardness increased from 30 HRB (as-cast) to approximately 70 HRB, while in the T7 condition, hardness decreased to approx. 25 HRB and was constant for analyzed SDAS.
4. Highest strength, i.e., UTS (330MPa (47.9ksi), YS (300MPa (43.5ksi)) with corresponding elongation of 1.5% was achieved for solution treatment and T6 aging conditions (at 510°C (950°F) for 9 hours).
5. Reduction in SDAS from 35 to 10µm resulted in approximately 10-25% improvement in YS and UTS respectively for alloys solution treated at 510°C (950°F) in the peak-aged (T6) and over-aged (T7) conditions.
6. Dilatometer analysis showed a clear correlation between strength and transformation kinetics indicating that alloys aged at 250°C (482°F) for 16 hours (T7 condition) showed good dimensional stability (i.e., no visible phase transformation during alloy re-heating).

Acknowledgments

The authors would like to acknowledge the financial support of the Advanced Structural Materials for Next Generation Vehicles (ASM-NGV) Program of Natural Resources Canada. The authors thank Dr. A. Javaid for wedge casting experiments and R. Zavadil B.Sc. and M. Aniolck M.Sc. Eng. from CANMET-MTL for their work on the microstructure analysis and thermal analysis data evaluation. Moreover, we thank Kevin Woo Seok Yeom from University of Waterloo and Lucas Nicodemou from McMaster University for their contributions to graphical data presentation done during their Co-op program.

References

1. W. Kasprzak, Z. Deng, J. Powell, M. Niewczas, "Aging Characteristics, Dimensional Stability and Assessment of High Temperature Performance of Cast Al-Si Alloy for Powertrain Applications", Proc. 12th Int. Conf. on Aluminium Alloys, September 5-9, 2010, Yokohama, Japan. *The Japan Institute of Light Metals*, pp. 669-674.
2. T. D. Donofrio, "New Efficiencies for Aluminum Engine Block and Cylinder Head Heat-Treatment Systems," *Industrial Heat Treating,* February 2010, pp. 47-49
3. M. Zeren, "Effect of Copper and Silicon Content on Mechanical Properties on Al-Cu-Si-Mg Alloys," *Journal of Materials Processing Technology,* Vol. 169 (2005), pp. 292-298.

4. M. Tash, F. H. Samuel, F. Mucciardi and H. W. Doty, "Effect of Metallurgical Parameters on the Hardness and Microstructural Characterization of As-Cast and Heat-Treated 356 and 319 Aluminum Alloys," *Materials Science and Engineering* A. Vol. 443 (2007) , pp. 185-201.
5. F. J. Tavitas-Medrano, A. M. A. Mohamed, J. E. Gruzleski, F. H. Samuel and H. W. Doty, "Precipitation-Hardening in Cast Al-Si-Cu-Mg Alloys," *Journal of Materials Science,* Vol. 45 (2010), pp. 641-651.
6. I. Alfonso, C. Maldonado, G. Gonzalez and A. Bedolla, "Effect of Mg Content and Solution Treatment on the Microstructure of Al-Si-Cu-Mg Alloys," *Journal of Materials Science,* Vol. 41 (2006), pp. 1945-1952.
7. J. Y. Hwang, H. W. Doty and M. J. Kaufman, "The Effects of Mn Additions on the Microstructure and Mechanical Properties of Al-Si-Cu Casting Alloys," *Materials Science and Engineering* A. Vol. 488 (2008), pp. 496-504.
8. "Standard Practice for Heat Treatment of Aluminum-Alloy Castings from All Processes," *ASTM*, Designation: B8917/B917M-01.
9. H. Moller, G. Govender and W. E. Stumpf, "Application of Shortened Heat Treatment Cycles on A356 Automotive Brake Calipers with Respective Globular and Dendritic Microstructures," *International Journal of Cast Metals Research,* Vol. 20 (2007), pp. 340-346.
10. M. Tiryakioglu, "The Effect of Solution treatment and Artificial Aging on the Work Hardening Characteristics of a Cast Al-7% Si-0.6%Mg Alloy," *Materials Science and Engineering*, A. Vol. 427 (2006), pp. 154-159.
11. G. Timelli, O. Lohne, L. Arnberg and H. I. Laukli, "Effect of Solution Heat Treatments on the Microstructure and Mechanical Properties of a Die-Cast AlSi7MgMn Alloy," *Metallurgical and Materials Transactions* A. Vol. 39 (2008), pp. 1747-1758.
12. Y. Zhou, J. Kang, Y. Rong, F. Yi, Y. Ma, J. Fang, H. D. Brody, X. Pan and J. E. Morral, "Integrated Numerical Simulation of Process Optimization for Solutionizing of Aluminum Alloys," *23rd ASM Heat Treating Society Conference,* 2005, pp. 1-8.
13. W. Kasprzak, J. H. Sokolowski, H. Yamagata, M. Aniolek, H. Kurita, "Energy Efficient Heat Treatments for Hypereutectic Al-Si Engine Blocks Made using Vacuum High Pressure Die-Casting," *Journal of Materials Engineering and Performance,* Published Online, April 30, 2010 (DOI: 10.1007/s11665-010-9658-5).
14. W. Kasprzak, H. Kurita, J. H. Sokolowski, H. Yamagata, "Energy Efficient tempers for Aluminum Motorcycle Cylinder Blocks," *Advanced Materials Processes (AMP),* March 2010, pp. 24-27.
15. R. N. Lumley, I. J. Polmear, P. R. Curtis, "Rapid Heat Treatment of Aluminum High-Pressure Diecastings," *Metallurgical and Materials Transactions,* A. Vol. 40 No. 7, pp.1716-1726.
16. Y. He, A. Javaid, E. Essadiqi and M. Shehata, "Numerical simulation and Experimental Study of the Solidification of a Wedge-Shaped AZ31 Mg Alloy Casting," *Canadian Metallurgical Quarterly,* Vol. 48 No. 2 (2009), pp. 145-156.

Proceedings of the 26th ASM Heat Treating Society Conference
B.L. Ferguson, R. Jones, D.S. MacKenzie, and D. Weires, editors
https://doi.org/10.31399/asm.cp.ht2011p0212

Effect of Heat Treatment on Fracture Toughness of Micro-Alloyed Steel

Mr. Joydeb Nag Chaudhury[1] and Prof. R.C. Prasad[2]
[1]Fr. C Rodrigues College of Engg.,University of Mumbai, Mumbai, India
[2]Indian Institute of Technology Bombay, Mumbai, India

Abstract

Micro-alloyed steels are being increasingly accepted by industry in various fields of application and are available with a wide variety of microstructures. Extensive literature is available on processing routes of Micro-alloyed steels. However, experimental information on their suitability to fracture mechanics based approaches towards design and manufacture is found to be relatively scarce in two distinct areas:

- Alteration of fundamental fracture mechanics properties of Micro-alloyed steels in presence of structural restraints in the form of pre-stress and pre-strain.
- Comparative study of the effect of heat treatment practice on fracture mechanics properties of Micro-alloyed steels relative to their as-rolled conditions.

It is in this context that the present work explores experimental determination of quasi-static initiation fracture toughness (J_{1c}) (as per ASTM E-1820) of low carbon (0.19%) micro-alloyed steel in as-rolled condition without any heat treatment. The study also explores the effect of normalizing, shot-peening, and cyaniding followed by shot-peening on fracture toughness parameter so obtained. The normalizing heat treatment, shot-peening and cyaniding followed by shot-peening indicate a positive influence on initiation fracture toughness of this Micro-alloyed steel. Results, when compared with as-rolled condition, show that cyaniding followed by shot-peening have led to a 2.7 times increase in J_{1c} of the Micro-alloyed Steel under study. Thus the surface hardening followed by shot-peening of (CT) specimens appears to have most positive influence in terms of enhancement of qualified J_{1c} values. Although fracture initiation toughness is in general known to decrease with increase in yield strength, the Micro-alloyed steel under study when normalized displayed simultaneous improvement in yield strength and J_{1c}. An attempt was made to explain this inter-relationship between yield strength and J_{1c}. Furthermore, the observed effects, in elastic-plastic regime, of heat treatment, mechanical pre-stress and chemical surface hardening on the initiation fracture toughness were also explained.)

1. Introduction

Modern high strength micro-alloyed steels are being extensively used in a wide spectrum of applications. For critical structures made of Micro-alloyed steels, experimental study of fundamental parameters of fracture mechanics, therefore, assumes significance; however, it has been observed from literature survey that the experimental information related to fracture mechanics based parameters (initiation fracture toughness, J_{1c} in particular) and the effects of heat treatment, mechanically induced pre-stress and surface hardening treatment on these parameters are inadequately documented in the literature. It was, therefore, decided to undertake the present study with following objectives:

- To estimate fracture toughness of as-rolled Low Carbon (0.2%) thick plate (16 mm) Micro-alloyed steel designated as SR BQT- 10 by the manufacturer, Essar Steel Limited, India and to study the effect, on fracture toughness of SR BQT 10, of:
  (a) heat treatment (normalizing),
  (b) compressive pre-stress in the form of shot-peening, and
  (c) surface hardening in the form of cyaniding followed by shot- peening.

The experiments include chemical analysis; optical, SEM and TEM study; phase analysis; grain size distribution analysis; normalizing heat treatment; shot-peening treatment; cyaniding (surface hardening) treatment; X-ray surface and depth profile analysis of residual stress distribution; tensile testing and hardness testing. Present experimental work also includes Fracture Mechanics-based experiments such as determination of Initiation Fracture Toughness (J_{1c}) as per ASTM E1820. It was possible to determine valid J_{1c} values (38.60 kJ/m^2, on an average) for SR BQT- 10 in as-rolled condition, on which the effects of normalizing, shot-peening and a combination of cyaniding and shot-peening were studied subsequently. Each of these processes was found to enhance valid J_{1c} values of as-rolled BQT-10 but to a varying degree. The best improvement in initiation fracture toughness value could be achieved through cyaniding and shot-peening applied consecutively that resulted in an average J_{1c} = 105.2 kJ/m^2.

Fracture initiation toughness is in general known to decrease with increase in yield strength; but the Micro-alloyed steel under study when normalized displayed simultaneous improvement in yield strength and J_{IC}. An attempt was also made to explain the observed effects of normalizing heat treatment, mechanically induced pre-stress in the form of shot-peening and chemical surface hardening in the form of cyaniding on the initiation fracture toughness of SR BQT- 10 in elastic-plastic regime.

2. Source & Chemistry of Research Materials

The material selected for study is a Low Carbon thick plate (16 mm) Micro-alloyed Steel designated as SRBQT- 10 by the manufacturer, Essar Steel Limited, India. The steel is of grade to the nearest equivalent of ASTM A 516 Gr. 70 and was received in the "As-rolled" condition without any mill-anneal or any other form of heat treatment.

Chemical analysis of the steel was done on Applied Research Laboratory (Switzerland) make Optical Emission Spectrometer – Model 3460, as per ASTM E-415: 1999. Specifically, Nitrogen analysis is done on LECO (USA) make 'Nitrogen & Oxygen Analyzer' – Model TC-136, as per ASTM E-1019: 2004. The results are listed in Table 1.

Table 1 Chemical Composition of BQT-10

C	Mn	S	P	Si	Cr	Ni	Cu
0.19	1.15	0.004	0.017	0.25	0.018	0.037	0.160
Ti	**V**	**Nb**	**Mo**	**Al**	**W**	**N**	
0.022	0.001	0.020	0.004	0.043	< 0.010	0.009	

3. Manufacturing Process History of Micro-alloyed Steel (BQT-10)

3.1. Production Process Sequence

The entire processing history of BQT-10 is outlined in the form of Process Flow Charts vide Figs. 1(a) and (b).

3.2. Rolling Schedule of Micro-alloyed Steels

The Controlled Rolling Schedule for BQT-10, as provided by the manufacturer, is highlighted in Table 2.

4. Experimental Details

4.1. Normalizing of BQT-10

Since the major factors contributing to strength properties of Micro-alloyed steels are grain-refinement and micro-precipitation, normalizing heat treatment was attempted to effect further grain-refinement without causing major alteration in the basic micro-constituents of the as-rolled material. The objective is to study the effect of normalizing on initiation fracture toughness in the elastic-plastic regime. With this objective, the (CT) specimens as well as metallographic specimens of BQT-10 were normalized.

While CT specimens of BQT-10 were utilized for experimental fracture mechanics study, metallographic specimens of BQT-10 in the normalized condition were studied for micro-structural details by optical microscopy, SEM and TEM. The methodology adopted for selection of normalizing variables is presented below.

4.1.2. Selection of Time of Holding at Normalizing Temperature

In order to determine the optimum time of holding at 920 °C, six cut-out plates of 285 x 285 x 16 mm size were held at 700 °C for 5 minutes. The temperature was raised to 700 °C at the rate of 5 °C /min. Thereafter the temperature was raised to normalizing temperature of 920 °C @ 1 °C per minute. All the 6 plates were held at 920 °C for different time intervals, followed by air cooling to room temperature. Finally, six uniaxial tension test specimens were machined from these 6 plates and tested for strength and elongation properties. The results are tabulated in Table 3 below:

4.1.1. Selection of Normalizing Temperature

Since in Micro-alloyed steels ferrite transformation starts at Ar3 and finishes at Ar1, the normalizing is practiced at Ar3 or below. The determination of Ar3 for specific Micro-alloyed steels is primarily a subject of shop-floor trial following use of empirical

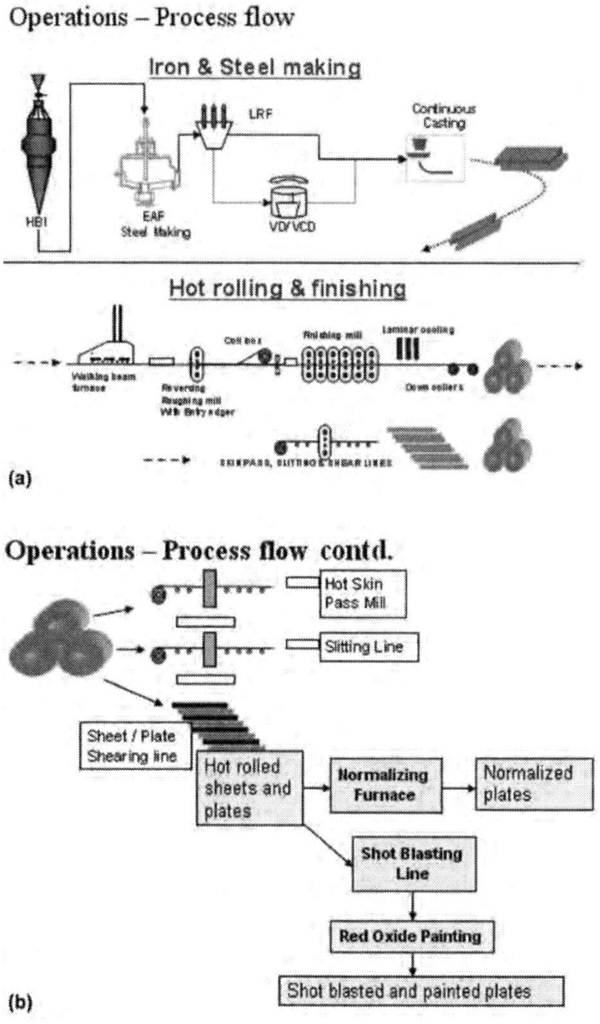

Figs. 1 (a) & (b): Generalized Production Process Sequence of Micro-alloyed steels. Process flow for Grade BQT-10: HBI-EAF-LRF-VD- CCM- HSM-DSC. (Abbrvs. used: HBI- Hot briquetted Iron; DRI- Direct reduced Iron; EAF- Electric arc furnace; LRF- Ladle refining furnace; VD- Vacuum Degassing; VCD- Vacuum carbon de-oxidation; CCM- Continuous casting machine; HSM- Hot strip mill; DSC- Down stream complex).

Table 2 Roll Pass Schedule of BQT-10.

Material ID No.	Thickness, mm	Reheating F/c, temp, °C	Draft pass 01	
136471	16	1253	14.86	
Draft pass 02	Draft pass 03	Draft pass 04	Draft pass 05	
17.6	19.4	21.7	27.7	
Draft pass 06	Draft pass 07	R/M Exit Temp, °C	Finish mill ave. finish temp, °C	Ave C.T., °C
35.1	34.06	1115	863	659

(Abbrvs. used: Matl- Material; Thk- thickness; F/c-Furnace; R/M- Rolling Mill; Ave.- average; C.T.- Coiling Temperature).

Table 3 Tensile Properties of BQT-10 Plates.

Specimen ID, (Normalized)	Holding Time at 920°C, (mins)	U.T.S., (N/mm2)	0.2% Offset Yield Strength (N/mm2)	Elongation, %
BQT 10-3H	10	505	276	46.2
BQT 10-2H	15	513	288	48.4
BQT 10-4H	25	516	308	45.6
BQT 10-7H	30	522	321	48.6
BQT 10-8H	30	518	326	45.6
BQT 10-6H	40	520	310	49.2
BQN-1*	-	471	270	40.8

*Without normalizing in the as-rolled condition.

equations. Boratto, E. et al developed empirical equations to simultaneously determine Tnr (no recrystallization), Ar3 and Ar1 temperatures.[1] Ouchi ,C. et al [2] had proposed the following equation for the determination of Ar3 and accordingly, the Ar3 temperature values calculated by Boratto et al showed reasonable correlation with Shop Floor trials.

$$A_{r3} = 910 - 310C - 80Mn - 20Cu - 15Cr - 80Mo + 0.35(h-0.8) \quad (1)$$

where h is the thickness of the sample being submitted to rolling.

However, Ouchi's formula does not consider effect of important Micro-alloying additions such as Ni and V. It was, therefore, decided to determine normalizing temperature as per the following empirical relationship (based on shop-floor trials):

$$A_{r3} = 910 - (203\%C) - (15.2\%Ni) + (44.7\%Si) + (104\%V) + (31.5\%Mo) + (13.1\%W) \quad (2)$$

The normalizing temperature, A_{r3}, for BQT-10, thus, works out to be about 905°C to 910°C as per this formula after considering an additional margin of 25 to 30° C.

4.1.3. Normalizing Variables and Hardness

Based on the above results, normalizing time of 30 minutes was selected and eight numbers of oversize (65mmx65mmx14mm) blocks of BQT-10 were machined out and heat treated by normalizing at 920°C. The entire procedure adopted was as follows:

All the BQT-10 blocks were first pre-heated to 200-300°C in a 500 Kg. muffle surface, thereafter loaded to 500Kg. oil-fired salt bath (40% BaCl2 + 60% BaCO3) at 700°C. The temperature was then raised to 920°C @ 70°C/hr and the blocks were held for 30 minutes at this temperature followed by forced air cooling using exhaust fan. Prior to the finish-machining, four of these blocks thus normalized were tested for hardness and the following results were observed (Table 4):

Table 4 Hardness Values of Normalized vs. As-rolled BQT-10 Blocks

Normalized Specimen ID	BHN Hardness (Normalized)	Equivalent Rockwell Hardness (Normalized)	As-rolled Specimen ID	BHN Hardness (As-rolled)	Equivalent Rockwell Hardness (As-rolled)
JNC_BQT(N)_J1c_8	153	R_B 81	BQT-10-1	145	72
JNC_BQT(N)_J1c_10	156	R_B 82	BQT-10-2	148	73
FCP1_BQT(N)_K(+)_MOD	156	R_B 82	BQT-10-3	148	74
FCP2_BQT(N)_K(+)_MOD	153	R_B 81	...	...	...

4.2. Optical, Scanning and Transmission Electron Microscopy of BQT-10

4.2.1. Optical Microscopy of BQT-10

The following Figs. 2 and 3 show the light optical microstructures of BQT-10 in as-rolled and in normalized condition respectively. It may be observed from these figures that there is an over-all grain refinement and increase in alloy-pearlite content achieved through normalizing.

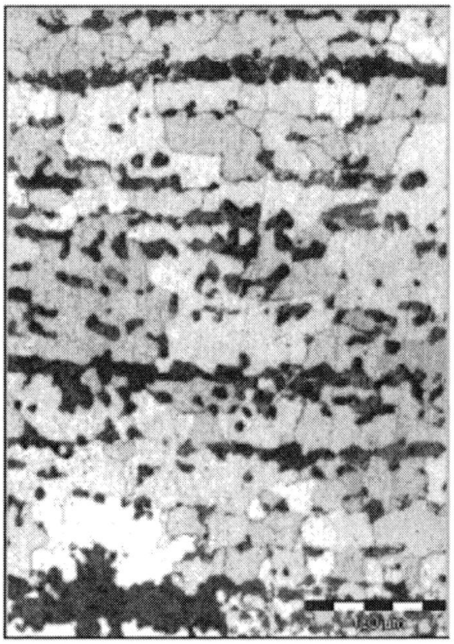

Fig.2: Optical Microstructure of BQT-10 As-rolled (RD-ST Plane), at 500X.

Fig.3: Optical Microstructure of BQT-10 Normalized (RD-ST Plane), at 500X

4.2.2. Grain Size Frequency Distribution of BQT-10

Figs. 4 and 5 show Grain Size Distribution of BQT-10 in as-rolled and in normalized conditions respectively. These figures indicate that there is an over-all reduction in grain size variation achieved through normalizing.

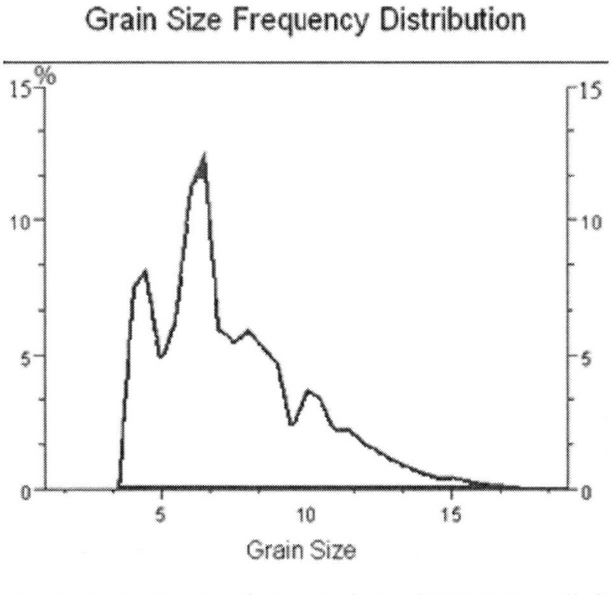

Fig. 4: Grain Size Distribution Analysis of BQT-10 (As-rolled) along the Principal Rolling direction (BQT10_RD_GM).

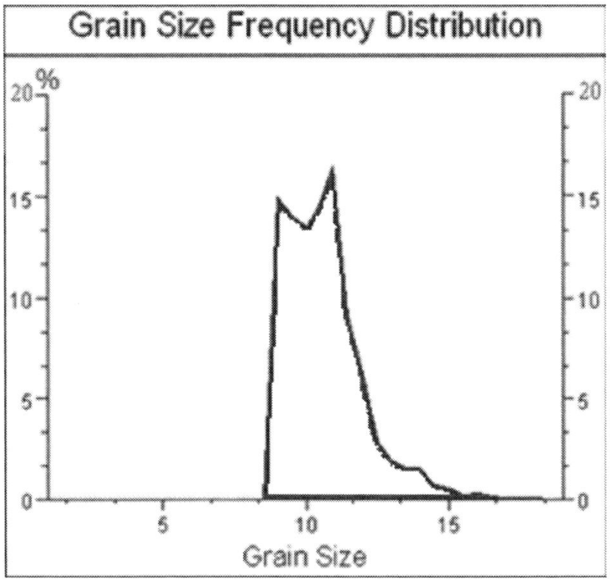

Fig. 5: Grain Size Distribution Analysis of BQT-10 (Normalized) along the Principal Rolling direction (BQT10_RD_N_GM).

4.2.3. SEM Study of BQT-10

SEM study was also conducted for BQT-10 both in the as-rolled condition as well as in the normalized condition. Frequent occurrences of ferrite-carbide type of aggregates surrounding ferrite grains could be noticed in BQT-10 in the as-rolled condition although the morphological details were not clear; however, in the normalized condition SEM micrographs reveal lamellar regions that are well defined and considerably refined as compared to their as-rolled counterparts and strongly resemble ferrite-pearlite type of microstructure (Figs.6 to 8).

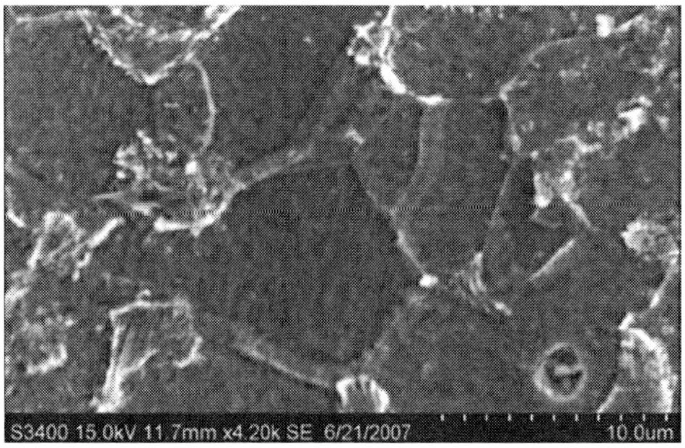

Fig. 6: SEM Micrograph of BQT-10 (As-rolled) at 4200X (Specimen Id. BQTRD_04)

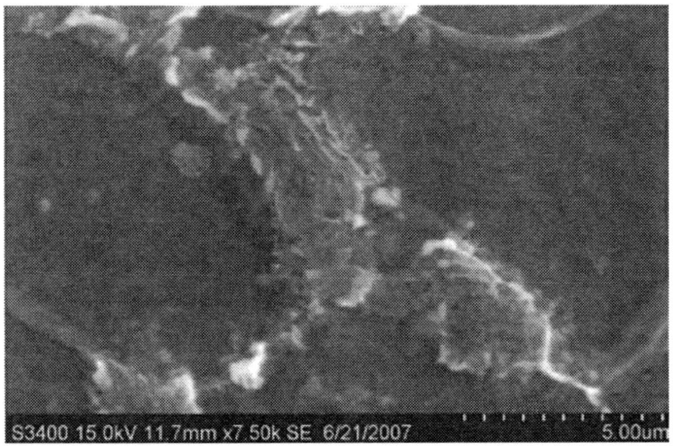

Fig. 7: SEM Micrograph of BQT-10 (as-rolled) at 7500X (Specimen Id. BQTRD_04)

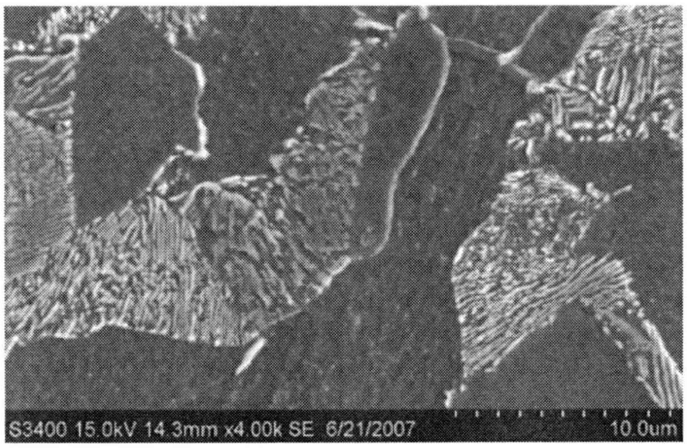

Fig. 8: SEM Micrograph of BQT-10 (normalized) at 4000X (Specimen Id. BQTRD_N_ 08)

4.2.4. TEM study of BQT-10

TEM study of BQT-10 in the as-rolled condition could only be performed on a limited scale since a number of specimens had to be discarded due to unacceptable chemical etching that occurred following jet polishing. Optical micrographs for BQT-10 in as-rolled condition typically resemble banded microstructure consisting of ferrite (bright) and pearlite (dark) bands. TEM study however indicates that the microstructure is rather complex and appears to consist of (i) quasi-polygonal ferrite with grain boundaries containing dislocation substructure and grain boundary triple points [Figs. 9(a) and 9(c)], (ii) grain boundary carbides at triple points [Figs. 9(a) and 9(c)], and (iii) grain boundary carbides at triple points that further extend, though discontinuously, along the grain boundaries [Fig.9(c)]. Figs. 9(b) and 9(f) show a lath substructure similar to upper bainite but with ferrite plates much broader and with considerable dislocation density. The laths appear to be slightly distorted and separated by discontinuous strings and isolated particles of carbides typical of upper bainite morphology in low carbon steels. Also, any evidence of retained austenite could not be revealed. The TEM micro-structures of Figs. 9(b) and 9(f), therefore, appear to indicate bainite type of lath substructure although it is difficult to ascertain the morphology in terms of upper bainite or lower bainite. However, similar morphological observation has been documented for lower bainite in the literature [3], [4].

While optical micrographs for BQT-10 in as-rolled condition also occasionally display nodules of dark phases rather than normal rounded type of pearlite colonies. TEM photo-micrograph [Figs. 9 (d) and (e)] gives the appearance of parallel ferrite plates with intervening carbides. Similar microstructure was identified by Prof. Bhadeshia [5] as pearlite with alloy carbide instead of cementite. There is also occasional evidence of microscopic features resembling very fine round/globular inclusions in dislocation substructures. BQT-10 in as-rolled condition has, therefore, a complex microstructure that appears to have been characterized by quasi-polygonal ferrite with discontinuously extended grain boundary carbides, ferrite-carbide aggregate with intervening carbides and bainitic type of ferrite lath substructure.

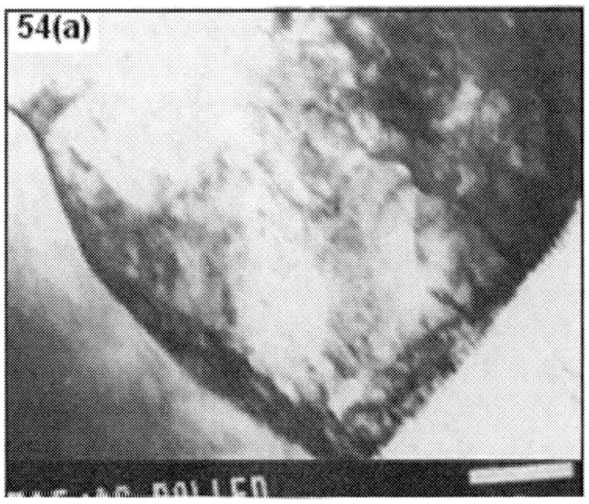

Fig. 9(a): TEM photo-micrograph 1 viewed at a magnification of 30K(BQT-10 As-rolled).

Fig. 9(b): TEM photo-micrograph 2 viewed at a magnification of 50K(BQT-10 As-rolled).

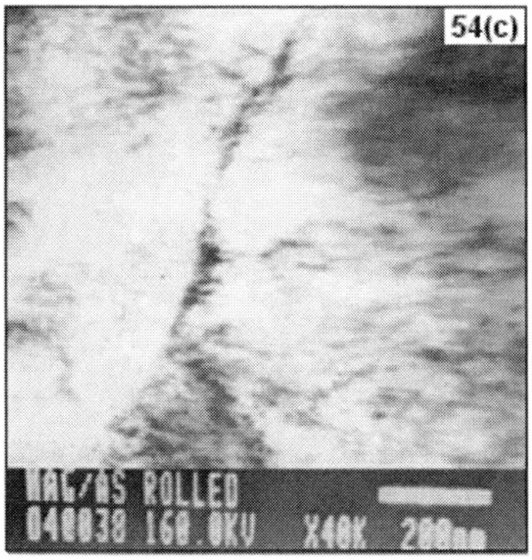

Fig. 9(c): TEM photo-micrograph 3 viewed at a magnification of 40K(BQT-10 As-rolled).

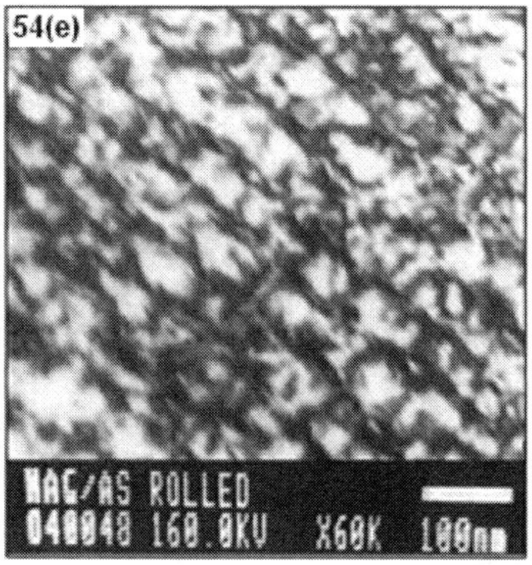

Fig. 9(e): TEM photo-micrograph 5 viewed at a magnification of 60K(BQT-10 As-rolled).

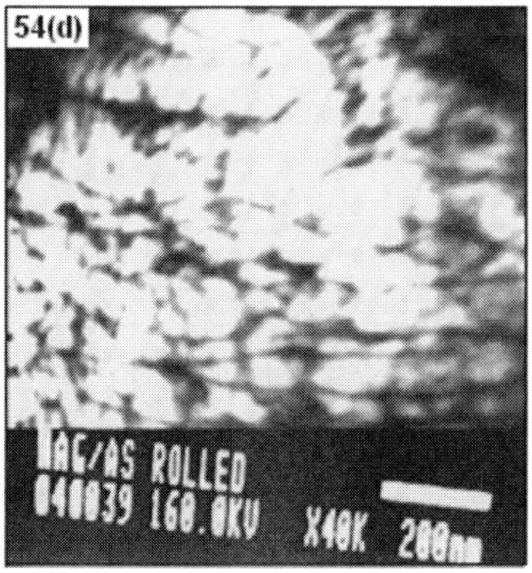

Fig. 9(d): TEM photo-micrograph 4 viewed at a magnification of 40K (BQT-10 As-rolled).

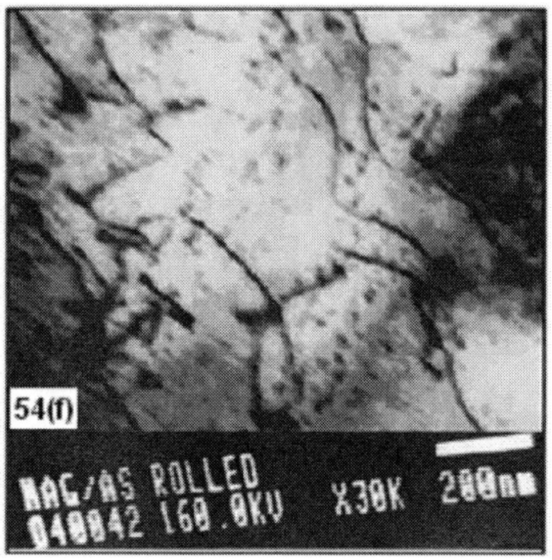

Fig. 9(f): TEM photo-micrograph 6 viewed at a magnification of 30K (BQT-10 As-rolled).

TEM study of BQT-10 in normalized condition could not reveal any region of polygonal ferrite or lath substructure unlike that of BQT-10 in the as-rolled condition. Rather, TEM photo-micrographs [Figs. 10(a) to 10(i)], in general, appear to represent a ferrite – carbide aggregate that significantly differs in morphological detail from classical pearlite and resembles so-called pearlitic bainite which, for steels containing carbide-forming elements, is a pearlite with carbide phase as an alloy carbide instead of cementite[5,6]. Since strong carbide forming elements such as Nb, Ti and V are capable of forming alloy carbides preferentially at alloying concentrations less than 0.1 wt%, the formation of ferrite - alloy carbide aggregate to constitute alloy pearlite during normalizing of BQT-10 appears feasible. These alloy carbides exhibit a 'rod'-like morphology [Figs. 10(b), (d), (e), (f), (g) and (h)]; they are frequently discontinuous giving a fragmented appearance to the microstructure [Figs. 10(a), (c) and (f)] and form in bands which are closely parallel to the interface.

Inadequate diffusion of strong carbide forming elements or insufficient availability of carbon at the growing interface at transformation temperature may be attributed to these frequent occurrences of discontinuity in carbide morphology. There is also evidence for spheroidizing tendency of these fragmented carbides [Figs. 10(b), (c), (f), (i) and (j), particularly Fig.10 (j)]. Furthermore, a closer inspection of a few TEM micrographs [Figs. 10(c), (f), (h) and (i)] appears to suggest the precipitation of alloy carbides on dislocations within the ferrite rather than adjacent to ferrite. The microstructure of BQT-10 in the normalized condition thus appears to be characterized by an alloy pearlite type of ferrite-alloy carbide aggregate as revealed by light microscopy and TEM.

Although TEM study was conducted based on comparison with TEM images reported by past researchers as detailed in preceding sub-paragraph, it was noted that availability of

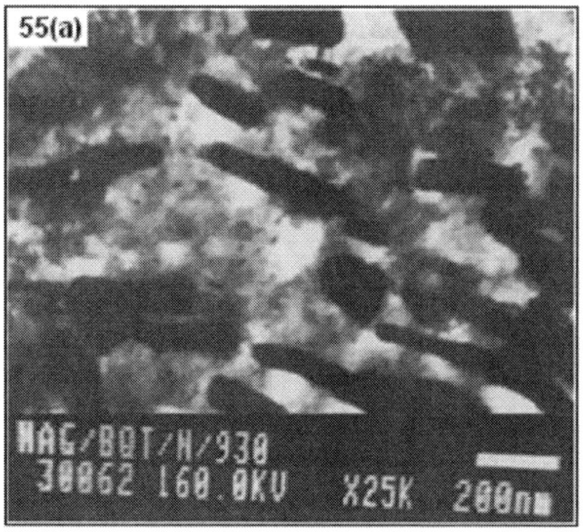

Fig. 10(a): TEM photo-micrograph1 viewed at a magnification of 25K(BQT-10 Normalized).

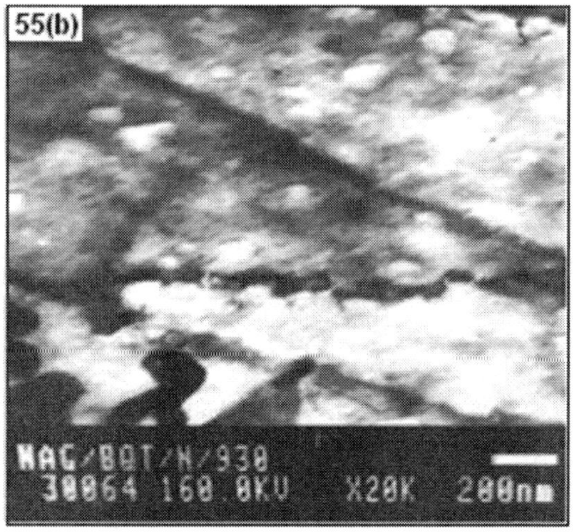

Fig. 10(b): TEM photo-micrograph2 viewed at a magnification of 20K (BQT-10 Normalized).

Fig. 10(c): TEM photo-micrograph 3 viewed at a magnification of 20K (BQT-10 Normalized).

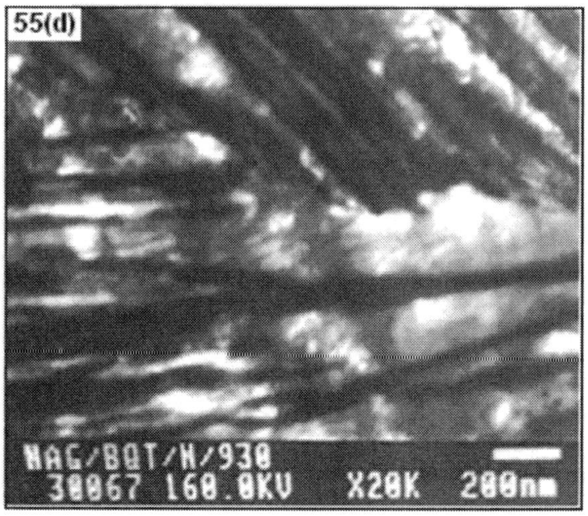

Fig.10 (d): TEM photo-micrograph4 viewed at a magnification of 20K(BQT-10 Normalized).

advanced techniques such as FE-TEM EDX Analysis, X-ray Absorption Spectroscopy (XAS) and Scanning Transmission Electron Microscopy (STEM) would have been of great benefit for authentic qualitative and quantitative identification of micro-constituents observed, which were not possible in the present case.

4.2.5. Phase Analysis

Phase analysis was performed for BQT-10 in as-rolled as well as in normalized condition using a systematic point counting procedure (ASTM E562). From the results of Manual Point Count, it was noted that BQT-10(as-rolled) appears to have % dark phase (pearlite) of the order of 40%. However, on normalizing BQT-10, the volume percentage of dark phase (pearlite) showed an increase by about 10%. The overall results obtained through systematic point counting are exhibited in Table 5 below:

4.3. Cyaniding of BQT-10

With due consideration given to factors such as the (CT) Specimen size, time requirement for different surface hardening methods, the expected benefits due to combined diffusion of carbon and nitrogen into the surface layer, low Al – content of BQT-10 and lack of literature availability on effects of cyaniding as a surface hardening process on fracture mechanics properties, it was decided to cyanide four CT-LLD specimens of BQT-10 for J-R Curve study, which were subsequently shot-peened to study the combined effect of cyaniding and shot-peening on initiation fracture toughness (J1c). In addition, two (CT) –MOD specimens

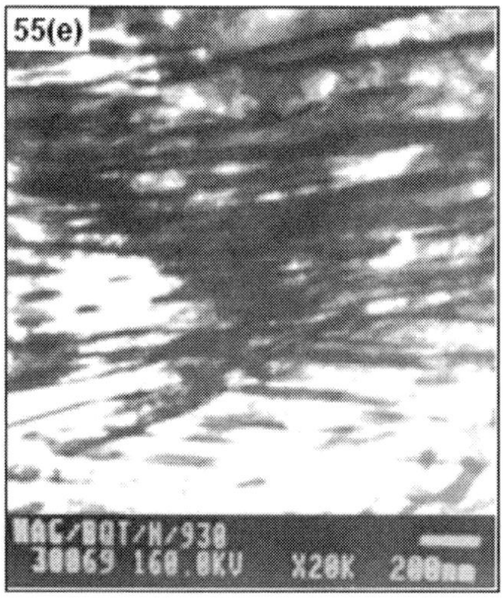

Fig. 10(e): TEM photo-micrograph5 viewed at a magnification of 20K (BQT-10 Normalized).

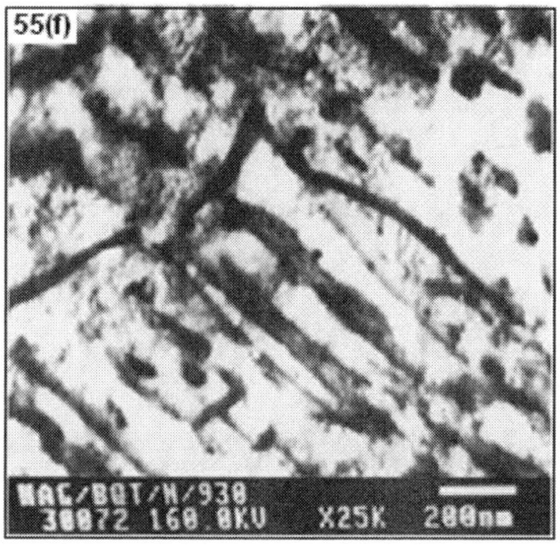

Fig. 10(f): TEM photo-micrograph 6 viewed at a magnification of 25K (BQT-10 Normalized).

were cyanided for Stress Depth Profile Analysis. It was decided to shot-peen one of them after cyaniding and thereafter, to subject both these specimens (cyanided and cyanided-cum-shot-peened) to Stress Depth Profile Analysis to determine the extent of compressive stress affected zone below the specimen-surface due to cyaniding and cyaniding-cum-shot-peening. In addition, two metallographic specimens of size 25x25x10mm were cyanided and one of the two specimens was then tested for Vickers Microhardness in a LECO make Vickers Microhardness Tester – Model LM 300 AT under an applied load of 25 gms and the following results (Table 6) were obtained:

The second specimen was then studied under Optical Microscope and the Microstructures, as shown in Figs. 11 and 12, clearly indicate a microscopically flat case of almost uniform depth and

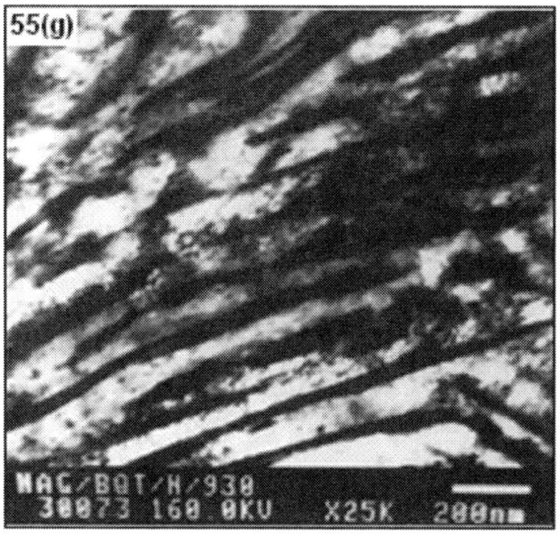

Fig. 10(g): TEM photo-micrograph 7 viewed at a magnification of 25K (BQT-10 Normalized).

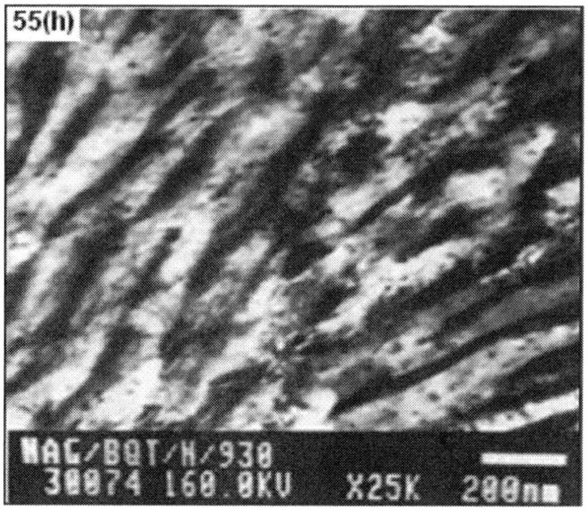

Fig. 10(h): TEM photo-micrograph 8 viewed at a magnification of 25K (BQT-10 Normalized).

a core that has significantly transformed from its characteristic banded structure to a martensitic microstructure.

4.3.1. Cyaniding Process Parameters for BQT-10

All specimens were first pre-heated at 650°C in an electric furnace for 45 minutes. The specimens were then transferred to a salt bath consisting of 30% NaCN + 40% BaCO3 + 30% BaCl2 at 850 - 880°C. The bath temperature was then raised to 880 – 900°C and the specimens were allowed to soak for 4½ hours with the objective of achieving 150-200μm case depth. The specimens were then mar-quenched in a Martempering Furnace [Molten NaNO2 (50%) + NaNO3 (50%)] maintained at 160 - 180°C, held there for 30 minutes and air cooled to room temperature. The cyanided specimens were immediately tempered at 130 - 140° C for 2 hours, followed by air cooling to room temperature.

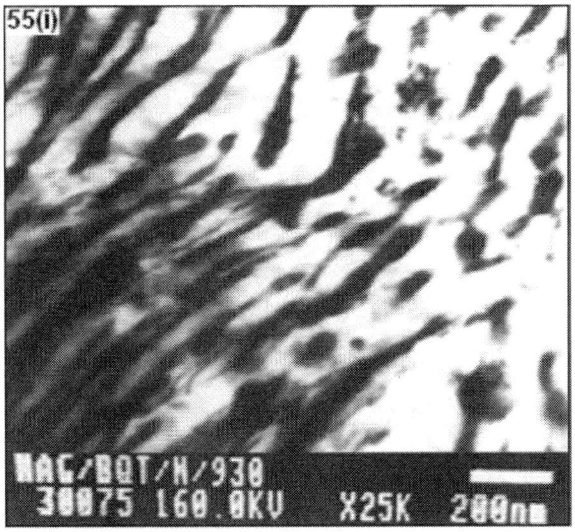

Fig. 10(i): TEM photo-micrograph 9 viewed at a magnification of 25K (BQT-10 Normalized).

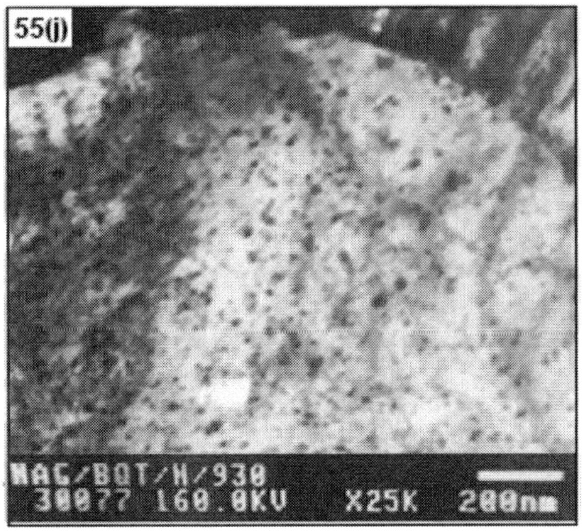

Fig. 10(j): TEM photo-micrograph 10 viewed at a magnification of 25K (BQT-10 Normalized).

Table 5 Percentage Phase Analysis of BQT-10 as rolled and BQT-10 normalized

Specimen ID.	Description	%Ferrite	%Pearlite
BQT10RD_04_500X	BQT10 As-Rolled.	57.79	42.21
BQT10LT_06_500X	BQT10 As-Rolled.	55.32	44.68
BQT10ST_07_500X	BQT10 As-Rolled.	64.97	35.03
BQT10RD_N_08_500X	BQT10 Normalized	45.21	54.71
BQT10LT_N_09_500X	BQT10 Normalized	47.82	52.18
BQT10ST_N_10_500X	BQT10 Normalized	48.96	51.04

Table 6 Effect of Cyaniding Cycle on Hardness of BQT-10

Cyanided Specimen Id.	Vickers Microhardness – Case	Vickers Micro hardness - Core
BQT10 - CN9	500	185
BQT10 - CN9	508	198
BQT10 - CN9	550	200

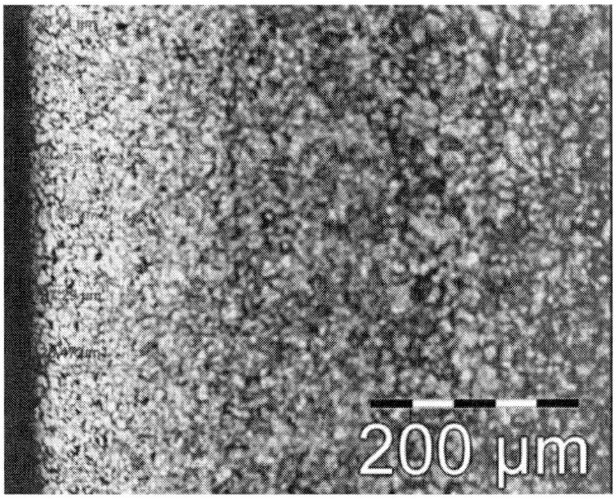

Fig. 11: The microstructure of cyanided BQT10 specimen at 200x indicates the formation of well-defined cyanided layer of uniform case depth (120µm on an average).

Fig. 12: The microstructure of core region (in the vicinity of mid-thickness) of the cyanided QT10 specimen at 1000x indicates almost complete absence of banded structure which has transformed to a predominantly acicular (martensitic) microstructure.

4.4. Shot-peening of BQT-10

The purpose of shot-peening is to introduce compressive stresses at surface levels of as-rolled as well as cyanided BQT-10 specimens. While shot-peening is known to have played a beneficial role in terms of fatigue resistance, much of the research publications in this direction are found to have followed Total Life Approaches. The main objective of shot-peening in the present investigation is to study the effect of residual compressive stresses at surface level on fracture toughness parameters as applicable to Micro-alloyed steels. Table 7 shows Chemistry and Properties of Martensitic Steel Shots used for shot-peening.

Table 7 Chemistry of & Properties of Martensitic Steel Shots

Typical Chemistry, %	C, 0.91	Si, 0.89	Mn, 0.91	S, 0.043	P, 0.044
Typical Density	7.32 gm/ml				
Hardness	400 to 500 HV				

4.5. Residual Stress Depth Profile Measurements using X-Ray diffraction

A total of 5 residual stress depth profiles were determined on five shot-peened specimens using a "XSTRESS 3000" diffractometer manufactured by Stresstech Oy, Finland. Material removal was made by electrochemical polishing (Movipol 3) and depth was measured with a dial indicator. The specimens were marked as I, III, III Repeat, VII & VIII. Table 8 below lists shot-peened specimens that were tested for residual stress depth profile measurements:

Table 8 List of Shot-peened Specimens subjected to Residual Stress Depth Profile Measurements

Specimen Id.	Micro-alloyed Steel Type	Processing Condition
CTS Specimen I	BQT-10 (RCP)	CN+SP (all around)
CTS Specimen III	BQT-10	SP (all around)
CTS Specimen III-Repeat	BQT-10	SP (all around)
CTS Specimen VII	BQT-10	CN(all around)
CTS Specimen VIII	BQT-10	SP (all around)

The directions Phi(Φ)=0° and Phi(Φ)=90° (Rolling Direction) can be seen from Fig. 14.

Figure 15 shows a typical Stress-Depth Profile of a BQT-10 (CT) specimen which was shot-peened all around below the machined notch. From the profile it may be seen that a state of residual compressive stress exists at the surface and below due to shot-peening. The value of compressive stress at the surface is of the order of 400 MPa and this state of compression continues to exist to a depth of approximately 0.25 to 0.3 mm below the surface. Fig. 16 shows a typical Stress-Depth Profile of a BQT-10 (CT) specimen which was cyanided and then shot-peened all around below the machined notch. The value of compressive stress at the surface in this case is of the order of 600 MPa and this state of compression continues to exist beyond the depth of 0.3 mm

Fig.13. An interior view of Shot Peening Chamber prior to commencement of peening process.

Fig. 14: The measurement directions of residual stress on shot-peened BQT10 (CT) specimens are marked with red arrows.

below the surface. It is thus evident that (CT) specimens when cyanided and subsequently shot-peened give the maximum depth of compressive residual stress distribution below the surface.

4.6. Determination of Initiation Fracture Toughness of BQT-10 in As-rolled, Normalized, Shot-peened and Cyanided-cum-Shotpeened Conditions

For BQT-10 in the as-rolled condition, J-R curve and J1c were determined from a 16mm BQT-10 plate using specimens with machined notch aligned along the rolling direction following the ASTM E 1820 procedure.. Chevron notch configuration was used as the fatigue crack starter notch configuration for BQT-10(LLD) specimens. The dimension 'W' selected was 50mm and the specimen thickness "B" selected was 12.5 mm. All the BQT-10 (CT) specimens were made in T-L (long transverse – longitudinal) orientation such that the intended crack propagation

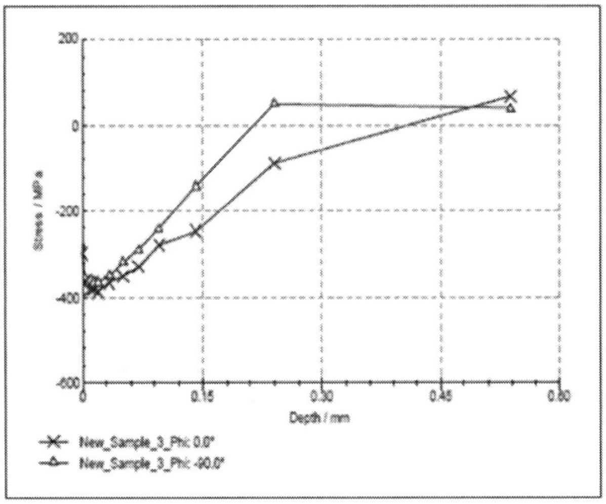

Fig. 15: Stress-Depth profile of 12.5 mm thick BQT-10 shot-peened all around (Spec.Id.: CTS Specimen III-Repeat)

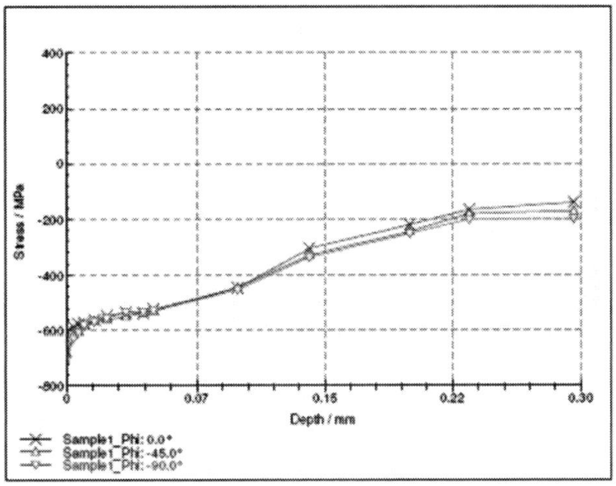

Fig. 16: Stress-Depth profile of 12.5 mm thick BQT-10 cyanided and shot-peened all around (Spec. Id.: CTS Specimen I).

is in longitudinal (rolling) direction throughout. Fatigue pre-cracking, followed by the test as per the Resistance Curve Procedure using single specimen technique for determination of J-R Curve, was performed using three BQT-10 (CT) specimens in as-rolled condition on a 50kN Horizontal Desktop Servo-Hydraulic Material Test System manufactured by Bangalore Integrated System Solutions (BISS), India. The fourth (CT) specimen was tested on a 100kN Vertical Servo-hydraulic Test System manufactured by MTS Systems Corporation, USA, for comparison purpose. All the specimens except Specimen ID JNC_BQT_J1C_6 generated qualified results as per ASTM E 1820. The qualified J1c values of BQT-10 Specimens in as-rolled condition are tabulated in Table 9 below:

Following Fig. 17 exhibits a typical J-R Curve of BQT-10 in as-rolled condition (Specimen Id. JNC_BQT_J1C_4).

Table 9: Qualified Initiation Fracture Toughness values (J1c) of BQT-10 in As-rolled condition

Specimen ID.	J1c (kJ/m²)	Average J1c (kJ/m²)
JNC_BQT_J1C_4	43.51	
JNC_BQT_J1C_5	35.8	38.6
JNC_BQT_J1C_7	36.4	

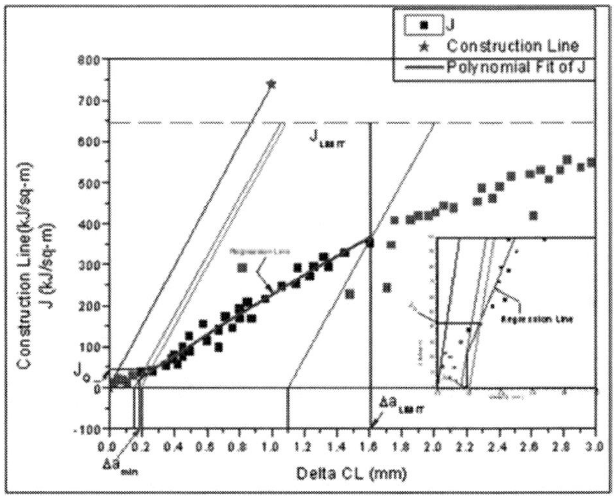

Fig.17: J-R Curve of Specimen Id. JNC_BQT_J1C_4 with Inset showing Intersection of Polynomial Regression Fit with 0.2 mm Offset Line for JQ determination.

Following the same procedure as per ASTM E 1820, it was also possible to obtain valid J1c values for BQT-10 in Normalized, Shot-peened and Cyanided-cum-Shotpeened Conditions.

Following Table 10 summarizes the results and Table 11 shows tensile properties of BQT-1 both in as-rolled condition and in normalized condition.

5. Results and Discussion

Test results of Table 10 indicate that each one of the three applications namely 'normalizing', 'shot-peening', and 'cyaniding-cum-shot-peening' improves initiation fracture toughness but to a varying extent; the best improvement in initiation fracture toughness value could, however, be achieved through a combined treatment of cyaniding and shot-peening applied consecutively that resulted in an average J_{Ic} value of 105.2 kJ/m².

The beneficial effect of normalizing heat treatment on J_{Ic} value may be explained in terms of micro-structural modification that occurred during heat treatment of banded micro-structure of BQT-10 in as-rolled condition. Pearlitic banding as observed in the micro-structure of BQT-10 in as-rolled condition (Fig. 2) is usually attributed to micro-segregation of manganese, precipitation of non-metallic inclusions or hot rolling at low finishing temperatures and cooling rates. Pearlitic banding has

Table 10 Summary Results-Initiation Fracture Toughness Values (J1c) for BQT-10 under Different Conditions

Specimen ID.	J1c (As-Rolled) (kJ/m²)	J1c (Normalized) (kJ/m²)	J1c (Shot-peened) (kJ/m²)	J1c (Cyanided and Shot-peened) (kJ/m²)
JNC_BQT_J1C_4	43.51			
JNC_BQT_J1C_5	35.8			
JNC_BQT_J1C_7	36.4			
JNC_BQT(N)_J1C_8		69.79		
JNC_BQT(N)_J1C_9		74.5		
JNC_BQT(N)_J1C_10		82.82		
JNC_BQT(PS)_J1C_11			82.42	
JNC_BQT(PS)_J1C_12			100.62	
JNC_BQT(PS)_J1C_13			105.67	
JNC_BQT(CN+PS)_J1C_14				85.46
JNC_BQT(CN+PS)_J1C_15				116.28
JNC_BQT(CN+PS)_J1C_16				113.76

(Abbrvs. used: N- Normalized; PS- Shot-peened; CN+PS- Cyanided and subsequently, Shot-peened).

been reportedly altered by heat treatment and changes to thermo-mechanical process schedules of the steel [7]. However, the literature available on the effect of normalizing heat treatment on elastic-plastic fracture toughness of Micro-alloyed steels appears to be rare. It is suggested that the improvement in J1c value as observed with normalized BQT-10 (Table 10) is related to the morphological changes of micro-constituents and refined grain size distribution that occurred due to normalizing. With reference to Phase Analysis Report (500X, manual point count method) of Table 5, an increase in average pearlite content by about 10 to 12% is noted in the normalized micro-structure (52.6%) when compared with as-rolled micro-structure (40.6%). Also, as observed from Figs. 2 and 3, the original banded structure of as-rolled BQT-10 is refined to a greater extent by normalizing. For example, the most frequently observed grain size along the principal rolling direction for BQT-10 in as-rolled condition is approximately ASTM Grain Size No.7 [Fig.4] whereas the same along the principal rolling direction for BQT-10 in as-normalized condition is almost ASTM Grain Size No.11 [Fig.5]. The lack of non-uniformity in grain size frequency distribution is also significantly reduced by normalizing, as evident from Fig.5 when compared with Fig.4. The range of grain size distribution is found to be approximately restricted to ASTM Grain Size No.8 to15 for BQT-10 in as-normalized condition as against a much wider variation in grain size distribution (from ASTM Grain Size No.3 to 15) for BQT-10 in as-rolled condition. TEM Micrographs [e.g. Figs.10 (f) and (i)] indicate that the pearlite morphology of as-rolled BQT-10 appears to have altered, due to normalizing, to frequently discontinuous and curvy pearlite giving a fragmented appearance to the microstructure. In classical pearlite (i.e. without micro-alloying additions) with directional orientation of ferrite-cementite lamellar aggregate , brittle cementite platelates and cementite-ferrite interfaces are expected to provide relatively easy path for crack propagation from fatigue pre-crack. In contrast, the alloy pearlite type of ferrite-alloy carbide aggregate, as observed in TEM Micrographs of normalized BQT-10, tends to be distributed with curvy and fragmented appearence and occasional spheroidization [Figs. 10(a), (i) and (j), for example]. From Table.11 displayed above, it is seen that that tensile yield strength and (%) tensile elongation of BQT-10 have concurrently improved by normalizing, which may be interpreted as an increased resistance to initiation of gross plastic deformation and an increased ability to plastically deform along the tensile direction (taken as rolling direction) once the plastic deformation initiates. These findings may be interpreted as experimental evidence for transformation of BQT-10, after normalizing, into more 'plastic' or 'ductile' state in terms of ability to undergo gross plastic deformation.

In LEFM arena, decreasing grain size produces greater fracture toughness (KIc) and greater tensile strength virtually in all metals (austenitic and ferritic steels, for example) [8]. But fracture toughness of high strength materials is almost always inversely proportional to their yield strength (σys). Increase in tensile strength is not always accompanied by an increase in yield strength (hot-rolled C-Mn structural steel, for example) [8]. However, present experimental investigation indicates otherwise in yielding fracture mechanics (YFM) or elastic-plastic fracture mechanics (EPFM) arena for the micro-alloyed steel considered. Normalizing was found to have caused an increase not only in initiation fracture toughness (J1c) but also in tensile yield strength (σys) and % tensile elongation of BQT-10.

Based on observed improvement in initiation fracture toughness (Table.10) and also in tensile ductility (Table.11) of BQT-10 in normalized condition, it may be suggested that the plastic zone ahead of the crack tip in pre-cracked (CT) specimens or local necking zone prior to tensile instability in uniaxial tension test specimens is characterized by void nucleation- and void growth-controlled ductile fracture mechanism under the action of far-field loading. Voids initiate at pearlite colonies, grain boundary

Table 11: Tensile Properties of BQT-10 in As-rolled and Normalized Conditions

Specimen Id and Condition of the Material under Study	Holding Time at 920 °C (mins.)	U.T.S (N/mm2)	0.2% Offset Y.S (σ_{ys}) (N/mm2)	Effective Y.S (σ_Y) =[1/2(U.T.S + Y.S)]	Elongation, %	Y.S / U.T.S RATIO	(U.T.S –Y.S)
BQT10- 7H Normalized	30	522	321	421	48.6	0.61	201
BQT10- 8H Normalized	30	518	326	422	45.6	0.63	192
BQN-1 As-rolled	-	471	270	370	40.8	0.57	201
BQT-10 -T2 As-rolled	-	474	276	375	41.64	0.58	198
BQT-10 -T3 As- rolled	-	476	276	376	41.08	0.58	200

carbides and MnS inclusions although void initiation at pearlite colonies is considered to be the dominant mechanism. It is possible that due to considerable grain refinement and uniformity in grain size distribution, nucleation of voids dominates at early stages of crack advancement over growth and coalescence of voids. If the rate of void growth is relatively less than the rate of void nucleation, the void may be expected to reach the critical size for plastic or ductile fracture at a relatively higher value of applied stress. It is then probable that the resultant resistance to initiation of crack advancement through local crack tip plastic zone in a pre-cracked (CT) specimen or to initiation of gross plastic deformation at the onset of necking in an un-notched tensile specimen should increase.

It may, therefore, be suggested that both phenomena (yield strength and J_{1c}) appear to be governed by finer grain size, more uniform grain size distribution and increased content of refined alloy-pearlite in the microstructure. Considerable dominance of void nucleation over void growth and coalescence in plastic zone ahead of macroscopic stress field or microscopic crack tip stress field may be one of the prime controlling factors that contribute to simultaneous improvement in 0.2% offset yield strength and J1c for the micro-alloyed steel considered (BQT-10 in normalized condition).The literature available on the experimental study of the effect of residual compressive stress on fracture toughness is rare. With reference to Table 10, it may be noted that the beneficial effect of 'shot-peening' and of 'cyaniding followed by shot-peening' are comparable but the effects are definitely more pronounced (e.g. J1c = 105.2 kJ/m2) than normalizing (J_{1c}= 75.70 kJ/m2) in terms of improvement in J1c values. This observation may be attributed to the resistance offered by residual compressive stress to propagation of a pre-crack that is open at surface and sub-surface levels in a shot-peened or a cyanided-cum-shot-peened fatigue pre-cracked (CT) specimen. The residual compressive stresses induced at surface and sub-surface levels due to cyaniding and shot-peening applied consecutively is expected to act as an effective barrier to initiation of crack advancement from the pre-crack. The increase in J1c by about10 kJ/m2 as obtained by consecutive application of cyaniding and shot-peening [in comparison to shot-peened (CT) specimens] may be attributed to the higher value of residual compressive stress at the surface (-600 MPa) and higher stress depth profile (Fig.16). Major transformation from banded pearlitic microstructure to martensitic microstructure due to cyaniding of BQT-10 and higher level of residual compressive stress at the surface and sub-surface levels due to combined effect of cyaniding and shot-peening may be assigned as the principal cause of increased resistance to crack initiation as observed with BQT-10 specimens that were cyanided and shot-peened consecutively [Figs.11 and 12].

An attempt was made to combine J-R Plots of BQT-10 in As-rolled, Normalized, Shot-peened and Cyanided-cum-Shot-peened conditions without constructional details (Fig. 18). The data plotted in Fig. 18 are collectively taken from experimental data upto 1.5 exclusion lines in order to make a comparative study of J-controlled crack growth processes in the valid regions under different conditions.

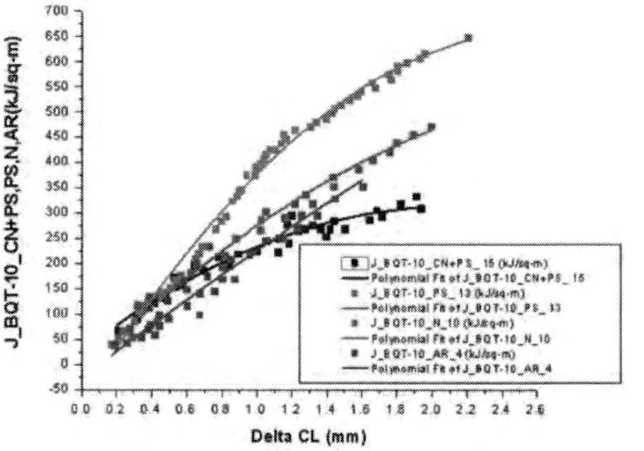

Fig. 18: J-R Curves of BQT-10 in As-rolled, Normalized, Shot-peened and Cyanided-cum-Shot-peened conditions. (Abbrvs. used: AR_4 → BQT-10 As-rolled Specimen Id. JNC_BQT_J1C_4; N_10 → BQT-10 Normalized Specimen Id. JNC_BQT (N) _J1C_10; PS_13 → BQT-10 Shot-peened Specimen Id. JNC_BQT (PS)_J1C_13; CN+PS_15 → BQT-10 Cyan. &-Shot-peened Spec. Id. JNC_BQT (CN+PS) J1C_15).

To understand the post-initiation crack growth process in the context of initiation fracture toughness under study, Table 12 is displayed below showing J1c values of specimens specifically considered in Fig. 18. It may be observed from Table 12 and Fig. 18 that although initiation fracture toughness was maximum for BQT-10 in cyanided-cum-shot-peened condition, J-resistance to crack growth was considerably reduced once the crack initiated, particularly as compared to BQT-10 in shot-peened condition with maximum J-resistance. The reason may be attributed to post-cyaniding heat treatment applied to cyanided specimen prior to shot-peening. The microstructure containing carbide network at surface and sub-surface levels along with acicular features developed at core regions may have reduced the crack tip plasticity after initiation that may be responsible for reduced J-resistance. The effect of shot-peening along the entire surface of "only-shotpeened" specimen ahead of the open crack tip was not influenced by any prior micro-structural modification due to cyaniding. Shot-peening was thus able to contribute in totality to J-resistance of the "only-shotpeened" specimen, leading to significant improvement in J-resistance. On the other hand, the normalized specimen showed improved J- resistance in tune with improvement in J_{1c} value when compared with as-rolled specimen. This relative improvement in J-resistance may be attributed to over-all micro-structural refinement that had taken place due to normalizing heat-treatment which, in turn, may have resulted in relatively slower rate of void growth during crack advancement.

6. Conclusion

Among normalizing, shot-peening, cyaniding and cyaniding followed by shot-peening it was observed that cyaniding followed by shot-peening is the most beneficial in terms of Initiation

Table 12 Initiation Fracture Toughness Values (J1c) for Specific Specimens of BQT-10 under Different Conditions

Specimen ID.	J1c (As-Rolled) (kJ/m²)	J1c (Normalized) (kJ/m²)	J1c (Shot-peened) (kJ/m²)	J1c (Cyanided and Shot-peened) (kJ/m²)
JNC_BQT_ J1C_4	43.51			
JNC_ BQT(N)_ J1C_10		82.82		
JNC_ BQT(PS)_ J1C_13			105.67	
JNC_BQT (CN+PS)_ J1C_15				116.28

Fracture Toughness, JIc for the Micro-alloyed Steel tested (BQT-10). Results, when compared with as-rolled condition, indicate that cyaniding followed by shot-peening have led to a 2.7 times increase in J1c. It may, therefore, be concluded that residual compressive stresses when applied to open pre-cracks in LLD (CT) specimens of micro-alloyed steel under study is likely to delay the onset of crack extension from a pre-existing fatigue crack which in turn improves the initiation fracture toughness of the material in elastic-plastic regime.

7. Acknowledgements

The authors acknowledge the support of Essar Steel, Hazira, in terms of material assistance and would like to make a special mention of Mr. Milind Patil, Dy. General Manager, Product Development & Research, Essar Steel, India, for his continued assistance and moral support. The authors would also like to gratefully acknowledge the help and support extended by Dr. G.K. Dey, Head, Material Science Division, BARC, Mumbai, India, for TEM study of Micro-alloyed Steel.

8. References

1. Boratto, F. et al. In: Thermec '88. Proceedings. Iron & Steel Institute of Japan, Tokyo, 1988 – pp. 383-390.
2. Ouchi, C. et al. Translation of the Iron & Steel Institute of Japan Iron & Steel Institute of Japan (ISIJ), 22(3): 214- 222, March, 1982.
3. "Bainitic Steels: Part One", Key to Metals Database. http://steel.keytometals.com/Articles/Art144.htm
4. Roger K. Steele, 'Steel Alloys with Lower Bainite Microstructures for Use in Railroad Cars and Track' (January2002), U.S. Department of Transportation, Washington, D.C.20590. http://www.fra.dot.gov www.volpe.dot.gov/sdd/docs/integrity/bainitealloys.pdf
5. H. K. D. H .Bhadeshia, 'Bainite in Steels-Transformations, Microstructures and Properties', Second Edition, IOM Communications Ltd., London SW1Y 3DB, ISBN 1-86125-112-2. www.msm.cam.ac.uk/phasetrans/2004/z/3750-011.pdf
6. H. K. D. H .Bhadeshia and Robert WilliamKerr Honeycombe, "Steels: Microstructures and Properties", 3rd Edition (2006), Elsevier Ltd., Oxford, U.K. ISBN-10:0-7506-8084-9
7. A. J. Duncan, K. H. Subramanian, R.L. Sindelar, B.J. Wiersma, "J- Integral Fracture Toughness Testing and Correlation to the Microstructure of A285 Steel for Fracture Analysis of Storage Tanks", WSRC-MS-2000-00282, Rev.1.
8. Campbell, F.C. (Editor) 'Elements of Metallurgy and Engineering Alloys, ASM International, 2008 –Science – page nos. 238 & 376

Proceedings of the 26th ASM Heat Treating Society Conference
B.L. Ferguson, R. Jones, D.S. MacKenzie, and D. Weires, editors
https://doi.org/10.31399/asm.cp.ht2011p0226

The Influence of Nitrocarburizing on Wear Behaviour of Forging Dies

J.B. Mane*, R.C. Prasad**, B. Radhakrishnan*

* Bharat Forge Ltd, Pune, Maharashtra, India, manejaya_18@yahoo.com, +91-9423236475, bradhakrishnan@bharatforge.com, 020-26702466.

**Indian Institute of Technology Bombay, Bombay, Maharashtra, India, rcp@iitb.ac.in, 022-25764604.

Abstract: Die life is an important ingredient in cost of forgings, particularly in hot forgings. A number of surface hardening techniques are used to improve the die wear life. Surface hardening mainly constitutes surface preparation and its treatment to obtain desired properties. In this investigation gaseous ferritic nitrocarburizing was carried on the DIN 1.2714(55NiCrMov7) steel that is used to manufacture dies for crankshafts and axle beams. The compound layer (White layer), diffuse layer, and core structure were identified using optical microscope. The effect of surface treatment (sand blasting) and process parameters like nitrogen and carbon activities on the formation of different layers during nitrocarburizing are reported and discussed in the paper. The wear rate with respect to sliding distance, sliding velocity and normal load are reported along with the analysis of wear mechanisms.

Key words: hot forging dies, Gaseous ferritic nitrocarburizing, white layer, Die wear, Wear mechanisms.

Introduction: One of the most important ingredients in cost of forgings is the cost of tooling involved. Die cost ranges from 10 to 15% of the cost of a forging. This includes cost of die material, machining of dies and subsequent heat treatment, if necessary. The indirect cost of dies is however, far more significant. Therefore identifying different modes of die failures and understanding dominant mechanism is the essential first step in the path to increase die life [1, 2].

Wear is one of the prominent mode of die failure. Techniques employed to improve the die wear life include surface hardening processes like carburizing, nitriding, carbonitriding, nitrocarburizing and boriding. These treatments are applied to the dies to achieve the desired properties in the material. Diffusion methods modify the chemical composition of the surface with hardening species such as carbon, nitrogen or boron. Surface hardening processes constitutes mainly surface preparation processes that help clean and prepare the die surfaces and surface treatment processes that give the desired properties to the material. Surface preparation is an integral part of diffusion process and it aids in easier and better treatment through removal of unwanted impurities, reduces interfacial surface tension, and removes scaling as well as condition of the surface for better bonding. Inadequate surface preparation is often the single greatest cause for failure of surface treatments [3]. In this investigation gaseous ferritic nitrocarburizing was carried on the DIN 1.2714(55NiCrMov7) steel that is used to manufacture dies for crankshafts and axle beams. This process alters the die surface structure in three layers. These layers are compound (white), diffuse and core structure layer. Out of these layers the compound layer is responsible for improving the tribological properties. Compound layer constitutes the epsilon (iron carbonitrides) and gamma prime (iron nitrides) composition. Monophase epsilon compound layer is more beneficial for improving wear resistance. Increase in wear resistance is reported to increase with volume percentage of epsilon phase and effective nitrocarburizing depth [4].

Experimental details.

For this study die steel DIN 1.2714 (55NiCrMov7) has been considered that is used to manufacture dies for crank shaft and axle beams. Chemical composition of die steel is given in Table 1. Before nitrocarburizing two pieces of the die steel surfaces were polished with 120G, 320G paper and 400G buffing wheel that is close to the die surface finish (0.75-0.95 micrometer) and they were marked as piece (1) and piece (2). The polished surface of the test piece(1) was sand blasted by holding the piece in magnetic vice and putting the load in blasting chamber. The sample was blasted with silicon carbide abrasive particle of 100G size at a operating pressure of 4.7-5.0 bar. The surface roughness values of polished and blasted samples are listed in Table 2.

Table.1. Chemical composition of die steel DIN1.2714.

C	Si	Mn	P	S	Cr
0.5-0.6	0.1-0.4	0.6-0.9	0.03+0.005	0.02+0.005	0.8-1.2
Mo	Ni	V			
0.35-0.55	1.5-1.8	0.05-0.15			

This material is in hardened and tempered condition.

Table.2. Roughness values for prepared samples.

Checking stage	Piece(1) Surface Roughness	Piece(2) Surface Roughness
Polished surface	0.83 micrometer	0.87 micrometer
Polished surface sand blasted	2.04 micrometer	

The nitrocarburizing was carried out by keeping the samples in closed furnace of SARLIN make with the surface facing upward. The gaseous atmosphere consists of nitrogen (N_2), ammonia (NH_3) and carbon dioxide (CO_2) in nitrocarburizing process. The samples were preheated at 400°c for 7 hrs.

The process is carried out in two stages at 580°c. Nitrogen gas is first purged for 60 min at 580°c at the rate of 26 m^3/hr. In first stage the volume percentage of the gases were maintained at nitrogen- 52% (6.6 cubic meters per hour), ammonia- 43% (5.5 cubic meters per hour) and carbon dioxide- 5% (0.5 cubic meter per hour) for 6 hrs. In second stage the volume percentage of the gases were maintained at nitrogen- 63% (6.6 cubic meters per hour), ammonia- 32% (4.0 cubic meters per hour) and carbon dioxide- 5% (0.5 cubic meters per hour) for 6 hrs. Controlled cooling was carried out in the furnace from 580°c to 280°c. During cooling N_2 flow was kept as 800 lit/hr. Then the pieces were allowed to cool at room temperature.

The two samples were mounted with Bakelite powder at 175°c as shown in Fig.1. Then transverse surfaces were ground to remove the cutting edges and polished by 320 and 400G papers followed by cloth grinding with aluminium oxide paste. After that the samples are cleaned with acetone and etched with 3% nital etchant.

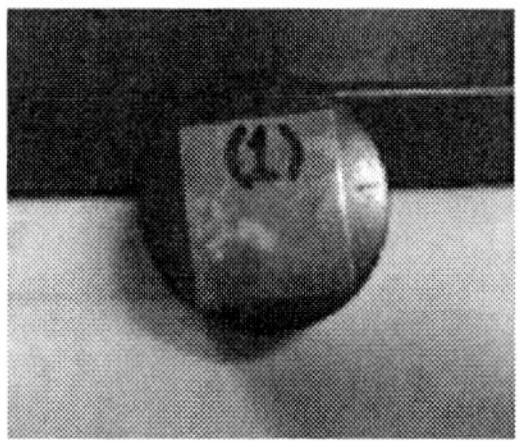

(Sand blasted sample) (Polished sample)

Fig.1. Prepared sample surfaces.

The effective case depth was measured with Rockwell hardness tester by applying the load of 200 gm. First point was taken at 0.02 mm and spacing between two indentations was kept at 0.05mm. The hardness was measured at 16 locations for both the samples. The measured Rockwell hardness value is converted in HV and listed in Fig.4. Limiting effective case depth was determined by adding the 50HV to the corresponding core hardness [5]. After that the samples were etched and examined under optical microscope.

In order to increase the effective case depth and the volume percent of epsilon phase (iron carbonitrides) as compared to gamma prime (iron nitrides) phase the activity of nitrogen and carbon was changed. To change these activities the volume percent of ammonia in the furnace was changed by adjusting the ingoing ammonia gas flow. This process is divided into two stages as it described earlier. The volume percentage of ammonia in first stage was changed from 43% to 50% to increase the activity of nitrogen and carbon by keeping the volume percentage of ammonia same in second stage [6].

Afterwards the wear resistance of these processed specimens was investigated by carrying out the wear test at high temperature. This test was carried out on pin on disc geometry tribometer TR20LE model DUCOM make as shown in Fig.2. From this test the specific wear rate, loss of weight as well as loss of volume has been analysed [7]. The coefficient of friction and frictional force was also calculated. The pins as shown in Fig.3 were nitrocarburized with existing and change in gas flow composition with and without sand blasting prior to nitrocarburizing process.

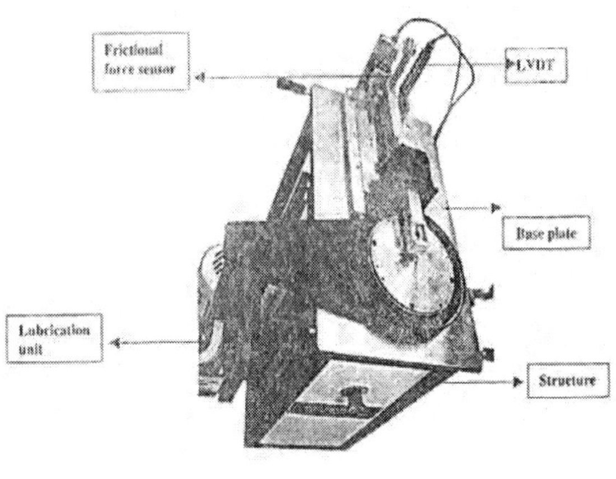

Fig.2. Schematic representation of pin on disc tribometer.

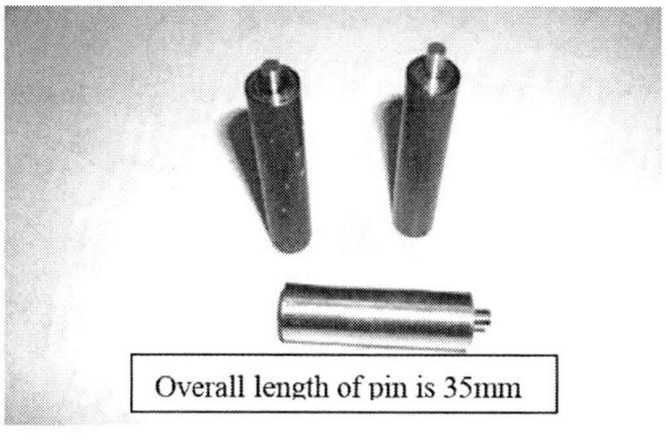

Fig.3. pins prepared for wear testing.

This particular tribometer is based on the principle of Archad equation [8].

$$k = \frac{\Delta V}{F_N L}$$

Where K- Wear rate in mm^3/Nm, Δv- worn volume (mm^3), FN- Normal force (N), L- Sliding Distance (m). Wear testing parameters were selected as stipulated by David Englund is given in Table.3 [9].

Table.3. Wear testing parameters.

Process Parameter	Process Value
Contact Pressure	15MPa
Pin rubbing diameter	4mm
Pin holding diameter	8mm
Sliding velocity	0.25m/s
Pin temperature	400 degree Celsius
Duration of test	4.0hrs

Disc material is En31 having 58-62 HRC hardness.

In order to assess the influence of nitrocarburizing on tribological behaviour of die steel, four categories of pins with case depth as zero mm effective case depth, 0.30-0.31mm, 0.35-0.36mm and 0.47-0.48mm were tested.

Results and Discussions.

Influence of surface preparation on microstructure and hardness.

The surface roughness of polished and sand blasted surface increases to 2.04 micrometer compared to polished surface 0.83 micrometer. The effective case depths formed in these two cases were 0.44mm and 0.39mm respectively as shown in Fig. 4 and 5. However increase in surface roughness beyond certain limit is not acceptable during forging because it may lead to cracking during impact loading due to notch sensitivity created by rough surface. Therefore sand blasting process is introduced before the nitrocarburizing without increasing surface roughness too much. The microstructure after nitrocarburizing consists of three different layers. The compound (white) layer, diffusion layer and core structure as shown in Fig.6 and 7. The compound layer consists of epsilon and gamma primee phases, diffuse layer consists of alloying element nitrides and core structure consists of tempered martensite.

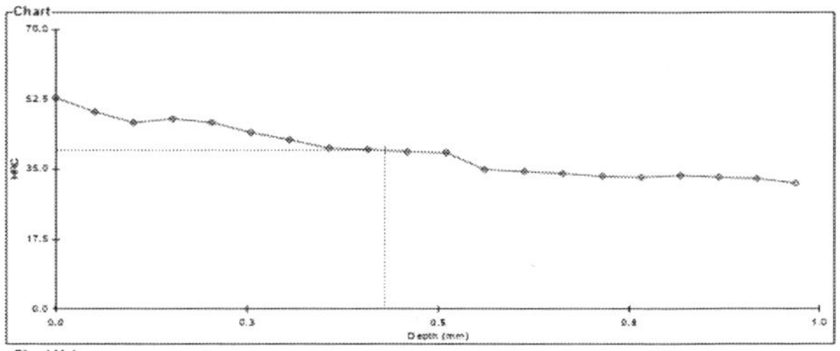

Chart Values

Field #	Hardness HRC (HV)	Depth mm	H Diag. µm	V Diag. µm
1	52.9 (558)	0.02	24.0	27.6
2	49.4 (504)	0.07	26.2	28.0
3	46.8 (468)	0.12	26.3	28.0
4	47.6 (479)	0.17	26.5	29.2
5	46.8 (468)	0.22	27.1	29.2
6	44.2 (436)	0.27	28.9	29.4
7	42.4 (416)	0.32	29.4	30.3
8	40.3 (395)	0.37	29.8	31.4
9	40.0 (392)	0.42	30.1	31.4
10	39.5 (387)	0.47	30.5	31.4
11	39.2 (384)	0.52	30.7	31.4
12	35.0 (345)	0.57	32.5	33.0
13	34.5 (341)	0.62	33.0	33.0
14	34.0 (336)	0.67	33.4	33.0
15	33.3 (329)	0.72	34.1	33.0
16	33.0 (327)	0.77	34.3	33.0
17	33.5 (332)	0.82	33.2	33.7
18	33.0 (327)	0.87	33.7	33.7
19	32.8 (325)	0.92	33.9	33.7
20	31.5 (314)	0.97	34.6	34.1

Sample Statistics

Minimum:	31.5 HRC (314 HV)	
Maximum:	52.9 HRC (558 HV)	
Count:	20 of 20	
	Hardness	Depth
Eff. Case Depth	39.8 HRC (390 HV)	0.44 mm

Fig.4.Hardness profile to check the effective case depth for sand blasted surface.

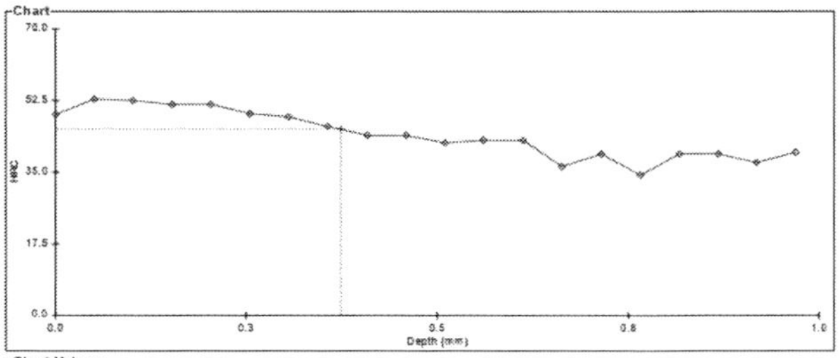

Chart Values

Field #	Hardness HRC (HV)	Depth mm	H Diag. µm	V Diag. µm
1	49.1 (499)	0.02	25.3	29.2
2	52.9 (558)	0.07	25.1	26.5
3	52.6 (553)	0.12	25.3	26.5
4	51.7 (539)	0.17	26.0	26.5
5	51.7 (539)	0.22	26.0	26.5
6	49.4 (504)	0.27	27.1	27.1
7	48.5 (491)	0.32	27.6	27.4
8	46.2 (461)	0.37	28.3	28.5
9	43.9 (433)	0.42	28.9	29.6
10	43.9 (433)	0.47	28.9	29.6
11	42.1 (413)	0.52	29.8	30.1
12	42.7 (420)	0.57	29.8	29.6
13	42.7 (420)	0.62	29.6	29.8
14	36.4 (358)	0.67	31.6	32.8
15	39.5 (387)	0.72	31.0	31.0
16	34.3 (338)	0.77	32.5	33.7
17	39.5 (387)	0.82	31.2	30.7
18	39.5 (387)	0.87	31.0	31.0
19	37.2 (365)	0.92	32.1	31.6
20	39.8 (390)	0.97	30.7	31.0

Sample Statistics

Minimum:	34.3 HRC (338 HV)	
Maximum:	52.9 HRC (558 HV)	
Count:	20 of 20	
	Hardness	Depth
Eff. Case Depth	45.5 HRC (451 HV)	0.39 mm

Fig.5. Hardness profile to check the effective case depth for polished surface.

Microstructural images:

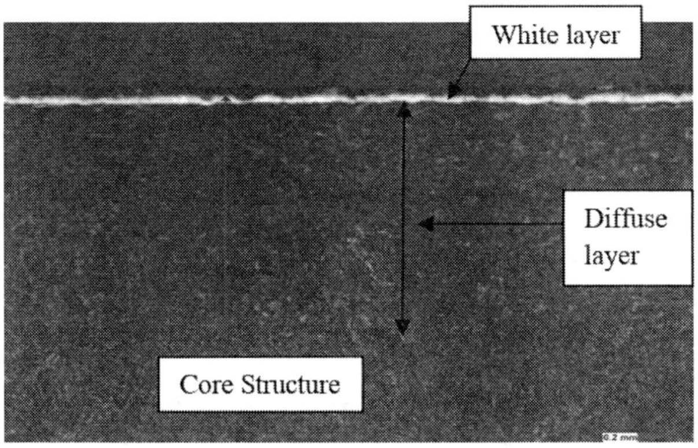

Fig.6. Optical microscope image showing case depth adjacent to sand blasted surface (Ra- 2.04 micrometer) at 100X.

Fig.7. Optical microscope image showing case depth adjacent to polished surface (Ra- 0.87micrometer) at 100X.

Comparison of effective case depth formed during nitrocarburizing for different treated samples (Fig.8).

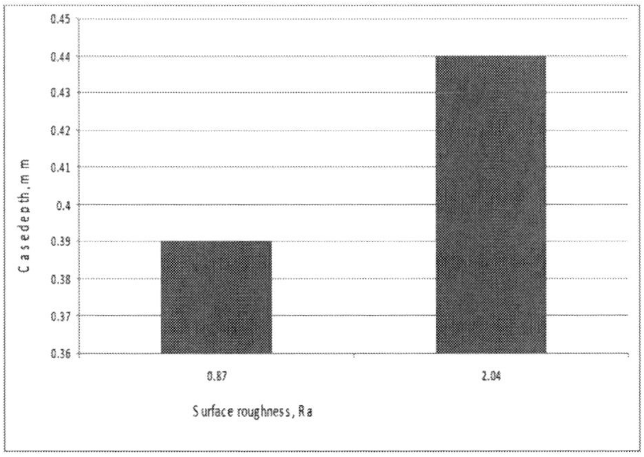

Fig.8. Variation of case depth for various treated surfaces.

The effective case depth increases with surface roughness as shown in Fig.8. Sand blasting introduces the uniform surface roughness and compressive residual stresses on the surface [10]. As the surface roughness increases the contact area for nascent carbon and nitrogen to diffuse increases i.e. there are more sites for nucleation.

Influence of change in activity of nitrogen and carbon on case depth and microstructure.

The volume percent of ammonia was increased from 43 to 50 to increase the activity of carbon without much increase in activity of nitrogen by adjusting the ingoing ammonia gas flow to increase the effective case depth. This also helps to increase the volume percent of epsilon phase as compared to gamma prime phase in compound layer as reported by other researchers and shown in Fig.9.

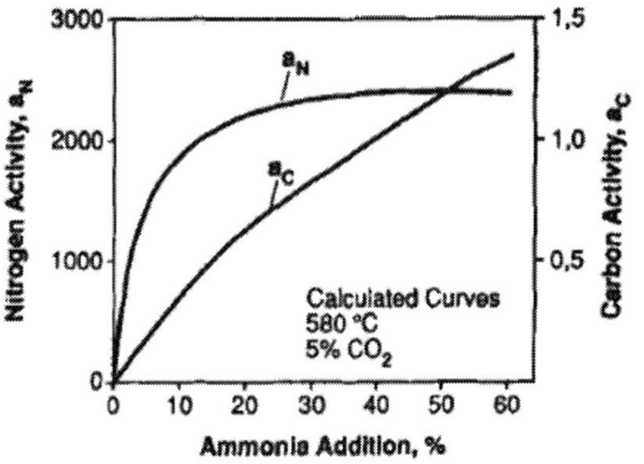

Fig.9. Change in activity of nitrogen and carbon with respect to change in ammonia variation [6].

The surface hardness and effective case depth for different gas flows are given in Table 4and 5.

Table.4. Surface hardness and depth for existing gas flows.

Existing gas flows (43% ammonia)			
Without sand blast		**With sand blast**	
Surface hardness (HRc)	**Effective case depth in mm**	**Surface hardness (HRc)**	**Effective case depth in mm**
57.31	**0.30 - 0.31**	52.43	**0.34 - 0.36**

Table.5. Surface hardness and depth for change in gas flow.

Change in gas flows (50% ammonia)			
Without sand blast		With sand blast	
Surface hardness (HRc)	Effective case depth in mm	Surface hardness (HRc)	Effective case depth in mm
53.60	0.35 - 0.36	51.66	0.47 - 0.49

With change in gas flow, the case depth increases both in case of sand blasted and without sand blasted samples. The surface hardness for both the cases is almost same. It shows that the increase in volume percentage of ammonia results in the formation of more epsilon phase in compound layer as compared to gamma prime phase. The variation in the weight percentage of carbon and nitrogen in the compound layer changes the phase formation in compound layer. Therefore, by measuring the weight percentage of nitrogen and carbon, the actual phase formed in compound layer can be identified. It can be analysed by Fe-N-C phase diagram as shown in Fig.10.

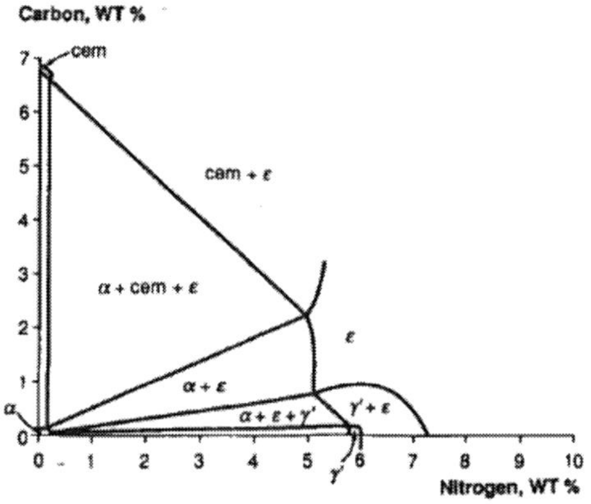

Fig.10. Fe-N-C phase diagram for 570-580 deg. Celsius [6].

For example, if the weight percentage of carbon is 1-2% and nitrogen is around 6% the monophase epsilon layer can be obtained. The volume percentages of epsilon and gamma prime phases formed in compound layer in both the cases were measured qualitatively by murakami reagent. The murakami reagent consists of 20gm KOH, 20gm $K_3Fe(CN)_6$ in 100ml aqua dist at 80 degree Celsius [11].

When nitrocarburized specimens were etched with 2% nital, and murakami reagent at 80 degree Celsius, the compound layer appears to be black and white. The black area shows the epsilon phase and white area shows the gamma prime phase. This was measured by the phase analysis technique available in optical microscope by selecting the gray scale in specific area in the compound layer and measuring the percentage of black and white areas. The black phase is shown by red colour and white phase is shown by blue colour respectively. Analysis was carried out on 3-4 locations in compound layer and the average value of those readings was taken. It is observed that the volume percentage of epsilon and gamma prime is 60.2 and 39.8 respectively for existing gas flow sample and volume percentage of epsilon and gamma prime is 76.7 and 23.3 respectively for change in gas flow sample as shown in Fig.11, 12 and 13. This may be due to the increase in chemical potential gradient of nascent nitrogen and carbon. The epsilon phase is more ductile than gamma prime phase.

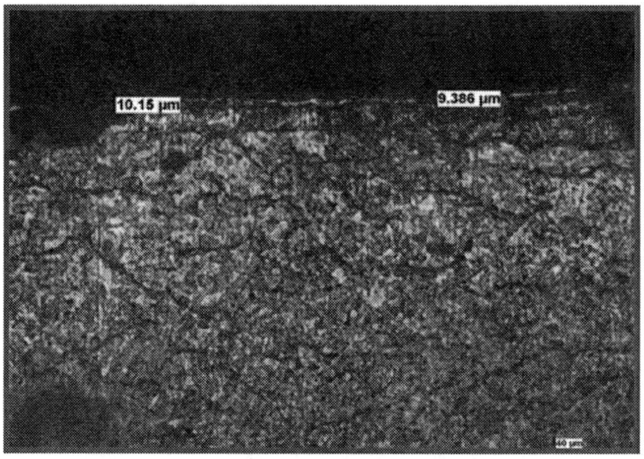

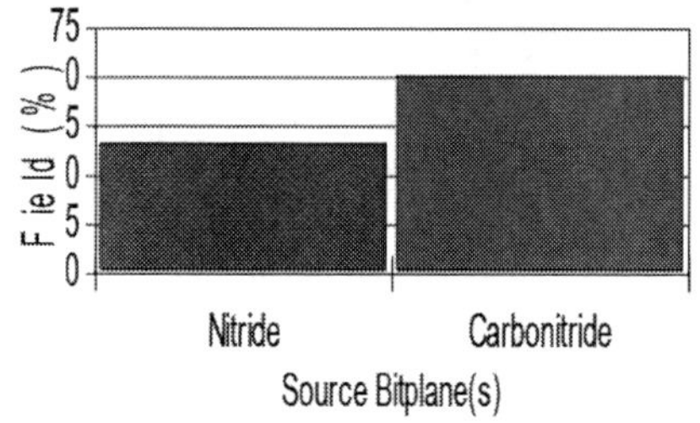

S. Bitplane(s)	Field (%)
Nitride	39.8
Carbonitride	60.2

Fig.11. Micro structural image at 500X when etched with murakami etchant shows volume percentage of epsilon and gamma prime phase in compound layer for 43% ammonia.

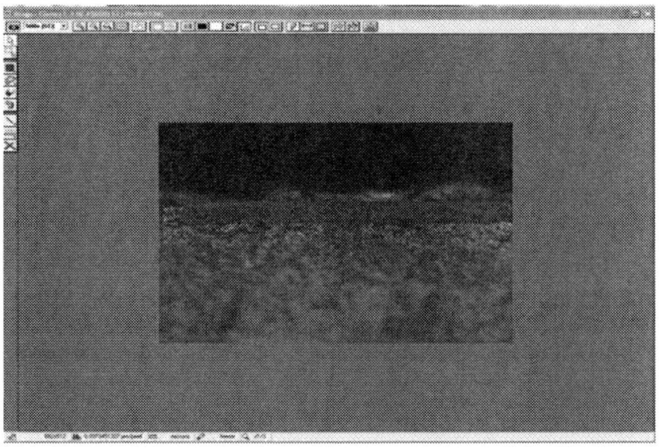

Fig.12. Selection of specific area on compound layer to measure the percentage amount of black and white phases.

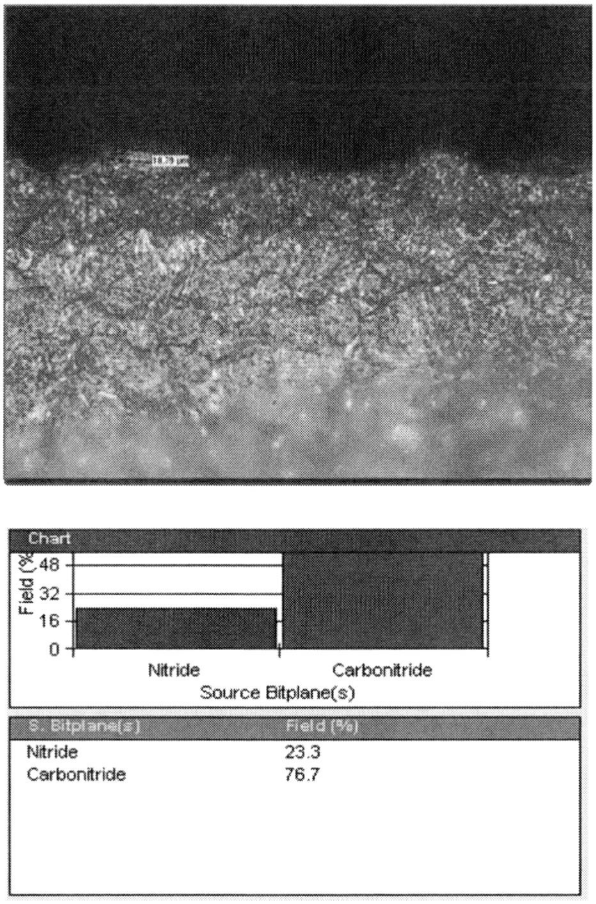

Fig.13. Micro structural image at 500X when etched with murakami etchant shows the volume percentage of epsilon and gamma prime phase in compound layer for 50% ammonia.

Influence of nitrocarburizing on wear resistance of die steel material.

Wear testing was carried out on eight numbers of pins and they were categorized as per the effective case depth observed. It is observed that the wear rate in case of zero case depth i.e. without nitrocarburized was 200-220 micron per hour. And in case of nitrocarburized pins the wear rate was 100-120 micron per hour. The pins without nitrocarburizing were tested for one hour and all other nitrocarburized pins were tested for four hour duration. The mean coefficient of friction was found by the WINDUCOM software through data acquisition system. The frictional forces were calculated with the help of normal load applied and mean coefficient of friction. It is observed that the frictional force is lower in case of non nitrocarburized specimens because at high temperature the softer material gets deformed and removed from its parent material by adhesive wear mechanism. The comparison of wear against time, sliding distance for different specimens is shown graphically in Fig.14 and 15. Also the comparison of coefficient of friction and frictional force against time is shown (Fig. 16, 17). The wear testing results are given in Table.6.

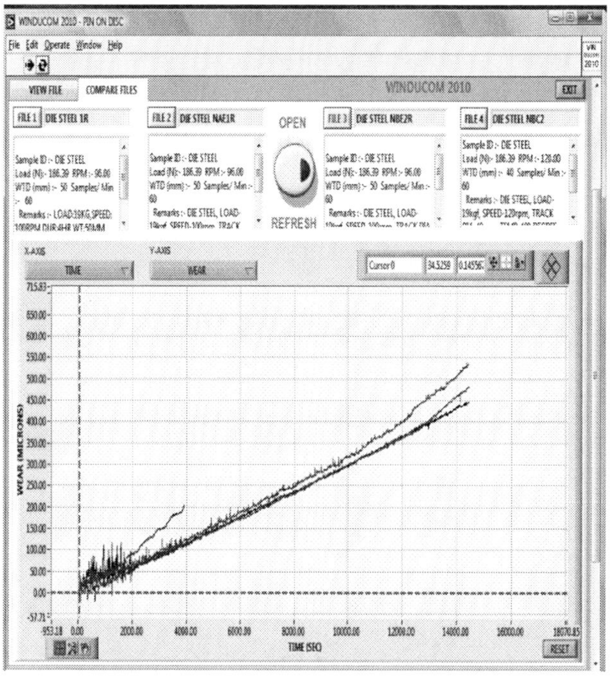

Fig.14.Wear in micron against time for different specimens having different effective case depth. Red line indicates zero mm depth pin, blue line indicates 0.31mm depth pin, green line indicates the 0.36mm pin and black line indicates 0.47mm depth pin.

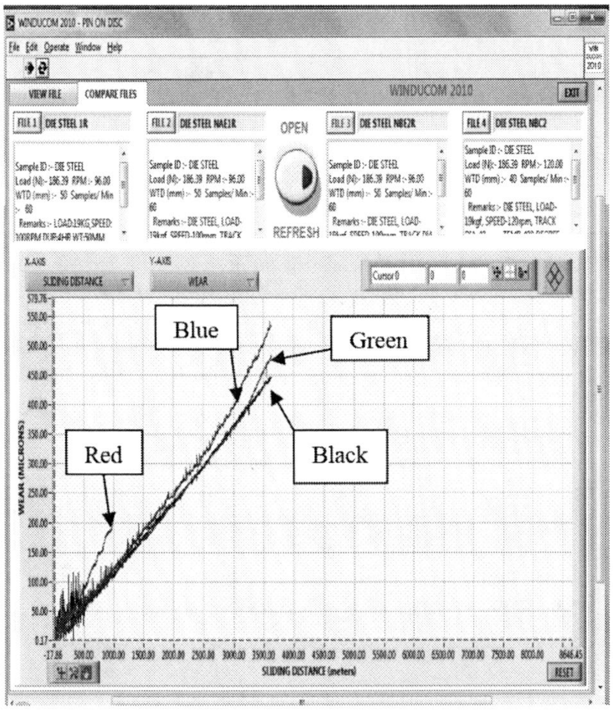

Fig.15.Wear in micron against sliding distance for respective pins.

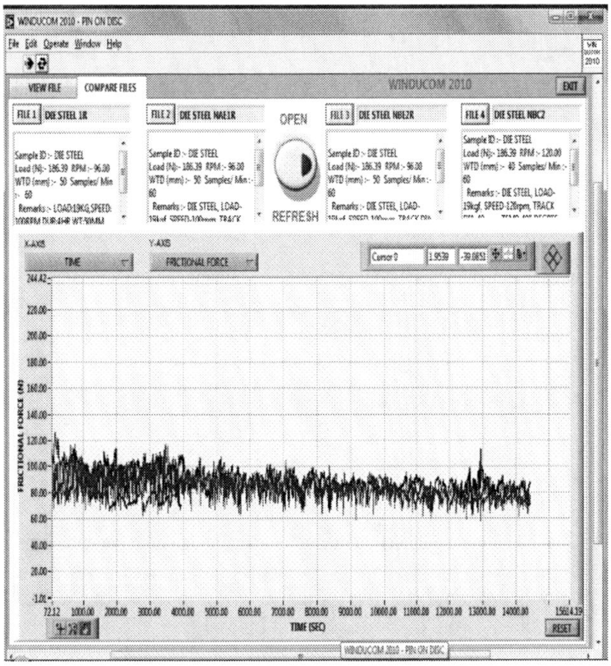

Fig.16. Frictional force against time.

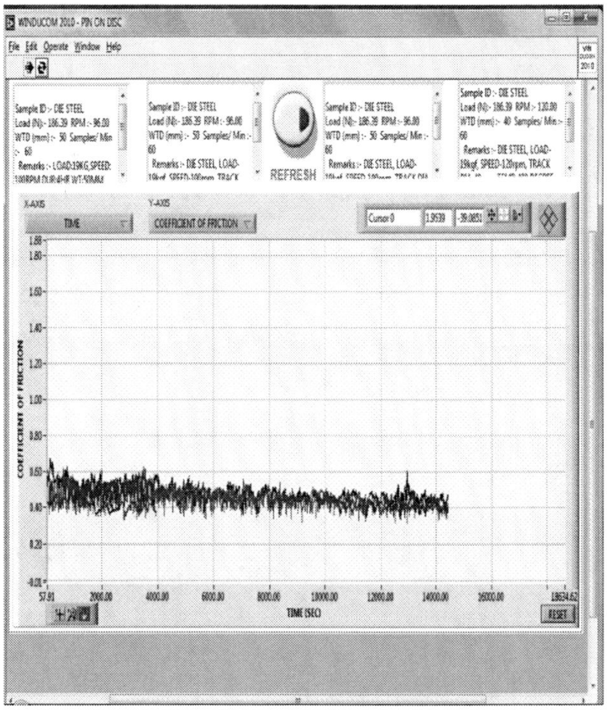

Fig.17. Variation of coefficient of friction against time.

Table.6. Wear testing results.

Effective case depth in mm	Process details	Wear in micron	coefficient of friction	Frictional force in (N)
Zero	W/N	*200/220	0.415 / 0.411	78.85 / 78.09
0.30-0.31	AE	533.99 / 736.27	0.459 / 0.469	87.25 / 89.11
0.35-0.36	BE/AC	480.63 / 453.11	0.456 / 0.444	86.64 / 84.36
0.46-0.48	BC	445.73 / 418.56	0.481 / 0.430	91.39 / 81.70

*- This wear value is of one hour test.

W/N – Without nitrocarburized. **AE** – Without sand blasted existing gas flow. **BE** – Sand blasted existing gas flow. **AC** - Without sand blasted change in gas flow. **BC** - Sand blasted change in gas flow.

Furthermore the weight and volume loss were calculated. First the weight loss was calculated by taking the initial weight and final weight after wear testing. And then by considering the density for this material as 7.80 Kg/dm^3 volume loss was calculated for respective pins. Those results are tabulated in Table.7.

Table.7. Loss of weight and volume during wear testing.

Effective case depth (mm)	Weight loss (gm)	Volume loss in cubic mm
Zero	0.039 / 0.042	5.0 / 5.38
0.30-0.31	0.054 / 0.065	6.92 / 8.33
0.35-0.36	0.043 / 0.045	5.51 / 5.77
0.46-0.48	0.042 / 0.041	5.26 / 5.38

The results of the wear test indicate that higher the case depth lower is the weight and volume loss.

The wear grooves were seen on pin surface and wear track on disc surface. The wear mechanism observed was abrasive as seen in Fig. 18and 19.

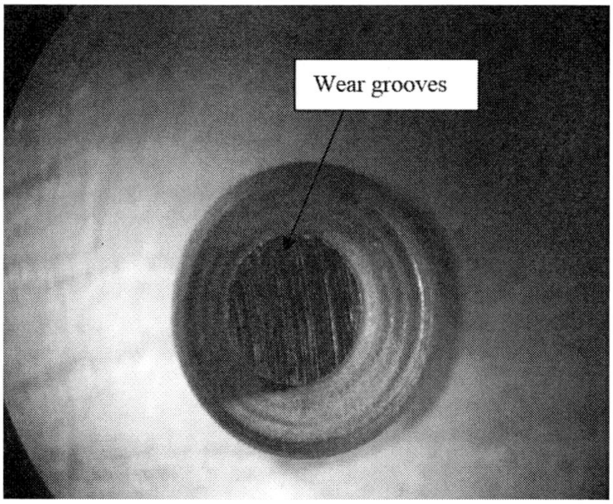

Fig.18. Wear groves on pin surface at 12.5X.

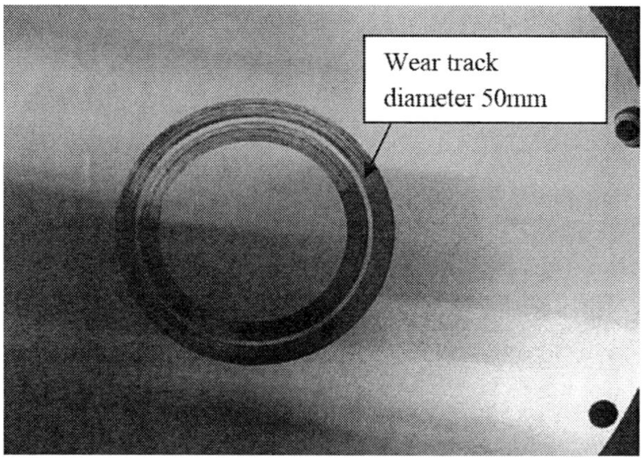

Fig.19. Wear track on disc at12.5X.

Conclusions

1. Surface treatment like sand blasting before nitrocarburizing helps to improve the uniform surface finish and it imparts the compressive residual stresses on the surface to be nitrocarburized. Therefore it helps to increase the effective case depth during nitrocarburizing.

2. Due to change in activity of nitrogen and carbon the thickness of compound layer is slightly increased but the surface hardness remains same. It is attributed to the formation of ductile epsilon phase. Epsilon phase has hcp structure, which has work hardening characteristics. Therefore it's hardness increases with the applied load that increases the wear resistance.

3. The nitrocarburized pin shows the constant wear rate up to effective case depth. Once the depth is over the wear rate increases to the level of core material wear rate.

4. The pins having more effective case depth show constant wear rate for longer time that subsequently increases the wear resistance. Also, increase in case depth results in reduction in weight and volume loss.

Acknowledgement:

The authors wish to thank the management and staff of Bharat Forge ltd. Pune for their cooperation in conducting the nitrocarburizing trials and to carry out the testing.

References:

[1] K.Lange, L.Cser, M.Geiger, J.A.G.Kals-"Tool life and tool quality in bulk metal forming", Annals of the CIRP Vol.41/2, 1992, pp. 667-675.

[2] Sailesh Babu, Dilmar Ribeiro and Rajiv Shivpuri-"Material and surface engineering for precision forging dies", Ohio state university, 1999.

[3] S.Lampman-"Introduction to surface hardening of steel", in Heat treating Vol.4, ASM handbook, 1991, pp. 607-626.

[4] J.R.Davis-"Surface hardening of steels", ASM, International materials information society, 2002, pp.178-209.

[5] J.Baranowska, K. Szczecinski, M.Wysiecki-"Increasing of gas nitriding kinetics via surface pretreatment", Institute of materials engineering, surface and coating technology 151-152, 2002, pp. 534-539.

[6] Lars Sproge and Jan Slycke-"Control of the compound layer structure in Gaseous nitrocarburizing", springer-Verlag New York, 1992, pp. 105-112.

[7] I.M.Hutching,Tribology: Friction and wear of engineering materials, chapter 5, 1992, pp. 77-78.

[8] G. Castro, A.Fernandez-vicente, J.Cid-"Influence of nitriding time on wear behaviour of an AISI H13 steel during a crank shaft forging process", Elsevier wear 263, pp. 1375-1385, 2007.

[9] David Englund and Dan Forsstrom-"Investigation into the failure of hot forming tools and their tribological behaviour",Lulea university of technology, 2010, pp.12.

[10] Jayant B. Mane-"Effect of nitride layer on wear life of hot forging dies", Master of Technology thesis at Indian Institute of Technology Bombay, 2011.

[11] M.A.J. Somers and E.J. Mittemeijer-"Formation and growth of compound layer on gaseous nitrocarburizing of iron: Kinetics and microstructural evolution", Laboratory of metallurgy, delft university of technology, 1989, pp. 191-194.

Proceedings of the 26th ASM Heat Treating Society Conference
B.L. Ferguson, R. Jones, D.S. MacKenzie, and D. Weires, editors
https://doi.org/10.31399/asm.cp.ht2011p0235

Heat Transfer Properties of a Series of Oxidized and Unoxidized Vegetable Oils in Comparison with Petroleum Oil-Based Quenchants

Ester Carvalho de Souza, Lauralice C.F. Canale
Universidade de São Paulo, Escola de Engenharia de São Carlo, São Carlos, SP, Brazil.
Email: esterfisica@yahoo.com.br; lfcanale@sc.usp.br, phone: 5516 3373 9579, fax: 5516 3373 9590.

G. Sánchez Sarmiento
Universidad del Salvador, Facultad de Ciencia y Tecnología, Buenos Aires, Argentina.
Universidad Austral, Facultad de Ciencia y Tecnología, Buenos Aires, Argentina.
Email: gustavo.ssarmiento@kbeng.com.ar, phone: 5411 4326 7542, fax: 5411 4326 2424

Eliana Agaliotis
Consejo Nacional de Investigaciones Científicas y Técnicas (CONICET), Buenos Aires, Argentina.

Juan C. Carrara, Diego S. Schicchi
Universidad Tecnológica Nacional, Facultad Regional Buenos Aires, Medrano Street Nº 951, Buenos Aires, Argentina

George E. Totten
Texas A&M University, Department of Mechanical and Materials Engineering, Seattle, WA, USA
Email: getotten@getottenassociates.com, phone: 206 788 0188, fax: 815 461 7344.

Abstract

Vegetable oils, especially soybean oil, exhibit substantially poorer thermal-oxidative stability than commercially available petroleum oil quenchant formulations. Therefore, to achieve any commercially interesting performance vegetable oils must be stabilized by the addition of antioxidant inhibitors. This work describes the ability of two commercially available antioxidants, Irganox L 57 and Irganox L 109 to stabilize soybean oil against thermal-oxidative degradation. In addition, the effect of antioxidant stabilization on quenching performance was evaluated by determining the profile of heat transfer coefficient variation throughout the quenching process at different times after being subjected to an accelerated thermal-oxidation ageing test. The results of this work are discussed here.

Keywords: Vegetable oils; Antioxidants; Quenchants; Cooling Curve Analysis

Introduction

Propererties, such as hardness and strength, of carbon and alloy steels are dependent on the cooling time-temperature profile (cooling curve) exhibited by the quenching medium relative to the TTT (time-temperature-transformation) or CCT (continuous cooling transformation) curves for the steel of interest. Quenchants not only control properties but also, depending on the quenching process and quenchant selection, control the residual stress profile with the steel and are critically important to optimize distortion control and prevent cracking. Depending on the steel and hardening process, the most common quenchants include: air, oil, water, brine, aqueous polymer solutions and high-pressure gas quenching.

Of the vaporizable quenchants, petroleum oil derived fluids continue to be the most commonly encountered throughout the industry globally. However, there is an on-going effort to replace petroleum oil derived industrial fluids because of: price and the environmental and toxicological properties hazards exhibited by petroleum products. Various quenching media have been evaluated as alternatives to petroleum oil including water, especially utilizing water in conjunction with time quenching [1-3] or by intensive quenching [4], water-gas [5], aqueous polymers [6, 7], and high-pressure gas [8]. However, petroleum oil derived quenchants continue to dominate the marketplace.

Currently, there is increasing interest in the use of biobased oils as replacements for petroleum oil basestocks for lubricant and quench oil formulation. Biobased oils are defined as "industrial products including fuels, but not food or feed, made from renewable agricultural or forestry resources". Seed oils are currently one of the most common sources of biobased oils and among the various seeds from which seed oils are derived. Soybeans are in greatest production worldwide and the United States has the highest soybean production rate, followed by Brazil [9]. Of the various seed oils produced in the USA in 2004, soybean oil has the largest production rate and is therefore one of the oils of greatest interest for use in replacing petroleum oil derived quenchants [10].

Tagaya and Tamura compared the quench severity of different vegetable oils including soybean, rapeseed and castor oils with mineral oils and fish/animal oils with respect to fluid source and viscosity and oxidative stability for various naturally derived fluids [11]. Fujimura and Sato followed this work in 1963 by examining the quenching performance of the vegetable oils reported by Tagaya and Tamura and different petroleum oil quenchants and concluded that the performance of soybean and rapeseed oil were essentially equivalent [12].

Currently, the most commonly cited vegetable oil basestocks used for quenchant formulation are canola oil [13], and soybean oil derivatives [14]. Although, vegetable oils exhibit cooling performance different from the petroleum oil standard used, in none of the references was any systematic comparison of vegetable oil structure with oxidative stability and its corresponding impact on quenching performance demonstrated. The cooling curves reported were obtained according to ISO 9950 [15] which is essentially equivalent to ASTM D6200 "Standard Test Method for Determination of Cooling Characteristics of Quench Oils by Cooling Curve Analysis" [16]. Both ISO 9950 and ASTM D6200 utilize a 12.5 mm dia x 60 mm INCONEL 600 cylindrical probe with a Type K thermocouple inserted to the geometric center.

Oxidative stability performance improvements of vegetable oils can be achieved by chemical or genetic modifications [17-19] or by process improvements such as winterization and partial hydrogenation [20, 21]. Winterization (fractionation) is performed to remove crystallized fats and improve the pour point of the base oil. The performance objective is to reduce the linolenic and linoleic ester content of the vegetable oil to increase the oxidative stability making the resulting vegetable oil more suitable for use in industrial applications [20]. Honary demonstrated that oxidative stability of soybean oil is substantially improved by partial hydrogenation. However, this also increases the level of saturation levels in the oil which produce crystallized fats which may be removed by a winterization process [21].

Genetic engineering has produced soybean oil with > 85% oleic acid ester contents [17, 18]. There are, however, a number of substantial problems with genetic engineering to reduce polyunsaturation including costs associated with regulatory approval.

Totten et al reported the results of cooling curve, hardenability, heat transfer, and wettability characterization studies conducted with crude and partially hydrogenated and winterized soybean oils provided by Honary [21, 22]. The soybean oils investigated showed no significant differences in cooling behavior and the rewetting conditions on the test specimen surface and the cooling rates were similar. Comparison of the cooling time-temperature and cooling rate curves showed that the vegetable oils exhibited faster cooling rates than the mineral oil used as a reference [22].

Although there are various references to the quenching performance of soybean oil, and other vegetable oils such as the work reported by Prabhu [23, 24], with the exception of Tagaya and Tamura [25] and Canale, et al [26], no one has reported the effect of the oxidative degradation on quenching performance and the ability to stabilize cooling performance, as determined by cooling curve performance with the addition of antioxidants. The objective of this work was to study the degradation of quenching performance of soybean oil with and without the addition of antioxidants.

Experimental Methods and Materials

The soybean oil used for this work was purchased at the local market in Sao Carlos, Brazil and was used in the "as-purchased" condition[1]. Quenching performance of these oils was compared to two commercially available petroleum quenching oils: MicroTemp 157 - a conventional "slow" oil and MicroTemp 153B - an accelerated "fast" oil.

The two antioxidants were used for this work are Irganox L 57 and Irganox L 109. Both were used as supplied by Ciba Geigy Corporation. The composition of the soybean oil blends prepared for this study are shown in Table 1.

Table 1: Composition of soybean oil/antioxidant blends and their designation codes.

Concentration of antioxidant	Name used
Soybean – As Purchased	SO
SO + 0,3% Irganox L 57 + 0,7% Irganox L 109	SO 1
SO + 1% Irganox L 109	SO 2
SO + 1% Irganox L 57	SO 3
SO + 0,5% Irganox L 57 + 0,5% Irganox L 109	SO 4

Viscosity was measured at 40 °C an 100 °C according to ABNT NBR 10441 – 10/02 [28].

The oxidative stability of the vegetable oils and quenching oil was determined using an accelerated oxidation test previously reported by Bashford and Mills [29] and equipment built for this purpose by Farah[2] [30]. A schematic illustration is provided in Figure 1. The oxidation test was conducted as

[1] Crude soybean oil is refined by a series of processes to remove impurities that affect the taste, smell, appearance, and stability of the oil. The refining processes involve de-gumming, alkali refining, bleaching, and deodorization. However, it is important to note that it is possible for the thermal treatment during the purification process may actually lead to peroxide formation and result in poorer oxidative stability when subjected to elevated temperatures during use [27].

[2] Historically, the earliest report of using a version of this test was by Tagaya and Tamura [25] who developed the test based on the Indiana Stirring Oxidation Test [31].

follows: into a 2300 mL glass vessel (12.5 cm diameter, 19 cm in height) was added 2000 mL of the vegetable oil which was heated to 150 ± 2 °C for 15 minutes with agitation. During this time, the fluid was aerated using a gas sparge tube at 4 L air/hour. (The agitation was provided by an electrically driven propeller mixer with speed settings of 0-10 and a setting of 6 was used.) After 12 minutes, the electrical resistance immersion heater, aeration system and agitation was automatically shut off for 15 minutes during which time the fluid was cooled to 125 ± 2 °C in 3 minutes. The fluid was reheated again with agitation and air sparging and these 15 minute cycles were repeated over the test duration of 48 hours.

Cooling curves were obtained under unagitated conditions according ASTM D6200-01 – "Standard Test Method for Determination of Cooling Characteristics of Quench Oils by Cooling Curve Analysis" at a bath temperature of 40 °C. This test method is based on the 12.5 mm dia. X 60 mm cylindrical INCONEL 600 probe assembly. After heating the probe in a furnace to 850 °C (1562 °F), it was then manually and rapidly immersed into 2000 mL of the oil to be tested which was contained in a tall-form stainless steel beaker. The probe temperature and cooling times are recorded at selected time intervals to establish a cooling temperature versus time curve.

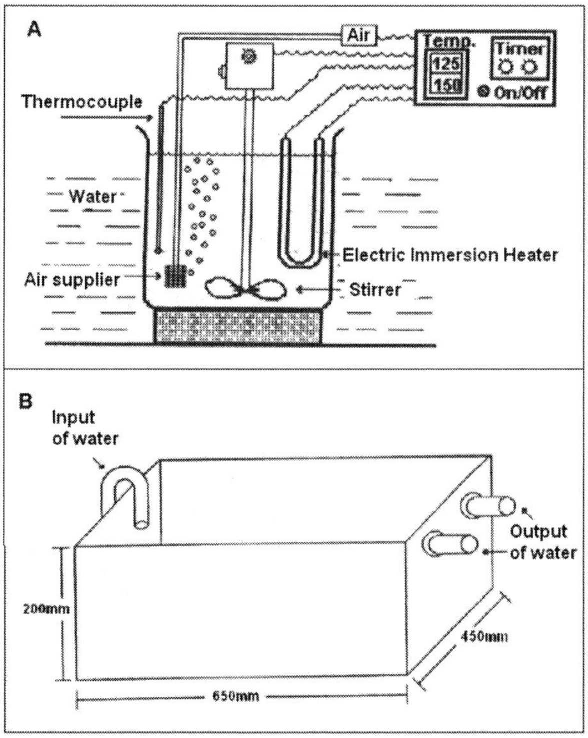

Figure 1: Schematic illustration of the accelerated oxidation system (A) heating, agitation and blow air into the oil sample (B) Refrigeration system [30].

Calculation of Heat Transfer Coefficients

Heat transfer processes are complex and the heat transfer coefficient is a complex function of variables describing this process. Generally, the heat transfer coefficient is a function of the fluid flow, component (probe) shape and dimensions, temperature and physical properties of the liquid: thermal conductivity, specific heat capacity, density and viscosity. The heat transfer coefficient can be defined as the quantity of heat transferred per unit time per unit area of a surface when the difference of temperatures between the surface and liquid equals one degree absolute. Since quenching processes are actually heat transfer processes, the heat transfer coefficient is an excellent single parameter for quenchant characterization.

In this work, the commercial code HT-MOD (Heat Treating Modeling)[3] was used to simulate the heat treatment process [32]. This code is also used to calculate heat transfer coefficients as function of time by solving an inverse heat transfer problem. The model is based on a numerical optimization algorithm which includes a finite element module for calculating, with respect to time and space, the temperature distribution and its coupled micro-structural evolution. In this case, since an Inconel 600 probe was used that does not undergo microstructural phase transformation, the differential problem solved is only the heat conduction equation (1) in a cylindrical probe:

$$\nabla\cdot\left(k(\vec{r},T)\cdot\nabla T\right)=c(\vec{r},T)\,\rho(\vec{r},T)\,\frac{\partial T}{\partial t} \qquad (1)$$

where $k(\vec{r},T)$, $c(\vec{r},T)$ and $\rho(\vec{r},T)$ denote respectively the thermal conductivity, the specific heat and the density of the material as dependent of the position $(\vec{r})$ and temperature T. $T(\vec{r},t)$ is subjected to the initial condition:

$$T\left(\vec{r},t=0\right)=T_0\left(\vec{r}\right);\quad \vec{r}\in\Omega \qquad (2)$$

and the following boundary conditions at the surface of the steel probe:

$$-k\frac{\partial T}{\partial n}=h_i\left(T-T_{qu}\right)\quad on\ \Gamma_i,\ i=1,\cdots,p \qquad (3)$$

where $h_i(T)$ are the heat transfer coefficients corresponding to different portions of the boundary Γ and T_{qu} the quenchant temperature. Each one of these p cooling zones has a time dependent heat transfer coefficient that varies strongly along each partial boundary, depending on the heat transfer mechanism (vapour blanket, nucleate boiling, convective cooling) that governs the energy flow.

Provided that the temperature change inside the component and on its surface are measured, it is possible to solve the inverse heat transfer problem to determine the time variation

[3] HT-MOD is a commercial code which is available from: KB Engineering S.R.L.; Florida 274, Piso 3, Of. 35 (1005) Buenos Aires – *Argentina;* Tel: 54-11 4326-7542; Fax: 54-11 4326-2424; Internet: http://www.kbeng.com.ar.

of the heat transfer coefficients, that best satisfies productions demands. The time dependence of the heat transfer coefficient can be approximated by polygonal functions, each one defined by a set of parameters $h_i^{(r)}(r = 1,\cdots,p\ ;\ i = 1,\cdots,q)$.

$$S = S\left(h_i^{(r)}\right) = \sum_{k=1}^{n}\left(T_k^m - T_k^c\right)^2 = min \qquad (4)$$

n being the total number of measured temperatures, i.e the number of points times the number of measurements at each point.

The selection of the initial values for these coefficients and of the quantity and length of the time intervals was sample-dependent. The mean square difference between the measured and calculated temperatures obtained after optimization of the heat transfer coefficients was about 1 °C. Table 2 summarizes the thermo-physical properties INCONEL 600 used for this work [33].

Table 2: Temperature-Dependent Thermo-Physical Values for INCONEL 600.

Thermal Conductivity		Specific Heat		Density	
Temp. [°C]	K [W/m.K]	Temp. [°C]	c [J/kg.K]	Temp. [°C]	ρ [kg/m3]
50	13.4	50	451	20	8400
100	14.2	100	467	100	8370
150	15.1	200	491	200	8340
200	16.0	300	509	300	8300
250	16.9	400	522	400	8270
300	17.8	500	533	500	8230
350	18.7	600	591	600	8190
400	19.7	700	597	700	8150
450	20.7	800	602	800	8100
500	21.7	900	611	900	8060
700	25.9				
900	30.1				

Discussion

A. Thermal Oxidation Test Results

In view of the relatively poor inherent oxidative stability of soybean oil and the commercial interest in using this commonly available basestock for industrial fluid formulation, it was selected for a subsequent preliminary screening test of using two antioxidants, Irganox L 57 and Irganox L 109. The thermal-oxidation test described in the Experimental Section was used and the degree of oil oxidation was followed throughout the duration of the test by measuring the resulting fluid kinematic viscosity with respect to oxidation time as shown in Figure 2. For hydrocarbon-based fluids, thermal-oxidative degradation will result in an overall viscosity increase with respect to time, although in some cases this increase may be preceeded by a small viscosity decrease due to initial hydrocarbon chain cleavage before an exponential increase in viscosity.

Although there was some variation of the fluid viscosity upon addition of the antioxidants, the differences were relatively minimal. Throughout the ageing test, the results obtained show that both Irganox L 109 and Irganox L 57 both exhibited better stabilization than their blends. For most of the test, Irganox L 109 exhibited somewhat better performance than did Irganox L 57. At the conclusion of the test, however, there was an inversion between 60 and 70 hours where Irganox L 57 seemed to stabilize with little additional increase in viscosity whereas the Irganox L 109 continued on its exponential viscosity increase. This is an important observation and unfortunately the test was not continued beyond 70 hours. To draw a definitive conclusion, longer test durations would be required.

One of the reasons for evaluating blends of Irganox L 57 and Irganox L 109 was to search for potential synergies. Based on these limited studies, no positive synergies by blending both antioxidants were observed.

Figure 2 also shows that the performance of neither Irganox L 57 or Irganox L 109 exhibited the much better performance obtained with both commercial petroleum oil quenchants: Micro Temp 157 or Micro Temp 153B. Clearly substantially greater stabilization of the soybean oil candidates is necessary if oxidative stability that is comparable to that exhibited by petroleum oils is to be achieved.

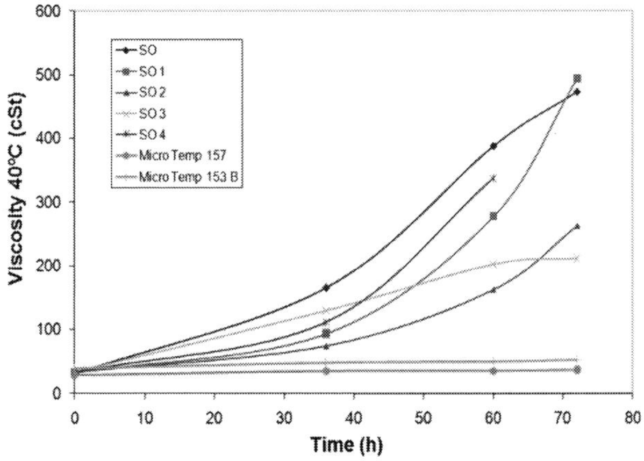

Figure 2: Comparison of the ability of Irganox L 57 and Irganox L 109 and their blends to stabilize soybean oil using the Bashford and Mills oxidative stability test. Also shown for comparison are two commercial petroleum oil formulations: Micro Temp 153B and Micro Temp 157.

B. Effect of Aging on Heat Transfer Coefficients During Quenching

Any hyrdrocarbon-containing quenchant will exhibit some finite propensity to oxidize over time when subjected to high interfacial temperatures such as those encountered upon immersion of austeniteized steel (≈ 850 °C) and oxygen (air). The degree and rate of oxidation will vary with respect to the chemical structure of the hydrocarbon moiety; presence, structure and concentration of antioxidants; presence of contaminants such as water and metals including copper and iron; total load size and frequency of use of the quench bath among other variables [34-36]. Therefore, some net increase in quenchant viscosity over time is expected and since viscosity is an important variable in heat transfer, some variation in heat transfer rates is also expected [37]. The question is how great is the variation and what would be the relative variation when different quench media are compared.

For this work, cooling curve analysis was performed according to the ASTM D6200 test procedure at a quenchant bath temperature of 40 °C with the different soybean oil formulations and petroleum oil-based quenchants shown in Table 1. The heat transfer coefficients as a function of time and surface temperatures data is shown in Figures 3-6.

As expected, the heat transfer coefficient vs time data showed that the conventional, unaccelerated Micro Temp157 petroleum oil-based quenchant exhibited significantly longer film-boiling behavior compared to the fast, accelerated Micro Temp 153B quench oil. This behavior is due to the addition of additives that rupture the vapor blanket that forms around the hot-metal surface upon immersion of the metal into the quenchant. Significantly, this work confirms earlier reported cooling curve behavior showing that vegetable oils do not exhibit this behavior [38].

Table 3 summarizes the maximum heat transfer coefficient, surface temperature and cooling time where this occurs. The maximum heat transfer coefficient for fresh, unoxidized soybean oil is significant less than soybean oil containing antioxidants and is comparable to the unaccelerated Micro Temp 157 quench oil which interestingly is less than the accelerated petroleum oil, MicroTemp 153B. Also, while the addition of an antioxidant or antioxidant mixture does significantly increase the maximum heat transfer coefficient relative to fesh, unstabilized soybean oil, the amount of increase with respect to different antioxidant additions is not very large.

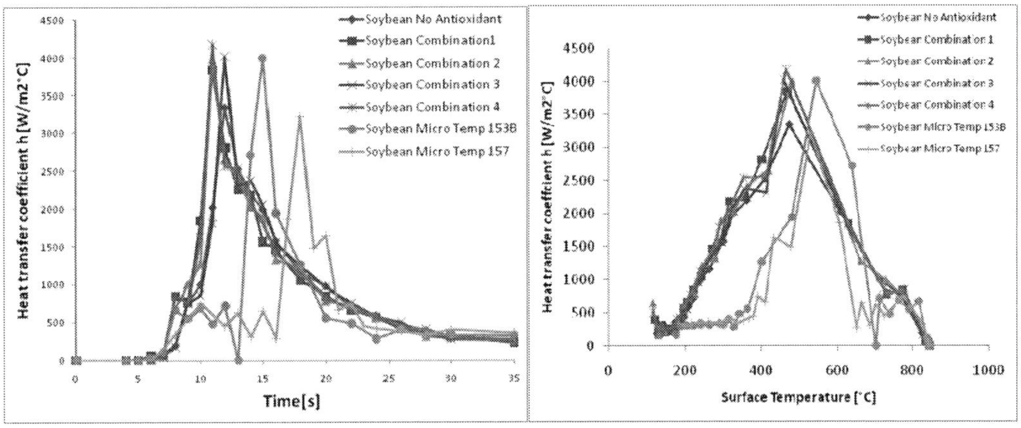

Figure 3: Heat transfer coefficient as a function of time (left) and as a function of surface temperature (right), calculated for the seven oils, time of degradation 0 h and bath temperature 40 °C.

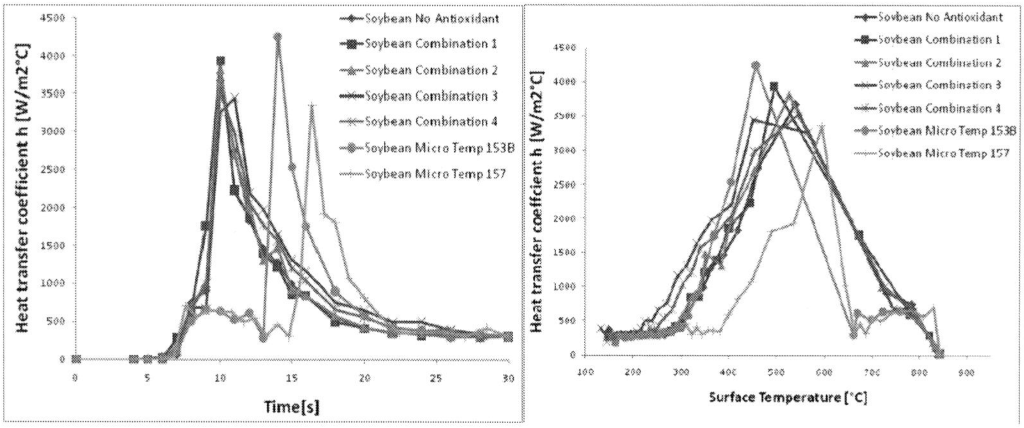

Figure 4: Heat transfer coefficient as a function of time (left) and as a function of surface temperature (right), calculated for the seven oils, time of degradation 24 h and bath temperature 40 °C.

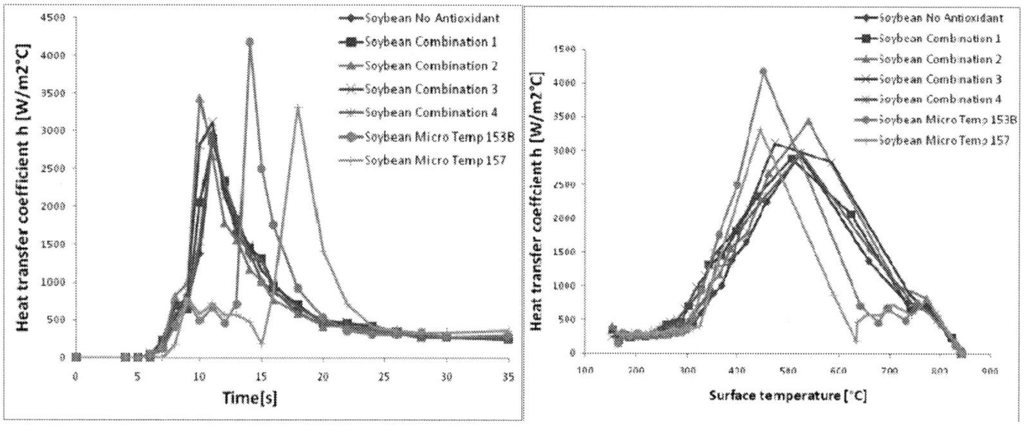

Figure 5: Heat transfer coefficient as a function of time (left) and as a function of surface temperature (right), calculated for the seven oils, time of degradation 48 h and bath temperature 40 °C.

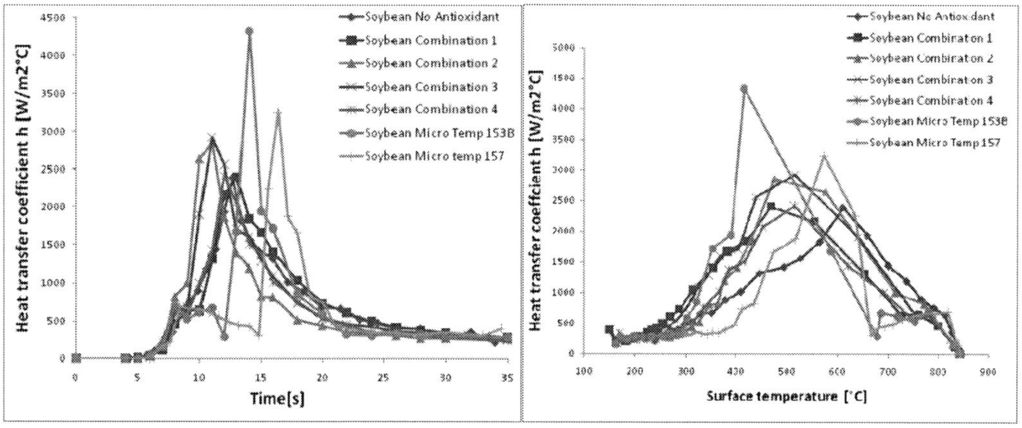

Figure 6: Heat transfer coefficient as a function of time (left) and as a function of surface temperature (right), calculated for the seven oils, time of degradation 60 h and bath temperature 40 °C.

The effect of anitioxidant addition on surface temperature and time where the maximum heat transfer coefficient occurs does not vary greatly for soybean oil. However, the surface temperature where the maximum heat transfer coefficient occurs is similar for both Micro Temp 153B and Micro Temp 157 but the time where this event occurs is somewhat greater for both oils than the soybean oil formulations with the unaccelerated Micro Temp 157 exhibiting the longest time – as expected.

Aging of the different oils showed a general decrease in the maximum heat transfer coefficient with respect to time as would be expected with the overall fluid viscosity increases illustrated in Figure 2. Generally, the amount of decrease was greatest for the least stable oils. There was not a clear correlation in the surce temperature and cooling time where the maximum heat transfer coefficient occurred which may be due to the overall reported variability of the ASTM D6200 method.

Compared to the variation of the maximum heat transfer coefficients observed for the soybean oil fluids with increasing ageing time, neither petroleum oil-based quenchant; Micro Temp 153B or Micro Temp 157 exhibited any significant difference in cooling behavior throughout the test which is reasonable in view of the minimal viscosity change observed throughout the ageing test for both oils as shown in Figure 2.

Conclusions

The effect of the addition of two commercial antioxidants, Irganox L 57 and Irganox L 109, on the thermal-oxidative stability of soybean oil was studied. The results showed that both L 57 and L 109 did inhibit the thermal-oxidative degradation process compared to uninhibited soybean oil. However, even though the concentration of these antioxidants was not optimized, neither produced a stabilized soybean oil that even roughly approximated the stability of a commercial, fully formulated petroleum-based quenchant.

Cooling curve analysis showed that soybean oil, with or without the addition of an antioxidant, exhibited any significant full-film boiling behavior whereas the petroleum oil quenchants did exhibit such behavior and, as expected, the accelerated petroleum oil (Micro Temp 153B) exhibited the less stable vapor blanker cooling behavior compared to the conventional, Micro Temp 157 quench oil.

Cooling curve studies also showed that the maximum heat transfer coefficient was greater (faster) for soybean oil containing antioxidants compared to uninhibited soybean oil. In addition, the maximum heat transfer coefficient decreased faster (due to fluid viscosity increase attributable to the thermal-oxidative degradation process) than the inhibited soybean oil fluids with Irganox L 109 and Irganox L 57 being the most effective. However, this performance fell considerably short of that observed for both fully formulated petroleum oil quenchants whose maximum heat transfer coefficients did not change significantly throughout the test.

As a result of this work, it is concluded that although the addition of antioxidants do stabilize soybean oil, substantially better performance is required if vegetable oils are to be effective functional equivalents to petroleum oil formulations. This mkay be done by selecting different vegetable oil composit ions with less unsaturation, by genetic modification of soybean seed oils or by chemically modifying and stabilizing the vegetable oil structure.

Table 3: Summary of quenching heat transfer parameters obtained from cooling curve analyses on the different soybean oil formulations and commercial petroleum-based quenchants.

Quenchant[1]	Ageing Time[2]											
	0 h			24 h			48 h			60 h		
	h_{max}	T_S	t_{hmax}	h_{max}	T_S	t_{hmax}	h_{max}	T_S	t_{hmax}	h_{max}	T_S	t_{hmax}
SO	3340	473	12	3672	538	10	2946	522	11	2391	610	13
SO1	3840	469	11	3940	494	10	2886	505	11	2402	469	13
SO2	3989	480	11	3817	526	10	3440	539	10	2859	475	11
SO3	4020	458	12	3450	451	11	3113	474	11	2920	515	11
SO4	4170	465	11	3518	544	10	2966	524	11	2416	515	12
MicroTemp 153B	4000	544	15	4250	456	14	4170	450	14	4330	415	14
MicroTemp 157	3210	529	18	3355	593	16.5	3310	444	18	3243	572	16.5

1.The quenchant designations indicate: SO – uninhibited soybean oil; SO1 – 0.3% Irganox L 57 + 0.7% Irganox L 109; SO2 – 1% Irganox L 109; SO3 – 1% Irganox L 57; SO4 – 0.5% Irganox L 57 + 0.5% Irganox L 109; Micro Temp 153B is a commercial accelerated petroleum oil quenchant and Micro Temp 157 is a commercial unaccelerated petroleum quenching oil.

2.The parameters shown here are: maximum heat transfer coefficient for the overall cooling process - h_{max} [W/m^2·°C]; Surface temperature at h_{max} – T_S (°C); and the time at h_{max} - t_{hmax}

Acknowledgements

To CAPES by the financial support and MSc Renata Leal for physical chemistry analysis.

References

[1] J.L. Burns and V. Brown, "The How and Why of Time Quenching", American Machinist, 1940, Vol. 84, No. 15, July 24, p. 523-526.

[2] M.S.P. Murthy, B.R. Ghosh, P.P. Sinha, M.C. Mittal, and B.K. Sarkar, "Studies on the Effects of Quenching Media and Quench Delays on the Properties of 12 mm Thick 15CDV6 Steel Plates", Transactions of the Indian Institute of Metals, Vol. 35, No. 1, February, p. 33-42.

[3] C.W. Finkl and N. Cerwin, "Method of Controlled Fluid Quenching of Steel", U.S. Patent 5,180,444, January 19, 1993.

[4] N.I. Kobasko, M.A. Aronov, J.A. Powell, L.C.F. Canale and G.E. Totten, "Intensive Quenching Process Classification and Applications", Heat Treatment of Metals, 2004, Vol. 31, No. 3, p. 51-58.

[5] H. Yu, J.A. Nicol, R.A. Ramser and D.E. Hunter, "Method of Heat Treating Metal with a Liquid Coolant Containing Dissolved Gas", US Patent 5,681,407, October 28, 1997.

[6] G.E. Totten, "Polymer Quenchants: The Basics", Adv. Mat. & Proc., 1990, Vol. 137, No. 3, p 51-53.

[7] G.E. Totten, G.M. Webster, S.W. Han and S.H. Kang, "Immersion Time Quenching Technology to Facilitate Replacement of Quench Oils with Polymer Quenchants for Production of Automotive Parts", The 1st International Automotive Heat Treating Conference, Eds. R. Colas, K. Funatani and C.A. Stickels, ASM International, Materials Park, OH, 1998, p. 449-455.

[8] Pritchard and S. Rush, "Vacuum Hardening of High-Strength Steels: Oil Versus Gas Quenching", Heat Treating Progress, 2007, May/June, p. 19-23.

[9] S. Serhan, "The Use of Vegetable Oils in Bio-Based Products", Presentation available from: National Center for Agricultural Utilization Research, USDA/ARS, 1815 N. University St., Peoria, IL, USA, E-Mail: erhansz@NCAUR.USDA.GOV

[10] Anon, "Corn Oil", Brochure Published by Corn Refiners Association, 1701 Pennsylvania Ave. N.W., Washington, DC. 20006-5805 (www.corn.org)

[11] M. Tagaya and I. Tamura, " Studies on the Quenching Media 3rd Report. The Cooling Ability of Oils," Technology Report, Osaka University, Vol 4, 1954, p 305-319.

[12] Y. Fujimura and T. Sato, "The Composition of Quenching oIl and Quenching Effects", The Iron and Steel Institute of Japan (ISIJ), 1963, Vol. 49, p. 1008-1015.

[13] Brennan, R.J., Faulkner, C.H. "A New Quenching Alternative", in Totten, G.E., Funatani, K., Howes, M.A.H., Sjostrom, S. (Eds.), 1996, Conf. Proceed. 2nd International Conference on Quenching and Control of Distortion. ASM International, Materials OH, pp. 423-428.

[14] Honary, L.A.T., "Performance of vegetable oils as a heat treat quenchant, in Totten, G.E., Funatani, K., Howes, M.A.H., Sjostrom, S. (Eds.), 1996, Conf. Proceed. 2nd International Conference on Quenching and Control of Distortion. ASM International, Materials OH, pp. 595-605.

[15] ISO 9950-1995 "Industrial quenching oils - determination of cooling characteristics - nickel-alloy probe test method".

[16] ASTM D6200-07 "Standard Test Method for Determination of Cooling Characteristics of Quench Oils by Cooling Curve Analysis", ASTM International, 100 Barr Harbor Drive, West Conshohocken, PA 19428 USA

[17] W. Castro, J.M. Perez, S.Z. Erhan and F. Caputo, "A Study of the Oxidation and Wear Properties of Vegetable Oils: Soybean Oil Without Additives", J. Am. Oil Chem. Soc., 2006, Vol. 83, No. 1, p. 47-52.

[18] S. Knowlton, "Soybean Oil Having High Oxidative Stability", U.S. Patent 5,981,781, November 9, 1999.

[19] E.B. Cahoon, "Genetic Enhancement of Soybean Oil for Industrial Uses: Prospects and Challenges", AgBioForum, 2003, Vo. 6, Nos. 1&2, p. 11-13.

[20] C. Tompkins and E.G. Perkins, "Frying Performance of Low-Linolenic Acid Soybean Oil", J. Am. Oil Chem. Soc., 2000, Vol. 77, No. 3, p. 223-229.

[21] L.A.T. Honary, "Soybean Based Hydraulic Fluid", U.S. Patent 5,972,855, October 26, 1999.

[22] Totten, G.E., Tensi, H.M., Lanier, K., "Performance of Vegetable Oils as a Cooling Medium in Comparison to a Standard Mineral Oil", J. Mat. Eng. and Perf., 1999, Vol. 8, No. 4, p. 409-416.

[23] K.N. Prabhu and P. Fernandes, "Determination of Wetting Behavior, Spread Activation Energy, and Quench Severity of Bioquenchants", Metallurgical and Materials Transactions B, 2007, Vol. 38, No. 4, August, p. 631-640.

[24] P. Fernandes and K.N. Prabhu, "Comparative Study of Heat Transfer and Wetting Behaviour of Conventional and Bioquenchants for Industrial Heat Treatment", International Journal of Heat and Mass Transfer, 2008, v 51, n 3-4, p. p 526-538.

[25] M. Tagaya and I. Tamura, "On the Deterioration of Quenching Oils," Technology Report, Osaka University, Vol 7, 1957, p 403-424.

[26] Canale, L.C.F., Fernandes, M.R., Agustinho, S.C.M., Totten, G.E. And Farah, A.F., "Oxidation of Vegetable Oils and Its Impact on Quenching Performance", Int. J. Materials and Product Technology, 2005, Vol. 24, No. 1-4, p. 101-125.

[27] T. Kinami, N. Horii, B. Narayan, S. Arato, M. Hosokawa, K. Miyashita, H. Negishi, J. Ikuina, R. Noda, S. Shirasawa, "Occurrence of Conjugated Linolenic Acids in Purified Soybean Oil", J. Am. Oil. Chem. Soc., 2007, Vol. 84, No. 1, p. 23-29.

[28] ABNT NBR 10441 – 10/02. Produtos de petróleo - Líquidos transparentes e opacos - Determinação da viscosidade cinemática e cálculo da viscosidade dinâmica.

[29] A. Bashford, A.J. Mills, "The Development of Improved Additives for Quenching Oils using Laboratory Simulations", Heat Treatment of Metals, 1984, No. 1, pp. 9-14.
[30] A.F. Farah, (2002) Caracterização de óleos vegetais como alternativa para meios de resfriamento utilizados no tratamento térmico de têmpera. Dissertação (Doutorado) - Interunidades em Ciência e Engenharia de Materiais, Universidade de São Paulo, São Carlos.
[31] G.G. Lamb, C.M. Loane and J.W. Gaynor, "Indiana Stirring Oxidation Test for Lubricating Oils", Ind. and Eng. Chem. Analytical Edition, 1941, Vol. 13, No. 5, p. 317-321.
[32] Sarmiento, G. S., Gastón, A., Vega, J. "Inverse heat conduction coupled with phase transformation problems in heat treating process". COMPUTATIONAL MECHANICS – NewTrends and Applications. E. Oñate and S.R.. Idelsohn (Eds.). CIMNE, (Barcelona, 1998). CD Book. Part VI, Section 1, Paper 16.
[33] Clark, J. and Tye, R., "Thermophysical Properties Reference Data for Some Key Engineering Alloys", High temperatures—high pressures, Vol. 35/36, (2003/2004), pp. 1–14.
[34] I. Debruyne, "Soybean Oil Processing: Quality Criteria and Flavor Reversion", Oil Mill Gazetteer, 2004, Vol. 110m July, p. 10-11.
[35] I. Paz and M. Molero, "Catalytic Effect of Solid Metals on Thermal Stability of Olive Oils", J. Am. Oil. Chem. Soc., 2000, Vol. 77, No. 2, p. 127-130.
[36] W.F. Bowman and G.W. Stochowiak, "Application of Sealed Capsule Differential Scanning Calorimetry – Part II: Assessing the Performance of Antioxidants in Base Oils", Lubrication Engineering, 1999, Vol. 55, No. 5, May, p. 22-29.
[37] S. Blaine and P.E. Savage, "Reaction Pathways in Lubricant Degradation. 2. N-Hexadecane Autoxidation", In. Eng. Chem. Res., 1991, Vol. 30, p. 218502191.
[38] Daniel Komatsu, Elki C. Souza, E. Carvalho de Souza, Lauralice F.C. Canale, George E. Totten, " Effect of Antioxidants and Corrosion Inhibitor Additives on the Quenching Performance of SoybeanOil", Strojniški vestnik (Journal of Mechanical Engineering – Slovenia), 2010, Vol. 56. No. 2, p. 121-130.

Proceedings of the 26th ASM Heat Treating Society Conference
B.L. Ferguson, R. Jones, D.S. MacKenzie, and D. Weires, editors
https://doi.org/10.31399/asm.cp.ht2011p0244

Simulation of Stress and Strain for Induction Hardening Applications

Dmitry Ivanov, Leif Markegård, John Inge Asperheim
EFD Induction a.s., Skien, Norway, www.efd-induction.com
{idg, lm, jia}@efdgroup.net, phone +47-35 50 60 00, fax +47-35 50 60 10

Abstract

The possibility to manage stress and strain in hardened parts might be beneficial for a number of induction hardening applications. The most important of these benefits is the improvement of fatigue strength, avoidance of cracks and minimization of distortion. An appropriate and powerful way to take the stress and strain into account during the development of a process is to make use of computer simulations. Own developed and commercial software packages have been coupled to incorporate the electromagnetic task into the calculations. The simulations have been performed with a following analysis of the results. The influences of two different values of quenching intensity, strength of initial material structure, strength of austenite, surface power density-frequency-time-combination, and work piece diameter on the residual stress are studied.

Introduction

The residual stress has decisive influence on the service properties of a steel component. In many cases, the compressive residual stress at the surface of a part allows improvement of fatigue strength (both initiation and propagation crack phase), corrosion resistance, etc. [1-3].

Induction surface hardening is known as a clean, efficient and precise process that could be integrated into a production line because of the high processing speed. Sometimes induction surface treatment allows improvement of fatigue life of the component regarding carburizing or shot-peening [4]. Normally, the compressive stress develops at the surface of the component [5, 6]. Stress levels vary in a wide range, and in some cases, the harmful tensile stress at the surface or in superficial layers can be obtained [5]. A number of investigations show that the induction surface hardening parameters have great influence on the residual stress [6-9]. In order to use the results of such investigations properly, a clear understanding of the influence nature is important. A good explanation of the results also allows development of new approaches to improve the residual stress distribution.

Even simplified physics of the stress and strain development during induction hardening includes a series of nonlinear and coupled phenomena, e.g., phase transformations, heat conduction, electromagnetism, elasto-plasticity and TRIP [10, 11]. Perhaps because of this, in most cases investigators concentrate only on a couple of the important parameters that influence the residual stress. Common explanations of the compressive residual stress at the surface of an induction hardened component are the dilatation associated with martensite formation and constraint of the relatively cold core [5, 12]. Some of the investigations take additionally the plastic strain into account [8].

A common approach to analysis of the stress and strain development during induction hardening must cover the variety of the most important phenomena and reflect the internal correlations of the process. The problem of identifying such phenomena is closely related to the simulation issues. Developers of mathematical models make the assumptions and simplifications sufficient for certain predictive accuracy. The results of such investigations could be partially used for analysis of the simulation results. In this paper such an attempt is made. The constitutive relationship of a series of developed stress-strain models is used for investigation of the cause-and-effect chains in typical steel components subject to induction surface hardening. Such relationship is the additive strain decomposition.

1. Approach to Analysis

a) Additive Strain Decomposition
The first step of process formalization is a choice of a sophisticated material model. The most popular metallurgic stress and strain formulations are plasticity, viscoplasticity and creep [13]. Plasticity represents the rate-independent inelastic behavior of solids, whereas viscoplasticity and creep are rate-dependent phenomena. For induction hardening of steel, the plasticity model with additive decomposition of either strain or strain rate as a constitutive relationship is the most common choice because of the simplicity, and it gives sufficient accuracy [10, 11].

Regarding strain rate, strain decomposition is more preferable as an analysis instrument because the integral values of strain components could be visualized. Taking into account the features of induction hardening processes, it can be stated for an arbitrary time of the process and point of the simulated component:

$$\varepsilon_{ij} = \varepsilon_{ij}^{e} + \varepsilon_{ij}^{p} + \varepsilon_{ij}^{Th} + \varepsilon_{ij}^{Ph}, \quad (1)$$

where ε_{ij} is the total strain, ε_{ij}^{e} elastic strain, ε_{ij}^{p} plastic strain (TRIP is included), ε_{ij}^{Th} thermal strain and ε_{ij}^{Ph} phase volumetric change strain.

Every component of strain tensor decomposition has its nature and could be interpreted from the physical point of view. The following properties of the strain components named above are important.

- Both thermal (prefix '*epsT_*' – designation on charts) and phase volumetric change ('*epsPh_*') strains are the initial sources and the cause of the effects in a hardened component. Strain has a local character and can be calculated with use of material properties, phase transformations and temperature at any point of a body.

- Elastic strain ('*epse_*') directly reflects the stress. Therefore, this component must be treated with the force balance of a body taken into account. As opposed to the source strains, elastic strain cannot be considered as a local value because it reflects the part's integrity, e.g., compressive elastic strain in a hardening zone is a cause of the compensating tensile elastic strain in the core of a hardened component. Elastic strain is always limited by the yield point of the actual phase mixture for the considered moment.

- Plastic strain ('*epsp_*') is an irrecoverable part of a strain. Since the residual phase volumetric change strain is dependent on the material properties and thermal strain is completely recoverable, plastic strain is the most important parameter that can be varied in order to manage the residual stress. Plastic strain can be calculated with the known laws of plastic flow, plastic criterion, etc. It is a continuation of a body reaction on the source strains and constraints after the elastic strain is yielded.

- Total strain ('*eps_*') is an observable quantity that can be measured with known techniques. The important property of this component is a possibility to reflect the constraints of the strain. The absence of constraints makes the total strain changed with the source strains. In case the zone with acting source strains is constrained, total strain is limited in reaction and elastic strain, with a possible continuation in plastic strain takes place.

With use of the above interpretations and Eq. 1, it is possible to build the cause-and-effect chains for an investigated hardened component. The tensor components could be considered separately; namely, the additive decomposition is justified also for certain directions such as axial or tangential in the cylindrical system of coordinates.

Since the main goal of this paper is the analysis of residual stress, elastic strain is of main interest and Eq. 1 can be rearranged for any of the strain tensor spatial components:

$$\varepsilon^{e} = \varepsilon - \left(\varepsilon^{p} + \varepsilon^{Th} + \varepsilon^{Ph}\right) \tag{2}$$

Thermal strain depends on temperature and the coefficient of thermal expansion. Phase volumetric change strain depends on the difference between the specific volumes of initial and resulting structures. The temperature curve of the complete induction hardening process usually is closed-loop. Therefore, the contribution of ε^{Ph} and ε^{Th} components to the residual stress value depends on the resulting structure only. With alternation of a structure's specific volume, the structure must be changed itself. This way of residual stress management is not flexible, but can be applied in practice.

According to above considerations, when a material is given, only total and plastic strains give theoretical possibility of flexible residual stress control. The difference between these values is crucial. To ensure the compressive residual stress, the plastic strain should be as high as possible, whereas the total strain should be minimized.

b) Geometrical Considerations

At first glance, the geometrical variety of hardened parts makes it impossible to apply general rules of stress development to an arbitrary region of a complex component. Indeed, the distribution of the stress is strongly dependent on the geometry of the component. Nevertheless, a big class of local regions in components could be represented as a convex or concave region. The simplest way to treat such regions is to consider them as cylindrical parts with or without an internal hole. An additional simplification and generalization of the task might be achieved by consideration of such cylinders as an infinite long height. It is possible if the axial force balance boundary condition is imposed. In this case, the following properties of geometry should be noted:

- The system can be represented in a cylindrical system of coordinates (Fig. 1). Hollow cylinder is described by two values, namely inner radius (IR) and outer radius (OR).

- There are only three non-zero spatial components of strain, namely radial (ε_r), axial (ε_z) and tangential (ε_φ). The postfixes '_r', '_z' and '_teta' are going to be used further to designate the spatial components on charts.

c) Design of simulations

The simulation procedure is as follows. First of all the induction heating process is calculated. The resulting distribution of temperature for a series of time steps is exported to a file that can be imported to our own developed simulation tool. The import of temperature distribution is possible for an arbitrary mesh because of field's interpolation. The considered procedure is a linear coupling of electromagnetic, thermal and phase transformation tasks. Nevertheless, the accuracy of such approach is sufficient for the goals of our investigation.

Design of computational experiments is represented in Table 1. The hardened surface for all the simulations except of No. 6 is external. The power and frequency of an induction heating convertor are adjusted to provide the equal austenitization depth conditions as much as possible. The maximal surface temperature after heating stage is the same for all the experiments and equal to 1000 °C. The detailed values of the variable parameters are described in corresponding sections of the article.

Material properties for steel Ck45 are used for simulations [14]. Ac1 and Ac3 temperatures are corrected according to the actual heating speed [15].

Table 1: Design of simulations

Sim. No	OR, mm	IR, mm	Current*/Freq. Surf. Power/Freq.	Heating time, s	Aus. depth**, mm	Quench dwell, s	Variable parameter
1	10	1	2360A/40kHz	1,9	2,0	0,5	Reference
2	20	2	1570A/40kHz	2,5	2,0	0,5	External radius
3	10	1	2360A/40kHz	1,9	2,0	0,5	Quench. Intensity
4	10	1	2360A/40kHz	1,9	2,0	0,5	Initial Material Strength
5	10	1	2360A/40kHz	1,9	2,0	0,5	Austenite Strength
6	20	2	1530 $W{\cdot}cm^{-2}$/40kHz	3,5	1,1	0,5	Convex/Concave Geometry
7	40	2	2050A/15kHz	2,5	2,0	0,5	External radius
8	40	2	648A/300 kHz	8,5	2,0	0,5	Surface power density-frequency-time-combination

*Current in a coil of 10x10mm Cu-conductor with core. Air gap 2mm
**Austenitization temperature 840 °C

2. Analysis Case

The application of the additive decomposition of the strain to analyze a real problem can be demonstrated with simulation No. 1.

The development of strain components at the external surface is shown in Fig. 1. The initial source of deformation is the thermal strain, which causes the compensating elastic and plastic strains because of the core's constraints (total strain is limited). During the austenitization, the phase volumetric change strain decreases because of the less specific volume of austenite.

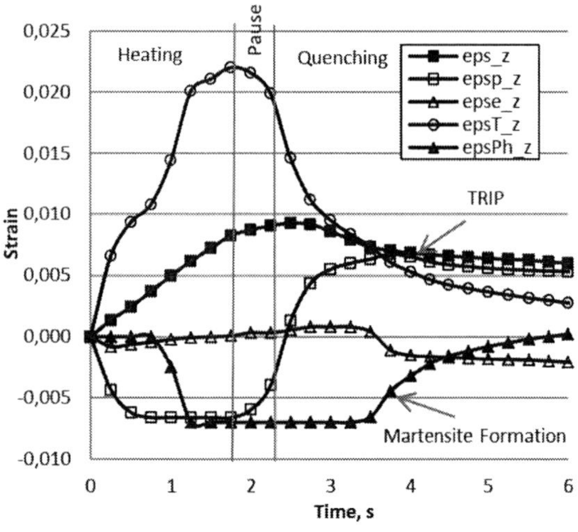

Fig. 1: Development of the strain components at the surface of a part (depth 0 mm) for simulation No. 1

After heating, the temperature at the surface starts to drop. Total strain is inert because of the adjusted layer constraints and has no possibility to react immediately. The surface is in tension and the inversion of the elastic strain with the continuation in plastic strain follows [8]. The possibility of the plastic elongation is conditioned by the properties of the soft austenite.

Martensite formation causes an increase of the phase mixture's specific volume. This increase is the source strain that is compensated by elastic compression. Because the hardening zone suffers the compression on this stage, TRIP slightly decreases the level of accumulated plastic elongation. The above explanation can be applied to an arbitrary point of the investigated component.

3. Quenching Intensity

In a real process, the quenching intensity can be altered in a number of the different ways, namely by use of the different quenching media, spray pressure, etc. Quantitatively, this parameter is estimated with a heat transfer coefficient (HTC) that is significantly dependent on the temperature at the cooling surface. The aim of the current investigation is to find the direction in which quenching intensity influences the results, rather than the exact residual stress distribution. Therefore, constant values of HTC have been used for an additional simplification.

Two values of HTC are considered: medium quenching intensity (MQI) for simulation No. 1 is equal to 10 kW/m^2K and low quenching intensity (LQI) for simulation No. 3 is equal to 5 kW/m^2K. The distribution of the residual stress as a function of radius is shown in Fig. 2. Here and further prefix '*sig_*' is used for the designation of stress.

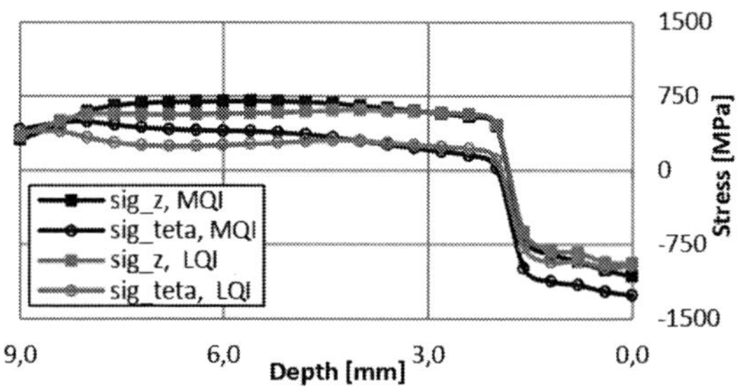

Fig. 2: Residual stress for the different quenching intensities

The axial compressive stress at the surface of the hardening zone for MQI simulation is about 1070 MPa, whereas for LQI it is about 940 MPa. Tangential compressive stress is also improved for the MQI simulation. Similar results have also been obtained by other researchers [12].

The total axial strain in an infinite long cylinder is uniformly distributed. Therefore, it is more convenient to build the explanation for this direction.

The development of the axial strain components at the hardened surface for both simulations is shown in Fig. 3. At the first stage, the more intensive quenching causes a drastic decrease in the thermal strain at the surface, whereas the total strain, dependent on the average temperature of the whole cylinder, has no possibility to change so fast. As a result, plastic elongation is prolonged for MQI (see plastic strain, 4s). Increased resulting plastic elongation raises the residual compressive stress.

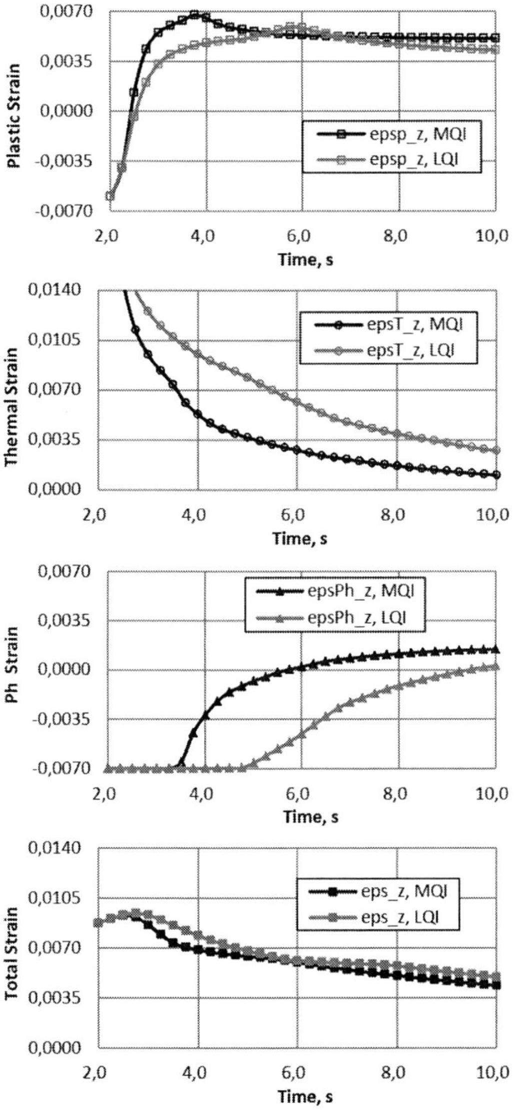

Fig. 3: Development of axial strain at the surface of the part during quenching for the different quenching intensities

For the considered case, the increased quenching intensity plays a beneficial role. High compressive stress at the surface of a component increases the fatigue strength if crack initiation site is located at the surface.

4. Initial Structure's Strength

In the previous case, improved compressive residual stress is achieved by an increase of the plastic elongation in the hardening zone. The basic increase is attained in the time period between the end of heating and the beginning of martensite formation (plastic elongation stage). To get the same effect, strength of the initial structure can be increased. In this case, the resulting compressive plastic strain in the hardening zone after the heating stage would be reduced due to the increased elastic compensation of thermal expansion. After plastic elongation stage, the residual plastic strain level would probably be higher because of that.

The strength of the initial structure can be improved by a pre-treatment such as Q&T. In the current investigation two cases are analysed. The first one corresponds to normalized structure of Ck45 steel (simulation No. 1). The material properties for the second case have been obtained by multiplying the temperature dependent yield strength of the normalized structure by 2 (simulation No. 4). Abbreviations LSIS (Low Strength Initial Structure) and HSIS (High Strength Initial Structure) are used. Temperature dependencies of the yield limit for both simulations are shown in Table 2.

Table 2: Yield limit of initial structure

Temperature, ℃	20	400	800
LSIS Yield limit, MPa	335	55	10
HSIS Yield limit, MPa	670	110	20

The comparison of the residual stress for these cases is shown in Fig. 4. Indeed, a small improvement of the compressive stress in the hardening zone can be observed for the HSIS-case. Another important factor for the HSIS case is an increase of the detrimental tensile stress at the core's border. Such behaviour has been already revealed by the experimental investigations [7].

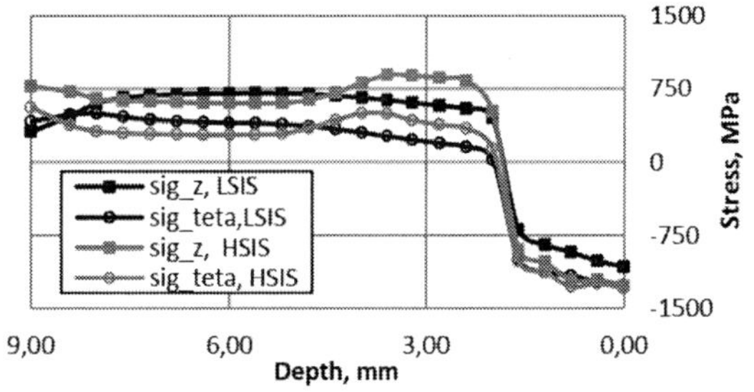

Fig. 4: Distribution of residual stress for the different strengths of the initial structure

Thermal and phase transformation conditions are equal for both simulations. Therefore, the examination of the development of total and plastic strains is enough for explanation (Fig. 5).

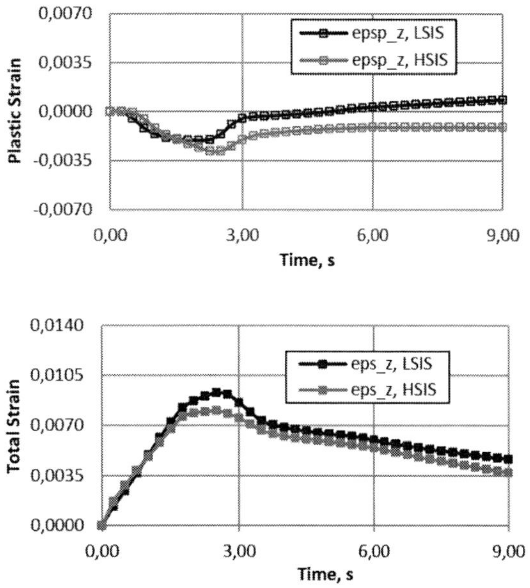

Fig. 5: The development of the plastic and total strains at the border of a core (depth 2,6 mm) for the different quenching intensities

The comparison of total strain developments during heating shows greater constraints for the surface in case of HSIS. Since the thermal and phase transformation conditions are the same, the elastic and plastic strains are responsible for this behaviour of the total strain. During heating, compressive stress in the heating zone is compensated by tensile stress in the central part of the core. HSIS allows higher stress (elastic strain). In the case of LSIS, the elastic strain is rapidly yielded and continued in a plastic elongation of the core.

The described process is shown in Fig. 6 with the distribution of the plastic strain just before quenching. The core's elongation weakens the constraint for the hardening zone. As can be seen in Fig. 6, the plastic compression at the surface is lower for LSIS before quenching.

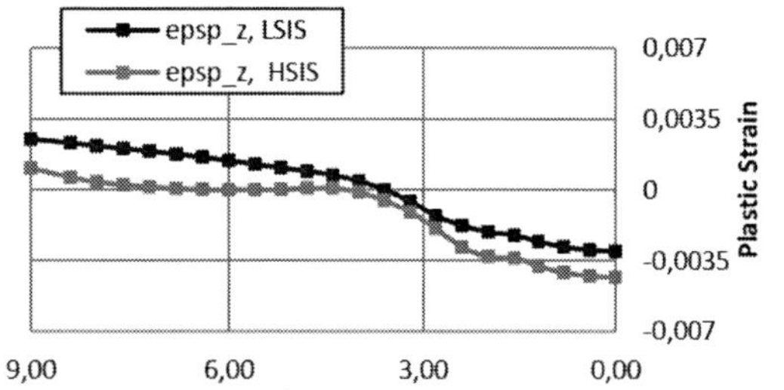

Fig. 6: Axial plastic strain distribution just before quenching (time 2,4s)

After the quenching has begun, the core's border, as well as the hardening zone, is subjected to thermal contraction that requires compensation according to Eq. 1.

For both the HSIS and LSIS cases, such compensation is again a reaction of the plastic and elastic strains, but because of a high yield limit of the core, the plastic elongation for HSIS is less compared with LSIS. At the same time the hardening zone has the same austenite properties. As a result, the residual stress in the hardening zone is improved, but the core's border suffers the high detrimental tensile stress.

5. Austenite Strength

Strength of austenite depends on the different factors such as a temperature, austenite's homogeneity and grain size, a presence of the undissolved carbides in the austenite's matrix, etc. Since the yield limit of a material influences the relation between the elastic and plastic strains, it is hypothetically possible to achieve a greater plastic elongation magnitude for the cases of weak austenite. To verify this assumption, the austenite properties of Ck45 steel are modified by a simple multiplication the temperature dependent yield strength by 2. The abbreviation LSA (Low Strength Austenite) is given to the reference simulation No. 1, and HSA (High Strength Austenite) for simulation No. 5 with modified austenite properties. The temperature dependencies of the yield limit of austenite for both experiments are shown in Table 3.

Table 3: Yield limit of austenite

Temperature, ℃	20	400	900
LSA Yield limit, MPa	243	39	7
HSA Yield limit, MPa	486	78	14

Distribution of the residual stress is quite unexpected. With a sufficient accuracy, residual stresses for LSA and HSA are equal (Fig. 7). The same result has been achieved for other geometries and austenite properties.

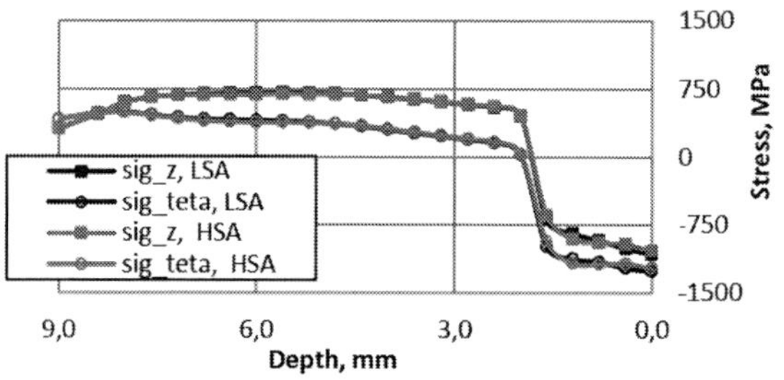

Fig. 7: Residual stress distribution for the different austenite's strength

To understand the cause of such stress' behaviour, we can investigate the strain development curves in depth 1 mm (Fig. 8). The stress evolution can be evaluated with elastic strain, whereas change volumetric strain can be used for identification of certain transformation moments.

Difference in stress for hot austenite is negligible because of the small level of yield limit (see Table 3).

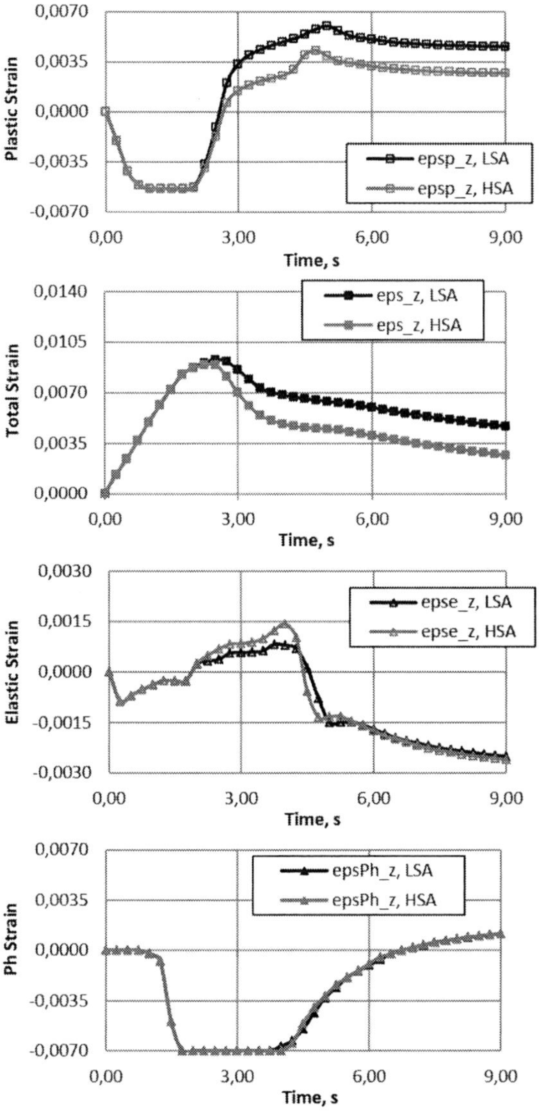

Fig. 8: Development of axial strains in hardening zone (depth 1mm) for the different austenite's strength

The temperature drops after 1,9s of heating and this causes the plastic elongation for both cases according to explanation in section 3. Cooling of the austenite zone leads to an increase of the yield limit difference and, therefore, the difference in plastic and elastic strain relations. As a result, stress is higher and plastic elongation is lower for HSA case. The stress difference is at a maximum just before martensite formation. This fact is in good agreement with the initial considerations of this section. But immediately after the martensite formation has started, the plastic strain in HSA case starts to grow significantly. The difference in elastic strain then quickly becomes negligible.

At first glance, such behaviour of plastic strain in HSA case is strange. Indeed, the austenite matrix starts to strengthen by martensite and plastic flow gets into difficulties. Another aspect of plastic flow in steels must be responsible for this, namely the TRIP phenomenon. Relatively high tensile stress during the early martensite formation in HSA case introduces the additional plastic strain increment into total elongation when the plastic flow in a common sense is impossible.

As it was already stated in section 1a, the difference between the total and plastic strains is crucial for residual stress in condition of equal thermal and phase volumetric change strains. The TRIP effect makes these differences equal in the two cases.

It can be concluded that the austenite yield limit does not influence the residual stress itself. Distortion related problem can be observed, namely the difference in total strain.

6. Part Outside Diameter

The geometrical factor has a great influence on the development of the stress and strain in a component. Three different cases have been investigated.

Simulation No. 1 represents a small outside diameter (SOD) and simulations No. 2 and No. 7 are for the medium (MOD) and large (LOD) outside diameters respectively.

The result of simulations is presented in Fig. 9. It can be concluded that with a reduced outside diameter, compressive stress in a hardening zone is improved. The explanation can be built on the comparison of SOD and LOD simulations (Fig. 10).

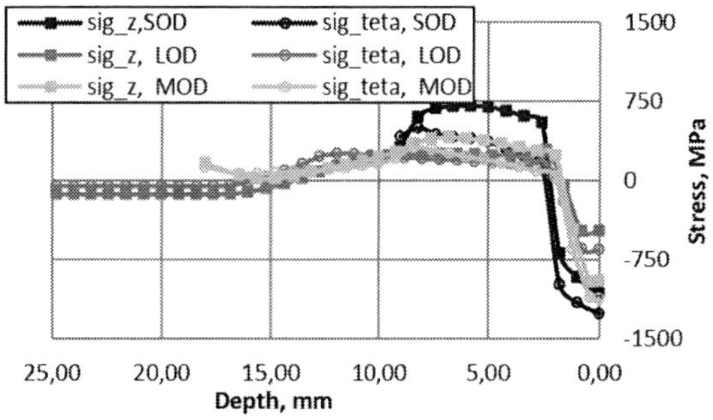

Fig. 9: Residual stress for the different outside diameters of a part

According to the total strain curve, the hardening zone during heating for SOD simulation is weak-constrained as opposed to LOD. As a result, the plastic contraction during heating is less and the residual plastic elongation is greater for SOD.

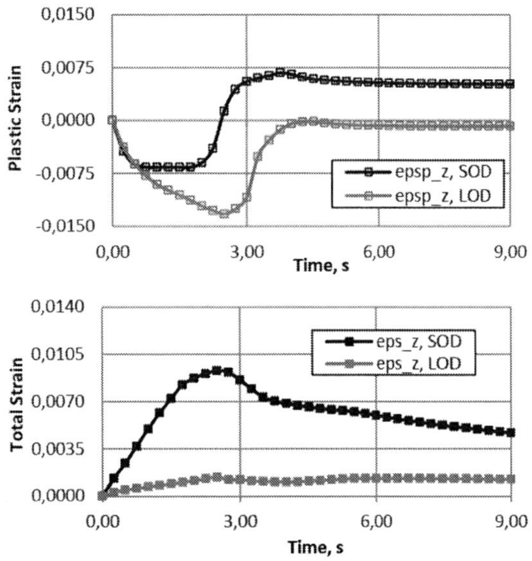

Fig. 10: The development of axial strains at the surface of a part for the different part outside diameters

The cause of the decreasing of constraints for small diameters is smaller differences in the temperature inside a part due to heat conduction. The whole core is heated and expanded more due to its higher average temperature, thereby making it possible for the hardening zone to expand more easily.

7. Convex vs. Concave Geometries

To compare the hardening of the convex and concave geometries, we can compare simulation No. 2 (External Hardening – EH) and simulation No. 6, in which the internal hardening (IH) is performed. Distributions of the residual plastic strain and stress are shown in Fig. 11.

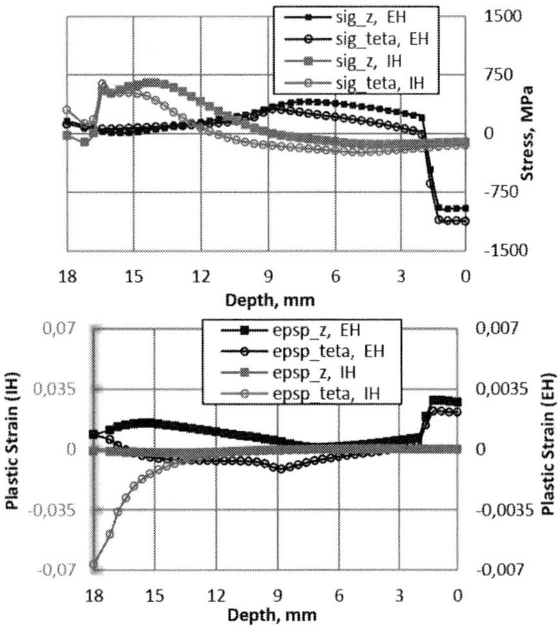

Fig. 11: Residual stress and plastic strain for external and internal hardening of hollow cylindrical part

The most important result is the tangential tensile residual stress for IH case. The magnitude of the tangential residual plastic strain differs by 10 times and is in contraction. Such "tangential anomaly" occurs when the heated zone is stiffly constrained by a relatively cold core (large OD) and the expansion has to go into the bore when the radius decreases. This process can be observed in Fig. 12.

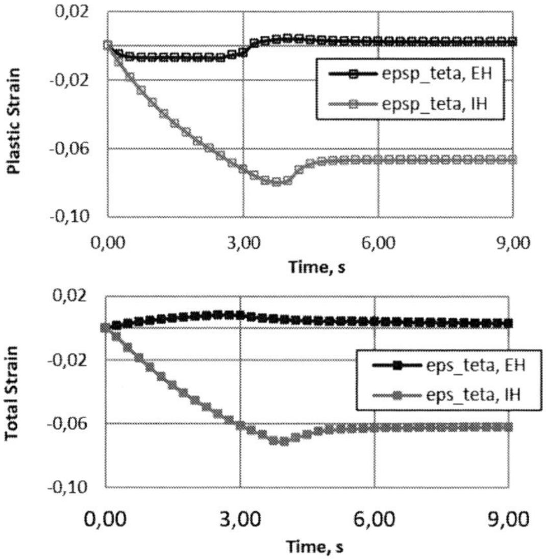

Fig. 12: The development of tangential strains at the surface of a part for external and internal hardening of hollow cylindrical part

Total strain at the surface for IH case is negative despite the thermal expansion on the heating stage. Such "over-constraint" causes the great plastic compression.

The effect of "anomaly" is magnified by small ID. It can be seen that this effect could be very harmful for the residual stress.

The big difference in axial component of plastic strain can be explained by the heat conduction in a cylindrical system. Heating the core through a small internal diameter is more difficult compared with the external heating. Therefore, the surface power density for simulation No. 6 is increased significantly. The constraints of the core in these conditions are stiffer since its average temperature is low. Hence, a greater plastic contraction during the heating stage develops for the IH case and the residual plastic elongation has a smaller magnitude.

8. Induction Heating Parameters

The distinctive characteristic of induction heating is the smooth variation of the induced current density versus depth. A penetration depth value depends on the frequency as well as the material properties [16]:

$$\delta_c = \sqrt{\frac{\rho}{\pi \cdot f \cdot \mu}}, \qquad (5)$$

where δ_c is the current penetration depth, ρ electrical resistivity, f frequency, μ magnetic permeability.

If the total power into the part is constant, the volume power density in the superficial layers will increase with frequency. To avoid overheating the surface and get the same hardening depth, the power should be reduced and the heating time increased with higher frequency. Beside this, the thermal properties of steel should be taken into account to ensure the equal austenitization conditions for the different heating frequencies [16].

In current work two different combinations of surface power density, frequency and time are used for comparison. First combination is based on the 15 kHz frequency and a heating time of 2,5s (simulation No. 7). The second one is adjusted to get the same austenitization conditions with high frequency 300 kHz and a heating time of 8,5s (simulation No. 8). The abbreviations MFH (Medium Frequency Heating) and HFH (High Frequency Heating) are used respectively.

The residual distributions of stress and plastic strain are shown in Fig. 13.

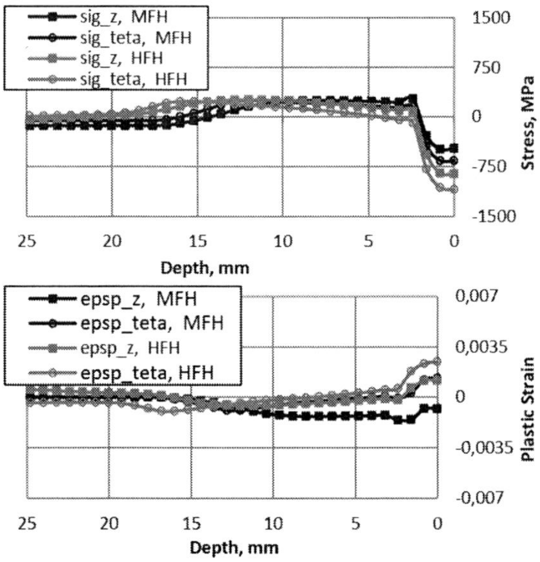

Fig. 13: Residual stress and plastic strain for the two surface power density-frequency-time combinations

The compressive stress is increased significantly for HFH case. As it can be seen in the plastic strain distribution, the cause of such stress improvement is the prolonged plastic elongation both for axial and tangential spatial components.

The development of axial plastic and total strain at the hardened surface is built to understand the nature of increased magnitude of plastic elongation (Fig. 14). It can be clearly seen, the total strain in HFH case is increased. It means that the hardening zone for this case is less constrained by core, and because of that plastic compression on the heating stage is deceased.

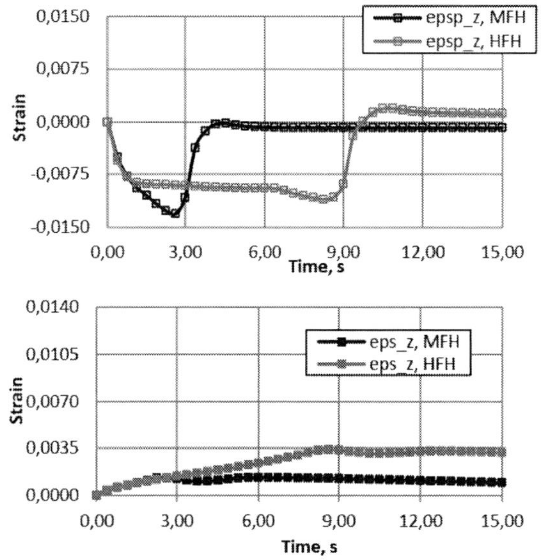

Fig. 14: The development of axial strains at the surface of a part for the different surface power density-frequency-time combinations

The resulting plastic elongation is higher for HFH case because of the expanded core due to the longer heating time. The cause of such expansion is higher average temperature of the core. The distribution of thermal strain just before quenching for both simulations is shown in Fig. 15. It can be clearly seen that thermal expansion of the hardening zones are almost equal, whereas the difference in core average temperature is significant. Heating time for HFH case is much longer and the temperature is more uniformly distributed along the radius because of the heat conduction.

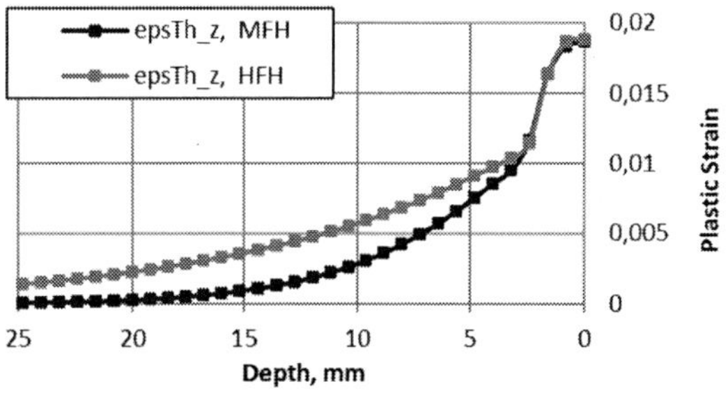

Fig. 15: Axial thermal strain distribution just before quenching for the different surface power density-frequency-time combinations (time 3s for MFH, 9s for HFH)

The described principle of compressive residual stress increase is related to one explained in section 6 (Part Outside Diameter). In both cases, the responsible for increased plastic elongation cause is the higher average core temperature.

Conclusions

The additive decomposition of strain is a convenient instrument for analysis of different stress and strain effects in an induction hardened steel part. Examination of the development of the strain component allows taking into account a wide range of cause-and-effect relations. The proposed explanations are based on calculation results from approved and verified mathematical models.

Simulations have shown the important role that the plastic and total strains have on residual stress formation. Understanding the behaviour of strain's development allows changes to the induction surface hardening parameters in order to get the best result.

Benefits of high average temperature of the core are demonstrated. Intensive quenching allows keeping the core's temperature at a relatively high level until the end of martensite transformation in the hardening zone. At the same time, a disadvantage of intensive quenching can be crack initiation during a quenching process [17]. Therefore, the use of high quenching intensity should be handled with care. A smaller outer diameter gives increased average core temperature due to heat conduction in the part. The same nature of the compressive residual stress improvement can be observed in case of longer heating with use of higher frequencies.

The influence of initial material strength is more complicated. A high level of tensile stress can be achieved at the core's border for the strengthened initial structure.

Changes in austenite's yield limit have negligible effect on the distribution of residual stress. Simulations reveal that this is due to the TRIP effect. Since the proper incorporation of TRIP into calculations is an open problem, it is difficult to estimate the relevance of this result to real cases. Despite the lack of difference in residual stress, the residual difference in total strain can be observed for the distinct austenite yield limit.

A big difference between the hardening of convex and concave geometries is demonstrated. In case of internal hardening (concave geometry), the "tangential anomaly" is of great interest because of the detrimental effect it has on the residual stress.

References

[1] Lu, J. Prestress engineering of Structural Material: A Global Design Approach to the Residual Stress Problem. book auth.] Howes M., Inoue T. Totten G. Handbook of Residual Stress and Deformation of Steel. Ohio: ASM International, 2001, pp. 11-26.

[2] Löhe, D., Lang, K.-H. and Vöhringer, O. Residual Stress and Fatigue Behavior. [book auth.] G. Totten, M. Howes and T. Inoue. Handbook of Residual Stress and Deformation of Steel. Ohio: ASM International, 2001, pp. 27-53.

[3] Webster, G. A. and Ezeilo, A. N., "Residual stress distributions and their influence on fatigue lifetimes", International Journal of Fatigue, No. 23 (2001), pp. S375-S383.

[4] Coupard, Dominique, et al., "Residual stresses in surface induction hardening of steels: Comparison between experiment and simulation", Materials Science and Engineering Vol. A, No. 487 (2008), pp. 328-339.

[5] Li, Z., Freborg, A. and Ferguson, B. L. "Applications of Modeling to Heat Treat Processes", Heat Treating Progress, May/June (2008), pp. 28-33.

[6] Grum, J. Induction Hardening. [book auth.] G. Totten, M. Howes and T. Inoue. Handbook of Residual Stress and Deformation of Steel. Ohio: ASM International, 2001, pp. 220-247.

[7] Kristoffersen, H. and Vomacka, P., "Influence of process parameters for induction hardening on residual stress", Materials and Design, No. 22 (2001), pp. 637-644.

[8] Markegård, L. and Kristoffersen, H., "Residual Stress After Surface Hardening - Explanation of How Residual Stress is Created", Dubrovnik 2009, pp. 359-366.

[9] Hossain, S., et al., "Application of quenching to create highly triaxial residual stress in type 316H stainless steel", International Journal of Mechanical Sciences, No. 48 (2006), pp. 235-243.

[10] Hoemberg, D., "A mathematical model for induction hardening including mechanical effects", Nonlinear Analysis: Real World Applications, No. 5 (2004), pp. 55-90.

[11] Şimşir, C., Gür, C. H., "A Mathematical Framework for Simulation of Thermal Processing of Materials: Application to Steel Quenching", Turkish J. Eng. Env. Sci., No. 32 (2008), pp. 85-100.

[12] Kobasko, N., "Optimal Quenched Layer at The Surface of Steel Parts", Dubrovnik 2001, pp. 25-33.

[13] Alberg, H. and Berglund, D., "Comparison of plastic, viscoplastic, and creep models when modeling welding and stress relief heat treatment", Comput. Methods Appl. Mech. Engrg., No. 192 (2003), pp. 5189–5208.

[14] Geijselaers, H.J.M. Numerical Simulation of Stresses due to Solid State Transformations. The Simulation of Laser Hardening. Hellendoorn: Ponsen & Looijen, 2003. p. 142. Thesis University of Twente, Enschede - with ref. with summary in Dutch.

[15] Orlich, J., Nose, A. and West, P. Atlas zur Warmebehandlung der Stahle. Dusseldorf: M.B.H., 1973.

[16] Markegård, L. and Asperheim, J. I. "New Nomographs for Induction Surface Hardening of Steel", Dresden 2003.

[17] Kobasko, N.I., Morganyuk, V.S. and Dobrivecher, V.V. Control of Residual Stress Formation and Steel Deformation during Rapid Heaing and Cooling. [book auth.] G. Totten, M. Howes and T. Inoue. Handbook of Residual Stress and Deformation of Steel. Ohio: ASM International, 2001, pp. 312-330.

Proceedings of the 26th ASM Heat Treating Society Conference
B.L. Ferguson, R. Jones, D.S. MacKenzie, and D. Weires, editors
https://doi.org/10.31399/asm.cp.ht2011p0253

Distortion control of transmission components by optimized High Pressure Gas Quenching

Dr. Volker Heuer, ALD, *Hanau, Germany (Tel.: +49 06181 307 3372; dr.volker.heuer@ald-vt.de)*
Donald R. Faron, General Motors, *Pontiac, MI (donald.r.faron@gm.com)*
David Bolton, ALD Thermal treatment, *Port Huron, MI (Tel.: +1 7652090870)*
Mike Lifshits, ALD Thermal treatment, *Port Huron, MI (Tel.: +1 8103570690)*
Dr. Klaus Loeser, ALD, *Hanau, Germany (Tel.: +49 06181 307 3366; dr.klaus.loeser@ald-vt.de)*

Abstract

The paper presents how "Dynamic Quenching" and "Reversing gas flow" - processes are successfully applied on Internal ring gears and planetary gears for a 6 speed automatic transmission. The specific challenge was to reduce distortion in such a way that subsequent machining operations are entirely eliminated. As a result of extensive development in the quenching process it was possible to meet the design metrological requirements.

The internal ring gears have been in continuous production since 2006. By utilizing the special CFC-fixtures and quench methodology of "Dynamic Quenching," the customer was able to achieve the design intent, while eliminating all machining operations of the ring gears following LPC/HPGQ.

Subsequent testing and monitoring over a two year period progressively demonstrated that conformance and quality inspection was reduced accordingly.

Introduction

Proper distortion control has become even more important than in previous days. To answer the demand for fuel-efficient vehicles, modern transmissions are built much lighter. Therefore the components of the transmission exhibit less wall-thickness which makes them more sensitive to distortion. Distorted gear components cause noise in the transmission, can require post heat treat machining processes and may even create problems during transmission-assembly. Therefore distortion control has become more important than ever.

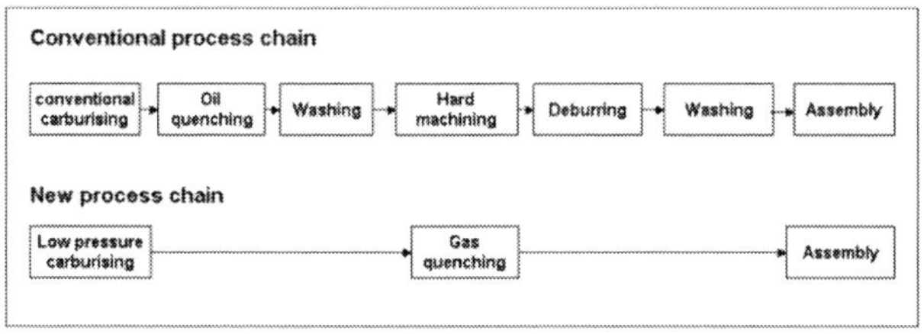

Fig.1 : Conventional and new process chain for the manufacturing of gear components

By applying the technology of Low Pressure Carburizing (LPC) and High Pressure Gas Quenching (HPGQ) heat treat distortion can be significantly reduced. LPC is a case hardening process which is performed in a pressure of only a few millibar using acetylene as the carbon source in most cases. During HPGQ the load is quenched using an inert gas-stream instead of a liquid quenching media. Usually nitrogen or helium are used as quench gas.

With an optimized distortion control it is possible to simplify the process chain significantly. Fig. 1 shows how the process chain can be simplified, if the specified geometrical values of the components can be guaranteed after gas quenching. If the simplified process chain can be applied, then this will result in lower costs per part, lower throughput times and lower energy consumption during production. Since there is no need to dispose any oil after the quench and since cleaning operations after the quench are unneeded, the simplified process chain is much more environmentally friendly as well.

Distortion Mechanisms and High Pressure Gas Quenching (HPGQ)

The relevant mechanisms that cause distortion of components during heat treatment have been described extensively in literature [Heeß 2007]. Three different types of stress in the material contribute to distortion: Residual stresses [Heeß, 2007], Thermal stresses and Transformation stresses. They are influenced by part-geometry, steel-grade, casting, forging, machining etc. and they depend on the heat treatment. If the total stress in the component exceeds the yield stress, then distortion of the component takes place. The numerous potential factors that are influencing distortion have been published by Walton [Walton, 1992].

The technology of High pressure gas quenching (HPGQ) offers a tremendous potential to reduce heat treat distortion. Conventional quenching-technologies such as oil- or polymer-quenching exhibit inhomogeneous cooling conditions. Three different mechanisms occur during conventional liquid quenching: film-boiling, bubble-boiling and convection. Resulting from these three mechanisms the distribution of the local heat transfer coefficients on the surface of the component are very inhomogeneous. These inhomogeneous cooling conditions cause tremendous thermal and transformation stresses in the component and subsequently distortion. During HPGQ only convection takes place which results in much more homogenous cooling-conditions [Stich, 1995]. Significant reductions of distortion by substituting Oil-quench with HPGQ have been published [Altena et al, 2005]. Another advantage of HPGQ is the possibility to adjust the quench-intensity exactly to the needed severity by choosing quench-pressure and quench-velocity. Typical quench pressures range form 2 bar to 20 bar. The gas-velocity is controlled by a frequency converter and typical gas-velocities range from 2 m/s to 15 m/s depending on the part-geometry and the steel-grade of the component. Typical gases applied for HPGQ are nitrogen and helium [Heuer et al, 2009].

Furnace Equipment and Fixturing for Distortion Control

The design of the gas quenching chamber is of key importance to minimize distortion. The chamber needs to provide a high gas-velocity to ensure that the core-hardness specification is met and the chamber needs to provide a very uniform distribution of the gas-velocity to minimize the spread of distortion within the load. Intensive numerical flow calculation (CFD-studies) and experimental studies led to the design of the quenching chamber of the ModulTherm® - system [Loeser et al, 2005], see Fig. 2.

Fig. 2: ModulTherm® heat treat system with gas quenching chamber

As in the case of liquid quenching, proper fixtures and optimized loading of the parts is important for gas quenching, too. As an alternative to alloy fixtures, carbon composite materials, e. g. CFC, were introduced for the use as fixtures in heat treating applications. Fig. 3 shows an example of a load on CFC fixturing. The major advantage is that CFC-fixtures do not show deformation during the heat treatment process thereby assuring optimum positioning of the parts. This has an significant, positive effect on distortion control.

Fig. 3: Load of Internal ring gears on CFC fixturing

Reversing gas flow

Modern gas quenching chambers offer the possibility to reverse the direction of the gas-flow during quenching. This "Reversing gas flow" means that the flow of gas is alternated back and forth from "top-bottom" to "bottom-top". By alternating the gas-flow direction, there is less difference in the cooling curves of parts placed in different layers. This reduces the spread of distortion inside the load.

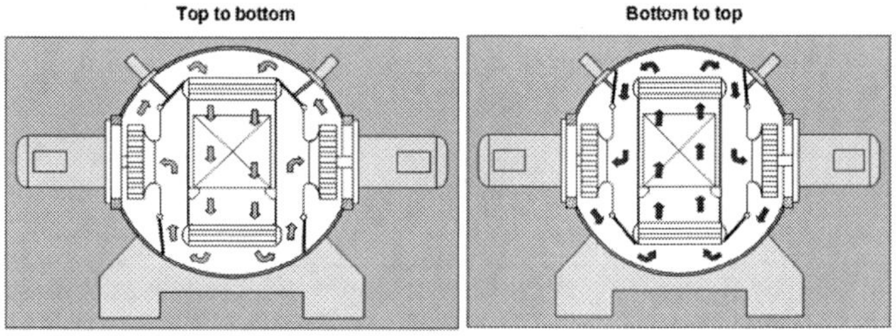

Fig. 4: "Reversing gas flow" as applied in the ModulTherm® – quenching chamber

Dynamic Quenching

To achieve optimum quenching results with respect to microstructure, hardness, and distortion, the gas quenching parameters need to be well adjusted. And to further reduce distortion, a quenching process has been developed where the quenching parameters gas-pressure and/or gas-velocity are stepwise varied during quenching, see Fig. 5. This process, so-called "Dynamic Quenching" is typically divided into three steps [Heuer, 2003]:

step 1: high quenching severity until a certain part-temperature is reached

step 2: quenching severity is reduced for a set time to allow for temperature equalization in the part.

step 3: quenching severity is increased again until the end of the quenching process.

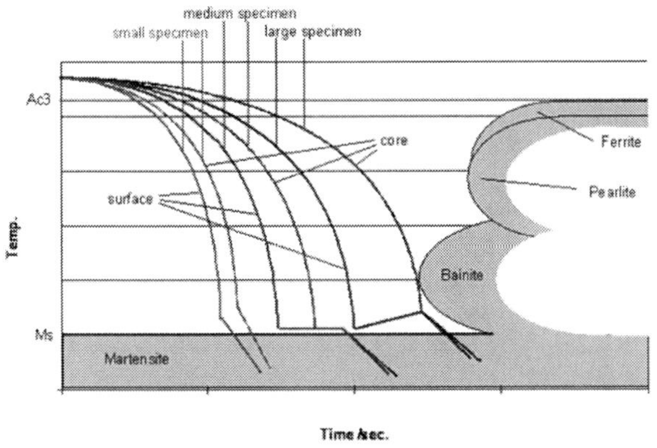

Fig. 5: Schematic illustration of dynamic quenching for specimen of different sizes

The control system of the quenching chamber allows to control the different quenching steps of "Dynamic Quenching" in a very accurate way with high reproducibility. Optimum results are achieved when using helium. The light quenching gas helium can be decelerated and accelerated very precisely for optimum distortion control.

Application of "Dynamic Quenching" in serial production

The "Dynamic Quenching" process was applied on the Reaction internal gears of a 6 speed automatic transmission. An intensive process optimization programme was started before the start of production. The specific challenge was to optimize distortion control of the gears to such extent that hard machining is completely eliminated and the process chain is simplified as postulated in Fig. 1.

The furnace supplier and transmission manufacturer worked in close cooperation on the optimization programme and successfully implemented a Dynamic Quench practice for the start of production in 6/2006. This low distortion process consists of LPC using acetylene as carburizing source and HPGQ using helium as quench medium with an optimized Dynamic quenching process. The application of this optimized process and the use of CFC-fixtures made subsequent machining-operations after heat treatment unnecessary as postulated in Fig.1.
The Reaction Internal gears have an outer diameter of 152 mm, 103 internal teeth and are made of 5130 material. A picture of the part is given in Fig. 6. The case hardening depth CHD after heat treat is specified as 0,3…0,6 mm and surface hardness is specified as 79…83 HRA. The geometry after heat treat is specified with a maximum circularity of 150 µm.

Fig. 6: Reaction Internal gear (d = 152 mm, 103 internal teeth)

The results of a detailed distortion-study from serial production are presented in the following. 48 pre-measured parts were equally distributed into different layers of a standard production load and were geometrically inspected after heat treat. One production load consists of 120 pieces treated in 10 layers, see Fig. 3. To cover all "extreme" positions in the load, it was made sure that parts from all 8 corners and parts from the middle of the load were geometrically inspected. All measurements were performed with a CNC analytical gear-checker.

The circularity of the 48 measured pieces before and after heat treatment can be found in Fig. 7. The helix angle variation (Vbf) for each part is shown in Fig. 8. As shown in Fig. 7 and 8, the achieved level of distortion is very small. The maximum change in circularity amounts to 41 µm and the average change in circularity is only 7 µm. All circularity- values after heat treat are well below the specified maximum of 150 µm. The helix angle variation after LPC and HPGQ is below the specified maximum as well. The average change during heat treat in helix angle variation is 10 µm for the left flank and 11 µm for the right flank. Measurements of two associated internal gears in the six speed transmission, the Output- and Input Internal gears, showed similar results.

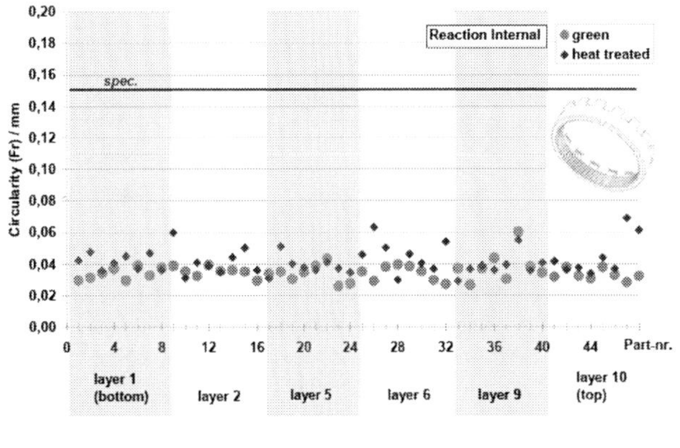

Fig. 7: Circularity of Reaction Internal gears before and after heat treat (LPC and HPGQ with Dynamic Quenching); spec. maximum after heat treat: 150 µm

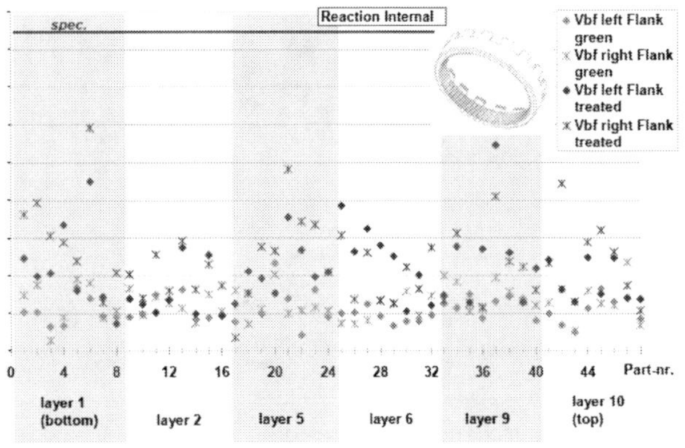

Fig. 8: Helix angle variation of Reaction Internal gears before and after heat treat (LPC and HPGQ with Dynamic Quenching)

Serial production of the 6 speed transmission started in 6/2006. Initially all Internal ring gears from each load were checked for excessive distortion with a so called "Roll checker". A "Roll Checker" is an automated measurement system utilizing a rolling master held by a pivoting yoke. The yoke enables the roll master spindle to move in the lead and taper directions as the rolling master rotates in tight mesh with the test component. Separate transducers located in the gimble head monitor lead and taper travel. As distortion proved to be very consistent, the transmission manufacturer decided to abandon the one hundred percent inspection of all parts. Since 10/2008 only two gears per load are being measured with a CNC analytical gear checker. One part from the top corner (TC) and one part from the bottom middle (BM) are being inspected in each load. These two positions were chosen to cover the "extreme" positions in the load.

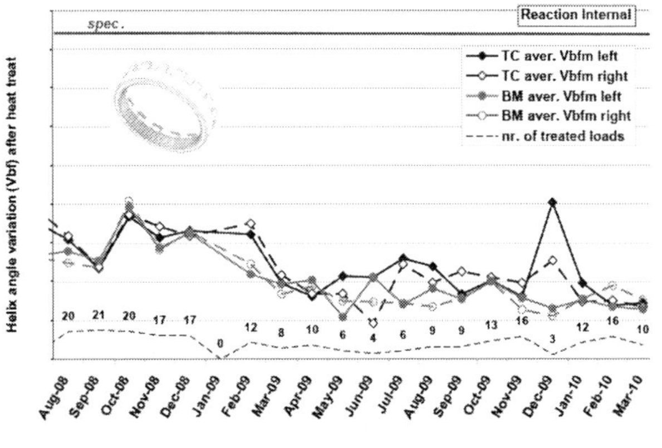

Fig. 9: Reaction Internal gears: helix angle variation after heat treat averaged per month of production (TC=sample from top corner of the load; BM=sample from bottom middle of the load; the number of inspected loads is indicated per each month)

Results from distortion monitoring since 8/2008 until 3/2010 are presented in Fig. 9. The diagram shows the helix angle variation of Reaction Internal gears after heat treat (Vbf) averaged per month of production. Additionally the number of inspected loads per month is indicated in the diagram. The total number of inspected loads from 8/08 until 3/10 was 229 loads. Fig. 9 demonstrates that the distortion-values in serial production are very stable. This was achieved with the help of the optimized LPC and HPGQ process in combination with the stable manufacturing process chain of the components.

Application of "Reversing gas flow" in serial production

The "Reversing gas flow" process was applied on the FD-Pinion planetary gears of another 6 speed automatic transmission. The gears are made of 5120 material, have an outer diameter of 31mm, a height of 32mm and 24 external teeth. One load consists of 1056 pieces treated in 9 layers, see Fig. 10.

Fig. 10: FD Pinion planetary gear (d=31mm, 24 teeth) and production load (9 layers with 1056 pcs total)

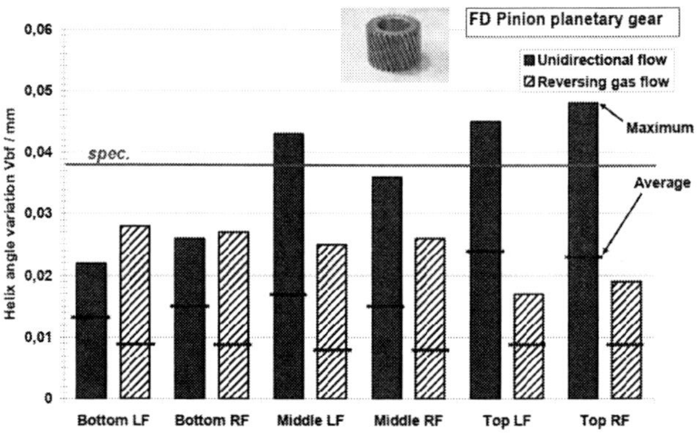

Fig. 11: Reduction of distortion by application of "Reversing gas flow": comparison between "unidirectional" and "Reversing gas flow" (helix angle variation of FD Pinion gears after heat treat in the bottom, middle and top layer; LF = left flank; RF = right flank; spec. maximum after heat treat: 38 μm)

Fig. 11 shows the improvement that was achieved when introducing the "Reversing gas flow" process. When applying "unidirectional gas flow", the gas is flowing only from top to bottom through the load. With "Reversing gas flow" the flow of gas is alternated back and forth from "top-bottom" to "bottom-top", as illustrated in Fig. 4. As shown in Fig. 11 with "unidirectional flow", the parts in the middle and top layer of the load exhibit excessive distortion. With "Reversing gas

flow" the helix angle variations were significantly reduced. E.g. for the right flank of the gears from the top layer, the maximum helix angle variation was reduced by 61%.
With the optimized "Reversing gas flow" process it is not necessary to machine the teeth after heat treatment. Only the bores and faces of the gear are machined after heat treat.

Necessary steps for low distortion heat treatment

The necessary steps that need to be accomplished to provide low distortion values are given in Fig. 12.
The design of the gas quenching cell is of key importance to minimize distortion. The chamber needs to provide a very uniform distribution of the gas-velocity to minimize the spread of distortion within the load. Another important factor is proper fixturing. Modern CFC-materials (carbon reinforced carbon) are well suited as fixture-material for gas quenching.

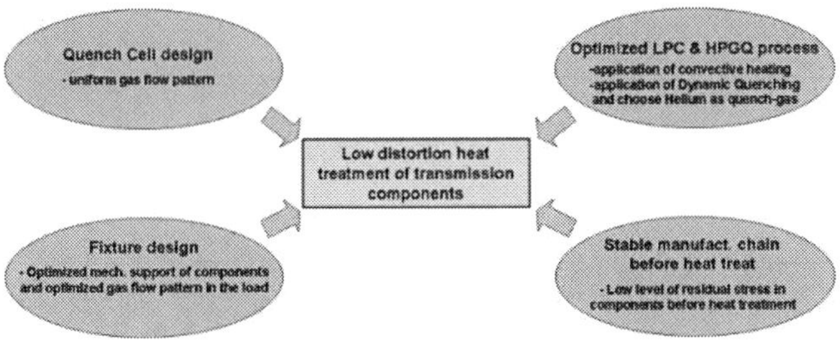

Fig. 12: Necessary steps for low distortion heat treatment with LPC and HPGQ

Further improvements can be achieved using "convective heating" for homogenous temperature distribution followed by carburizing and the application of "Dynamic Quenching"-processes. However, if the components contain high residual stress from prior manufacturing processes, it is impossible to achieve low distortion-values during heat treat. Therefore an optimized and stable manufacturing process chain including melting, casting, cutting, soft machining etc. is mandatory to create low levels of residual stress in the components before heat treating.

Conclusions

LPC and HPGQ processes were successfully applied on gear-components of 6 speed automatic transmissions. Helium was chosen as quench-gas and optimized "Dynamic quenching" and "Reversing gas flow" technologies were applied for distortion control.
For relatively large diameter/thin wall Internal Ring gears, "Dynamic Quenching" is used in serial production. A distortion study was conducted to analyze the geometrical stability of Reaction Internal gears during heat treat. The maximum change of circularity during heat treat was 41 μm and the average change of circularity was as low as 7μm. All values after heat treat are well below the specified maximum. The level of distortion stays at such a low level that subsequent machining operations are entirely eliminated. This led to enormous cost savings for the transmission manufacturer. Production of the gears started in 6/2006. Results from continuous distortion monitoring during production prove that the distortion of the LPC and HPGQ process is very stable. Therefore in 10/2008 it was decided to reduce the quality inspection drastically which resulted in significant cost savings.
For a planetary gear (FD-Pinion gear), densely packed for geometry, the "Reversing gas flow"-technology is used in serial production. Compared to "unidirectional gas flow" the helix angle variations were substantially reduced. With this optimized "Reversing gas flow" process it is not necessary to machine the gear - teeth after heat treatment.

References

[Altena et al, 2005] Altena H., Schrank, F. und Jasienski, W.: Reduzierung der Formänderung von Getriebeteilen in Gasaufkohlungs-Durchstoßanlagen durch Hochdruck-Gasabschreckung. In: HTM 60 (2005)1, S. 43-50
[Heeß, 2007] Heeß, K.: Maß- und Formänderungen infolge Wärmebehandlung von Stählen. Expert Verlag. 3. neu bearbeitete Auflage (2007). ISBN-10: 3-8169-2678-9
[Heuer, 2003] Heuer, V., Löser, K.: Entwicklung des dynamischen Abschreckens in Hochdruck-Gasabschreckanlagen. In: Mat.-wiss. u. Werkstofftech. 34 (2003), page 56-63
[Heuer et al, 2009] Heuer, V.; Loeser, K.: Low pressure carburizing for transmissions. In: Gear solutions July 2009
[Loeser et al, 2005] Loeser, K.; Stueber, G.; Welzig, G.; Heuer, V.: United States Patent No. US 6,913,449 B2 „Apparatus for the treatment of metallic workpieces with cooling gas". July 5th 2005
[Stich, 1995] Stich; Tensi. In: HTM 50 (1995)
[Walton, 1992]: Walton, H.: Dimensional changes during hardening and tempering of through-hardened bearing steels. Quenching and Distortion control (Conference Proceedings). ASM International , 1992, S. 265-273

Proceedings of the 26th ASM Heat Treating Society Conference
B.L. Ferguson, R. Jones, D.S. MacKenzie, and D. Weires, editors
https://doi.org/10.31399/asm.cp.ht2011p0258

Effect of the Oxidation Stability of Soybean Oil and Palm Oil on Steel Quenching Performance

Diego Said
Universidad Tecnológica Nacional, Facultad Regional Buenos Aires,Medrano Street N° 951, Buenos Aires, Argentina

Gabriela Belinato, Rosa L. Simencio Otero, Lauralice C.F. Canale
Universidade de São Paulo, Escola de Engenharia de São Carlos, São Carlos, SP, Brazil.
E-mail: rosa_simencio@yahoo.com.br; lfcanale@sc.usp.br, phone: 5516 3373 9579, fax: 5516 3373 9590.

Gustavo S. Sarmiento, Analía Gastón,
Universidad Nacional de Rosario, CIUNR-FCEIA, Rosario, Argentina
Universidad del Salvador, Facultad de Ciencia y Tecnología, Buenos Aires, Argentina

George E. Totten
Portland State University, Department of Mechanical and Materials Engineering, Portland, OR USA

Abstract

The potential use of vegetable oil derived industrial oils continues to be of great interest because vegetable oils are: relatively non-toxic, biodegradable and they are a renewable basestock alternative to petroleum oil. However, the fatty ester components containing conjugated double bonds of the triglyceride structure of vegetable oils typically produces considerably poorer thermal oxidative stability than that achievable with petroleum basestocks under typical use conditions where furnace loads of hot steel (850 °C) are typically rapidly immersed and cooled to approximately 50-60 °C bath temperatures. This is especially true when a vegetable oil is held in an open tank with agitation and exposure to air at elevated temperatures for extended periods of time (months or years). Unfortunately, as vegetable oils degrade, their fluid viscosity is expected to increase resulting in decreased quench severity. Clearly, thermal-oxidative stability is essential. This paper will review thermal-oxidative stability and quenching performance of soybean oil and palm oil and the resulting impact on the heat transfer coefficient.

Introduction

Properties of steels and alloys vary with their composition and microstructures which are dependent on the heat treatment and quenching process used. Typically, steel is heated to its austenitization temperature and then cooled sufficiently fast to avoid pearlite transformation in order to obtain maximum hardness and strength. The critical cooling rate is dependent on both the specific heat capacity and thermal conductivity of both the steel and the quenchant in addition to quenching bath temperature and agitation.

Among the most common quenchants used is water. Petroleum oil (mineral oil) derived quenchants are used when lower cooling rates and more uniform cooling is desired for better distortion control and crack prevention of alloy steels. However, petroleum oil possesses a number of significant disadvantages including: 1.) poor biodegradability [1], 2.) toxicity [2], 3.) petroleum oil is not a renewable basestock [3] and it is flammable. Therefore, it is of continuing interest to identify viable alternatives to petroleum oil as a basestocks for industrial oil formulation such as quenchants for heat treating applications [3].

Currently vegetable oils are one of the most commonly identified alternative basestocks to petroleum oils primarily because they are biodegradable and are obtained from renewable sources. Brazil has shown interest in the behavior of soybean oil for use as coolants (and quenchants) and for lubricant formulations because it is a biodegradable renewable basestock and it represents 95% of all seed oil production in the country. Soybean oil has shown excellent performance in various commercial applications [4]. However, its oxidative instability and its narrow viscosity range potentially limit its use as a quenchant.

While palm oil (dende oil) occupies a prominent position on the world market, it is the second most produced seed oil behind only soybean oil, it is currently not produced at a high level in Brazil. Its use is mainly devoted to cooking but palm oil has great potential for use in the formulation of various industrial oils and its use is expected to grow in the future [5].

Relative to soybean oil, palm oil contains substantially higher saturated fatty ester content and significantly lowers polyunsaturated acid content in the overall triglyceride structure of the two seed oils and is therefore relatively more resistant to oxidation. This is because molecular double bonds, especially double bonds in conjugation, react more easily with oxygen to form free radicals leading to faster degradation when subjected to elevated temperatures and air as would be encountered during quenching of steel.

There have been various investigations on the use of vegetable oils (and animal oils) as quenchants. One of the earliest studies involving cooling curve and heat transfer analysis of quenching properties was conducted by Rose in 1940 with rapeseed oil [6]. Higher cooling rates were observed for rapeseed oil compared to petroleum oil which was attributed to the relatively poor stability of the vapor blanket formed by the rapeseed oil [6].

Tagaya and Tamura subsequently compared the quench severity of different vegetable oils including: soybean, rapeseed and castor oils with mineral oils and fish/animal oils with respect to fluid source, viscosity and oxidative stability for various naturally derived fluids [7]. These data showed that, although the Grossmann quench severity factors were comparable for both castor oil (H = 0.199) and soybean oil (H = 0.200), the cooling times from 700 to 300°C were significantly faster for castor oil than for soybean oil using a JIS K 2242 silver probe test which utilizes a 10 mm dia x 30 mm length cylindrical silver probe with a surface thermocouple [8].

Fujimura and Sato also studied the quenching performance of vegetable oils reported earlier by Tagaya and Tamura and also added castor oil to their study in addition to ethyl esters of oleic, palmitic and stearic acid versus different petroleum oil quenchants, and concluded that the performance of soybean and rapeseed oil were essentially equivalent [9]. Quenching performance of the ethyl esters were also equivalent to each other and exhibited greater quench severity than the vegetable oils. Castor oil, however, was found to be thermally unstable. The quenching performance of castor oil was also studied by Farah, et. al. [10] and the thermal instability of this oil was subsequently confirmed [11].

Currently, the most commonly cited vegetable oil basestocks used for quenchant formulation in the USA are based on canola oil [12,13] and soybean oil [14]. Recently, a crambe oil based fluid has been reported as a potential quenchant [15].

Prabhu studied the heat flux properties of palm oil [12] and extensive heat transfer and wetting studies of coconut, sunflower, groundnut, palm and castor oils [16-19]. With the exception of palm oil, these vegetable oils are not commonly considered as basestocks for quenchants in North and South America. Interestingly, with the exception of castor oil, relatively little difference in surface wetting properties were observed in Prabhu's work and quench severities comparable to a conventional, non-accelerated mineral oil were observed.

Totten, et. al. reported the results of cooling curve, hardening performance, heat transfer, and rewettability characterization studies conducted with crude and partially hydrogenated and winterized soybean oils [20] provided by Honary [21]. Because of the strong influence of the test specimen surface on the wetting behavior (and therefore on the quenching behavior), unalloyed (Ck 45) steel probes were used. This work provided a simultaneous examination of the martensitic transformation with local differences, and dependencies on the rewetting time (and dependencies on the metallurgical condition of the steel).

Cooling curve analyses showed that immersion quenching was primarily influenced by the sequence of the cooling phases and different heat transfer rates. Heat transfer on the specimen surface was primarily determined by the rewetting conditions and depended on the cooling characteristics of the quenching medium used, bath temperatures, and agitation rates. The soybean oils investigated showed no significant differences in cooling behavior and rewetting conditions on the test specimen surface and cooling rates were similar. Comparison of the cooling time-temperature and cooling rate curves showed that vegetable oils exhibited faster cooling rates than the mineral oil used as a reference.

Most recently, de Souza, et.al. [22, 23] and Kobasko, et.al. [23] performed extensive cooling curve analysis and heat transfer coefficient characterization of a series of unoxidized vegetable oils including: canola oil, soybean oil, corn oil, sunflower oil, and cottonseed oil and showed that while different, their quenching performance was not only similar but the quenching process occurred predominantly by convective heat transfer.

This paper will describe the quenching performance of uninhibited soybean oil and palm oil and the effect of oxidation on quenching performance compared to fully formulated, commercially available petroleum oil quenchants.

Materials and Experimental Methods

A total of 30 liters of soybean oil and 18 liters of palm oil was purchased at a local market in Brazil. The fatty ester content of soybean oil and palm oil was determined by gas chromatography [24].

The kinematic viscosity of soybean oil was determined using an AVS - 350 Schott-Geräte viscometer with a glass capillary (Ubbelohde 2C tube) according to the technical standard ABNT NBR 10441-10 / 02 [25]. The dynamic viscosity of palm oil was determined using a Brookfield DV-III Ultra rheometer with a No. 51 spindle. The kinematic viscosity (η) of the palm oil was determined from the dynamic viscosity (ν) by dividing by the density (ρ) at the same temperature: $\nu = \eta/\rho$ [26].

The accelerated aging equipment used for this work was described previously by Farah [11] which was based on the procedure reported earlier by Bashford and Mills [27]. For this test, a 2.7 L oil sample is placed in a metal beaker which was immersed in a constant temperature water bath where the temperature is cycled between 150 °C and 120 °C every 15 min. The accelerated oxidation was performed under agitation which was provided by using an air flow of 4 L per hour through the vegetable oil being tested. The oxidation test was conducted for 48 h for soybean oil and 60 h for oil palm (since palm oil was found to be significantly more stable) and interrupted every 12 h to determine the fluid viscosity at which time the fluid was returned to the aging system without adding more oil.

Cooling curves were obtained at 60 °C using a data acquisition system that includes a standard Wolfson probe (ISO 9950) with a Type-K thermocouple inserted to the geometric center to capture the change of temperature during the cooling process. This procedure was performed according to ASTM D6200-01 [28]. The data was then transferred to a computer using an A / D converter card to obtain real-time cooling curves which were subsequently used for numerical computational work.

Determination of Heat Transfer Coefficients

Heat transfer processes are complex and the heat transfer coefficient is a complex function of variables describing this process. Generally, the heat transfer coefficient is a function of the fluid flow, component (probe) shape and dimensions, temperature and physical properties of the liquid: thermal conductivity, specific heat capacity, density and viscosity. The heat transfer coefficient can be defined as the quantity of heat transferred per unit time per unit area of a surface when the difference of temperatures between the surface and liquid equals one degree absolute. Since quenching processes are actually heat transfer processes, the heat transfer coefficient is an excellent single parameter for quenchant characterization.

In this work, the commercial code HT-MOD1 (Heat Treating Modeling) [29] was used to simulate the heat treatment process. This code is also used to calculate heat transfer coefficients as function of time by solving an inverse heat transfer problem. The model is based on a numerical optimization algorithm which includes a finite element module for calculating, with respect to time and space, the temperature distribution and its coupled micro-structural evolution. In this case, since an Inconel 600 probe was used that does not undergo microstructural phase transformation, the differential problem solved is only the heat conduction equation (1) in a cylindrical probe with the boundary condition (2) at all the external surface in contact with the quenchant:

$$\frac{1}{r}\frac{\partial}{\partial r}\left[rk\frac{\partial T}{\partial r}\right]+\frac{\partial}{\partial z}\left[k\frac{\partial T}{\partial z}\right]+Q=c\rho\frac{\partial T}{\partial t} \qquad (1)$$

$$-k\frac{\partial T}{\partial n}=h_i\left(T-T_{am}\right) \qquad (2)$$

[3] HT-MOD is a commercial code which is available from: KB Engineering S.R.L.; Florida 274, Piso 3, Of. 35 (1005) Buenos Aires – ***Argentina;*** Tel: (54-11) 4326-7542; Fax: (54-11) 4326-2424; Internet: http://www.kbeng.com.ar/en/

where:

T	Temperature;
k	Thermal conductivity;
c	Specific heat;
ρ	Density;
Q	Heat source by enthalpies of phase transformations;
n	Normal direction at the probe surface;
h_i	Heat transfer coefficient on the probe surface;
T_{am}	Room temperature.

These calculations were performed using a finite element mesh containing 11 nodes along the radial direction and 21 nodes along the longitudinal direction. The total time of each process was divided into a sufficient number of time intervals where the linear variation of the heat transfer coefficient can be assumed [30]. The selection of the initial values for these coefficients and of the quantity and length of the time intervals was sample-dependent. The mean square difference between the measured and calculated temperatures obtained after optimization of the heat transfer coefficients was about 1°C. Table 1 summarizes the thermo-physical properties INCONEL 600 used for this work [31].

Table 1:Temperature-Dependent Thermo-Physical Values for INCONEL 600

Thermal Conductivity		Specific Heat		Density	
Temp. [°C]	K [W/m.K]	Temp. [°C]	c [J/kg.K]	Temp. [°C]	ρ [kg/m3]
50	13.4	50	451	20	8400
100	14.2	100	467	100	8370
150	15.1	200	491	200	8340
200	16.0	300	509	300	8300
250	16.9	400	522	400	8270
300	17.8	500	533	500	8230
350	18.7	600	591	600	8190
400	19.7	700	597	700	8150
450	20.7	800	602	800	8100
500	21.7	900	611	900	8060
700	25.9				
900	30.1				

Results and Discussion

1. Fluid Composition and Viscosity

Gas chromatographic analysis was performed on unoxidized (fresh) soybean oil and palm oil. The fatty acid composition of both vegetable oils is shown in Table 2. These data clearly show that there are large compositional differences between soybean oil and palm oil. Soybean oil has substantially greater amounts of polyunsaturation (linoleic and linolenic acid esters) than does palm oil. Also, palm oil contains much higher concentrations of saturated fatty acid esters (palmitic and stearic esters). The relatively higher amounts of unsaturation present in the soybean oil would be expected to produce substantially greater oxidative instability than palm oil. Palm oil viscosity and soybean oil viscosity both measured at 40°C and values of 61 cSt and 32 cSt respectively were obtained. These viscosities should be compared to the conventional petroleum oil (Microtemp 157) and an accelerated quench oil (Microtemp 153B) which yielded viscosities of 29 cSt and 40 cSt at 40 °C respectively.

Table 2: Fatty acid composition of soybean oil and palm oil.

Fatty Acid	Soybean Oil (%)	Palm Oil (%)
Palmitic	10.5	45.3
Stearic	2.7	5.9
Oleic	20.8	40.5
Linoleic	58.6	7.1
Linolenic	5.0	Not Detected

2. Oxidative Stability

Kodali showed previously that the relative rates of oxidation increased as the number of double bonds in the triglyceride ester groups increased and they followed the order [32]:

Stearic (1) < oleic (10) < linoleic (100) < linolenic (200)

The higher amounts of polyunsaturation in soybean oil relative to palm oil shown in Table 2 would be expected to result in substantially greater oxidation rates which are consistent with the data shown in Fig.1. Therefore, faster viscosity increases with respect to time would be expected for soybean oil. Importantly, both petroleum oil quenchants shown in Fig. 1 are considerable more stable than either uninhibited soybean oil or uninhibited palm oil which indicates the degree of the challenge confronting the quenchant formulator in developing vegetable oil-based quenchants with oxidation stability comparable to currently available petroleum oil-based quenchants.

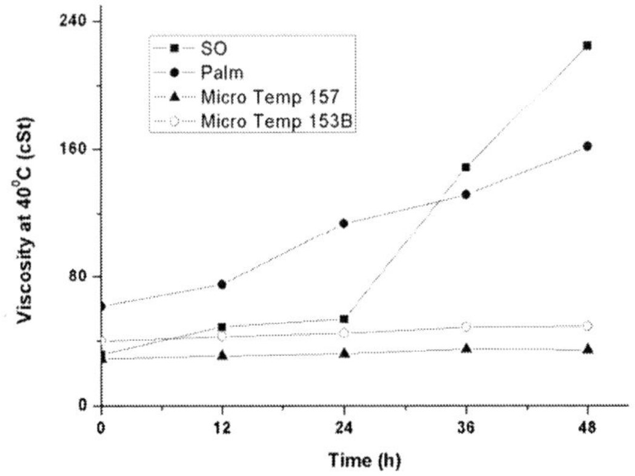

Fig. 1: Comparison of the oxidative stability of soybean oil (SO) and palm oil (Palm) and two commercially available petroleum oil-based quench oils: Micro Temp 153B and Micro Temp 157.

3. Quenching Performance

Quenching performance of the soybean oil, palm oil and the two petroleum oil quenchants used for comparison; Micro Temp 157, a conventional, unaccelerated petroleum oil, and Micro Temp 153B, an accelerated petroleum oil were evaluated by cooling curve analysis conducted according to ASTM D6200 under unagitated conditions. The cooling curves obtained are shown in Fig.2.

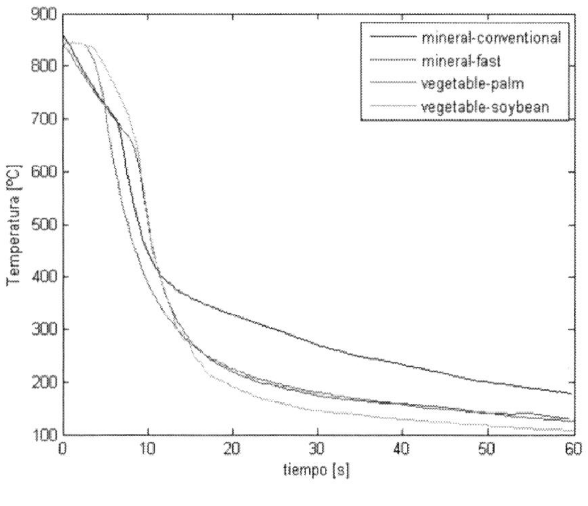

Fig. 2: ASTM D6200 cooling curve comparison of soybean oil, palm oil, a conventional and a fast petroleum oil quenchant at 60 °C bath temperature and no agitation.

During the course of these investigations, it was readily apparent by visually observing the metal-fluid interface during the quenching process that vegetable oils do not "boil" under these conditions. Only convective cooling was observed.

Actually, given the relatively high molecular weight, triglyceride molecular structure full-film boiling and nucleate boiling would not be expected when quenching into a vegetable oil. This behavior should be compared to petroleum oil derived quenchants since, as a class, they do exhibit full-film boiling and nucleate boiling in addition to convective heat transfer. These differences are apparent when viewing the cooling curves obtained for the vegetable oils in this study with the conventional oil (Microtemp 157) and an accelerated quench oil (Microtemp 153B) shown in Figure 1.

Micro Temp 157 and Micro Temp 153B did behave as expected. The fast oil, Micro Temp 153B exhibited shorter film-boiling times than did Micro Temp 157 and both oils exhibited similar lower temperature cooling rates, although the Micro Temp 153B was somewhat slower which was probably due to a higher bulk fluid viscosity.

Interestingly, the cooling curves exhibited by palm oil and soybean oil were noticeably different. The lower viscosity palm oil did appear to cool significantly more quickly than did soybean oil. Also, as expected, the conventional petroleum oil also cooled more slowely than the fast oil. Furthermore, both petroleum oil quenchants exhibited noticeable film-boiling and nucleate boiling behavior not observed with either palm oil or soybean oil. As expected, the fast oil exhibited somewhat shorter film boiling times than the slow oil. This cooling curve profiles suggest that vegetable oils generally require no cooling rate accelerator such as that used in the fast oil to reduce full-film cooling. This type of cooling behavior would be desirable for difficult to harden, crack-sensitive carbon steels. The cooling rates, however, were somewhat faster for the vegetable oils as a class when contrasted to petroleum oil quenchants, especially in the critical martensitic temperature transition range (300 °C), which suggests that they may not be as desirable for higher alloy, crack sensitive steels.

The quenching behavior of these oils was also examined by comparing the cooling time-heat transfer coefficient and surface temperature–heat transfer coefficient profiles which were calculated as described previously and are shown in Figures 3-6. Fig. 3 illustrates the cooling behavior of palm oil. By comparison, the heat transfer coefficient behavior of soybean oil shown in Fig. 4 appeared to show some incipient film boiling which may be due to the lower oxidative stability which would result in some potential formation of sufficiently low molecular weight volatile components at the hot metal interface, especially under unagitated conditions which would increase somewhat the residence time of the fluid at higher temperatures to cause this effect. Interestingly, the time to reach the maximum heat transfer coefficient was longer and the surface temperature where this occurs was lower for soybean oil than palm oil. Furthermore, the maximum heat transfer coefficient was substantially greater for soybean oil which may also be explained by the interfacial evolution of volatile components leading to lower interfacial viscosity ies due to the degradation process.

The quenching behavior of both palm oil and soybean oil was compared to the accelerated (fast) (Fig. 6) and unaccelerated (slow) mineral oil quenchants (Fig. 5). Fast mineral oil quenchants contain additives to destabilize the vapor blanket surrounding the hot metal which typically results in shorter time durations of film boiling behavior than that observed for the unaccelerated (slow) petroleum oil quenchants. This was observed in this study as well. The accelerated mineral oil did exhibit significantly shorter film boiling times than did the unaccelerated mineral oil. Both the accelerated and unaccelerated mineral quenchants exhibited comparable maximum heat transfer coefficients but maximum heat transfer coefficient occurred at a significantly lower surface temperature for the unaccelerated mineral oil than did the accelerated mineral oil.

When comparing the heat transfer coefficients for palm oil (Fig. 3) and soybean oil (Fig. 4) with the unaccelerated mineral oil (Fig. 5) and the accelerated mineral oil (Fig. 6), it was observed that the maximum heat transfer coefficient for palm oil occurred at approximately the same surface temperatures. A similar comparison showed that the maximum heat transfer coefficients for soybean oil also occurred at approximately the same surface temperatures. Soybean oil exhibited the highest maximum heat transfer coefficient of the four fluids studied. The maximum heat transfer coefficient for palm oil was less than either mineral oil quenchant. Only palm oil showed no film boiling and soybean oil exhibited an incipient film boiling behavior as discussed above whereas both mineral quenchants exhibited a clearly observable film boiling behavior with the accelerated mineral oil exhibiting somewhat shorter film boiling t imes than the unaccelerated mineral oil.

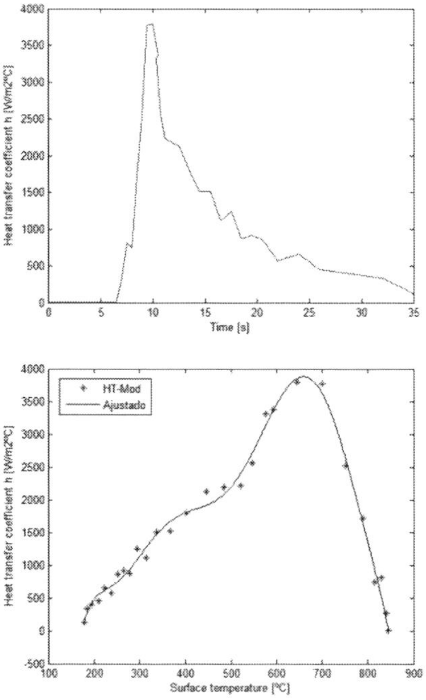

Fig. 3: Heat transfer coefficient as a function of time (top) and as a function of surface temperature (bottom), calculated for palm oil.

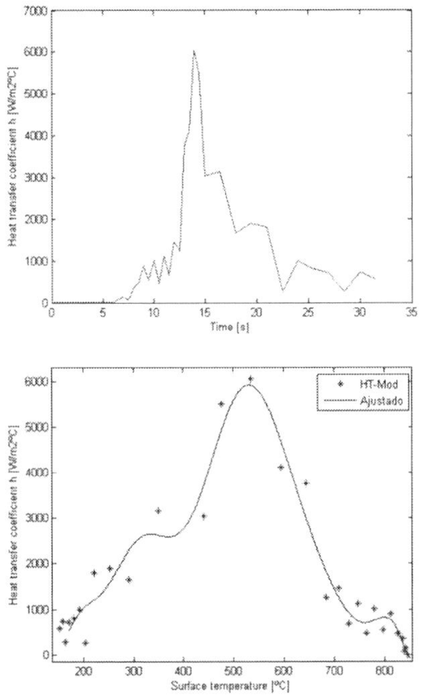

Fig. 4 Heat transfer coefficient as a function of time (top) and as a function of surface temperature (bottom), calculated for soybean oil.

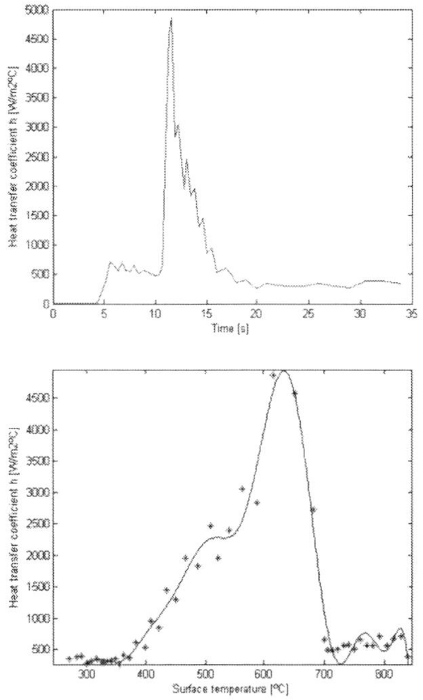

Fig. 6: Heat transfer coefficient as a function of time (top) and as a function of surface temperature (bottom), calculated for an accelerated (fast) mineral oil.

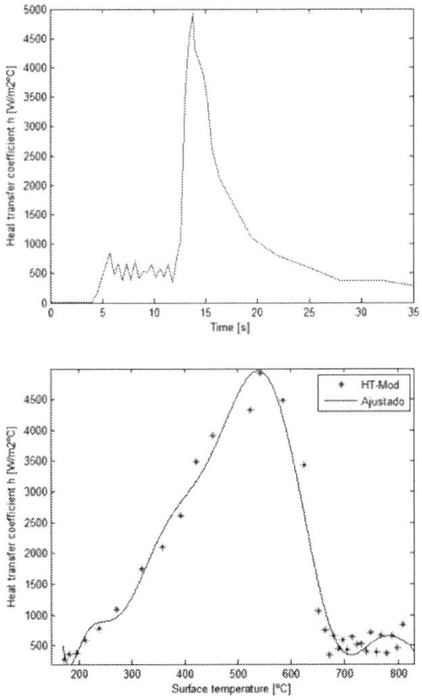

Fig.5: Heat transfer coefficient as a function of time (top) and as a function of surface temperature (bottom), calculated for an unaccelerated, (conventional or slow) mineral oil.

Conclusions

The oxidative stability of two uninhibited vegetable oils, palm oil and soybean oil, with substantially different molecular compositions was compared. In this study, palm oil was shown to be substantially more stable than soybean oil which was due to the much lower degree of unsaturation. Of the ester component structure of the palm oil. However, even though palm oil was significantly more stable than soybean oil, it was still much less stable than the fully formulated mineral oil quenchants used for comparison. Clearly, effective antioxidants must be identified for palm oil if it is going to successfully compete with mineral oil quenchants.

The quenching behavior of palm oil and soybean oil was also evaluated by comparing the cooling curve time-core temperature curves and the cooling time - heat transfer coefficient and cooling time-surface temperature behavior. These data showed that palm oil exhibited essent ially no vapor blanket cooling which is consistent with the interfacial behavior observed visually during the data acquisition process. Soybean oil, however, did appear to exhibit some incipient vapor blanker cooling as observed by some possible full-film boiling in the cooling curves and heat transfer coefficient curve behavior. The mineral oil quenchants used for comparison did exhibit full-film cooling behavior with the accelerated oil exhibiting shorter full-film boiling durations than did the unaccelerated mineral oil. Interestingly, the soybean oil exhibited the greatest heat transfer coefficient followed by both the unaccelerated and the accelerated mineral oils (which were comparable in this regard and the lowest heat transfer coefficient was observed with the palm oil.

This work is continuing with the objective of evaluated effective antioxidants and to compare the residual stress profiles after quenching. These results will be reported when complete.

Acknowledgement

The authors acknowledge their appreciation to CNPq (Conselho Nacional de Desenvolvimento Científico e Tecnológico - "National Counsel of Technological and Scientific Development") for the financial support for this work.

REFERENCES

[1]. Jones, N., "Managing used oil", Lubes 'n' Greases, 1996, Vol. 2, No. 6, p.20-21.

[2]. Henry,J.A., "Composition and toxicity of petroleum products and their additives", Human and Experimental Toxicology, 1998, Vol. 17, No. 2, p.111-123.

[3]. Erdman, K. D., Kling, G. H., Tharp, D. E., "High-performance biodegradable fluid requirements for mobile hydraulic systems", SAE Technical Paper Series, Paper Number 981518, 1998.

[4]. Rose, B., Honary, L., "A report on the field test performance of a soybean-based hydraulic oil", SAE, Technical Paper Series, Paper Number 982005, 1998.

[5]. Kodali, D. R., "High performance ester lubricants from natural oils", Industrial Lubrication and Tribology, 2002, Vol. 54, No 4, p.165-170, 2002.

[6]. Rose, A., "Das Abkühlungsvermögen von Stahlabschreckmitteln", Archiv.für das Eisenhüttenwesen, 1940, Vol. 13, No. 8, p. 345-354.

[7]. Tagaya, M., Tamura, I., Studies on the quenching media 3rd report. The cooling ability of oils.Technology Report, Osaka University, 1954, Vol. 4, p. 305-319.

[8]. JIS K 2242-1980, "Japanese Industrial Standard – Heat Treating Oil".

[9]. Fujimura, Y., Sato, T., "The composition of quenching oil and quenching effects", The Iron and Steel Institute of Japan (ISIJ), 1963, Vol. 49, p. 1008-1015.

[10].Farah, A.F., Crnkovic, O.R., Canale, L.C.F., "Castor oil: Studies for heat treatment applications", in 19th Heat Treating Society Conference Proceedings, Eds. S. Midea and G. Pfaffmann, ASM International, Materials Park, OH, 1999, p. 251-254.

[11].Canale, L.C.F., Fernandes, M.R., Agustinho, S.C.M., Totten, G.E., Farah, A.F., "Oxidation of vegetable oils and its impact on quenching performance", Int. J. Materials and Product Technology, 2005, Vol. 24, No. 1-4, p. 101-125.

[12].Brennan, R.J., Faulkner, C.H., Massuda, D., "A quenchant based on canola oil", Advanced Materials & Processes, August 1997, Vol. 152, No. 2, p. 32S – 32U.

[13].Brennan, R.J., Faulkner, C.H., "A new quenching alternative", in Totten, G.E., Funatani, K., Howes, M.A.H., Sjostrom, S. (Eds.), Conf. Proceed. 2nd International Conference on Quenching and Control of Distortion. ASM International, Materials OH, 1996, p. 423-428.

[14].Honary, L.A.T., "Performance of vegetable oils as a heat treat quenchant", in Totten, G.E., Funatani, K., Howes, M.A.H., Sjostrom, S. (Eds.), 1996, Conf. Proceed. 2nd International Conference on Quenching and Control of Distortion, ASM International, Materials OH, 1996, p. 595-605.

[15].Lazerri, L., De Mattei, F. Bucelli, F., Palmieri, S., "Crambe oil: A potentially new hydraulic oil and quenchant", Ind. Lubr. Tribol., 1999, Vol. 49, No. 2, p. 71-77.

[16].Prabhu, K.N., Fernandes, P., "Determination of wetting behavior, spread activation energy, and quench severity of bioquenchants", Metallurgical and Materials Transactions B, August 2007, Vol. 38, No. 4, p. 631-640.

[17].Prabhu, K.N., Prasad, A., "Metal/quenchant interfacial heat flux transients during quenching in conventional quench media and vegetable oils", J. Mat. Eng. and Perf., 2003, Vol. 12, No. 1, p. 48-55.

[18].Prabhu,K.N., Fernandes,P., "Effect of surface roughness on metal/quenchant interfacial heat transfer and evolution of microstructure", Materials and Design, 2007, Vo. 28, p. 544-550.

[19].Fernandes,P., Prabhu,K.N., "Comparative study of heat transfer and wetting behaviour of conventional and bioquenchants for industrial heat treatment", International Journal of Heat and Mass Transfer, 2008, v 51, n 3-4, p. p 526-538.

[20].Totten, G.E., Tensi, H.M., Lanier, K., "Performance of vegetable oils as a cooling medium in comparison to a standard mineral oil", J. Mat. Eng. and Perf., 1999, Vol. 8, No. 4, p. 409-416.

[21].Honary, L.A.T., Soybean based hydraulic fluid", U.S. Patent 5,972,855, October 26, 1999.

[22]. Ester Carvalho de Souza, Mauro R. Fernandes, Sylvana C.M. Augustinho, Lauralice de Campos Franceschini Canale, George E. Totten, "Comparison of Structure and Quenching Performance of Vegetable Oils", J. ASTM International, 2009, Vol. 6, No. 9, Paper Number JAI 102188.

[23]. Kobasko, N.I., Carvalho de Souza, E., Canale, L.C.F., Totten, G.E., Vegetable Oil Quenchants Calculation and Comparison of the Cooling Properties of a Series of Vegetable Oils, Strojniški vestnik (Journal of Mechanical Engineering – Slovenia), 2010, Vol. 56, No. 2, p. 131-142.

[24]. Souza, E.C.," Estudo da oxidação do óleo de soja com diferentes concentrações de aditivos anti-oxidantes, para uso em tratamentos térmicos de tempera", 2007. 160p. Dissertação (Mestrado em Ciência e Engenharia de Materiais) – Instituto de química de São Carlos, Instituto de física de São Carlos, Escola de engenharia de São Carlos, Universidade de São Paulo, São Carlos, 2007.

[25].Associação Brasileira de Normas Técnicas. NBR 10441 – 10/02., "Produtos de petróleo - Líquidos transparentes e opacos - Determinação da viscosidade cinemática e cálculo da viscosidade dinâmica", Rio de Janeiro: 2002.

[26].Manning, R.E.; Hoover, M.R.; "Chapter 32 - Flow properties and Shear Stability", in Manual 37 – Fuels and Lubricants Handbook: Technology, Properties, Performance and Testing; Eds. TOTTEN, G.E., WESTBROOK, S.R., SHAH, R.J.; 2003, ASTM International, West Conshohocken, PA, USA, p. 833-883; ISBN-10: 0803120966.

[27].Bashford, A.; Mills, A. J., "The development of improved additives for quenching oils using laboratory simulations", Heat Treatment of Metals,1984, Vol. 11, No. 1, p.9-14.

[28]. ASTM D6200 - 01(2007), "Standard Test Method for Determination of Cooling Characteristics of Quench Oils by Cooling Curve Analysis", ASTM International, West Conshohocken, PA, USA.

[29]. Sarmiento, G. S.; Gastón, A.; Vega, J. "Inverse heat conduction coupled with phase transformation problems in heat treating process". COMPUTATIONAL MECHANICS – New Trends and Applications. E. Oñate and S.R. Idelsohn (Eds.). CIMNE, Barcelona, 1998. CD Book. Part VI, Section 1, Paper 16.

[30].Penha, R.N., Canale, L.C.F., Totten, G.E., Sarmiento, G.S., "Simulation of Heat Transfer Properties and Residual Stress Analyses of Cooling Curves Obtained from Quenching Studies" J. ASTM International, 2006, Vol. 3, No. 5, Paper ID No. 13614.

[31].Clark, J. and Tye, R., "Thermophysical Properties Reference Data for Some Key Engineering Alloys", High temperatures—high pressures, 2003/2004, Vol. 35/36, pp. 1–14.

[32]. Kodali, D. R. High performance ester lubricants from natural oils. Industrial Lubrication and Tribology, 2002, Vol. 54,No. 4, p.165-170.

Proceedings of the 26th ASM Heat Treating Society Conference
B.L. Ferguson, R. Jones, D.S. MacKenzie, and D. Weires, editors
https://doi.org/10.31399/asm.cp.ht2011p0266

Epoxidized Soybean Oil: Evaluation of Oxidative Stabilization and Metal Quenching/Heat Transfer Performance

Rosa L. Simencio Otero, Lauralice C.F. Canale
Universidade de São Paulo, Escola de Engenharia de São Carlo, São Carlos, SP, Brazil.
E-mail: rosa_simencio@yahoo.com.br; lfcanale@sc.usp.br, phone: 5516 3373 9579, fax: 5516 3373 9590.

Diego Said Schicchi
Universidad Tecnológica Nacional, Facultad Regional Buenos Aires, Buenos Aires, Argentina.
Instituto Nacional de Tecnología Industrial, Parque Tecnológico Miguelete, Buenos Aires, Argentina.

Eliana Agaliotis
Consejo Nacional de Investigaciones Científicas y Técnicas (CONICET), Buenos Aires, Argentina.

George E. Totten
Texas A&M University, Department of Mechanical and Materials Engineering, Seattle, WA, USA
E-mail: getotten@getottenassociates.com, phone: 206 788 0188, fax: 815 461 7344.

Gustavo Sánchez Sarmiento
Universidad del Salvador, Facultad de Ciencia y Tecnología, Buenos Aires, Argentina.
Universidad Austral, Facultad de Ciencia y Tecnología, Buenos Aires, Argentina.
E-mail: gustavo.ssarmiento@kbeng.com.ar, phone: 5411 4326 7542, fax: 5411 4326 9176

Abstract

Vegetable and animal oils as a class of fluids have been used for hundreds of years, if not longer, as quenchants for hardening steel. However, when petroleum oils became available in the late 1800s and early 1900s, the use of these fluids as quenchants, in addition to their use in other industrial oil applications quickly dimished. This was primarily, but not exclusively, due to their generally very poor thermal-oxidative instability and the difficulty for formulating fluid analogs with varying viscosity properties. Interest in the use of renewable fluids, such as vegetable oils, has increased dramatically in recent years as alternatives to the use of relatively non-biodegradable and toxic petroleum oils. However, the relatively poor thermal-oxidative stability has continued to be a significant reason for their general non-acceptance in the marketplace. Soybean oil is one of the most highly produced vegetable oils in Brazil. Currently, there are commercially produced epoxidized versions of soybean oil which are available. The objective of this paper is to discuss recently obtained results showing the dramatic improvement in thermal-oxidative stability of epoxidized soybean oils and to discuss their potential use and heat transfer properties as viable alternatives to petroleum oils for hardening steel.

Introduction

Hardening steel occurs by first heating the steel to the specified austenitizing temperature, typically in the range of 750-1100°C and cooling it in such a way that the desired microstructure is formed obtaining required properties such as hardness, strength and toughness. The main objective of any quenchant is to produce the desired metallurgical transformations. In addition, the quenchant must prevent cracking and minimize distortions due to non-uniform heat transfer over the surface of the steel.

The microstructures that are formed during quenching depend on the heat transfer properties at the hot metal interface. The cooling time-temperature pathway is designated as the cooling curve. One method of illustrating the steel transformation microstructures during the quenching process is to superimpose the cooling time-temperature curve for steel cooled in a specific quenching medium over a TTT (Time-Temperature-Transformation) or CCT (Continuous-Cooling-Transformation) curve for the steel of interest.

The most common quenchants, depending on the steel and hardening process, include: air, petroleum oil, water, brine, aqueous polymer solutions and high-pressure gas quenching. Of the vaporizable quenchants, petroleum oil derived fluids are the most commonly encountered throughout the industry.

The challenge to replace petroleum basestocks due to potential problems with long-term availability in addition to the relative toxicity and poor biodegradability has done the vegetable oils as interesting alternative because their biodegradable, environmentally friendly and non-toxic renewable basestocks. Beyond these characteristics vegetable oil have good anti-friction properties, low volatility, high viscosity index and good miscibility with other fluids [1].Although the potential use of vegetable oils as basestocks for industrial oil formulation continues to be of interest, they possess a number of very substantial limitations, not the least of which is relatively poor thermal-oxidative stability relative to petroleum oil-derived formulations. For example, vegetable oils typically cannot withstand reservoir temperatures in excess of 80°C due to the onset of oxidation, although the use of antioxidants can partially offset this notable limitation [2, 3]. Oxidation limits the useful life vegetable oil-derived fluids because of the increased viscosity that results which is further accelerated by elevated temperatures and contact with metals such as iron and copper [4]. Adhvaryu *et al.* [5] have been concluded that soybean oil oxidizes at a rate that is at least an order of magnitude greater than that of petroleum oil.

Thermo-oxidative stability of a vegetable oil is dependent on the fatty ester composition of the triglyceride structure. Increasing amounts of unsaturation in the fatty ester structure leads to increased oxidative instability. Schneider [6] and Kodali [7] have reported that the relative rate of oxidation increases as the number of double bonds in conjugation with each other increases in the following relative order: stearic (1) < oleic (10) < linoleic (100) < linolenic (200) shown in Fig 1.

Knothe [8] reported that the initiation step of oxidative degradation involves hydrogen abstraction from fatty ester of the lipid by an initiator. The most favored position for hydrogen abstraction occurs with the lowest activation energy which is an allylic methylene, the (CH_2) adjacent to the CH=CH moiety, and the allylic radical that is formed is stabilized by resonance over the double bond structure and the resonance stabilization increases with the number of double bonds in conjugation. Bis-allylic positions in polyunsaturated fatty acids such as linoleic acid (double bonds at $\Delta 9$ and $\Delta 12$, giving one bis-allylic position at C-11) and linolenic acid (double bonds at $\Delta 9$, $\Delta 12$, and $\Delta 15$, giving two bis-allylic positions at C-11 and C-14), are even more prone to autoxidation than the allylic position of oleic acid as indicated in Fig. 1.

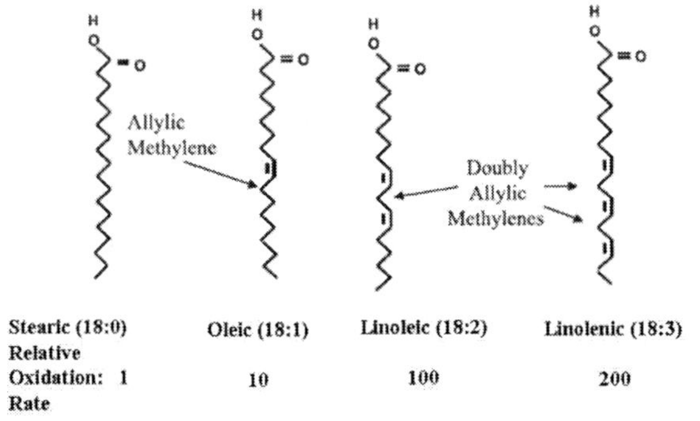

Fig. 1: Allylic methylene functionality: 18:3>18:2>18:1

Souza *et al.* [9] reported that although selection of the most effective antioxidants do provide substantially improved inhibition to oxidation, this is insufficient to rival the oxidative stability possible with the use of petroleum oil-based fluids. Clearly, something significantly more effective is needed to provide the necessary oxidative stability for applications where the fluid will be subjected to relatively high, even if only for a short-time, thermal excursions.

Others possibilities to improve the thermo-oxidative stability of the vegetable oils are genetic and chemical modification. Modification of the chemical structure of vegetable oils has been proposed for the development of environmental friendly vegetable-oil derived basestocks. One modification of vegetable oil structure that has proposed is epoxidation. Fig. 2 provides a generic illustration of a fully epoxidized soybean oil. Double bond epoxidation has been utilized for over 50 years as described by Julian *et al.* [10] and Findley *et al.* [11]. Epoxidized vegetable oils are of potentially great interest for commercial applications as lubricants, synthetic detergents, and for the production of polyurethane foams.

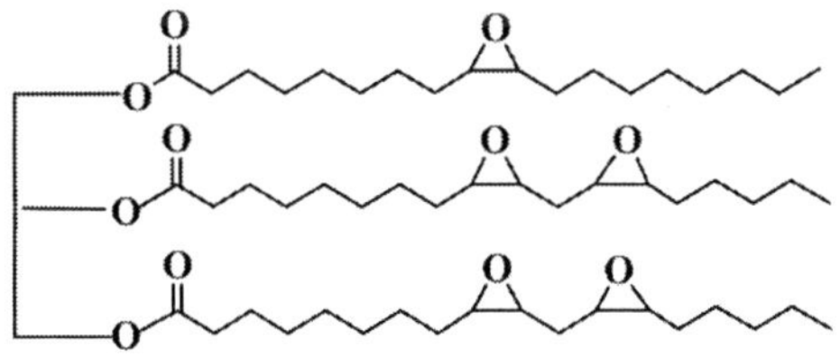

Fig. 2: Generic simplified illustration of one potential component in epoxidized soybean oil, adapted from Hwang and Erhan [12].

Wu *et. al.* [13] showed that epoxidized rapeseed oil exhibited greater oxidative stability than unepoxidized rapeseed oil without chemical modification. Moreover, the epoxidation treatment did not modify the biodegradability of the base stock.

Adhvaryu and Erhan [1] and Erhan *et. al.* [14] reported that epoxidized soybean oil demonstrated improved thermal and oxidative stability relative to unepoxidized soybean oil and genetically modified high oleic soybean oil in certain high temperature lubricant applications.

Epoxidation has been shown to significantly improve oxidative stability relative to unepoxidized triglyceride structure. Only limited data is available comparing the resulting oxidative stability of the epoxidized triglycerides with their functionally equivalent petroleum oil basestocks.

The objective of this paper is to discuss recently obtained results showing the dramatic improvement in thermal-oxidative stability of epoxidized soybean oils and to discuss their potential use and heat transfer properties as viable alternatives to petroleum oils for hardening steel.

Experimental Procedures

The epoxidized soybean oil used in this work was obtained from Inbra (Indústrias Químicas Ltda) and was designated as ESBO. The soybean oil, produced by Cargill Agrícola S/A, was purchased at a local market in São Carlos/SP, Brazil and was commercially designated as Liza and classified as "pure" soybean oil (SO). In addition, one Fatty Acid Methyl Ester (FAME) derivative of soybean soapstock designated as FAME 3P was obtained from Cognis do Brasil Ltda. (Soybean oil soapstock is a by-product of the caustic refining process of soybean oil.) A reaction schematic illustrating the general synthesis of FAME from a triglyceride such as soybean oil is shown in Fig. 3. The FAME 3P was added with stirring into ESBO producing the formulations designated in Table1.

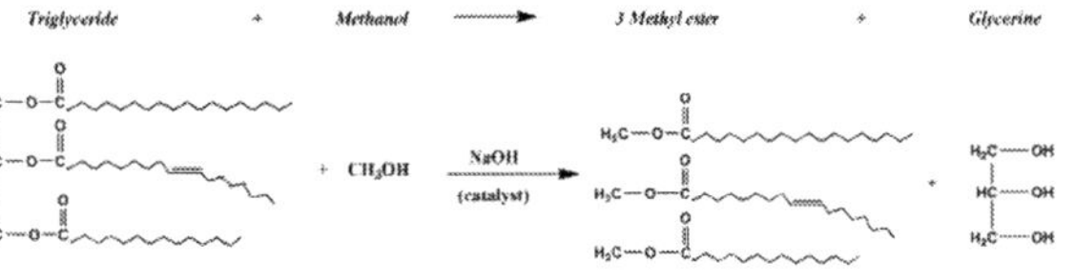

Fig. 3: General illustration of the synthesis of a FAME derivative from a triglyceride.

Table 1: ESBO_1 and FAME formulations using for thermal-oxidation study.

Designation	% ESBO	%FAME
ESBO	100	0
EF 30	70	30
EF 38	62	38
EF 60	40	60
FAME	0	100

Two fully-formulated, commercially available, petroleum-based quench oils used for comparison were: Lubrifort Temp 4 (conventional "slow" oil) and Lubrifort Temp 2 (accelerated "fast" oil). These petroleum-based quench oils were obtained from Quimifort Indústria e Comércio Ltda., São Carlos, SP, Brazil.

Viscosity was measured at 40°C and 100°C using calibrated Cannon-Fenske viscometer tubes according to ASTM D445-06 "Standard Test Method for Kinematic Viscosity of Transparent and Opaque Liquids (and Calculation of Dynamic Viscosity)". The viscosity index was calculated according to ASTM D2270-10 "Standard Practice for Calculating Viscosity Index from Kinematic Viscosity at 40° and 100°C". All viscosity measurements were run in duplicate and the average value was reported.

Cooling curves were obtained at 60°C, non-agitated condition, according to ASTM D6200 "Standard Test Method for Determination of Cooling Characteristics of Quench Oils by Cooling Curve Analysis". This method utilizes a 12.5 mm dia x 60 mm INCONEL 600 cylindrical probe with a Type K thermocouple inserted to the geometric center. After heating the probe in a furnace to 850 °C, it was then manually and rapidly immersed into 2000 mL of the oil to be tested which was contained in a tall-form stainless steel beaker. The probe temperature and cooling times were recorded at selected time intervals to establish a cooling temperature versus time curve. All measurements were performed in duplicate and averaged. The averaged data are reported here.

Results and Discussion

Viscosity Properties

The viscosity-temperature relationship is an important physical property for any potential basestock that may be used for steel quenchant formulations since heat transfer is exponentially related to the viscosity of the heat transfer medium. As steel cools, the viscosity at the interface between the cooling metal surface and the bulk fluid decreases and the interfacial heat transfer coefficient which is partly determined by the structurally dependent, viscosity-temperature relationship of the basestock will control the time-temperature (cooling curve) exhibited by the quenchant which, in turn, controls the overall hardening properties of the quenching medium. According to Santos *et. al.* [15], the viscosity of the triglyceride components of the vegetable oil increases with the fatty acid ester chain length and decreases with the amount of unsaturation in the fatty acid ester alkyl chain.

The results of kinematic viscosity measurements and the calculated viscosity index (VI) obtained for each formulation and the two petroleum oil quenchants are shown in Table 2. The viscosity of epoxidized soybean oil (ESBO), at both temperatures, is higher than the other formulations and as reported in previous study described in literature (Adhvaryu and Erhan, 2002). FAME exhibited the lowest viscosity compared with the other formulations, as expected since it is a relatively low molecular weight monoester. Also, increasing amounts of FAME addition to the ESBO resulted in a progressive viscosity decrease. The reason for performing this study was to examine the potential use of FAME addition to decrease the otherwise extraordinarily high viscosity of ESBO which would inhibit its potential use as a quenchant. These data show that, as expected, the addition of FAME to ESBO can be performed to obtain reasonably similar viscosity properties relative to unadulterated soybean oil. However, the viscosity data of the ESBO/FAME blends do not compare well with either petroleum oil quenchant evaluated.

The viscosity of the slow oil was considerably greater than either the soybean oil or most of the ESBO blends. The viscosity of the fast oil was substantially lower than either the slow oil or soybean oil. The viscosities of EF 60 blend was similar to the fast petroleum oil quenchant (Temp 2) at 40 °C, and the viscosity of EF 30 blend approximated the slow petroleum oil quenchant (Temp 4). Interestingly, while it is possible to blend ESBO/FAME to obtain the viscosity of either the fast or slow oil at 40 °C, it is not possible to match the viscosities at both 40 °C and 100 °C.

The reason for the inability to match the viscosity properties of these blends with the petroleum oil quenchants is due to the substantially different VI properties of these fluids. Table 2 shows that the VI of soybean oil (229) was significantly higher than either the fast (143) or slow (99) petroleum oil and the VI of the fast oil was nearly 50% greater than that of the slow oil. These values represent the expected change in viscosity of the oil with change in temperature and the lower the VI value, the greater the change that is expected. FAME had the highest VI (347) of the fluids evaluated, even greater than soybean oil, and the value for ESBO (141) was similar to that published previously (142) [16]. Increasing amounts of FAME in an ESBO/FAME blend resulted in a progressively increasing VI. It should be noted that a comparison of the viscosity properties of the fast and slow petroleum quench oils indicates that the basestocks are different for these two oils. The slow oil possesses a substantially greater fluid viscosity which would be expected to result in generally slower cooling rates throughout the quenching process relative to the fast oil since heat transfer decreases with increasing viscosity.

Table 2: Kinematic viscosity and viscosity index (VI) determined for every formulations and petroleum oil based quenchants.

Physical Property	Soybean Oil, Epoxidized Soybean Oil (ESBO), FAME and ESBO/FAME Blends						Petroleum Oil Quenchant	
							Slow Oil	Fast Oil
	SO	ESBO	EF 30	EF 38	EF 60	FAME	Temp 4	Temp 2
Viscosity (cSt at 40°C)	31.71	162.77	48.65	37.43	16.25	4.76	52.86	16.96
Viscosity (cSt at 100°C)	7.74	19.91	9.48	8.11	4.57	2.02	7.38	4.05
Viscosity Index	229	141	183	199	220	347	99	143

Quenching/Heat Transfer Performance

Quenching cooling times and cooling rates and heat transfer coefficients for each fluid evaluated in this work were determined by cooling curve analysis under unagitated conditions according to ASTM D6200 at a bath temperature of 60°C. Two petroleum-based quenchants were used for comparison: Lubrifort Temp 2 (fast oil) and Lubrifort Temp 4 (slow oil). The "slow" and "fast" oil designations typically refer to the film-boiling/nucleate boiling properties of a quenchant. When hot (≈ 850 °C) steel (or Inconel) is immersed into a vaporizable fluid such as a petroleum oil, the hot surface is surrounded by a vapor blanket and heat transfer occurs by a full-film boiling mechanism. Heat transfer through this vapor blanket is typically the slowest encountered in the process. Upon further cooling, the film collapses and nucleate boiling results. The transition between full-film boiling and nucleate boiling is designated as the Leidenfrost temperature and heat transfer is typically fastest in this region. When the temperature decreases to a temperature less than the boiling point of the components of the oil, heat transfer occurs predominantly by convection which is considerably slower than nucleate boiling but faster than full-film boiling. "Fast" quench oils typically contain additives that facilitate the rupture of the vapor blanket.

Since nucleate boiling is facilitated at a higher temperature (shorter time), such oils are used to harden otherwise difficult to harden steels such as carbon steels and low-hardenability alloy steels. Because nucleate boiling occurs sooner than would be observed for a slow oil (with an otherwise identical composition and physical properties) it is designated as a "fast" quenching oil. Figs. 5 and 6 show the cooling time-temperature curves and cooling rate curves for the fluids evaluated as part of this work. The cooling curve parameters for the fluids evaluated in this work are summarized in Table 3.

The first comparison to be discussed is for the "fast" and "slow" petroleum oil. When evaluating these results, it is important to consider the viscosity and viscosity index relationships shown in Table 2 which suggest that these two quench oils are formulated using substantially different basestock with the "fast" quench oil (Temp 2) being formulated with a relatively low viscosity (16.06 cSt) versus 52.86 cSt viscosity for the "slow quench oil (Temp 4). This complicates the cooling curve comparison of the two quench oils. Table 3 shows that the cooling rate at the higher 700 °C temperature ($CR_{700°C}$) is much faster for the slow oil and the time to cool to 700 °C ($t_{700°C}$) is somewhat faster. Furthermore, inspection of the cooling time-temperature and cooling rate curves do not exhibit the substantially reduced full-film boiling region more conventionally expected for fast oil compared to a slow quenching oil. In fact, the slow quenching oil, even at the higher viscosity, did not exhibit the expected extended full-film boiling. The cooling rate at 300 °C ($CR_{300°C}$) was essentially equivalent for both oils. However, the cooling times to 300 °C ($t_{300°C}$) and 200 °C ($t_{200°C}$) are substantially slower for the slow quench oil relative to the fast oil which is the expected behavior.

Table 3: Cooling parameters obtained by ASTM D6200 at 60°C bath temperature and with no agitation.

Cooling Curve Parameters obtained at 60°C bath temperature	Soybean Oil, Epoxidized Soybean Oil (ESBO), FAME and ESBO/FAME Blends						Petroleum Oil Quenchant	
							Slow Oil	Fast Oil
	SO	ESBO	EF 30	EF 38	EF 60	FAME	Temp 4	Temp 2
$C_{R700°C}$ (°C/s)	94.87	90.35	27.47	23.26	24.23	25.53	74.96	32.98
$C_{R300°C}$ (°C/s)	15.60	8.18	11.89	12.90	18.97	19.31	5.50	5.81
$C_{R200°C}$ (°C/s)	3.00	3.03	2.67	2.77	3.32	3.57	2.98	2.70
$t_{700°C}$ (s)	4.80	5.14	6.58	7.47	6.82	6.29	6.08	7.37
$t_{300°C}$ (s)	13.53	15.83	16.31	16.80	14.66	16.70	23.22	18.72
$t_{200°C}$ (s)	28.94	40.64	28.94	36.22	29.84	31.15	51.78	42.20

The next comparison to be performed relative to the fast and slow petroleum oil quenchants is soybean oil. Inspection of the soybean oil cooling time-temperature and cooling rate curves, shows that it does exhibit a minimal vapor blanket cooling region which typically accompanies full-film boiling. However, the cooling curve parameters for soybean oil in Table 3 shows that it exhibits substantially faster values for both $CR_{700°C}$ and $t_{700°C}$ relative to either petroleum oil quenchant suggesting that cooling process for soybean oil is probably predominated by convection. The $CR_{300°C}$ and $t_{300°C}$ is also much faster than either the fast or slow oil. While the $CR_{200°C}$ and $t_{200°C}$ is also faster than either petroleum oil, the difference is considerably less. Overall, the soybean oil is clearly the fastest of the quenchants evaluated.

When comparing the cooling time-temperature and cooling rate curves for FAME (which is the methyl ester of fatty acid components derived from methanolysis of soybean oil) it is important to realize that FAME is composed of monoesters which represent a much lower molecular weight and consequently is correspondingly more volatile than soybean oil which is a triglyceride (triester). Comparison of the cooling time-temperature curve and cooling rate curve for FAME shows that it exhibits a substantially longer vapor blanket (full-film boiling) region than either soybean oil or either petroleum oil quenchant. The data in Table 3, shows that the $CR_{700°C}$ is much slower than soybean oil and the cooling rates at $CR_{300°C}$ and $CR_{200°C}$ are faster than soybean oil, although the differences may not be significant. Although the cooling rate 700 °C is slower due to the full-film boiling process, the cooling rates at 300 °C and 200 °C are faster for FAME than those exhibited by either the fast or slow petroleum oil. The cooling time at 700 °C is very similar for FAME and the petroleum oil quenchants although the cooling times are faster for FAME at 300 °C and 200 °C.

The $CR_{700°C}$ for ESBO is approximately the same as soybean oil and like soybean oil, it exhibits a very minimal apparent vapor blanket region. In view of the expected very high boiling points for both soybean oil and ESBO and the corresponding low vapor pressure, it is likely that the cooling process for ESBO is also predominated by convection. The $CR_{300°C}$ is much less for ESBO than soybean oil which is likely due to the much higher viscosity of ESBO (162.8 cSt) compared to soybean oil (31.7 cSt). The $t_{700°C}$ and $t_{300°C}$ is slightly longer for ESBO than soybean oil although the $t_{200°C}$ is much longer for ESBO. Compared to FAME, the $CR_{700°C}$ is much faster because FAME exhibits a prolonged vapor blanket cooling region which is not observed for ESBO. The $t_{700°C}$ and $t_{300°C}$ is faster for ESBO but ESBO has a longer $t_{200°C}$. ESBO cools much faster to 700 °C than either petroleum oil and is also faster at 300 °C, although it is only marginally faster at 200 °C. The $t_{700°C}$ and $t_{300°C}$ are faster for ESBO than either petroleum oil. Although the $t_{200°C}$ is faster for ESBO than the slow petroleum oil, the values for $t_{200°C}$ are approximately comparable for ESBO and the fast petroleum oil.

The reason for doing this work was to determine if it is possible to reasonably match the quenching performance of a petroleum oil, whether fast or slow, or both, by blending FAME and epoxidized soybean oil (ESBO). The approached used to address this question was to blend different ratios of FAME and ESBO and compare the quenching properties with the fast and slow petroleum oil quenchants used for this study. Careful inspection of the cooling time-temperature and cooling rate curves shows that the vapor blanket cooling region was significantly extended relative to soybean oil and ESBO but somewhat less than that observed for FAME. However, the vapor blanket regions exhibited by EF 30, EF 38 and EF 60 were intermediate between FAME and ESBO but quite similar to each other. The $CR_{700°C}$ of the FAME and EF 30, EF 38 and EF 60 blends were comparable and were also slower than ESBO and both the fast and slow petroleum oil. The $CR_{300°C}$ followed the trend:

$$ESBO < EF\ 30 < EF\ 38 < EF\ 60 < FAME$$

This trend was directly proportional to decreasing fluid viscosities. No similar identifiable trend was observed for $CR_{200°C}$. In addition, none of the cooling times followed any significant trend. Finally, and importantly, none of the ESBO/FAME blends produced cooling curve time and rate parameters that matched well with the fast and slow petroleum oils used for this work. The reasons for this are varied but perhaps one of the most important is the inherent film boiling properties that seemed to be exhibited by FAME and its blends. This was further complicated by the varying viscosities exhibited by the blends.

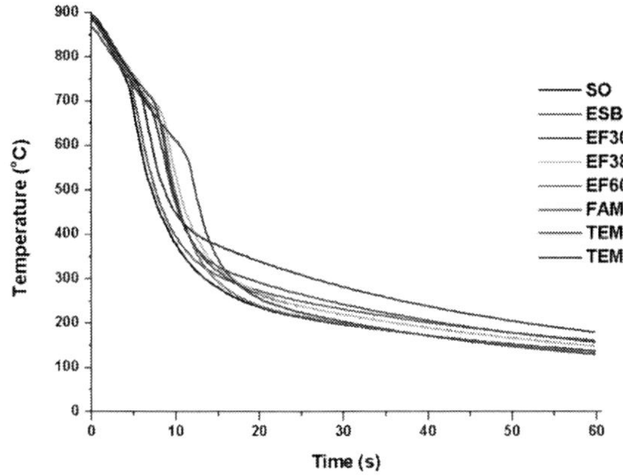

Fig. 4: Cooling curves data at 60 °C bath temperature with no agitation.

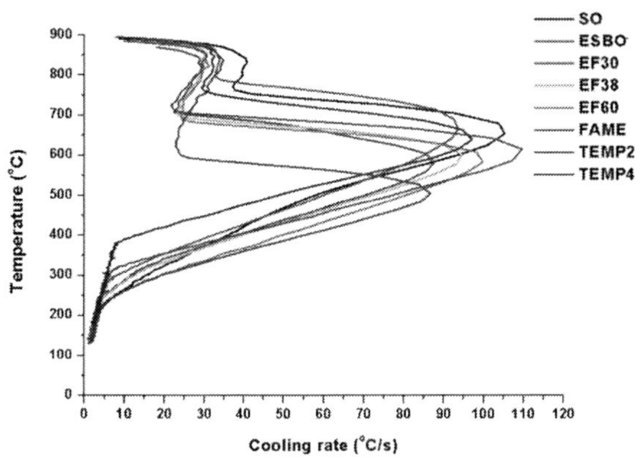

Fig.5: Cooling rates data at 60 °C bath temperature with no agitation.

Most conventional cooling processes involving vaporizable quenchants possess four cooling mechanisms: 1) shock boiling, 2) film boiling, 3) nucleate boiling and 4) convection cooling processes. Since the ASTM D6200 standard 12.5 mm dia x 60 mm cylindrical Inconel 600 probe provides cooling rate and temperature vs. time at the core of probe, it is only possible to evaluate "average effective heat transfer coefficients" which are used in the heat treating industry. During quenching, the heat transfer coefficient is dependent on: surface temperature of the steel part (probe), mass and flow velocity of the quenchant. The variation of the heat transfer coefficient during film boiling is sufficiently small to permit the use of average values (α). During nucleate boiling and convective cooling, average effective heat transfer coefficients can be determined. In this paper, heat transfer coefficients were calculated according to the theory of regular conditions and the calculation procedure described previously by Kobasko, et. al. [17] was used and will not be discussed further here.

Calculation of Heat Transfer Coefficients

Heat transfer processes are complex and the heat transfer coefficient is a complex function of variables describing this process. Generally, the heat transfer coefficient is a function of the fluid flow, component (probe) shape and dimensions, temperature and physical properties of the liquid: thermal conductivity, specific heat capacity, density and viscosity. The heat transfer coefficient can be defined as the quantity of heat transferred per unit time per unit area of a surface when the difference of temperatures between the surface and liquid equals one degree absolute. Since quenching processes are actually heat transfer processes, the heat transfer coefficient is an excellent single parameter for quenchant characterization.

In this work, the commercial code HT-MOD1 (Heat Treating Modeling) [18] was used to simulate the heat treatment process. This code is also used to calculate heat transfer coefficients as function of time by solving an inverse heat transfer problem. The model is based on a numerical optimization algorithm which includes a finite element module for calculating, with respect to time and space, the temperature distribution and its coupled micro-structural evolution. In this case, since an Inconel 600 probe was used that does not undergo microstructural phase transformation, the differential problem solved is only the heat conduction equation (1) in a cylindrical probe:

$$\nabla\cdot\left(k(\vec{r},T)\cdot\nabla T\right)=c(\vec{r},T)\,\rho(\vec{r},T)\,\frac{\partial T}{\partial t} \qquad (1)$$

where $k(\vec{r},T)$, $c(\vec{r},T)$ and $\rho(\vec{r},T)$ denote respectively the thermal conductivity, the specific heat and the density of the material as dependent of the position $(\vec{r})$ and temperature T. $T(\vec{r},t)$ is subjected to the initial condition:

$$T\left(\vec{r},t=0\right)=T_0\left(\vec{r}\right);\quad \vec{r}\in\Omega \qquad (2)$$

and the following boundary conditions at the surface of the steel probe:

$$-k\frac{\partial T}{\partial n}=h_i\left(T-T_{qu}\right)\quad on\ \Gamma_i,\ i=1,\cdots,p \qquad (3)$$

where $h_i(T)$ are the heat transfer coefficients corresponding to different portions of the boundary Γ and T_{qu} the quenchant temperature. Each one of these p cooling zones has a time dependent heat transfer coefficient that varies strongly along each partial boundary, depending on the heat transfer mechanism (vapour blanket, nucleate boiling, convective cooling) that governs the energy flow.

Provided that the temperature change inside the component and on its surface are measured, it is possible to solve the inverse heat transfer problem to determine the time variation

[1] HT-MOD is a commercial code which is available from: KB Engineering S.R.L.; Florida 274, Piso 3, Of. 35 (1005) Buenos Aires – ***Argentina;*** Tel: 54-11 4326-7542; Fax: 54-11 4326-2424; Internet: http://www.kbeng.com.ar.

of the heat transfer coefficients, that best satisfies productions demands. The time dependence of the heat transfer coefficient can be approximated by polygonal functions, each one defined by a set of parameters $h_i^{(r)}(r = 1,\cdots, p \,;\, i = 1,\cdots, q)$.

On calling T_k^m , the measured temperature, and T_k^c, the numerically calculated temperature at those points, one can pose the problem of obtaining the values of the heat transfer coefficients h_i that minimize the function:

$$S = S\left(h_i^{(r)}\right) = \sum_{k=1}^{n}\left(T_k^m - T_k^c\right)^2 = min \qquad (4)$$

n being the total number of measured temperatures, i.e the number of points times the number of measurements at each point.

The selection of the initial values for these coefficients and of the quantity and length of the time intervals was sample-dependent. The mean square difference between the measured and calculated temperatures obtained after optimization of the heat transfer coefficients was about 1°C. Table 4 summarizes the thermo-physical properties INCONEL 600 used for this work [19].

Table 4: Temperature-Dependent Thermo-Physical Values for INCONEL 600

Thermal Conductivity		Specific Heat		Density	
Temp. [°C]	K [W/m.K]	Temp. [°C]	c [J/kg.K]	Temp. [°C]	ρ [kg/m3]
50	13.4	50	451	20	8400
100	14.2	100	467	100	8370
150	15.1	200	491	200	8340
200	16.0	300	509	300	8300
250	16.9	400	522	400	8270
300	17.8	500	533	500	8230
350	18.7	600	591	600	8190
400	19.7	700	597	700	8150
450	20.7	800	602	800	8100
500	21.7	900	611	900	8060
700	25.9				
900	30.1				

Summary of Heat Transfer Coefficient Calculation Results

The calculated heat transfer coefficients as a function of cooling time and as a function of mean surface temperature for each of the fluids in Table 3 are shown in Figs. 6-13. The heat transfer coefficients for the two petroleum oil-based quenchants are shown in Fig. 6 (Fast Oil – TEMP 2) and Fig. 7 (SLOW Oil – TEMP 4). Once again, it is important to note that the fluid viscosity at 40 ºC for TEMP 2 is 16.96 cSt and is substantially lower than the 40 ºC for TEMP 4 which is 52.86 cSt. Based on the fluid viscosities alone, the maximum heat transfer coefficient for TEMP 2 would be expected to be greater than that of TEMP 4 which was, in fact, observed. The film boiling region for TEMP 2 was significantly longer than

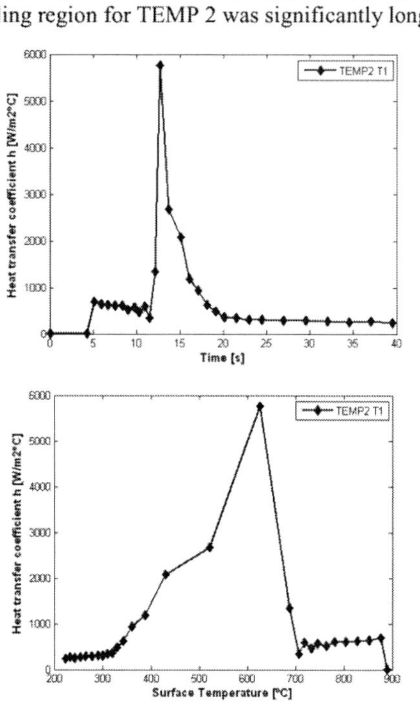

Fig. 6: Heat transfer coefficient calculated for TEMP2 oil as a function of time (top) and of mean surface temperature (bottom).

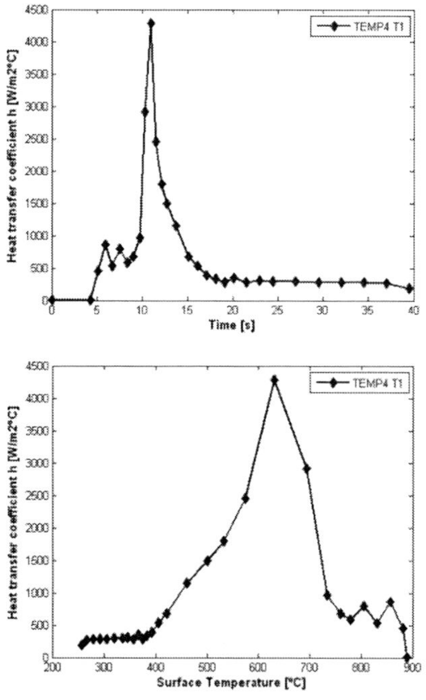

Fig. 7: Heat transfer coefficient calculated for TEMP4 oil as a function of time (top) and of mean surface temperature (bottom).

that exhibited for TEMP 4 which was not expected since fast (accelerated) petroleum oil quenchants are typically formulated with additives to destabilize film boiling behavior. However, these data suggest that the terms "FAST" and "SLOW" are more reflective of the maximum heat transfer rate and fluid viscosity than and behavior induced by additives.
The heat transfer coefficient data for soybean oil is shown in Fig. 8. An apparent film boiling region was observed but was considerable shorter than the petroleum oil quenchant designated as TEMP 4. The maximum heat transfer coefficient was intermediate between that observed for the two petroleum oils; TEMP 2 in Fig. 6 and TEMP 4 in Fig. 7. The maximum heat transfer coefficient for soybean oil occurred at approximately the same surface temperature as TEMP 4 but at a faster cooling time.

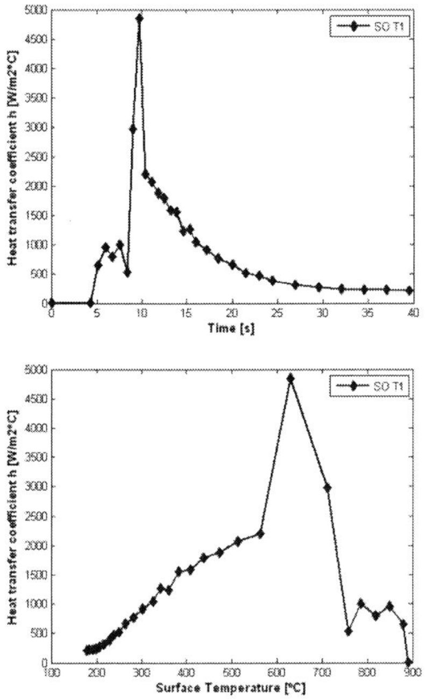

Fig. 8: Heat transfer coefficient calculated for SO oil as a function of time (top) and of mean surface temperature (bottom).

Comparing heat transfer coefficients exhibited by soybean oil (SO) shown in Fig. 8 with epoxidized soybean oil (ESBO) in Fig. 9 during quenching showed that ESBO exhibited the same apparent film boiling behavior as SO but the maximum heat transfer coefficient was much lower than either petroleum oil or SO. In addition, the maximum heat transfer coefficient occurred at the highest surface temperature of the fluids evaluated.

The calculated heat transfer results obtained for FAME during quenching are shown in Fig. 10. Although the maximum heat transfer coefficient for FAME was comparable to that obtained for soybean oil (SO), the total duration of the full-

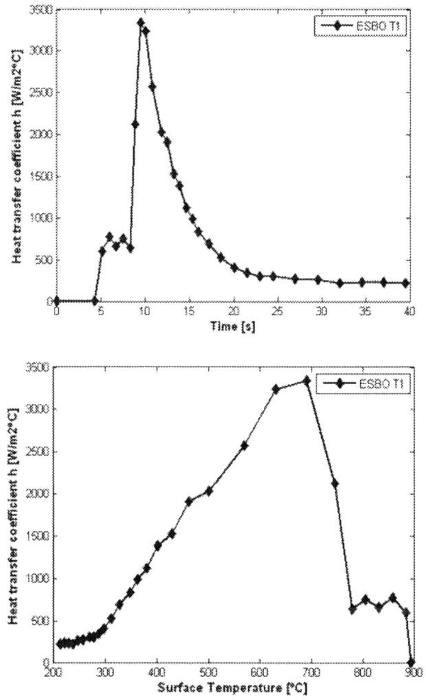

Fig. 9 - Heat transfer coefficient calculated for ESBO oil as a function of time (top) and of mean surface temperature (bottom).

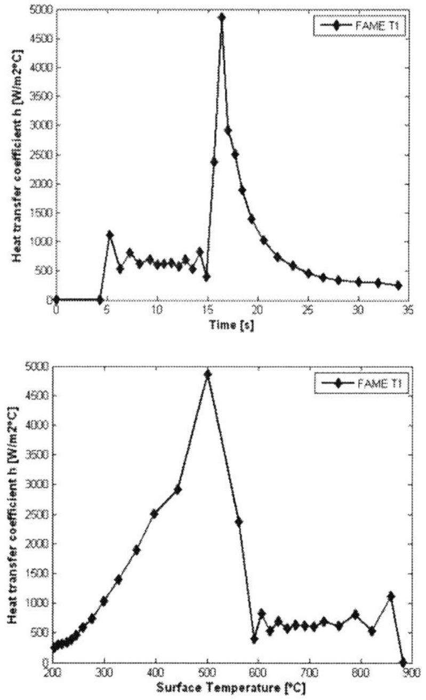

Fig. 10: Heat transfer coefficient calculated for FAME oil as a function of time (top) and of mean surface temperature (bottom).

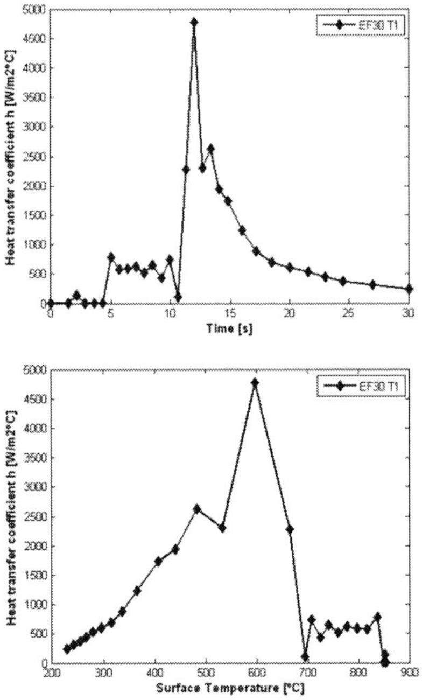

Fig. 11: Heat transfer coefficient calculated for EF30 oil as a function of time (top) and of mean surface temperature (bottom).

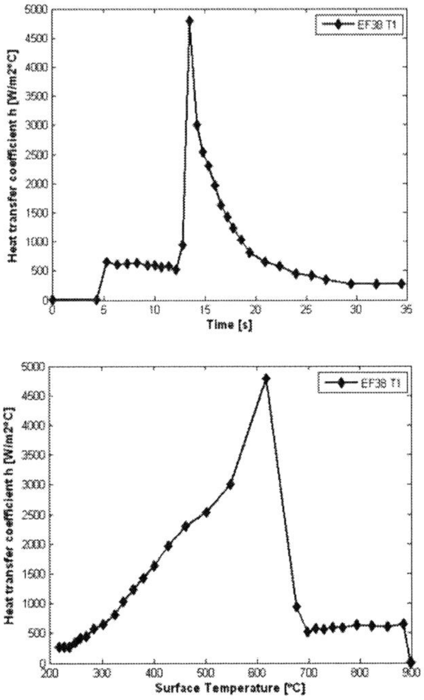

Fig. 12: Heat transfer coefficient calculated for EF38 oil as a function of time (top) and of mean surface temperature (bottom).

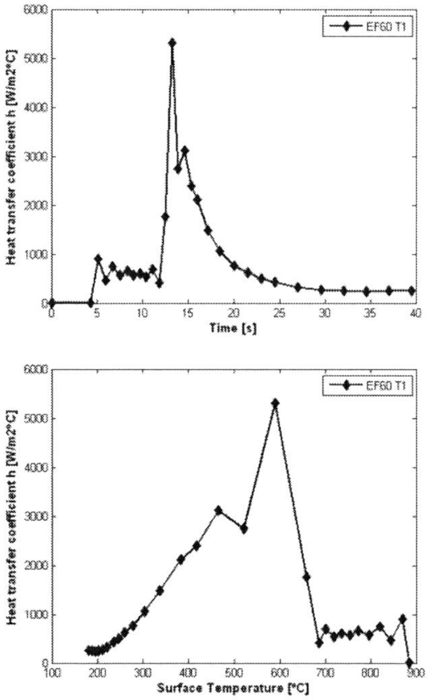

Fig. 13 - Heat transfer coefficient calculated for EF60 oil as a function of time (top) and of mean surface temperature (bottom).

film boiling region was much longer than either SO or any of the ESBO/FAME blends. In addition the maximum heat transfer coefficient occurred at a surface temperature substantially lower than either petroleum oil, SO or ESBO.

Figs 11, 12, and 13 illustrate the heat transfer coefficients obtained for EF30, EF38 and EF60 which are blends of ESBO and 30%, 38% and 60% FAME respectively. Interestingly, while there are some differences in these data, they tend not to be large. The cooling profiles, as indicated by the heat transfer coefficients, is dominated by the presence of FAME by exhibiting substantial durations of film-boiling behavior and somewhat lower surface temperatures where the maximum heat transfer coefficients are obtained. Not surprisingly, EF30 and EF38 exhibited similar maximum heat transfer coefficients with respect to each other and to FAME and slightly less than that value exhibited by EF60. However, further work may show that these differences are not significant. In all cases, the cooling behavior of these blends appear to be most similar to the TEMP2 petroleum oil quenchant than either ESBO or SO.

Conclusion

Soybean oil (SO), epoxidized soybean oil (EBSO), FAME and blends of ESBO containing 30%, 38% and 60% of FAME where prepared and the viscosity properties determined and cooling curve analysis performed using the ASTM D6200 test procedure. For comparison, two fully formulated, commercially available, petroleum-based "fast" and "slow" quench oils were also studied. The very high viscosity of ESBO would preclude its use as a quench oil due to excessive drag-out. Therefore, the potential use of FAME derived from soybean oil to reduce the viscosity of ESBO to model a petroleum oil-based quenchant was evaluated. The data obtained showed that the viscosity of ESBO was much more sensitive to temperature than either soybean oil or FAME. Increase amounts of FAME in the ESBO/FAME blend resulted in progressively less sensitivity of viscosity change relative to ESBO. The viscosity of the fast petroleum oil quenchant was more sensitive to temperature change than the slow petroleum oil quenchants showing that it is formulated using a different petroleum oil basestock.

The quenching performance of soybean oil suggests that heat transfer occurs predominantly by convection as does ESBO. However, FAME exhibits a prolonged film-boiling region suggesting that heat transfer occurs by three mechanisms: film-boiling, nucleate boiling and convection. The addition of FAME to ESBO imparts these characteristic heat transfer mechanisms to the overall cooling process and the duration of the film-boiling region of the ESBO/FAME blends is relatively insensitive to the amount of FAME in the blend.

The characteristic film-boiling process that typically accompanies petroleum oil-derived quenchants was not so clearly evident with the two petroleum oils evaluated here. This is most likely due to two factors. The first is that temperature change is measure in the center of the probe whereas heat transfer is controlled by the mechanism of the interfacial cooling process at the surface. Therefore, the probe may not be sufficiently sensitive to observe a short-lived cooling process exhibited by these petroleum oils. The second reason is that it is likely that both petroleum oils contain additive(s) to either eliminate or reduce the duration of film boiling. Since the ESBO/FAME blends did not contain such additives, it was not possible, from the work conducted thus far, to conclude if the an overall cooling process equivalent to these two particular petroleum oil quenchants is possible. This will be the subject of further work.

Acknowledgements

The authors gratefully acknowledge Inbra (Indústrias Químicas Ltda), Cognis Brasil Ltda and Quimifort Indústria e Comércio Ltda. for donating the epoxidized soybean oil, fatty acid methyl ester and mineral oil quenchants fluids, respectively, and Tecumseh do Brasil Ltda for allowing the use of their Chemical Laboratory for viscosity measurements. The authors also acknowledge CAPES - Coordenação de Aperfeiçoamento de Pessoal de Nível Superior) and Universidade de São Paulo (USP).

References

[1] Adhvaryu, A., Erhan, S.Z., 2002. "Epoxidized Soybean Oil as a Potential Source of High-Temperature Lubricants", Industrial Crops and Products, Vol. 15, pp. 247-254.

[2] Goyan, R., Melley, R.E., Wissner, P.A., Ong, W.C. "Biodegradable Lubricants", Lubrication Engineering, July, (1998), pp. 10-17.

[3] Zeeman, A., Sprengel, A., Niedermeier, D., Späth, M., "Biodegradable Lubricants – Studies of Thermo-oxidation of Metal-working and Hydraulic Fluids by Differential Scanning Calorimetry", Thermochimica Acta, Vol. 268, (1995), pp. 9-15.

[4] Ruger, C.W., Klinker, E.J., Hammond, E.G. "Abilities of Some Antioxidants to Stabilize Soybean Oil in Industrial Use Conditions", J. Am. Oil. Chem. Soc., Vol. 79, No. 7, (2002), pp. 733-736.

[5] Adhvaryu, A., Erhan, S.Z., Liu, Z.S., Perez, J.M. "Oxidation Kinetic Studies of Oils Derived from Unmodified and Genetically Modified Vegetables Using Pressurized Differential Scanning Calorimetry and Nuclear Magnetic Resonance Spectroscopy", Thermochimica Acta, Vol. 364, (2000), pp. 87-97.

[6] Schneider, M. "Government Industry Forum on Non-Food Uses of Crops (GFGNFG 7/7), (2002) – Case Study: Plant Oil Based Lubricants in Total Loss and Potential Loss Applications", Final Report Prepared for P. Smith, Network Manager, University of York, United Kingdom.

[7] Kodali, D.R.,. "High-Performance Ester Lubricants from Natural Oils", Ind. Lube Tribol., Vol. 54, No. 4, (2002), pp. 165-170.

[8] Knothe, G. "Dependence of Biodiesel Fuel Properties on the Structure of Fatty Acid Alkyl Esters", Fuel Processing Technology, Vol. 86, (2005), pp. 1059– 1070.

[9] Souza, E. C., Canale, L.C.F., Totten, G.E. "Vegetable Oil Structure and Antioxidants", Conf. Proceed. New Challenges in Heat Treating and Surface Engineering – Conference in Honor of Božidar Liščić, June 9-12, Cavtat Croatia, Publ. by Croatian Society for Heat Treatment and Surface Engineering, Zagreb, Croatia, (2009), pp. 45-52.

[10] Julian, P.L., Iveson, H.T., Radlove, S.B. "Hydroxylation of egetable Oils and Products Thereof", U.S. Patent 2,752,376, (1956).

[11] Findley, T.W., Ohlson, J.L., Kuester, F.E. "Treatment of Epoxy Fatty Compositions", U.S. Patent 3,035,069, (1962).

[12] Hwang, H., Erhan, S.Z. "Synthetic lubricant basestocks from epoxidized soybean oil and Guerbet alcohols", Industrial Crops and Products, Vol. 23, (2006), pp 311-317.

[13] Wu, X., Zhang, X., Yang, S., Chen, H., Wang, D. "The Study of Epoxidized Rapeseed Oil Used as a Potential Biodegradable Lubricant", J. Am. Oil. Chem. Soc., Vol. 77, No. 5, (2000), pp. 561-563.

[14] Erhan, S.Z, Adhvaryu, A., Liu, Z. "Chemically Modified Vegetable Oil-Based Industrial Fluid", U.S. Patent 6,583,302 B1, (2003).

[15] Santos, J. C. O., Santos, I. M. G., Souza, A. G. "Effect of Heating and Cooling on Rheological Parameters of Edible Vegetable Oils", J. Food Eng., Vol. 67, (2005), pp. 401–405.

[16] Doll, K.M., Sharma, B.K., Erhan, S.Z. "Friction Reducing Properties and Stability of Epoxidized

Oleochemicals", Clean, Vol. 36, No. 8, (2008), pp.700-705.

[17] Kobasko, N.I., Carvalho de Souza, E., Canale, L.C.F., Totten, G.E. "Vegetable Oil Quenchants: Calculation and Comparison of the Cooling Properties of a Series of Vegetable Oils", Strojniški vestnik - Journal of Mechanical Engineering, Vol. 56, No. 2, (2010), pp. 131-142.

[18] Sarmiento, G. S., Gastón, A., Vega, J. "Inverse heat conduction coupled with phase transformation problems in heat treating process". COMPUTATIONAL MECHANICS – NewTrends and Applications. E. Oñate and S.R. Idelsohn (Eds.). CIMNE, (Barcelona, 1998). CD Book. Part VI, Section 1, Paper 16.

[19] Clark, J. and Tye, R., "Thermophysical Properties Reference Data for Some Key Engineering Alloys", High temperatures—high pressures, Vol. 35/36, (2003/2004), pp. 1–14.

Proceedings of the 26th ASM Heat Treating Society Conference
B.L. Ferguson, R. Jones, D.S. MacKenzie, and D. Weires, editors
https://doi.org/10.31399/asm.cp.ht2011p0277

Effect of Bath Temperature on Surface Heat Flux during Quenching in CNT Nanofluids

K. Babu
Department of Mechanical Engineering, SSN College of Engineering, Kalavakkam – 603110, Chennai, India
babuk@ssn.edu.in, +91-44-24982656(236)

T.S. Prasanna Kumar
Department of Metallurgical & Materials Engineering, IIT Madras, Chennai – 600036, India
tsp@iitm.ac.in, +91-44-22574761

Abstract

In this paper, the effect of nanofluid temperature on the heat transfer rate has been studied during immersion quenching in Carbon Nanotube (CNT) nanofluid. For this purpose, CNT nanofluid was prepared by suspending 0.25wt% of chemically treated CNTs (TCNT) in de-ionized (DI) water without any surfactant. Quench probes with a diameter of 20 mm and a length of 50 mm were machined from 304L stainless steel and quenched in CNT nanofluid when maintained at 30 °C, 40 °C and 50 °C. The heat flux and temperature at the quenched surface were estimated based on the Inverse Heat Conduction (IHC) method using the measured temperature data during quenching as input. The computation results showed that the peak heat flux increased when the CNT nanofluid was maintained at 40 °C than at 30°C and decreased with further increase in the CNT nanofluid temperature. The enhanced heat transfer performance of CNT nanofluid during quenching at a slightly higher temperature is attributed to the increase in the Brownian motion of CNTs in nanofluid. The reduced heat transfer rate when the CNT nanofluid maintained at 50 °C during quenching was due to the predominant effect of quenchant temperature on quenching heat transfer rate.

Introduction

Immersion quenching is one of the most widely used processes for achieving martensitic and bainitic steels, where the rate of cooling has a decisive influence on the material properties. Heat transfer during quenching is complex and controlled by different cooling stages viz. i) the vapor phase ii) nucleate boiling and iii) the convective stage. The parameters that affect the heat transfer during quenching may include surface temperature, surface condition of the specimen, agitation of the quenchant, thermo-physical properties of the quenchant and the materials being quenched. The literature concerned with these various quenching parameters can be found in plenty. The heat transfer rate at the part surface is characterized by the surface heat flux or the heat transfer coefficient of the quenching medium. The heat flux at the quenched surface is unknown during quenching and can be calculated with the help of measured temperature data at an interior location in the part being quenched.

Estimation of the surface heat flux or heat transfer coefficient from the measured temperature data during quenching is based on the Inverse Heat Conduction (IHC) method. Prasanna Kumar [1] described a serial solution method for 2-D inverse heat conduction problem to estimate the multiple heat flux components and used it to estimate the heat flux components at the metal/mold interface during casting [2]. The same algorithm was used by Arun Kumar et al. [3] to study the spatial variation of heat flux at the metal-mold interface in gravity die casting. Babu and Prasanna Kumar estimated the surface heat flux and temperature using the IHC method for different initial soaking temperature [4] during quenching in water and also studied the heat transfer rates during quenching in CNT nanofluids [5, 6]. Sarmiento et al. [7] used the IHC method to estimate the temperature dependent heat transfer co-efficient during quenching.

Intensive Quenching (IQ) [8, 9], known as IntensiQuench^{SM}, is an alternate way of hardening steel parts developed by Nikolai Kobasko. IntensiQuench^{SM} process can be defined as cooling usually at a rate several times higher than the rate of normal or conventional quenching. In contrast to the conventional heat-treating practices, IQ calls for very high cooling rates for parts within the martensite phase. Kobasko's research showed that very fast and uniform part cooling actually reduced the probability of part cracking and distortion, while improving the surface hardness and durability of steel parts. In this perception, we wanted to achieve higher cooling rate during quenching possibly by the use of nanofluids. Nanofluids are the new class of engineering fluids and have attracted many researchers [10-25] because of their better heat transfer performance. Nanoparticles are suspended in the base fluids usually at very low concentration to improve their heat transfer performance. By suspending nanosized metallic particles, whose thermal conductivities are many folds higher than that of conventional liquids in the cooling or heating media, the resultant thermal conductivity of the base fluid is improved. It was thought that the use of nanofluids for quenching application might evolve as an alternate quenching technique and yield the benefits of IQ process [5, 6].

Choi [10] projected the possibility of using nanofluids for different practical applications including quenching. Narayan Prabhu and Peter Fernades [11] first used water and water based nanofluid containing 20 wt% alumina nanoparticles as quenchants and reported that the peak heat transfer coefficient of the nanoquenchant was 10% lower as compared to water. Kim et al. [12] quenched small metallic spheres in water and water based nanofluids having alumina, silica and diamond nanoparticles and showed that the quenching behavior of nanofluids was nearly identical with that of pure water. Lotfi and Shafii [13] reported a considerable reduction in the quenching ability of nanofluids prepared with Ag and TiO_2 nanoparticles as compared to that of pure water during quenching of silver spheres. These results on the use of nanofluids for quenching were not encouraging. However, Chopkar et al. [14, 15] reported the higher quenching efficiency of water and ethylene glycol based nanofluids.

Among the different nanoparticles, CNTs have been the main attraction of many researchers due to their higher thermal conductivity and very high aspect ratio for preparing nanofluids. Park and Jung [16] showed that the presence of CNT improved the nucleate boiling heat transfer coefficients of water and R22, a refrigerant. Ding et al. [17] reported that CNT nanofluids produced a maximum heat transfer enhancement of 350% by considering the flow conditions through a horizontal tube. They also studied the effect of CNT concentration on the thermal conductivity and viscosity of CNT nanofluids at different temperatures. Xie et al. [18] measured the thermal conductivity of nanofluids containing treated CNTs and proved that the thermal conductivity of nanofluids increased with an increase in CNT concentration. The different mechanisms for the improved heat transfer performance of nanofluids were elaborated by Keblinski et al. [19]. The effect of various parameters like nanoparticle size, concentration, temperature, pH value of base fluid, ultrasonication on the thermal conductivity enhancement and stability of nanofluids was investigated by several researchers [20-25].

Based on the above literature, it was decided to use CNT nanofluids for quenching application and we assessed the heat transfer rate during quenching of stainless steel (SS) probes in CNT nanofluids [5]. The results published in reference [5] were interesting that the heat transfer rate during quenching in CNT nanofluids was significantly affected by the preparation method. Heat transfer enhancement in nanofluids is affected by several parameters like its concentration, nanofluid temperature, etc. The effect of CNT concentration and agitation during quenching of SS probes in CNT nanofluids was studied and reported by us in reference [6]. The objective of this work is to study the effect of CNT nanofluid temperature on the heat transfer rate during quenching. Chemically treated CNTs were dispersed in de-ionized (DI) water at the concentration level of 0.25wt%. Quench probes with a diameter of 20 mm and length of 50 mm were machined from SS 304L and quenched in CNT nanofluid maintained at 30 °C, 40 °C and 50 °C. Time-temperature data during quenching were recorded using a data logger. The boiling curves during quenching in these conditions were obtained by following the procedure as reported in references [4-6]. The results showed that CNT nanofluid at 40 °C had a higher heat transfer rate than at 30 °C and the heat transfer rate had been poor when the CNT nanofluid was maintained at 50 °C.

CNT Nanofluids

Nanofluids refer to the fluids with suspended nanoparticles. By suspending nanoparticles at very low concentrations, usually 2vol% or less in heating or cooling fluids, the heat transfer performance is proved to be significantly improved by the researchers. The basic idea is to increase the effective thermal conductivity of the heat transfer fluids to achieve better heat transfer performance. The primary reason for the better heat transfer performance of nanofluids is attributed to the higher effective thermal conductivity of nanofluids than that of their base fluids. The nanoparticles that were used to produce nanofluids in the literature includes: aluminum oxide, copper, copper oxide, silver, silica and CNTs. CNT nanofluids have been reported to have higher heat transfer performance [10] because of the unique properties of CNTs such as the higher thermal conductivity which is in the order of ~3000W/mK (multi-walled CNT) and very high aspect ratio [17].

a) Synthesis and characterization of CNT

Multi-walled CNTs were synthesized through an arc discharge method with a DC power supply as described by Ravi Joshi *et al.* [26]. A rotating graphite disc at very low speed (10 rpm) was used as cathode and graphite rods of diameter 11 mm (99.7% purity) were used as anode. The power supply used was an AC/DC inverter TIG power source which supplied a current density of 150 A/cm^2 at 36 V. The arcing was carried out manually in an open air atmosphere. The soot deposited on the cathode was collected and crushed using mortar and pestle to get a fine powder of CNT. CNT powder was washed with distilled water and then soaked in toluene for 5 hours. It was washed again with distilled water and dried in an oven. Finally, the CNT powder was oxidized at 550 °C in a closed furnace for 2 hours to remove any amorphous graphite present. The yield of this method was reported to have ~55% by Ravi Joshi *et al.* [26] with a purity of ~60% CNT and the remaining being graphite spherulites.

The synthesized CNT powder was characterized in a X-ray diffractometer (Bruker Discover D8) with Cu-Kα radiation. The XRD pattern of CNT powder is shown in Fig. 1 (a). It has a peak at 26.29° which corresponds with (002) plane and confirms the highly graphitic structure of the CNT powder. The crystallite size was calculated to be 11.4 nm from the XRD pattern using Scherrer's formula. The high resolution transmission electron micrograph of CNT powder in Fig. 1(b) was taken by a Transmission Electron Microscope (Philips CM12) and shows CNT bundles which are entangled and embedded in graphite spherulites. The minimum outer diameter of multi-walled CNTs observed was 9.73 nm and the average diameter ranged from 9 to 15 nm with a length ranging from few hundreds of nm to microns. The average crystallite size was calculated to be 9.3 nm using Scherrer's equation and this value confirms the observed range of tube diameter in TEM picture. The details of CNT synthesis through arc discharge method and characterization of CNTs have already been presented by us in references [5, 6].

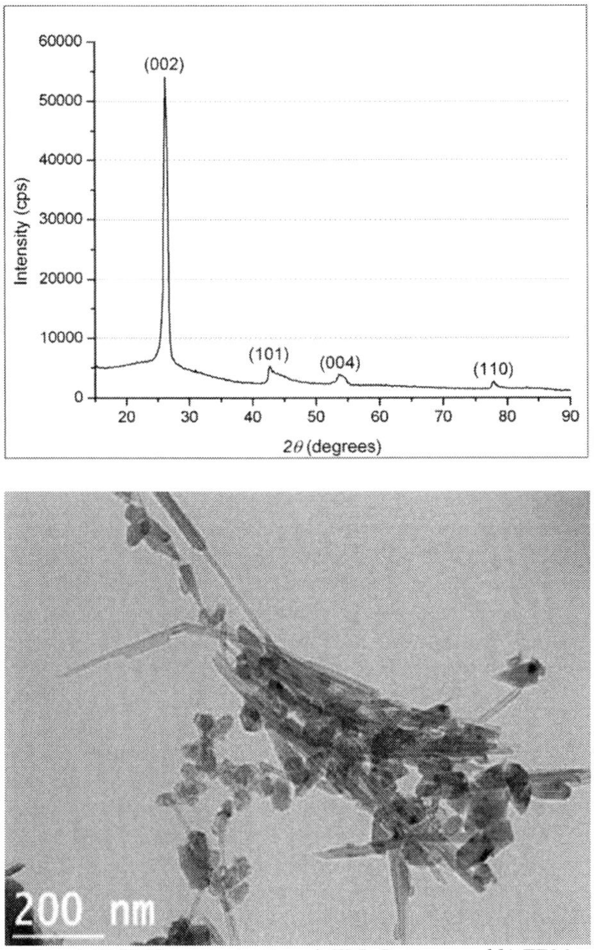

Fig. 1 CNT Characterization a) XRD Pattern and b) TEM Picture of CNT Powder

b) Preparation of CNT nanofluid

Nanofluids were prepared by the two-step method. The synthesized CNT powders were directly dispersed in DI water to prepare CNT nanofluid. CNTs are known to be hydrophobic and entangled. To disperse CNTs successfully in polar liquids such as water, either surfactant should be used or CNTs should be functionalized by chemical treatments to make them suspended in water. It has been reported that the use of surfactant in CNT nanofluids severely hampered the overall heat transfer rate during quenching by promoting a lengthy film boiling stage [5]. Thus, chemically treated CNTs (TCNT) were used for preparing the CNT nanofluid. The chemical treatment of CNTs involved soaking of CNTs in acid mixture containing HNO_3 and H_2So_4 in the volumetric ration of 1:3 for 5 h [2, 3]. A sample (30 ml of DI water + 0.5 wt% of TCNT) of CNT nanofluid was prepared by ultrasonication in a probe type sonicator (ROOPTELSONIC, Model No. USP-1000) and tested for its stability through visual observation. The settlement of the nanofluid sample after 1 h and 1 day is shown in Fig. 2 (a) and (b) respectively. The CNT nanofluid was stable enough for the quenching application as the quenching process itself does not last more than a minute. Thus, for the quenching application, 2 liter of CNT nanofluid was prepared with 0.25wt% of TCNTs in DI water. The CNT nanofluid was sonicated for 1 hour at 20 kHz to disperse the CNTs well in the same probe type sonicator.

Fig. 2 Stability of CNT nanofluid after (a) 1 h and (b) 1 day

Experimental Setup

Quench probes with a diameter of 20 mm and a length of 50 mm were prepared from SS 304 L which has no transformations during quenching. The selection of SS material for quench probes has avoided the non-linearity arising from solid-solid phase transformation part during inverse estimation of the surface heat flux. The schematic of the quench probes is given in Fig. 3(a). Manufacturing details of the quench probes can be found in references [4-6]. Since the IHC analysis is very sensitive to the input temperature data, even small fluctuations in the input temperature data that results from measurement errors would get amplified in the estimated surface heat flux. Two mineral insulated K-type thermocouples with a sheath diameter of 1 mm were used for temperature measurements during quenching. One of the thermocouples was calibrated at different temperatures ranging from 50 °C to 600 °C and the maximum deviation in temperature measurements was 0.34% at 600°C. COREFIX (a commercial adhesive paste) was used for fixing the thermocouple TC2 after placing it carefully in its position. It was allowed to cure at room temperature for 24 hours. Another thermocouple (TC1) was tied along with the connecting stem such that its tip was at the mid plane but ~2 mm away from the probe surface. The thermocouple TC1 was positioned diametrically opposite to TC2, so that its presence would not have any influence on the heat transfer near TC2 during quenching. The photograph of the quench probe with thermocouples is shown in Fig. 3(b). A fresh quench probe was used for each quenching experiment conducted in the CNT nanofluid maintained at 30 °C, 40 °C and 50 °C.

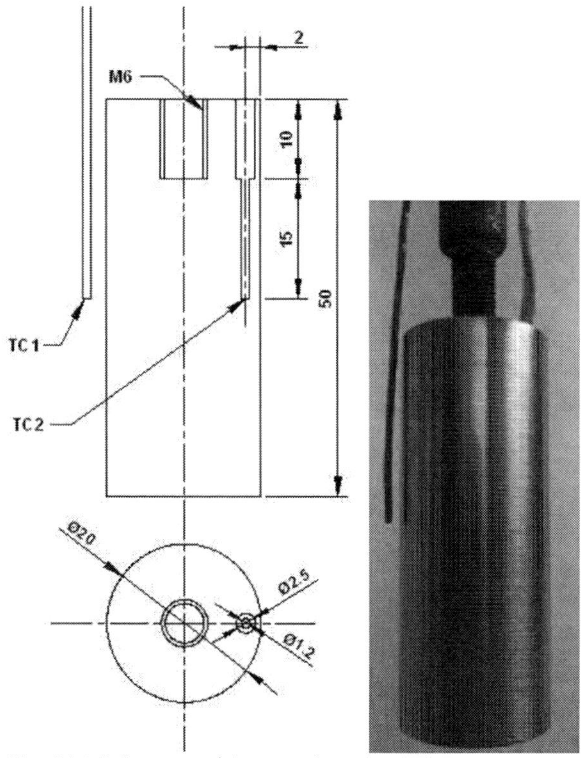

*Fig. 3(**a**) Schematic of the quench probes (All dimensions are in mm) (**b**) Photograph of the quench probe showing thermocouples*

A vertical tubular electrical resistance furnace was used for heating the quench probe. The probe was heated to 850 °C and soaked for 30 minutes. The probe was then quickly transferred and immersed into a beaker containing 2 liter of the CNT nanofluid at room temperature of 30 °C with the probe axis held vertical. The experiment was repeated by maintaining the CNT nanofluid at 40 °C and 50 °C using an external water bath. The external water bath was heated by a filament type immersion electric water heater. The external water bath was stirred manually at frequent intervals, so that the entire water bath was at uniform temperature. The bath temperature was monitored by another K-type thermocouple connected through a temperature controller, so that the quenchant temperature could be maintained at the required temperature with an accuracy of ± 1 °C. The time-temperature data ware recorded for every 0.1 s using a computer assisted data acquisition system (Agilant – Model No. 34970A). The temperature recording was started even when the probe was in the furnace and continued until the probe temperature reached about 50 °C during quenching.

Inverse Determination of Heat Flux and Surface Temperature

The unknown heat flux and temperature at the quenched surface were estimated using the IHC method from the measured temperature at TC2 during quenching as input. The serial IHC algorithm developed by Prasanna Kumar [1] was used in this work to estimate the heat flux at the probe-quenchant interface. The estimation of the unknown heat flux and temperature at the probe surface was exactly the same way as followed in references [4-6] and is briefly explained in the following sub-sections. The repeatability data of this quenching experiment has been presented in reference [5].

a) Finite Element (FE) Model for the Inverse Method

A single unknown heat flux boundary was assumed along the probe-quenchant interface, since the quench probes were quickly immersed during quenching. The mid section of the quench probe, where the tip of the thermocouples were positioned, was taken as the model domain and neglected the end effects for the estimation of unknown heat flux at the probe surface. Due to the circular geometry of the quench probes, only a quadrant was modeled and discretized in the finite element (FE) inverse solver. The effect of grid size and time step on the estimated surface heat flux during inverse calculation was studied by us in reference [4]. The suggested grid size of 1089 elements (3 node triangular elements) with the total number of nodes being 595 and the time step of 0.1 s were used for all the finite element calculations in this work. The accuracy and reliability of the inverse solver employed in this work has already been proven by us in reference [4]. The FE mesh used in this work is shown in Fig. 4 along with the different boundary conditions and is exactly the same as used in references [4 - 6]. The material properties of 304L stainless steel used in reference [4] were assigned to the FE model. The material properties at different temperatures within the temperature range given were treated as step-wise linear and were not extrapolated for the temperatures spanning outside the given ranges, rather it was taken as a constant. The initial temperature of all the nodes was set to 850 °C from which the probes were quenched in CNT nanofluid.

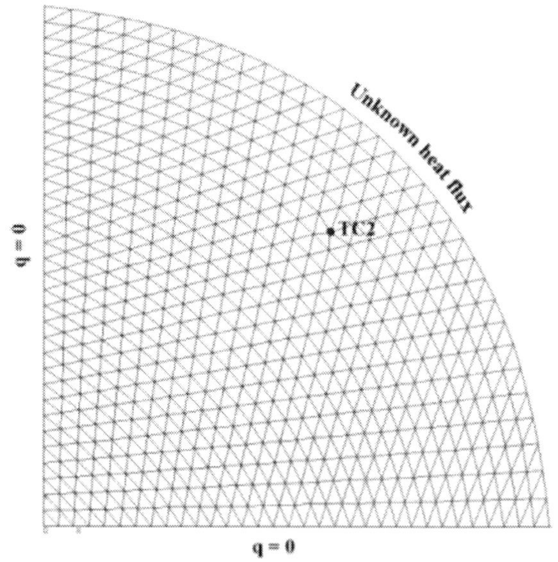

Fig. 4 Solution domain for the inverse problem of heat flux estimation during quenching

Results and Discussion

The temperatures recorded at TC1 and TC2 during quenching of the cylindrical quench probes in the CNT nanofluid maintained at different temperatures are plotted against time in Fig. 5. The initial flattened portion of the cooling curves corresponds with the temperatures recorded while the probes were in the furnace. The instant of probes' removal from the furnace is characterized by the deviation in TC1 and TC2 data. The instant of full immersion is exactly located with the help of TC1 data as it drops instantaneously down to the quenchant temperature. Upon full immersion, the film boiling or vapor blanket stage of cooling commences and the end of this stage is characterized by a small change in the slope of the cooling curves recorded at TC2 which can be noticed immediately after the instant of full immersion. The next stages are the transition boiling followed by the nucleate boiling and convective stages of cooling. The exact instant of start of these different cooling stages would not be accurate if the calculation is made at this stage, as these temperatures were recorded at 2 mm below the probe's surface. There would be a temperature gradient from the surface to core of the quench probes, and the temperature recorded at TC2 at any instant would be higher than that at the probe's surface.

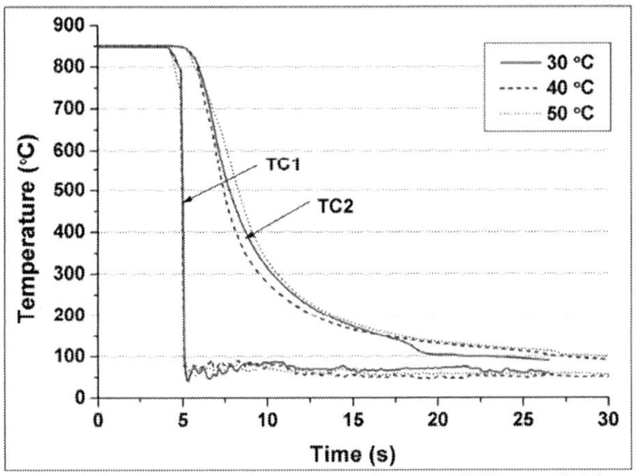

Fig. 5 Cooling curve during quenching in CNT nanofluids

a) Temperature and cooling rate at the quenched surface

The inverse estimation of heat flux and temperature at the surface was carried out using TC2 data from the instant of full immersion as input to the inverse solver and until the temperature at TC2 reached 90 °C. The calculated surface temperature for different cases of quenching is plotted in Fig. 6 as a function of time. Corresponding cooling rates are also calculated and plotted in Fig. 7 against time. The computed cooling curves at the probe surface clearly indicate that the drop in the surface temperature is faster when the CNT nanofluid was maintained at 40 °C. The cooling rate plots also clearly show the highest peak cooling rate in CNT nanofluid maintained at 40 °C followed by 30 °C and then by 50 °C. The reason for the variation in the cooling rate provided by CNT nanofluid at different temperatures is discussed in the next sub-section with respect to the surface heat flux.

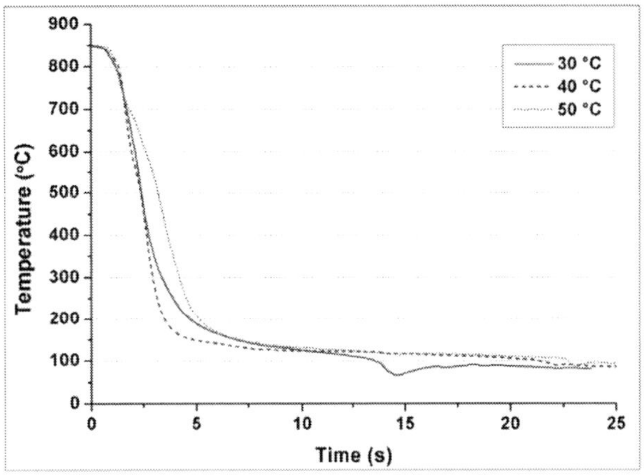

Fig. 6 Computed cooling curve at the probe surface

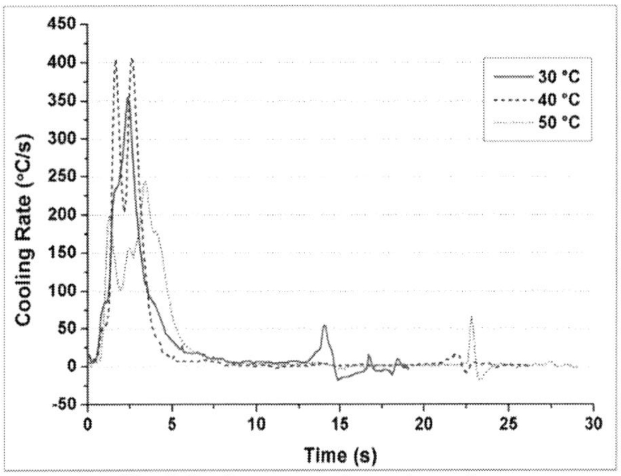

Fig. 7 Cooling rate at the probe surface

The mechanisms of heat flow in nanofluids have been discussed by Choi [10], Keblinski [19] and Das et al. [20] with respect to (i) Brownian motion (ii) molecular level liquid layering at liquid/particle interface (iii) the nature of heat transport (ballistic phonon effects) and (iv) the effect of nanoparticle clustering. Heat transfer rates during quenching in CNT nanofluids have been discussed with respect to these different mechanisms and can be found in references [5, 6].

b) Heat flux during quenching in CNT nanofluids

The estimated surface heat flux is plotted as a function of time and temperature in Fig. 8 and Fig. 9 respectively. The plot of surface heat flux as a function of surface temperature is referred to as conventional boiling curve during quenching. The heat flux starts from zero representing the instant of probes' removal from the furnace and rises slowly to a value of approximately 51000 W/m^2 within 0.5 s. This increase in the heat flux is caused by the heat transfer in the air due to natural convection and radiation during the probes' traverse in air before its full immersion in the quenchant [4 - 6]. A slight variation in these values for different cases is due to the fact that the probes were transferred manually from the furnace to the quenchant.

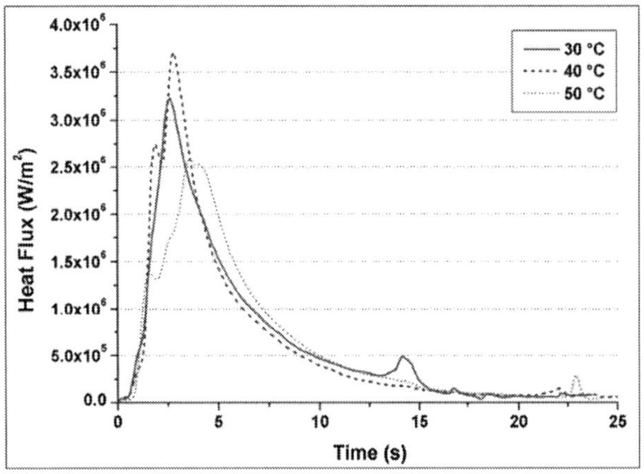

Fig. 8 Transient heat flux at the quenched surface

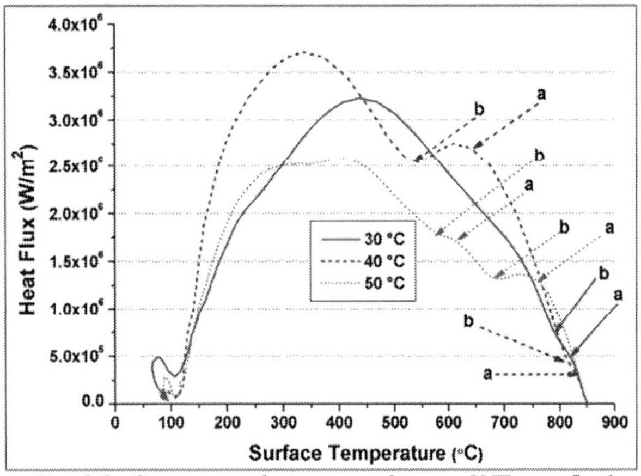

Fig. 9 Boiling curves during quenching in CNT nanofluid

The heat flux increases rapidly from this value to reach a peak value (peak heat flux) and then drops down as the probe tends to attain equilibrium temperature with the quenchant. The boiling curve during quenching has two peaks for all the cases; (i) first peak heat flux within the nucleate boiling stage and (ii) the second peak at the near end of quenching within the convective stage of cooling [4]. The start of the second peak is taking place when the surface temperature of the probes has reached ~100 °C which is the boiling point of base fluid, water. The different stages of cooling during quenching in CNT nanofluid maintained at different temperatures are discussed in the following paragraphs based on Fig. 8 and 9.

Upon full immersion, the film boiling stage takes place and the length of this film boiling stage in different cases of quenching is different. The start and end of film boiling stage on the boiling curves of different cases are marked on Fig. 9 as 'a' and 'b' respectively. When the nanofluid temperature was increased, the film boiling stage takes place in more than one distinct stage. For eg., the first film boiling stage in 40 °C case has happened immediately after the full immersion and the second film boiling has taken place when the surface temperature reached ~650 °C. There were some crackling sounds during quenching in CNT nanofluid maintained at 40°C and 50°C cases which confirmed the rupturing of film of vapors. The multiple stages of film boiling in 40 °C and 50 °C cases are not clearly understood and could not be captured because of the opaque nature of CNT nanofluid. However, the effect of nanofluid viscosity and wettability in causing such multiple film boiling stages can not be ruled out and needs to be studied further. In general, viscosity of any fluid is strongly influenced by the fluid temperature and the same effect can be expected in this experimental work too.

During the nucleate boiling stage of cooling, the surface heat flux reaches a maximum and then the slope of the cooling curves decreases as the probe cools continuously. The peak heat flux and cooling rate reached during quenching in different cases are presented in Fig. 10 as a bar chart. It is clearly evident that the peak heat flux and cooling rate have both increased when the CNT nanofluid was maintained at 40°C compared to 30 °C. When the CNT nanofluid was maintained at 50 °C, both these quantities have dropped down to a very low value than in the case of 30 °C. The increase in the peak heat flux and cooling rate in 40 °C compared to 30°C are 15% and 14% respectively. The decrease in the peak heat flux and cooling rate in 50 °C compared to 30°C are 20% and 31% respectively.

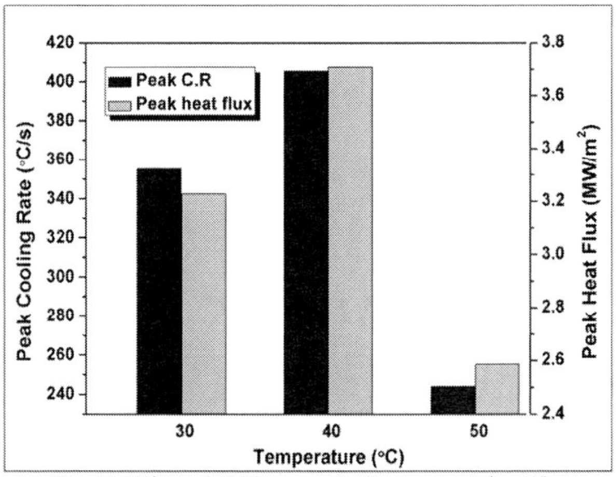

Fig. 10 Effect of CNT concentration on peak heat flux

The heat transfer performance of CNT nanofluid at slightly increased temperature is better and it becomes worse when its temperature is raised further. It is expected that the Brownian motion of CNTs in CNT nanofluid will get enhanced at elevated temperature. Thus, the increased heat transfer performance of CNT nanofluid can be attributed to the enhanced Brownian motion of CNTs. With further increase in the CNT nanofluid temperature, the Brownian motion of CNTs should still be higher and result in higher heat transfer performance of CNT nanofluid during quenching. But, the effect of bath temperature on the quenching heat transfer rate seems to have become more predominant than the effect of enhanced Brownian motion at very high temperatures. This finding suggests that the Brownian motion of nanoparticles in nanofluids is an important factor in the contribution of enhanced heat transfer performance of nanofluids in heat transfer applications.

Summary and Conclusions

Quenching experiments were conducted using SS quench probes in CNT nanofluid to study the effect of CNT nanofluid temperature on quenching heat transfer rates. The temperature and heat flux at the quenched surface have been calculated through the IHC method with the measured temperature history during quenching. During quenching, the CNT nanofluid was maintained at different temperatures using an external water bath. The effect of CNT nanofluid temperature on the heat transfer rate during quenching in CNT nanofluid has been studied from the estimated surface heat flux. The conclusions that are drawn from this experimental and numerical work are summarized as follows:

1. The heat transfer rate during quenching in CNT nanofluid is significantly affected by the quenchant temperature.
2. The peak heat flux during quenching has increased with a slight increase in the bath temperature, but dropped down with further increase in the bath temperature.
3. The increased heat transfer rate during quenching in CNT nanofluid at slightly elevated temperature is attributed to the enhanced Brownian motion of CNTs at higher temperature.
4. However, the drop in the heat transfer rate when the nanofluid temperature was further increased is attributed to the bulk effect of the bath temperature on the quenching heat transfer. In other words, the effect of bath temperature becomes predominant than the effect of enhanced Brownian motion of the nanoparticles.
5. Brownian motion of nanoparticles remains the important factor that contributes to the enhanced heat transfer performance of nanofluids in heat transfer applications.

Acknowledgment

The authors would like to acknowledge the support extended by Dr. Prathap Haridoss, Associate Professor, Department of Metallurgical & Materials Engineering, Indian Institute of Technology Madras, Chennai during CNT synthesis by Arc discharge method.

References

1. Prasanna Kumar, T.S., "A serial solution for the 2-D inverse heat conduction problem for estimating multiple heat flux components," *Numer. Heat Transfer Part B,* Vol. 45 (2004), pp. 541-563.
2. Prasanna Kumar, T.S. and Kamath, H.C., "Estimation of multiple heat flux components at the metal/mold interface in bar and plate aluminium alloy castings," *Metall. Mater. Trans. B,* Vol. 35, No. 3 (2004), pp. 575-585.
3. Arunkumar, S., Sreenivas Rao, K.V., and Prasanna Kumar, T.S., "Spatial variation of heat flux at the metal–mold interface due to mold filling effects in gravity die-casting," *Int. J. Heat and Mass Transfer,* Vol. 51 (2008), pp. 2676-2685.
4. Babu, K. and Prasanna Kumar, T.S., "Mathematical modeling of surface heat flux during quenching," *Metall. Mater. Trans. B,* Vol. 41, No. 1 (2010), pp. 214-224.
5. Babu, K. and Prasanna Kumar, T.S., "Estimation and analysis of surface heat flux during quenching in CNT nanofluids," *J. Heat Transfer,* Vol. 133, No. 7 (2011), 071501 (8 pp.).
6. Babu, K. and Prasanna Kumar, T.S., "Effect of CNT concentration and agitation on surface heat flux during quenching in CNT nanofluids," *Int. J. of Heat Mass Transfer,* Vol. 54 (2011), pp. 106-117.
7. Sarmiento, G.S., Chen, X., Vega, J., Totten, G., Raynoldson, R., Huynh, L., and Meekisho, L., "A comparison of cooling curve analysis using Inc-Phatran and Winprobe," *Proc 20th ASM Heat Treating Society Conference and Show,* St.Louis, U.S., 2000.
8. Aronov, M.A., Kobasko, N.I., and Powell, J. A., "Practical application of intensive quenching process for steel parts," *Proc 20th ASM Heat Treating Conference and Show*, St. Louis, U.S, 2000.
9. Andrew M. Freborg, B. Lynn Ferguson, Michael A. Aronov, Nikolai I. Kobasko, Joseph A. Powell, "Intensive quenching theory and application for imparting high residual surface compressive stresses in pressure vessel components," *J. Pressure Vessel Technology*, Vol. 125 (2003). pp. 188-194.
10. Stephen U.S. Choi, "Nanofluids: from vision to reality through research," *J. Heat Trans.* Vol. 131 (2009), 033106 (9 pp).
11. Narayan Prabhu, K., and Peter Fernades, "Nanoquenchants for Industrial Heat Treatment," *J. Mater. Engg. Perfor.*, Vol. 17 (2008), pp. 101-103.
12. Hyungdae Kim, Gregory DeWitt, Thomas McKrell, Jacopo Buongiorno, and Lin-wen Hu, "On the quenching of steel and zircaloy spheres in water-based nanofluids with alumina, silica and diamond nanoparticles," *Int. J. Multiphase Flow,* Vol. 35 (2009), pp. 427–438.
13. Lotfi, H., and Shafii, M.B., "Boiling heat transfer on a high temperature silver sphere in nanofluid," *Int. J. Therm. Sci.*, Vol. 48 (2009), pp. 2215-2220.
14. Manoj Chopkar, Prasanta K. Das, and Indranil Manna, "Synthesis and characterization of nanofluid for advanced heat transfer applications," *Scripta Materialia,* Vol. 55 (2006), pp. 549–552.
15. Manoj Chopkar, Kumar, S., Bhandari, D.R., Prasanta K. Das, and Indranil Manna, "Development and characterization of Al_2Cu and Ag_2Al nanoparticle dispersed water and ethylene glycol based nanofluid," *Mater. Sci. Engg. B,* Vol. 139 (2007), pp. 141-148.
16. Ki-Jung Park, Dongsoo Jung, "Enhancement of nucleate boiling heat transfer using carbon nanotubes," *Int. J. Heat Mass Trans.*, Vol. 50 (2007), pp. 4499–4502.
17. Yulong Ding, Hajar Alias, Dongsheng Wen, and Richard A. Williams, "Heat transfer of aqueous suspensions of carbon nanotubes (CNT nanofluids), " *Int. J. Heat Mass Trans.*, Vol. 49 (2006), pp. 240–250.

18. Huaqing Xie, Hohyun Lee, Wonjin Youn, Mansoo Choi, "Nanofluids containing multiwalled carbon nanotubes and their enhanced thermal conductivities," *J. Applied Physics,* Vol. 94, No. 8 (2003), pp. 4967-4951.
19. Keblinski, P., Phillpot, S.R., Choi, S.U.S., and Eastman, J.A., "Mechanisms of heat flow in suspensions of nano-sized particles (nanofluids)," *Int. J. Heat Mass Transfer,* Vol. 45 (2002), pp. 855-863.
20. Sarit Kumar Das, Nandy Putra, Peter Thiesen and Wilfried Roetzel, "Temperature dependence of thermal conductivity enhancement for Nanofluids," *J. Heat Trans.,* Vol. 125 (2003), pp. 567-574.
21. Chopkar, M., Sudarshan, S., Das, P.K., and Manna, I., "Effect of particle size on thermal conductivity of nanofluid," *Metall. Mater. Trans. A*, Vol. 39, No. 4 (2008), pp. 1535-1542.
22. Min-Sheng Liu, Mark Ching-Cheng Lin, I-Te Huang and Chi-Chuan Wang, "Enhancement of thermal conductivity with carbon nanotube for nanofluids," *Int. Comm. Heat Mass Trans.*, Vol. 32 (2005), pp. 1202-1210.
23. Amrollahi, A., Hamidi, A. A., and Rashidi, A. A., "The effects of temperature, volume fraction and vibration time on the thermo-physical properties of a carbon nanotube suspension (carbon nanofluid)", *Nanotechnology,* Vol. 19 (2008), 315701 (8 pp)
24. Lifei Chen, Huaqing Xie, Yang Li and Wei Yu, "Nanofluids containing carbon nanotubes treated by mechanochemical reaction," *Thermochimica Acta,* Vol. 477 (2008), pp. 21-24.
25. Paritosh Garg, Alvarado, J.L., Charles Marsh, Thomas A. Carlson, David A. Kessler, and Kalyan Annamalai, "An experimental study on the effect of ultrasonication on viscosity and heat transfer performance of multiwall carbon nanotube-based aqueous nanofluids," *Int. J. Heat Mass Trans.,* Vol. 52 (2009), pp. 5090-5101.
26. Joshi, R, Engstler, J., Nair, P.K., Haridoss, P., and Schneider, J.J., "High yield formation of Carbon Nanotubes using a roatating cathode in open air", Diamond Rel. Mater., Vol. 17 (2008) pp. 913-919.

Proceedings of the 26th ASM Heat Treating Society Conference
B.L. Ferguson, R. Jones, D.S. MacKenzie, and D. Weires, editors
https://doi.org/10.31399/asm.cp.ht2011p0285

Development of Low-Cost, Rapid Case Hardening Treatments for Austenitic Stainless Steels

Xiaolan Wang[1]; Zbigniew Zurecki[2]; and Richard D. Sisson, Jr.[1]
[1]*Worcester Polytechnic Institute, Worcester, Mass, USA*
evawang824@hotmail.com and sisson@wpi.edu, 508-831-5335
[2]*Air Products& Chemicals, Inc., Allentown, Pennsylvania, USA*
zureckz@airproducts.com,

Abstract

Austenitic stainless steels are critical in the modern economies with applications ranging from food processing and cryogenic machinery to medical implants and aerospace instrumentation. Tough, resistant to low-temperature embrittlement and many forms of corrosion, these steels are, nevertheless, prone to scratching and galling in service. Case hardening was found to be effective when combined with corrosive surface treatments or in low-pressure, direct plasma-ion discharges, but inhibited in simple atmospheric-pressure furnaces. This paper presents preliminary evaluation of new, rapid (3-4 hrs) nitriding and carbonitriding treatments at low- (500-565°C) and high- (1100°C) temperature ranges involving injection of high-voltage, electric arc-activated, N2-based, NH3 and hydrocarbon gas mixes to the conventional box furnace. Reported data includes characterization of stainless steel product layers using SEM-EDS, XRD, OM, Leco elemental analysis, and microhardness profiling, as well as laser gas analysis of the residual furnace atmosphere.

Introduction

Stainless steel is a widely used material where corrosion resistance is needed. In 2010, the production of stainless steel is 30.7 million metric tons worldwide. And among that, 57.7% is 300 series austenitic stainless steels.[1] Austenitic stainless steels are highly valued for their corrosion, oxidation and thermal resistance, toughness and ductility, even at cryogenic temperatures. These steels contain high level chromium (Cr), as well as nickel (Ni) and/or manganese (Mn) that help stabilize their austenitic structure.[2]

Many processes have been developed to improve the stainless steel hardness and wear resistance.[3-14] Three, largely proprietary processes are best known in the US at present: ion-nitriding and carburizing in partial vacuum, plasma furnaces[3-5]; low-temperature (350-550°C) nitriding involving metal surface pre-etching using corrosive and toxic gases[6-11], and solution nitriding at high temperatures assuring the presence of austenitic phase during diffusion treatment (1050-1200°C)[12-14].

It is desired to treat and harden the steel surface using nitriding, an inexpensive, thermochemical-diffusional process, well proven in the field of low-alloy and carbon steels. But unfortunately, the passive oxide films forming on metal surface act as dense diffusion barriers and prevent the conventional nitriding for stainless steel. Many methods have been developed recently in order to overcome the problem of passive oxide films. Thus, the metal surface could be dry-etched at elevated temperatures in halide gases such as hydrochloric acid (HCl) [15] or nitrogen trifluoride (NF_3) [16], or low-pressure (vacuum furnace) nitriding using plasma ion glow discharges [3-5]. The method requires a prolonged, multi-hour processing time, and necessitates significant capital, safety equipment, and maintenance expenditures. Due to the complex surface activation steps before the process or low pressure which normally used, the hardening step for stainless steel is often expensive.

Second problem is chromium can react with nitrogen and carbon and form nitrides which will cause loss of corrosion resistance. In 1985, Zhang and Bell observed that at temperatures below 450°C, large quantities of nitrogen or carbon can be dissolved in the stainless steel to form expanded austenite phase (S-phase) which can improve wear resistant and corrosion resistance of steel simultaneously.[6] But this technique often requires plasma/implantation based techniques, which limited production capacity.[17] Expanded austenite is metastable and tends to form chromium nitrides. The high content of N is obtained, because of the relatively strong affinity of Cr atoms for N atoms, to short range ordering of Cr and N. Due to the low mobility of Cr atoms as compared to N atoms at low treatment temperatures, chromium nitrides do not precipitate until after long exposure times. And high hardness (up to 1700 HV) can be obtained by nitriding austenitic stainless steel.[9] The high concentration of interstitial atom in solid-solution and the occurrence of an enhanced stacking fault density may contribute to strengthen the austenite stainless steel. But the production rate of S-layer is extremely slow due to the low process temperature. Normally 10-20μm S-layer is expected after 22 hours treatment, with the hardness above 1000HV.[8]

To making austenitic stainless steels case hardening process more affordable, modified conventional, atmosphere pressure process was investigated in this research. Activation of nitrogen or nitrogen-ammonia atmospheres was used, depending on the treatment type and temperature desired, hydrocarbon(HC) gas additions [18] and/or electric discharge (cold plasma) injection.[19-20] With small percentage active hydrocarbon gas in the ammonia, HCN can formed and it can improve the evenness of nitriding treatment and the nitrogen penetrating. El-Rahman [21] also investigated the effects of high percentage C_2H_2 used in r.f. plasma carbonitriding for austenitic stainless steel. CH_4 is also used in plasma immersion ion implantation X5CrNi189 steel (AISI 304 stainless steel).[22] The results show both improvement in element concentration, larger case depth and higher hardness.

Overall, case hardening stainless steel is difficult and expensive due to the native passive layer which blocks the nitrogen and carbon penetration. It is desired to find an inexpensive, efficient way to treat and harden the stainless steel using minimum modified atmospheric process which is well proven in the field of low-alloy and carbon steels. This study is mostly focused on exploring the feasibility of case hardening stainless steels, the objectives of the paper are to: (1) develop the conventional, atmospheric-pressure, low-cost surface hardening treatments for the case hardening of stainless steel at different temperature range; (2) explore the possibility of activation of atmosphere and metal surface to accelerate processing time and save energy and time; (3) generate S-layer in short time period with plasma or/and hydrocarbon activation atmosphere.

Experimental procedure

Atmosphere nitriding experiments were conducted on a semi-production scale, electrically heated box furnace, ATS 3350. Gas analyses were performed by laser gas analyzer (manufactured by ARI, model LGA-4ENAPBT) for CO, CO_2, H_2, HC and NH_3. The 203μm AISI 301 stainless steel shim stock and 762μm AISI 304 stainless steel plate used in this study were cut into 76 mm by 102 mm (3' by 4') square and degreased by acetone prior to the nitriding process. The cylindrical AISI 304 stainless steel bars were machined into Φ25mm*7mm (Φ1"*1/4") discs. Element concentration was test by X-ray fluorescence (XRF), listed in Table 1.

Table 1 Composition of the alloys tested by XRF (wt%)

	Ni	Mn	Cr	Mo	Cu	Fe
301 ss shim	7.23	0.71	16.54	0.24	0.34	Bal
304 ss plate	8.24	1.91	18.40	0.34	0.374	Bal
304 ss disk	8.37	1.57	18.16	0.39	0.30	Bal

The cold AC plasma, stream-activating injector was used in the test, equipment details were described elsewhere.[18-19] For low temperature test series, the atmosphere used were NH_3 (25 vol%) in N_2-stream with small amount of CH_4 (<2 vol%) addition, a total flowrate of 100 scfh, 1.5 volume change per minute in the box furnace. The nitrogen gas used to balance total gas stream has 99.995% purity. Each nitriding cycle involved 30min long heating period from room to treatment temperature, 4 hours nitriding step at 500 or 565°C and slow cooling in the furnace, listed in Table 2. For high temperature test series, process temperature was maintained at 1100°C, and N_2 process gas was used with small addition of other active gas of total 100scfh flowrate, details listed in Table 3. Each cycle involved 60min long heating period from room to treatment temperature, 4 hours holding time at 1100°C and quenched in water.

The specimens were weighted before and after the nitriding cycle with a conventional microbalance, accuracy of 0.1mg. Microhardness and SEM test were performed on coupons afterwards. Knoop hardness on the cross-sections with 25g for 10s, was taken for the coupons and shims on Shimadzu HMV-2000 Micro Hardness tester. Metallographic cross-sections of the coupons were etched with 2% Nital, oxalic acid or 50 vol% HCl + 25 vol% HNO_3 + 25 vol%H_2O solution prior to optical microstructural and SEM examination. LECO combustion test for through average element concentration was also performed on shim stock afterwards for selected test conditions. X-ray diffraction (XRD) was used to identify the nitrides phase near the surface. The sample was scanned on the MPD from 30-85° using X'Celerator optics, Cu-Kα radiation, a 0.03° step size and 800 sec count time.

Table 2 Low temperature nitriding tests for stainless steel

	T1	T2	T3	T4
Material	762μm 304ss plate		203μm 301ss shim	
Nitriding temp. (°C)	500	500	565	565
Time(hr)	4			
NH_3% - inlet	25			
CH_4% - inlet	2	2	1.25	1.25
'CH_4 on' time	Non-stop			Heatup only
Activation used	None	AC 80%	AC 20%	AC 20%
'Activation on' time	-	Non-stop	Non-stop	Heatup only
NH_3 activation	N	N	Y	Y

Table 3 High temperature tests for stainless steel

	T5	T6	T7	T8
Material	304 ss disk			
Treatment temp.(°C)	1100			
Time(hr)	4			
Atmosphere	N_2		1.5%CH_4+N_2	
Activation used	None	AC 80%	None	AC 80%
'Activation on' time	Non-stop			

Result and discussion

Initial stage of nitrides growing

At initial stage of 565°C low temperature nitriding, microscopic crystallites were observed growing on the surface of stainless steel coupons after the first few minutes of nitriding treatment, shown in Figure 1(a). The graph indicates fresh metal surface and the first crystallites on the surface. As the treatment time progresses, the entire metal surface becomes covered with the crystallites and connected, shown in Figure 1(b).

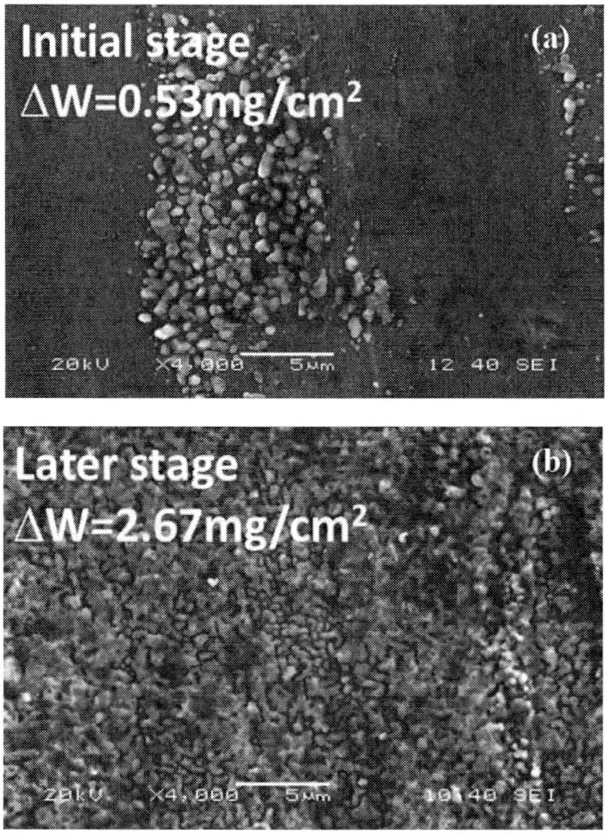

Figure 1 Top view of nitrided layers produced on AISI 301stainless steel by gas nitriding at 565 ℃, (a) initial stage; (b)after 4 hours.

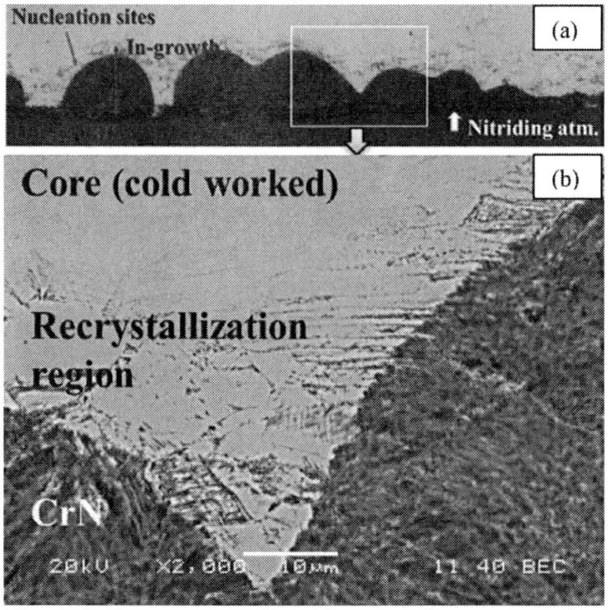

Figure 2 Cross-section of nitrides crystallites generated at 565 ℃on AISI 301 stainless steel, etched by 2% Nital acid.

Figure 2 shows 2% Nital acid etched cross sections of the metal surfaces covered by the crystallites under an increasing magnification, identified in Figure 1(a). The micrographs suggest that the nitriding process starts with a few selected nucleation sites rather than uniformly, and these surface nuclei, once formed, grow into the parent metal, joining together at a later stage. The initial absence of a planar growth front is interpreted by applicants as the consequence of the N_2-NH_3-CH_4 atmosphere used and its site-activating effect on metal surface. It's also observed that the recrystallization was initialed by nitrogen diffused into metal core, shown in figure 2(b) and (c).

Nitrides layers

After 4 hours treatment at 565°C (T3-4), a continuous nitrides layer formed in AISI 301 stainless steel shim. 301 stainless shim developed an 80micro "dark" nitrided layer which is 4 times larger than alloy steel (shown in Figure 3) with the hardness of 1000 ~1100 HK (shown in figure 4). With hydrocarbon gas injected for whole process time or light power plasma activation used, T3-T4 have identical layer thickness for all samples.

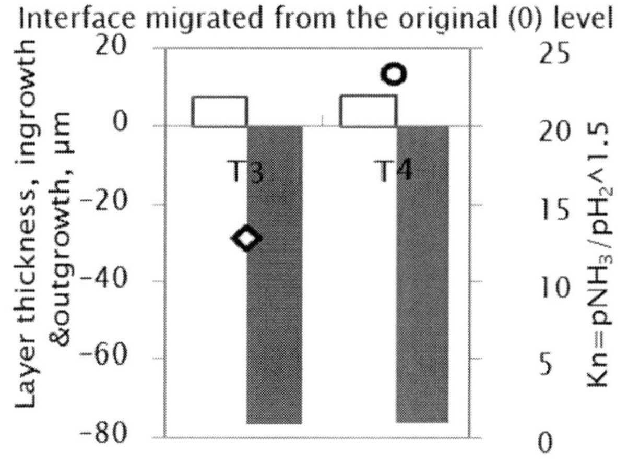

Figure 3 The ingrowth & outgrowth of 301 stainless steel nitrided layers (565 ℃, 25%NH_3, 4hrs)

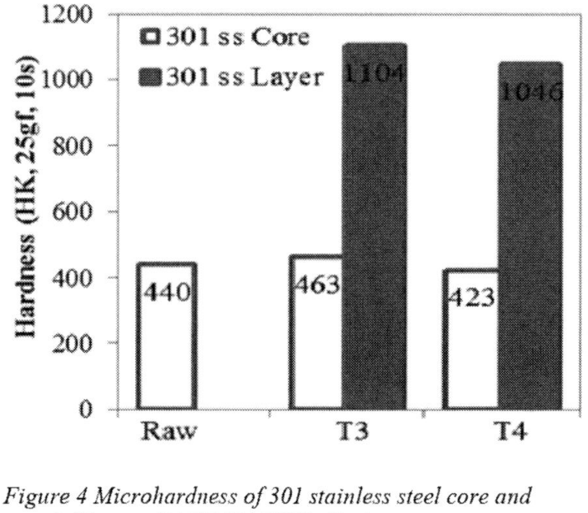

Figure 4 Microhardness of 301 stainless steel core and nitrided layers (565 ℃,25%NH3, 4hrs)

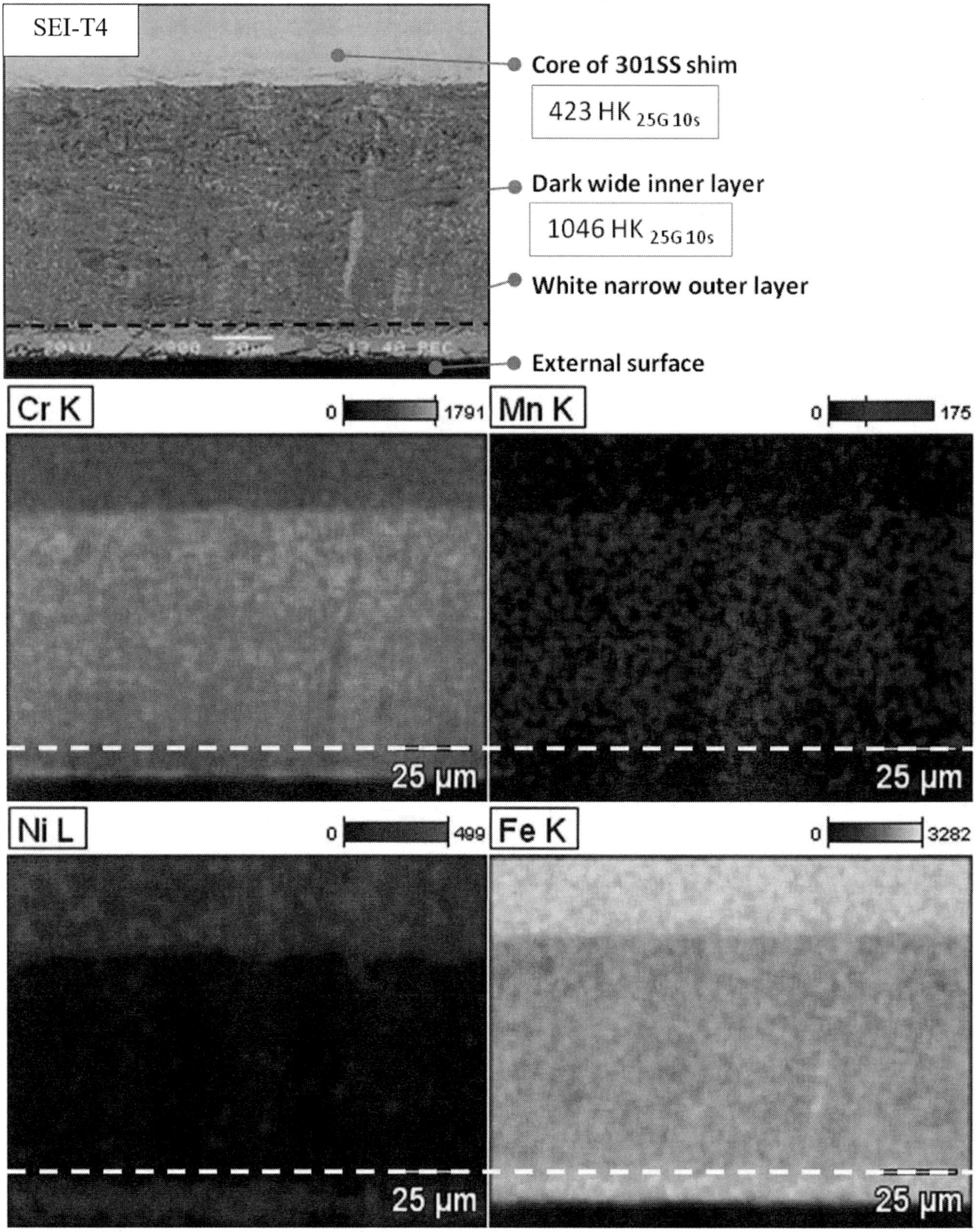

Figure 5 SEM-SEI-EDS cross-sectional images of subsurface regions of the nitrided 301 stainless steel shim

Secondary electron images (SEM-SEI) of the nitrided shim (T4) cross-sections, and the corresponding elemental maps of Mn, Cr, Ni and Fe were acquired using energy-dispersive X-ray spectroscopy (EDS) probe. 7-8μm outgrowth is observed for all the 301 ss samples, which has the same thickness of the white narrow outer layer observed under SEM. In this layer, higher Ni, Fe% and lower Cr, Mn% is observed by EDS, compared to dark wide inner layer. Cr-Mn rich layer catalyzing and nucleating the growth of "dark, wide, inner layer" below. The dark layers (shown in Figure 5) growing from the surface into the core of the specimen indicate the depth of nitriding; the white strip in the core is the unaffected parent metal. The difference in color response may be the consequence of different etching rates –the nitrided stainless steel is less resistant to etching than the parent stainless steel. The layers growing into 301 stainless steel are over 4-times thicker than the layers growing into low carbon steel. This finding is unexpected and suggests that the N_2-NH_3-CH_4 atmosphere described herein is uniquely suited for nitriding of highly-alloyed metals which tend to resist the conventional nitriding methods due to the presence of Cr-rich, passive oxide films.

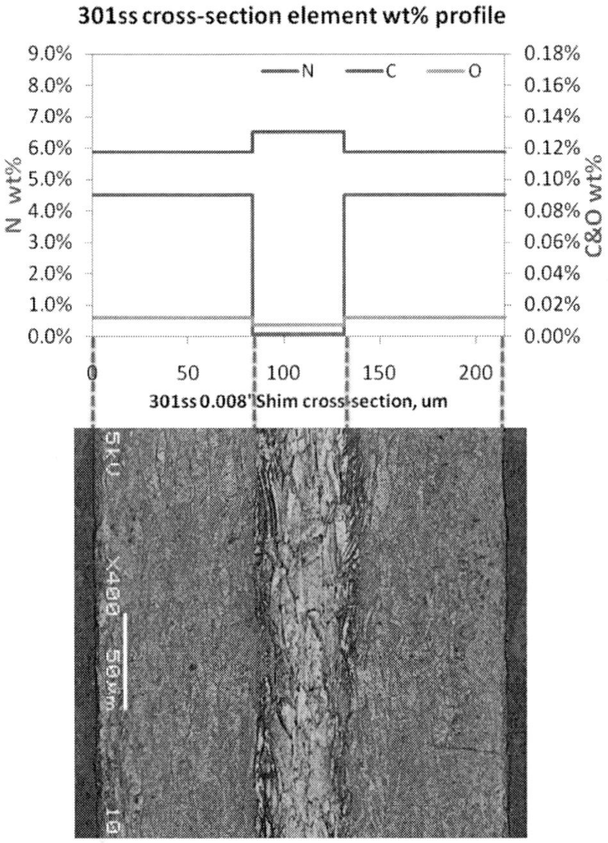

Figure 6 The estimated distribution of interstitial of N, C and O wt% in nitrided layer, plotted per elemental Leco analysis of treated and untreated shims (up) and SEM micrograph of cross section of the 301 SS shim after the nitriding treatment, etched by oxalic acid (bottom).

The same nitrided 301 stainless steel specimen (T4) was etched with oxalic acid in order to reveal grains in the nitrided layers and in the unaffected, parent metal core, here visible as a narrow strip in the center of the microscopic image, shown in Figure 6. Elemental chemical analyses were carried out on raw and nitride shims for nitrogen (N), carbon (C) and oxygen (O) using a Leco combustion gas extraction analyzer. The results are plotted directly above the image of the crosssection. It is apparent that the nitrided layers contain about 4.5 wt% of nitrogen while the N-content in the parent metal is zero. The O-level in the nitrided layers is very low, about 0.01 wt%, not much more than in the parent metal. Finally, the C-level in the nitride layers is below 0.12 wt%, less than in the parent metal. This confirms that the CH_4- containing atmosphere of this invention does not carburize the metal treated but, only, accelerates the nitriding. Comparison of the nitride layer and parent metal hardness suggests that the presence of the CH4 addition throughout the entire, 4-hour long nitriding cycle is important for metal hardening, even though there is no carburizing effect involved.

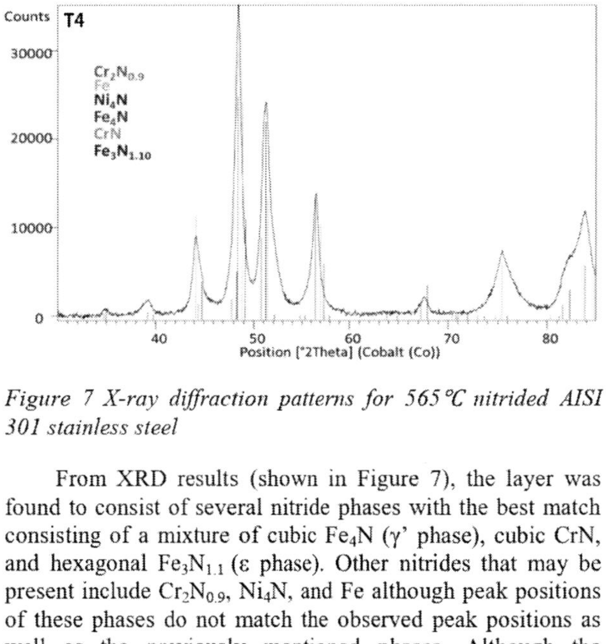

Figure 7 X-ray diffraction patterns for 565°C nitrided AISI 301 stainless steel

From XRD results (shown in Figure 7), the layer was found to consist of several nitride phases with the best match consisting of a mixture of cubic Fe_4N (γ' phase), cubic CrN, and hexagonal $Fe_3N_{1.1}$ (ε phase). Other nitrides that may be present include $Cr_2N_{0.9}$, Ni_4N, and Fe although peak positions of these phases do not match the observed peak positions as well as the previously mentioned phases. Although the nitrocarburizing atmosphere contains both NH_3 and CH_4, at low temperature, only Cr and Fe nitrides are formed, no carbides can be detected by XRD.

S-layer

The Cr in the stainless steel reacted with N, formed CrN, and lost the corrosion resistance due to adequate amount of Cr left in the steel. So a further reduced- temperature nitriding test was performed at 500°C for 4hr to form an expanded austenitic s-layer, a thermally metastable layer of austenitic (FCC) structure containing large quantities of N dissolved in metallic matrix. In this thin hard layer, the expanded austenitic phase can be detected, and the corrosion resistance is even better than base metal. By producing this phase, surface hardening without sacrifice corrosion resistance can be archived. But to form this layer, proper control of the process parameter is critical. The s-layer can only form at a low temperature (<530°C) with limited nitriding time. At the low temperature, the N diffusion rate is low and the whole process takes extremely long time to form a usable nitrided layer. A lot of research has been done in this area, with expensive ion sputtering, low-temperature plasma surface alloying and etc to accelerate the process.

Another group of tests was performed to target generating S-phase, by using modified conventional nitriding process. AISI 304 stainless steel coupons were treated for 4 hours in the N_2-NH_3-CH_4 atmosphere at a temperature reduced to 500°C. Due to an apparently too long treatment time and/or too high treatment temperature, the S-layer produced in the 1st stage became decorated with small nuclei of Cr-nitrides growing from the outer surface in. An important finding of this reduced-temperature, 500°C tests, is that the S-layer grown, and the coupon weight gain, ΔW, were one magnitude larger for the N_2- 25vol%NH_3-1.25vol%CH_4 atmospheres activated with AC electric arc at the inlet to the furnace. The electric activation is important especially during nitriding more alloyed stainless steels and/or during nitriding at lower temperatures.

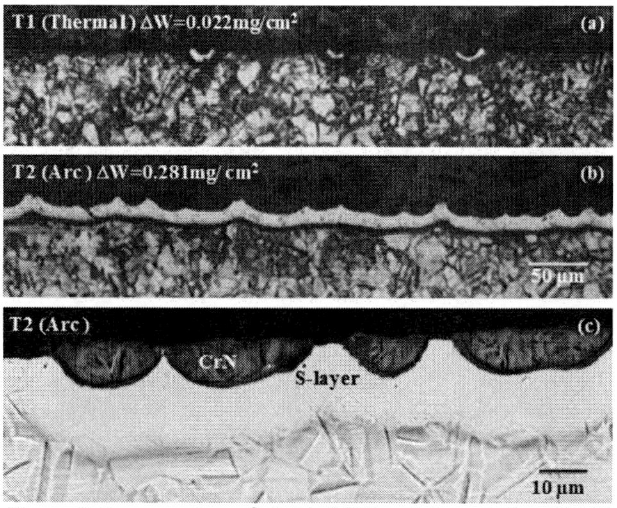

Figure 8 Cross-section of nitrided 304 stainless steel at 500℃ for 4hr, (a)T1:non-activated atmosphere run: (b)T2: activated atmosphere run; (c) SEM graphs of T2 cross section.

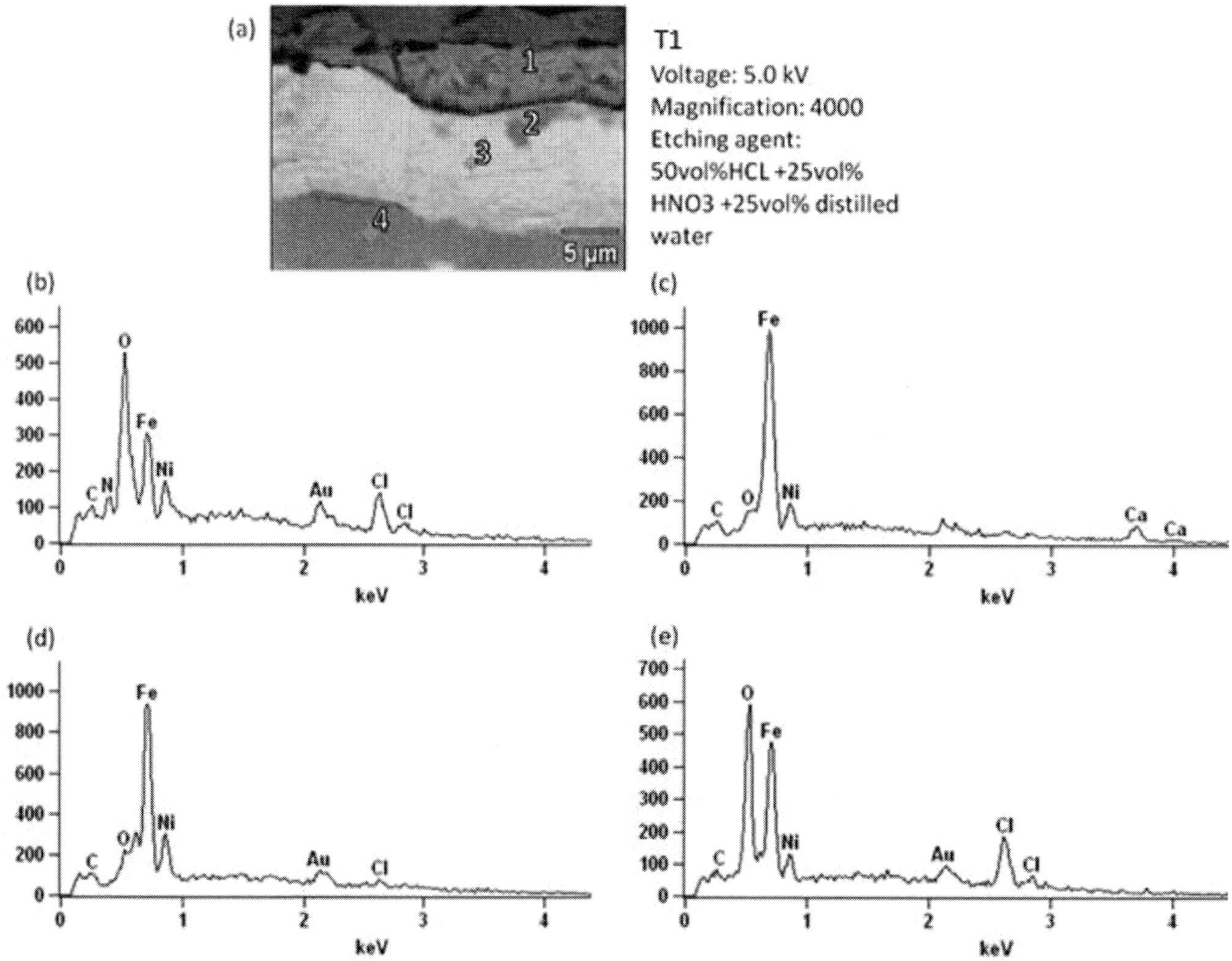

Figure 9 Cross-section of nitrided 304 stainless steel at 500℃ for 4hr, (a) Cross section by SEM; (b) Composition of cross section (position 1); (c) Composition of cross section (position 2); (d) Composition of cross section (position 3); (e) Composition of cross section (position 4).

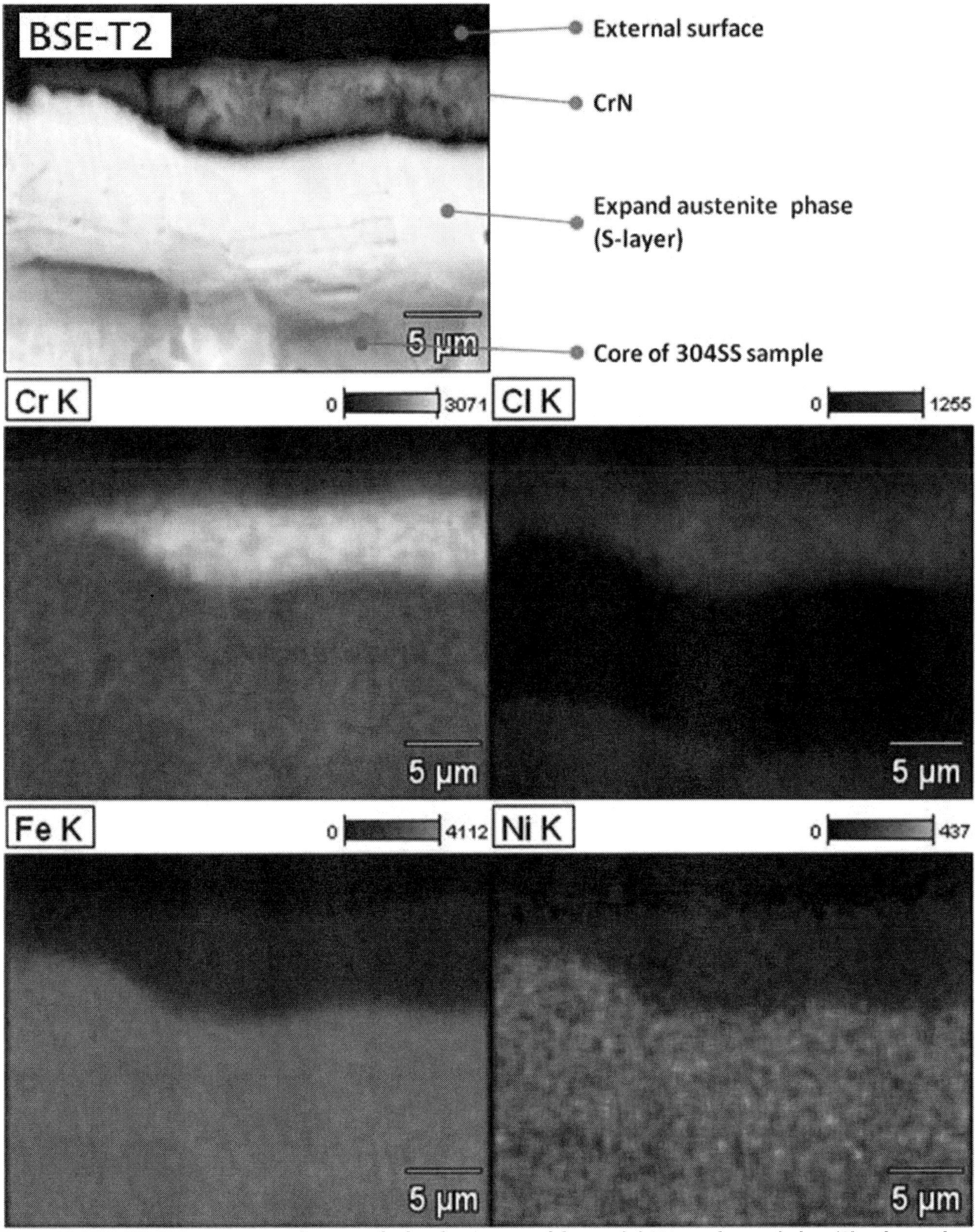

Figure 10 SEM-SEI-EDS cross-sectional images of subsurface regions of the nitrided 304 stainless steel plate (Cl is from etching agent HCl)

For N_2 and CH_4 activated in plasma discharge and NH_3 entering furnace atmosphere via separate inlet, arc discharge enhances nitriding of stainless. In this s-phase layer, fine grain structure can be observed under SEM, the element concentration didn't change from the bulk material. It is speculated that the difference can be related to a faster removal of passive film from stainless steel with arc-activated gases. Also, plasma-activation promotes the growth of S-layer. Elemental analysis of the typical S-layers decorated with nitrides, etched by acid is shown in Figure 9-10. It shows the topography of the nitride, the S-layer and the parent metal, the Cr-enrichment and the absence of a relatively non-reactive nickel (Ni) in the top nitride phase, the absence of chlorine (Cl) in the S-layer indicating its increased resistance to acid attack, and the uniform distribution of iron (Fe) across the material, except the Cr-enriched nitrides. It suggests that after adjusting the time and temperature of the treatment, it is possible to grow corrosion resistant S-layers without the use of expensive and toxic etchants or vacuum plasma ion nitriding chambers.

Solution nitriding and carburizing

High temperature treatments were also conducted on the AISI 304 stainless steel samples. Nitrogen gas is used as a nitrogen source not a protective atmosphere. At temperatures above 1050°C (1920°F), stable nitrogen molecules can dissociate at metallic surfaces into active nitrogen atoms. The dissociated nitrogen can despite the passive chromium-oxide layer, penetrate the surface and diffuse into metal piece. AISI 304 stainless steel disks were used in this group of test, process temperature and time were kept constant at 1100°C for 4 hour and only variable is atmosphere condition. Comparing the non-activated (T5-thermal) nitrogen atmosphere run with electric-arc activated (T6-arc) run, more nitrogen is picked up by parent metal, shown in Figure 11. The reaction is clearly been accelerated and higher surface hardness and deeper case depth were produced by arc-activated run. From the cross-section hardness profile, hardness increased from 200 to 350 HK and several hundred micron case depth was generated, shown in Figure 12.

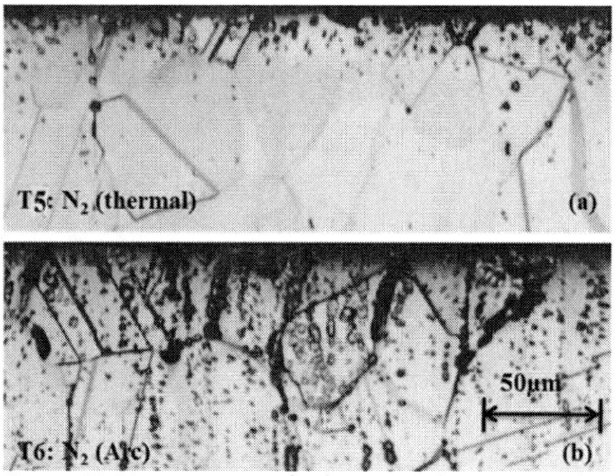

Figure 11 Microstructure of solution nitriding 304 stainless steel, (a) non-activated nitrogen atmosphere run, (b) electric-arc activated nitrogen atmosphere run.

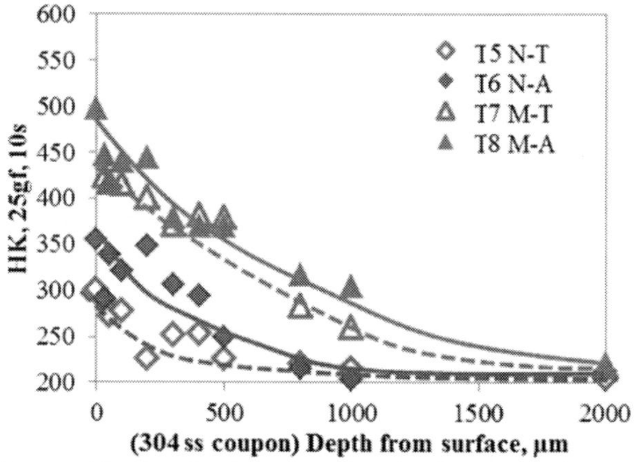

Figure 12 Microhardness profile of solution nitridng and carburizing 304 stainless steel (1100°C, 4hrs).

Small amount methane was added into process gas for T7 and T8. The test runs with hydrocarbon gas have significantly larger weight gain and the extra weight pickup during the process is due to the carbon diffusion. Pearlite was observed in the effective case region (in Figure 13), and much higher hardness can be achieved by carbon diffused into the base metal with larger case depth. From the Figure 13(b) microstructure photos, 1st layer (0~150µm): occupied with fine grains and nitrides, no carbides exist; 2nd layer (150~600µm): carbides and pearlite shape grains start to exist; 3rd layer (600~1500µm) large grains formed with pearlite. Nitrides and carbides seem don't precipitate in the same location. And more than 1mm hardening case was generated after 4 hour process. From the hardness result, methane runs have the highest hardness, which have 450-500 HK surface hardness. According to OES results (shown in Figure 14), carbon diffusion rate is higher than nitrogen, so carbon diffused into metal first, and then blocked the nitrogen penetration. This phenomenon resulted into high carbon pickup and little nitrogen pickup in the process.

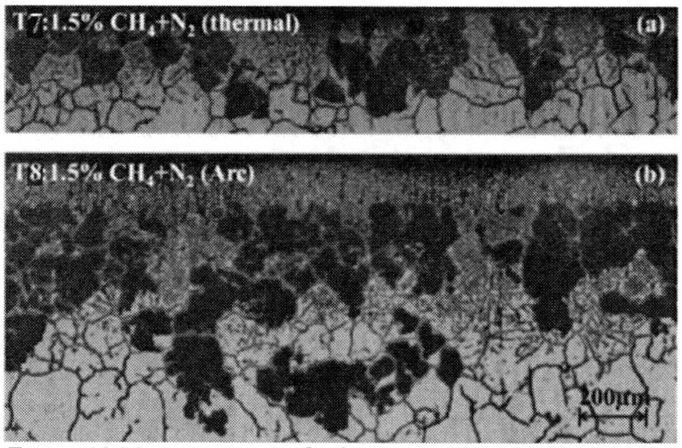

Figure 13 Microstructure of solution carburizing 304 stainless steel, (a) non-activated nitrogen atmosphere run, (b) electric-arc activated nitrogen atmosphere run.

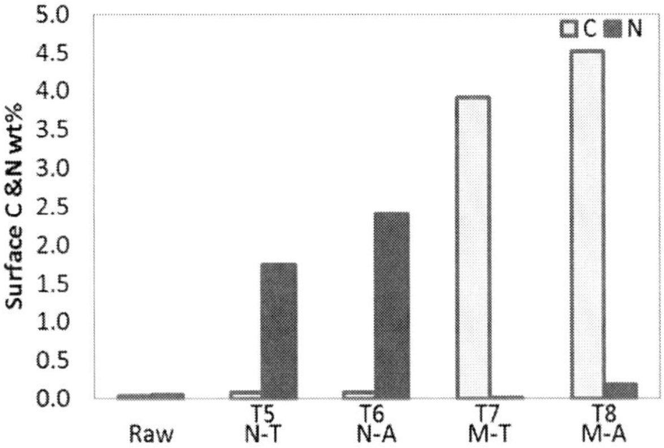

Figure 14 Surface concentrations of N and C for solution treatment samples.

Conclusions

In this study on the behavior of stainless steel under atmospheric heat treating condition with nitrogen based process gas, the following key observations are made:

1. 80 µm thick nitrided layer was successfully generated using a novel, 4-hr long, atmospheric-pressure gas treatment on 300-series austenitic stainless steel. High hardness of 1000 ~1100 HK was observed at this layer.
2. Electric arc activation of nitrocarburizing atmosphere increases the growth rate of S-layers by one order of magnitude. 15-20 µm S-layer was successfully formed after 4 hour process. But Cr-nitrides formed on top of the s-layer and lost corrosion resistance. Lower temperature or shorter time should be used in the future test to grow S-layer without CrN precipitates using arc activation.
3. Solution nitriding and carburizing at 1100°C with electric arc activation have improvement in hardness and larger case. And much higher hardness can be achieved by carbon addition.
4. Simple and cheaper methods to case hardening stainless steel are successfully demonstrated. With some optimizations, those processes can be used in commercial heat treating facility in near future.

Acknowledgments

The authors would like to thank J.L. Green for laboratory support and Air Products for funding and the permission to publish this study.

References

[1] ISSF announces, "Global stainless steel output figure for 2010", url: http://www.worldstainless.org/News/Media+releases/Stainless+steel+production+in+2010.htm, last accessed: March, 2011

[2] Coates, G. and Jenkinson, D., "What is Stainless Steel?", url: http://www.nickelinstitute.org/index.cfm/ci_id/11021.htm, last accessed: April, 2011

[3] Wei, R *et al*, "A comparative study of beam ion implantation, plasma ion implantation and nitriding of AISI 304 stainless steel", *Surface and Coatings Technology*, Volume 83, Issues 1-3, Pages 235-242, September 1996

[4] Blawert, C. *et al*, "Plasma immersion ion implantation of stainless steel: austenitic stainless steel in comparison to austenitic-ferritic stainless steel", *Surface and Coatings Technology*, Volume 85, Issues 1-2, Pages 15-27, November 1996

[5] Collins, G. A. *et al*, "Nitriding of austenitic stainless steel by plasma immersion ion implantation", *Surface and Coatings Technology*, Volumes 74-75, Part 1, Pages 417-424, September 1995

[6] Zhang, Z.L. and Bell, T., "Structure and corrosion resistance of plasma nitrided stainless steel". *Surf. Eng.* Vol 1 (2), 1985

[7] Christiansen, T. and Somers, M. A.J., "Characterization of low temperature surface hardened stainless steel", *Struers Journal of Materialography,* Sept., 2006

[8] Bell, T. and Li, C.X., "Stainless Steel: Low-Temperature Nitriding and Carburizing", *Advanced Materials & Processes*, Volume 160, Issue 6, 2002

[9] Buhagiar, J., Li, X. and Dong, H., "Formation and microstructural characterisation of S-phase layers in Ni-free austenitic stainless steels by low-temperature plasma surface alloying", *Surface and Coatings Technology*, Volume 204, Issue 3, Pages 330-335, October 2009

[10] Gontijo, L.C. *et al*, "Study of the S phase formed on plasma-nitrided AISI 316L stainless steel", *Materials Science and Engineering: A*, Volume 431, Issues 1-2, Pages 315-321, September 2006

[11] Tsujikawa, M. *et al*, "Effect of molybdenum and copper on S-phase layer thickness of low-temperature carburized austenitic stainless steel", *Surface and Coatings Technology*, Volume 202, Issues 22-23, Pages 5488-5492, August 2008,

[12] Schmalt, F., Berns, H., and Zaugg, R., "Solution Nitriding—A New High Temperature Nitriding of Stainless Steel", *Surface Engineering Coatings and Heat Treatments 2002: Proceedings of the 1st ASM International Surface Engineering and the 13th IFHTSE Congress,* Pages: 88-97, Jan 2003

[13] Edenhofer, B., "Solution Nitriding - A Cost Effective Case Hardening Process for Stainless Steels", *ASM Heat Treating Society Conference and Exposition*, 2007, COBO Center, Detroit, Michigan, USA

[14] Berns, H., "Case hardening of stainless steel using nitrogen", *Industrial heating,* vol. 70, no5, pp. 47-50, 2003

[15] Kochmanski, P. and Nowacki, J., "Influence of initial heat treatment of 17-4 PH stainless steel on gas nitriding kinetics", *Surface and Coatings Technology*, Volume 202, Issue 19, Pages 4834-4838, June 2008,

[16] Tahara, M. *et al*, U.S Patent 5,792,282

[17] Du, H. and Agren, J., "Gaseous Nitriding Iron - Evaluation of Diffusion Data of N in γ' and ε Phases", *Solid Phase Transformations*, 1994

[18] P.N. Kogyo, U.S. Patent 2007/0204934 A1
[19] Z. Zurecki *et al*, U.S. Patent 2008/0283153
[20] Zurecki, Z and Wang, X, "Atmosphere carburizing using electric discharge-activated nitrogen-natural gas mixtures," Heat Treating Conference and Exposition, Indianapolis, Indiana, Oct 2009.
[21] El-Rahman, A.M. Abd *et al*, "Effect of N2 to C_2H_2 ratio on r.f. plasma surface treatment of austenitic stainless steel", *Surface and coatings technology*, vol.183, p268-274, 2004
[22] Blawert, C. *et al*, "Characterisation of duplex layer structures produced by simultaneous implantation of nitrogen and carbon into austenitic stainless steel X5CrNi189", *Surface and coatings technology*, vol.128, p219-225, 2000

Proceedings of the 26th ASM Heat Treating Society Conference
B.L. Ferguson, R. Jones, D.S. MacKenzie, and D. Weires, editors
https://doi.org/10.31399/asm.cp.ht2011p0295

Effect of Process Time on Low-temperature Nitrided Austenitic Stainless Steels Layer Structure

D. Koshel, J. Kalucki
Nitrex Metal Inc., 3474 Boul. Poirier, St Laurent, QC, Canada H4R 2J5
jack.kalucki@nitrex.com

Abstract

Considerable interest has been expressed in nitriding austenitic stainless steels under condition that nitrogen is present mostly in solid solution. The effect of Supersaturation with nitrogen is to significantly increase hardness and develop high residual compressive stresses. This paper will expand on the topic by studying the relationship between the nitrogen content, layer morphology and properties. It will examine Nitrex's Nano-S™ nitriding processes performed at various nitriding times to achieve a range of nitrogen contents in austenitic AISI 304 stainless steel. Research also involved 410 grade martensitic stainless steel as well as 17-4 precipitation hardened steel, and a final analysis of these metals will be provided in October 2011. From these results, optimum nitriding conditions to achieve the desired properties may be recommended.

Keywords: Austenitic stainless steel, low temperature nitriding, nitriding potential control, corrosion resistance, wear resistance, galling resistance

Introduction

The main advantage of stainless steels for many applications is their high resistance to corrosion. However, they have mechanical properties inferior to many other steels, and consequently they may require high performance coatings which allow them to achieve a high surface hardness and resistance to galling and wear. Of the various approaches to surface hardening stainless steels, controlled gas nitriding is economical, practical and reliable in terms of control and repetition of results.
Where it concerns austenitic stainless steels, these have slower nitrogen diffusivities and consequently need more precise K_N control to prevent excessive white layer growth compared to conventional steel grades.
In the recent years, Nitrex Metal Inc. has developed new hardware and software which allows an extremely precise control of the gas nitriding process. This has lead to the possibility of new applications for which traditional nitriding technologies were insufficient to meet those technical requirements. In particular, the new advances allow for a more precise control of the atmosphere and the nitriding potential, K_N.
As a result of these newly developed control systems, we have been able to develop robust technologies to offer a wider range of applications for stainless steels as well as other grades such as tool steels or high alloy steels.

This paper will present examples of such nitriding using low-temperature gas nitriding at different process times, showing the optimum nitriding conditions which yield a surface layer with exceptional hardness, no white layer or other metal nitrides. The very high hardness is a result of the supersaturation of the austenitic crystal structure of the material with nitrogen. An additional benefit of this type of process is that the corrosion resistance is maintained.

1. Process controls

The Nitrex gas analyzer is a sampling system designed to measure hydrogen content in nitriding and nitrocarburizing atmospheres. The system is engineered to provide high accuracy and to calculate the parameters necessary for process control. An integrated sampling pump with variable output and a flow monitoring circuit with a filter saturation warning and flow alarm insure reliable sampling. The pump provides a continuously controlled flow despite sampling line obstructions or filter contamination, assuring accurate measurements and consequently better process control.
The Nitrex analyzer meets the industry's highest standards with an accuracy of ±0.5% of reading and ±0.2% of full scale. The linearity is <0.5% of full scale; the repeatability and drift are below 0.5% of full scale. The combination of the analyzer and flow controllers yields a closed loop control system. Consequently, gas nitriding can be performed with sufficient precision to ensure not only the thickness but the required structure of the nitrided layer as well.

2. Low temperature nitriding of austenitic stainless steels

Austenitic stainless steels show excellent corrosion resistance in many environments, but have the disadvantage of poor tribological properties. For applications where wear-resistance is an important factor, this class of steels may be inadequate or require a very conservatively designed part. Such a design would naturally incur higher material costs and more frequent maintenance. Examples of such applications include valves, injectors, latches, etc. which are exposed to corrosive environments.
One approach to enhancing wear resistance is to apply a surface treatment which offers a high hardness. Gas nitriding has been successfully applied to a wide range of steel grades, and can provide higher surface hardness than carburized layers. At the same time it insures an effective closed-loop control of process

chemistry, resulting with surface uniformity and repeatability better than plasma nitriding processes.
The difficulty in applying nitriding to stainless steels is relatively slow diffusivity of nitrogen in austenite. Increasing process temperature will speed up diffusion, but the increase is limited by the fact that chromium has high affinity to nitrogen. At temperatures higher than ~450°C (842°F) the development of chromium nitrides is thermodynamically favorable. However, the bounding of chromium to nitrides impairs the corrosion resistance of material. For this nitriding at lower temperatures has become a treatment of choice for stainless steels [1]. At low temperatures (below ~450 °C) the kinetics of nucleating and growing these nitride phases are much slower than the diffusion times to achieve the desired case depth. Very high nitrogen contents of up to 30 atomic percent are achievable [2].
The supersaturated solid solution is generally termed "expanded austenite" for austenitic stainless steels and has sometimes been used to describe the equivalent condition for the martensitic structures (expanded martensite). Christiansen and Somers have measured the solid solubility of nitrogen in two austenitic steels (AISI 304 and 316) and have shown interstitial occupancies as high as 61% at temperatures below 445 °C (833 °F), leading to the description of this as "colossal dissolution" [2]. High supersaturations have also been reported for low temperature carburizing of stainless steels by Collins and Williams who noted that superficial hardnesses exceeding 1000HV could be achieved on austenitic stainless steels [3]. In contrast, various research groups have shown that a surface hardness in the range of 2000HV can be achieved by low temperature nitriding, resulting in a supersaturated solid solution [4]. Esfandiari and Dong reported a maximum surface hardness of 1300HV with a 50 g load on a precipitation hardenable A286 stainless steel [5]. They further reported a significant increase in the wear resistance commensurate with the hardness. Spies et al. [6] further suggest that the nitriding temperature should be kept below 420 °C (788 °F) to maintain good corrosion resistance for the X5CrNi18-10 stainless steel while developing a high hardness and wear resistance.
Moreover, a number of studies have reported significantly better corrosion resistance of low-temperature nitrided stainless steels compared to untreated material [6, 7]. To explain the enhancement of the corrosion resistance of nitrogen supersaturated structures such as expanded austenite (or expanded martensite for martensitic steels) could be ascribed to their high nitrogen content. Speidel and Uggowitzer [8] interpret such an action of nitrogen as a result of multiple effects:

- Nitrogen dissolves during the chemical attack to form NH^+ which locally increases pH (a local passivating effect).
- Nitrogen plays a role in the formation of a dense oxynitrided passivating layer.
- Finally the formation of nitrites and nitrides was reported near the fresh corrosion pits, which depicts a local inhibiting effect.

The disadvantage of treating steels at low temperatures is the resulting shallow case depth (typically less than ~30 microns (1.18 mil)) which is strongly limited by the practical process times. In cases where a large case depth is required, nitriding must be performed at higher temperatures and the potential disadvantages of having metal nitride precipitations must be accepted. There are many applications where this is the case, including aerospace ones where the risk of exposure to highly corrosive atmospheres is reduced and the presence of the white layer yields other desirable properties such as reduced friction.
The primary interest of this work was to explore the safe limits of processing times before the formation of chromium nitrides at the surface.

3. Experimental procedure

At Nitrex Metal Inc., we have chosen the trade name of Nano-S™ for our low-temperature gas nitriding technology using controlled ammonia gas decomposition.
AISI 304 grade stainless steel 5x5x20 mm (~0.2x0.2x1″) samples were nitrided at 425 °C (797 °F) for times ranging from 120 to 2400 minutes using the proprietary Nano-S™ technology in a Nitrex NX-815 furnace. To achieve a suitable surface condition prior to nitriding, a special proprietary surface activation is performed as part of the nitriding heat treatment process.
Cross section specimens were electrolytically nickel plated in Wood's reagent and mechanically polished under low load. The samples used for the optical metallography were subsequently electrolytically etched in 10% oxalic acid (current density 0.5 A/cm^2). This etchant reveals the carbides/nitrides and the grain boundaries in the austenite. The hardness was measured using a Vickers indenter. The nitrogen concentration profiles were acquired by Energy-dispersive X-ray spectroscopy (EDX) using *JEOL JSM 6300F* Scanning Electron Microscope (SEM) at 20 kV acceleration voltage.
The arithmetic mean roughness, R_A was measured before and after the Nano-S™ treatment. The roughness was measured using a *Mitutoyo Surftest 201* profilometer.

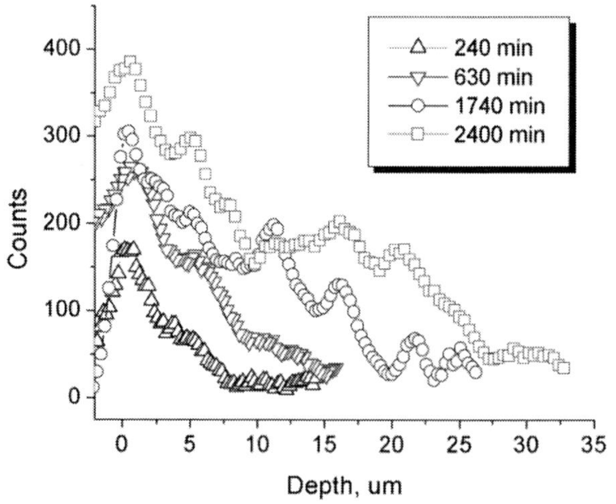

Fig. 1. EDX measured nitrogen profiles for different nitriding times

4. Results and discussion

The nitrogen concentration profiles estimated by EDX technique for the samples, nitrided for different times, are presented in Fig. 1. The signal counts were normalized using the level of K_α Fe signal measured in the bulk. Nitrogen concentration at the very top of the layer seems to increase with the process time. However, the probe beam diameter approaches the nitrided layer thickness for the shortest times of treatment, so the direct comparing of absolute values for such profiles is not reliable. The nitrided case depth follows approximately the square root of processing time law typical for the diffusion processes. Fig. 2 shows the nitrided case depth evolution in relation to the treatment duration, estimated from the EDX nitrogen profiles and from the optical microscope evaluation. In the first case the number corresponds to the distance between the surface and the depth where the nitrogen concentration drops to the core level. Under the optical microscope there is a threshold with regards to the nitrogen content and the etching response. The visible case depth therefore appears shallower than the nitrogen diffusion range.

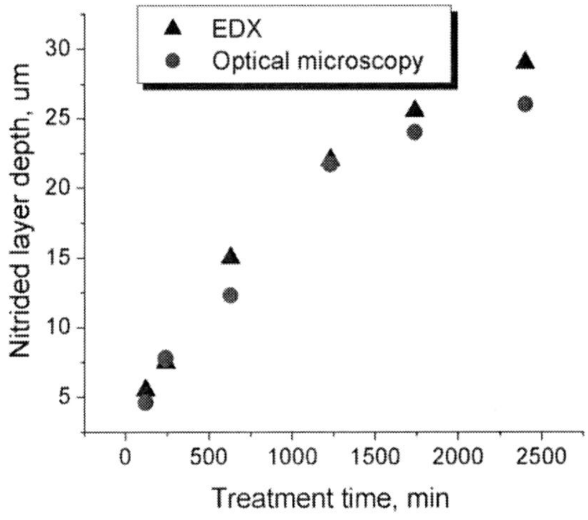

Fig. 2 Nitrided layer depth as a function of nitriding time.

Fig. 3 shows the cross sectional morphology evolution of a nitrided layer with the time of nitriding. The supersaturated austenite remains visible as a brighter feature under an optical microscope.The fact that many etchants do not attack the expanded austenite layer phase has been offered as evidence suggesting that it has superior corrosion resistance to untreated stainless steel. The nitrided layer consists of two distinct regions: the upper dark coloured zone and the lower bright zone. The separation line between these zones is rather sharp. It corresponds roughly to the locations of 'steps' in the EDX profiles (Fig. 1), probably due to the difference in relief caused by polishing. (Such 'steps' at the same positions were observed also for K_α Fe and Cr signals).

Egawa et al. [9] attribute the presence of the 'dark phases' in the nitrided layer to the precipitations of chromium nitrides which deteriorates the resistance of steel to etchant. They found such 'dark phases' to be distributed uniformly in the nitrided layer and along the grain boundaries. The authors reported the existence of a critical nitriding temperature, above which the dark zone appears (430-440 °C (806-824 °F) for the AISI 304 grade in 4 hours plasma process, 80% of nitrogen and 20% of hydrogen mix).

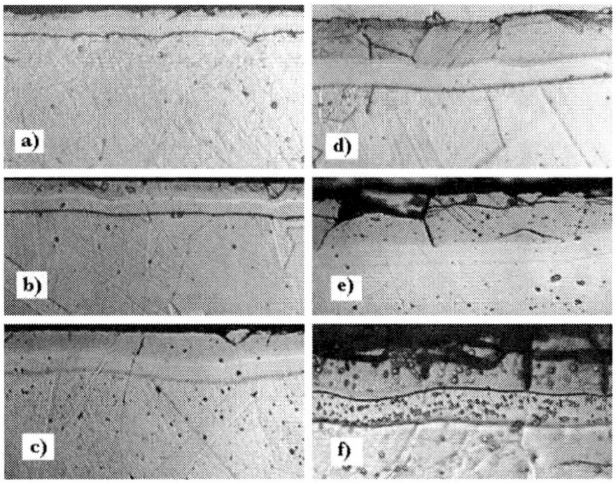

Fig. 3. Optical micrographs of structures, nitrided for a) 120 min b) 240 min c) 630 min d) 1230 min e) 1740 min f) 2400 min.

In our experiment the dark layer appeared at 120 min of nitriding on the very top of the nitrided case (Fig. 3a). Increasing the process time to 1230 min results in further growth of this layer as well as revealing grain boundaries by etching (Fig. 3d). Nitride precipitations were evident in both dark and bright zones under the optical microscope for the longest process time 2400 min (Fig. 3f). Moreover, the longest process times induce a visible cracking near the top of the dark zone (Fig. 3e, f). The SEM scan (Fig. 4) has revealed a cracking of the nitrided layer, parallel to the surface, as early as 240 min of nitriding.

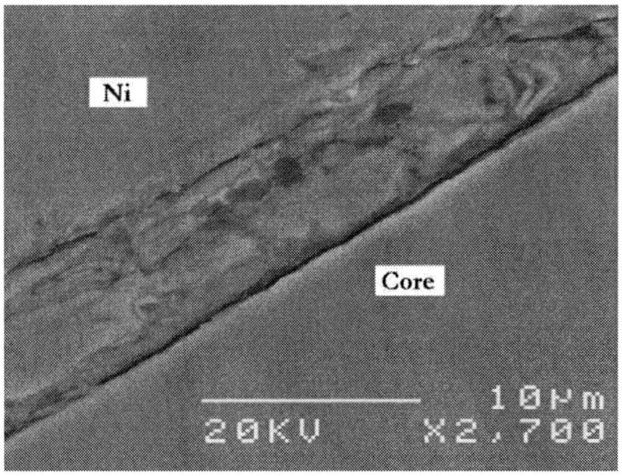

Fig. 4. SEM image of the sample, nitrided for 240 min.

Latella and Short studied the expanded austenite layer cracking under monolithic tensile loading, and noted that cracking of the supersaturated layer starts at a strain of 1.85 %, which is well within the plastic deformation region of austenite. Such

deformation corresponds to tensile stress of ~4.3 GPa [10]. The stress limit under compressive load is generally higher.
Christiansen and Somers have done an X-ray diffraction-based determination of supersaturated austenite lattice strains. They reported compressive stresses as high as 8 GPa, and associated the cracking of the expanded austenite layer with the compressive stress relaxation [2].

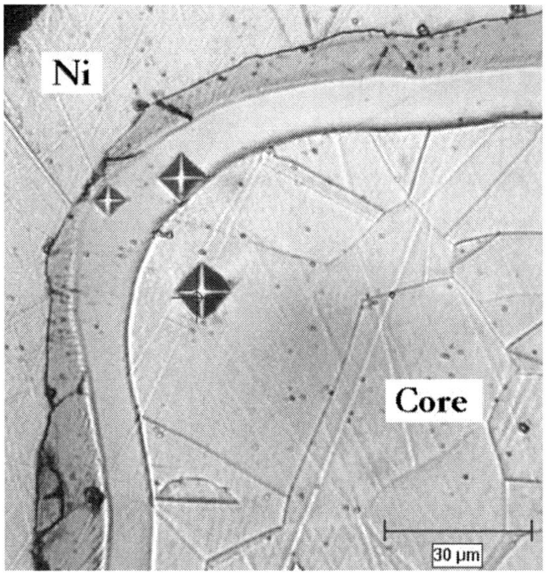

Fig. 5. Reflected light micrograph of the corner of sample nitrided for 1230 min.

There is a difference in the compressive stress field distribution in the flat nitrided surface and near a corner; one could expect a less stressed lattice in the latter case. Indeed, the examination of the nitrided layer near the corner (Fig. 5) has revealed a difference in the dark/bright layer relative thicknesses, comparatively to a flat surface (Fig. 3d). The non-attacked bright zone is significantly larger around the corner. This fact suggests that the high degree of stress of an expanded austenite does not only result in an extremely brittle top layer, but also decreases its resistance to the etchant. One could expect higher compressive stresses and possibly higher corrosion rates near features such as holes. Parts of a complex shape should therefore be nitrided to prevent corrosion damage near 'critical' surfaces. That could be done by a tight control of the nitriding potential, as used in Nano-S™.
The microhardness profile of nitrided samples follows the nitrogen concentration profile, reaching ~1500HV next to the surface of the sample, nitrided for 2400 min (Fig. 6). Such very high hardnesses have been reported previously for expanded austenite structures [4].
It is problematic to measure the hardness very close to the surface on a cross-section of a sample. We have estimated the hardness of the top of the nitrided layer by superficial microhardness testing.
The soft austenitic core has a pronounced influence on superficial hardness measurements. This is particularly true for the shortest processes, where the case depth is too shallow (Fig. 7). The effect became more noticeable with the indentation load increase. The purpose of this experiment was to illustrate the contact load properties of nitrided parts only, without providing accurate hardness numbers. As expected, the difference in measurements between different loads became less important for extended nitriding times.
The measured superficial hardness increases with the process time yielding as high as 1900HV at 1740 minutes for a 25 g load. At such a low load the indenter penetration depth is less than 1 micron, which is far below the nitrided layer thickness [~25 microns (0.98 mil) at 1740 minutes], thus the hardness should be close to the real value. The 1740-minute process resulted in maximal measured hardness for all four loads. The process of nitrides formation and compressive stress relaxation became more important with increased time, as it has been reported by Jegou *et al* [11]. That could explain the drop in the superficial hardness for the samples, nitrided for 2400 min.
One of the problems typically associated with nitriding steels is an increase in the surface roughness. For applications such as valves a low surface roughness is essential to maintain a good seal. To determine the effect of the Nano-S™ process on the surface roughness, polished specimens were prepared using conventional metallographic techniques to a final 0.05 micron (0.02 mil) Al_2O_3 polishing media.

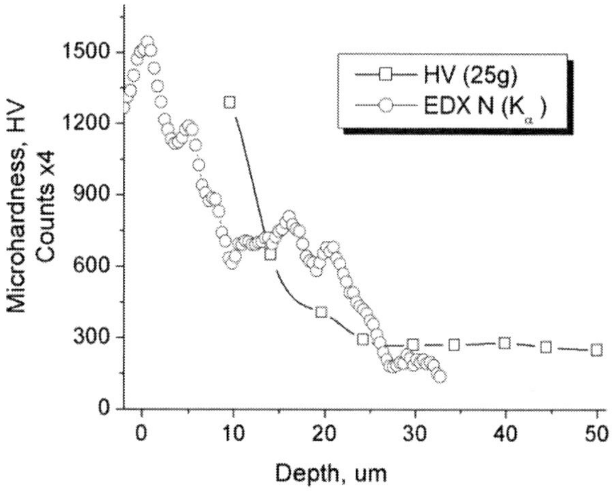

Fig.6. Vickers (25 g load) hardness depth profile compared to nitrogen distribution, measured by EDX method. Sample is nitrided for 2400 min.

For the purpose of this paper, we considered only the arithmetic mean roughness, R_A. This was measured before and after the Nano-S™ treatment on all samples. The roughness is the average of four profiles of 12.5 mm (0.5″) length on each specimen and the results are listed in Table 1.
The roughness of the austenitic stainless steel AISI 304 is essentially unchanged by the nitriding until 1740 min of treatment. The increase in surface roughness of the sample nitrided for 2400 min is most probably a result of the surface cracking, as mentioned previously.

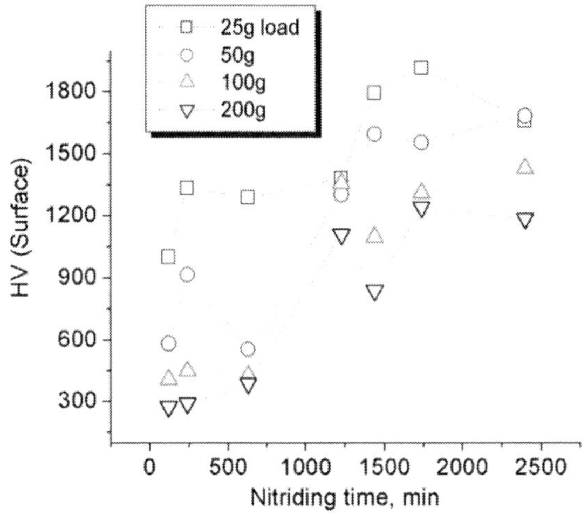

Fig. 7. Superficial hardness of the nitrided samples as a function of processing time.

These results demonstrate that choosing the appropriate Nano-S™ process can maintain a high quality surface finish on austenitic stainless steels. This would be of significant value to applications such as valves, which, as mentioned previously, must be polished to ensure a good seal.

Table 1. Arithmetic mean roughness R_A before and after Nano-S™ nitriding.

Nitriding time, min	R_A before Nano-S™	R_A after Nano-S™
120	0.08±0.01	0.08±0.01
240	0.07±0.02	0.08±0.01
630	0.08±0.01	0.08±0.02
1230	0.08±0.01	0.08±0.02
1740	0.08±0.02	0.09±0.02
2400	0.07±0.01	0.1±0.02

Conclusions

This paper has confirmed the parabolic trend of the nitrided layer buildup on AISI 304 austenitic stainless steel as a function of treatment time for the low temperature nitriding process. The presence of a hard layer has been attributed by other authors to a supersaturated solid solution of nitrogen in the austenite matrix. The layer nitrided with Nano-S™ consists of a top 'dark' etched zone and a bright etched internal sub layer. The harder 'dark' zone contains micro cracks resulting from the relaxation of the stress, induced by nitrogen supersaturation. The brighter internal zone's thickness is practically independent of the nitriding time. It appears as an interface zone between a highly stressed top layer and the soft core. This relatively soft 'bright' zone is most probably responsible for the frequently reported increase of corrosion resistance, as it is extremely resistant to chemical etching. Therefore, special care should be taken to nitride parts of complex geometry or that have some degree of residual stress. Proper selection of treatment time at a given temperature may result in a very high surface hardness, 1900 HV combined with a relatively deep case layer [~25 microns (0.98 mil)]. Although relatively thin, the layer has a negligible effect on the dimensions or surface topography of the parts. The high hardness may be positively correlated to enhanced wear resistance. This treatment is potentially interesting for a wide range of industrial applications of stainless steels where good tribological properties are required.

Nitriding time extended beyond the optimal will induce excessive cracking on the top of the hardened layer and nitride precipitations, impairing the mechanical and anticorrosive properties of the surface. In our experiment conditions, the 'critical time' was 1740 min.

Finally, while there exists in the market high-performance surface treatments for stainless steels, a host of new applications are being discovered and developed as a result of recent advances in process controls. At Nitrex Metal Inc. we successfully apply the Nano-S™ to a range of applications (molds for plastic/rubber, recreational all-terrain vehicle parts, gun barrels, surgical tools, marine fasteners) made from a variety of materials, including martensitic and precipitation-hardening steels. Advanced studies on the 'critical' treatment times for such materials are ongoing.

Acknowledgments

The authors are grateful to Nitrex Metal Inc. for permission to publish this work and to C. Chabanier (INRS-EMT) for EDX measurements.

References

[1] Fewell M.P. *et al*, "The nature of expanded austenite", *Surf. Coat. Techn.*, vol. 131 (2000) pp.300-306.

[2] Christiansen, T., Somers, M.A.J., "Controlled dissolution of colossal quantities of nitrogen in stainless steel", *Met. & Mat. Trans A*, 37A (2006), pp.675–682.

[3] Collins, S., Williams, P., "Low Temperature Colossal Supersaturaion", *Adv. Mat. & Proc.*, September (2006), pp.32–33.

[4] Fernandes, F.A.P., *et al.* "Stainless Steel Property Improvement by Ion Nitriding and Nitrocarburizing", Heat Treating Progress, July/August (2008), pp.41–43.

[5] Esfandiari M., Dong H., "Improving the surface properties of A286 precipitation-hardening stainless steel by low-temperature plasma nitriding", *Surf. & Coat. Technol.* Vol. 201 (2007) pp.6189–6196.

[6] Spies H.-J., Eckstein C., Zimdars H., "Structure and corrosion behaviour of stainless steels after plasma and gas nitriding" *Surf. Eng.*, Vol. 18 (2002), pp.459–461.

[7] Manova D. *et al.* "Influence of microstructure on nitriding properties of stainless steel", *IEEE Prans. Plasma Sci.*, Vol.34, No.4 (2006,) pp.1136-1140.

[8] Speidel M O, Uggowitzer P J "Ergebnisse der werkstoff", *Forschung*, Vol. 4, (1991) Verlag der Schweizerischen Akademie derWerkstoffwissenschaften.

[9] Egawa M. *et al.* "Effect of additive alloying element on plasma nitriding and carburizing behaviour for austenitic

stainless steels", *Surf. & Coat. Technol.* Vol. 205 (2010) pp.246-251.
[10] Latella B.A., Short K.T. "Tension-driven cracking of an expanded austenite layer", *J..Mat. Sci.* Vol. 39 (2004) pp.4321-4324.
[11] Jegou S. *et al.* "Evolution of residual stress in the diffusion zone of a model Fe-Cr-C alloy during nitriding", *J.Treatm.Mat* Vol. 66 (2011) , 3 pp.135-142.

Proceedings of the 26th ASM Heat Treating Society Conference
B.L. Ferguson, R. Jones, D.S. MacKenzie, and D. Weires, editors
https://doi.org/10.31399/asm.cp.ht2011p0301

THE VACUUM HEAT TREATMENT OF TITANIUM ALLOYS FOR COMMERCIAL AIRFRAMES

Robert Hill
Solar Atmospheres, USA

Abstract

Aeronautical engineers are consistently searching for new and optimal materials to achieve specific applications throughout an airframe. There are a multitude of considerations affecting the structural design of an aircraft such as the complexity of the load distribution through a redundant structure, the large number of intricate systems required in an airplane and the operating environment of that airframe. All of the above criteria is governed primarily by weight savings. Thus, the optimal materials selected today and for the future of airframes are composite material and titanium.

Introduction

Today's airframe is comprised of nearly half carbon fiber reinforced plastic and other composites. Composites, which are excellent in handling tension, greatly reduce maintenance due to fatigue when compared to an aluminum structure. When loading and environmental factors were analyzed, it was determined aluminum was a poor choice; whereas, titanium was an excellent low maintenance high strength material. In the new Boeing 787 Dreamliner, titanium has been expanded to approximately 15% of the total airframe.

Within the structure of the Boeing 787, the majority of the heat treatable titanium is in the following:

1. Landing gear and structural fittings (Ti5553)
2. Floor structure (Ti-6Al-4V)
3. Extrusions (Ti6Al-4V)
4. Nacelles (Ti-6Al-4V)

While Ti-6Al-4V has been the workhorse alloy for the industry, Ti5553 (Ti5Al-5V-5Mo-3Cr) is being used in a number of key flight-critical parts in the 787 aircraft. Ti5553 is a new near beta alloy that exhibits excellent hardening ability characteristics with superior strength combined with high fracture toughness and excellent high cycle fatigue behavior properties when compared to Ti-6Al-4V. As a result of these superior properties, Ti5553 forgings are being used in highly loaded locations such as flap tracks, pylon, side of body chords or landing gear applications.

Heat Treatment of Titanium

Without the proper heat treatment of this exotic material, specific critical metallurgical properties could not be attained. The heat treatment of Titanium 6Al-4V typically involves a solution treatment above the beta transus temperature followed by fast cooling (water quenching). The part is then age hardened at a prescribed temperature for a period of time. This treatment works well for most wrought products. However, the aeronautical engineer today must design for improved "buy to fly" ratios. (The "buy to fly" ratio is the mass of material that is required to machine a part compared to the mass of the original part.) Due to the excessive distortion from heat treating 6Al-4V during the water quenching, today's engineers need to develop a newer alloy that could be "controlled cooled" to control distortion. The alloy developed to control this distortion to enable the engineer to design to near net shapes is Ti5553. The BASCA (beta anneal slow cool age) heat treatment of 5553 not only allows for better ultimate properties but also better "buy to fly" ratios over its competitor Ti-6Al-4V.

Shortfalls with Current Titanium Heat Treating Specifications

Currently, there are a multitude of heat treat specifications that engineers design to. These may include their own "in-house" specifications (e.g. BAC5613). There is a heat treating specification for titanium parts only (e.g. AMS2801B). There are specifications that exist for wrought mill products (e.g. AMS4928Q or AMS-H-81200). Since there is not one universal heat treating specification it is understandable why contradiction and nebulous statements exist within all of these documents. I will summarize the top five problems as I see them from a heat treating point of view.

1. Heat treating specifications need to describe critical cooling rates for quenching per units of measure (F or C). What is the exact cooling rate of "water", "oil" or "air"? Cooling rates of liquid quenchants vary when considering immersion temperature of the media, agitation of the media and the heat exchange rate of the media. This problem is not inherent for titanium only. This problem exists for all materials and, therefore, we are continually distorting parts at times more than necessary.

2. All titanium heat treating specifications assume subsequent thermal process will produce an oxygen rich layer of alpha case. With today's vacuum technology this is not necessarily true. There are many methods available to metallurgists today to help minimize alpha case in vacuum. Specifications could define pre-cleaning methods, bake out frequencies of furnace and fixturing, use of pre-holds for vacuum levels, specify slower treating rates, multiple argon sweeps during pump down and prior to heating, and more accurate minimal vacuum levels which must be attained during the entire thermal cycle.

3. Severe problems can arise when titanium alloys come in contact with hydrogen rich environments. Typically, this hydrogen pickup occurs during the manufacturing and processing of the metal; especially at non-inert elevated temperatures. Deep vacuum processing along with elevated temperature depletes this hydrogen down to single digit part per million levels. Specifications need to state a maximum ppm level of hydrogen allowed to help eliminate hydrogen embrittlement and risk of failure in the material.

4. Specifications need to acknowledge the use of graphite for fixturing in vacuum. Graphite fixturing enables the engineer to design to nearer net shaped components. Graphite's attributes include excellent heat transfer characteristics, coefficient of thermal expansion (CTE) which mimics titanium, remains dimensionally stable and strong at temperature, and graphite can be easily machined to hold tight tolerances during thermal cycling.

5. Specifications currently do not clearly define and differentiate between work and control thermocouples and their individual intricacies. All values in the future must be recorded and reported on the <u>work piece thermocouples</u> only.

Looking to the longer term, Boeing and Airbus project the world fleet will double by the year 2029. This fleet will have airframes comprised of ever increasing composites and titanium. Currently, the Boeing/Airbus titanium usage is at 60 million pounds per year. Projections indicate that production will exceed 100 million pounds of metal by the year 2012 and maintain at that level through 2015! In order to properly thermally process this material downstream, metallurgists need to interact with Boeing, AMEC and SAE committees to discover and remedy these specification shortfalls. These actions will not only enhance the materials world, but also provide more safety to the military and commercial aerospace industries.

Proceedings of the 26th ASM Heat Treating Society Conference
B.L. Ferguson, R. Jones, D.S. MacKenzie, and D. Weires, editors
https://doi.org/10.31399/asm.cp.ht2011p0303

The Evolution of High "Tech" Vacuum Furnaces

Robert M. Huckins
G-M Enterprises, Corona, CA, USA
rhuckinsgm@earthlink.net, Phone: 630-762-1750, Corporate phone: 951-340-4646

Introduction

Today's "High Tech" vacuum furnace did not just appear on the heat treating market but has evolved as materials, controls & processes technologies have improved. "High Tech" vacuum furnaces were developed to improve parts metallurgical quality, braze quality, reduced inter-granular oxide formation and better control of the heating and quenching process. Ever since the creation of the first vacuum furnace for brazing in the late 1950's & early 1960's, about 50 years ago, continuous improvements have brought us to today's most advanced vacuum furnaces. A "High Tech" Vacuum Furnace is a furnace that utilizes the most advanced materials, designs, control systems and features that provide high performance, efficiency and productivity along with a low cost of ownership in utilities and maintenance.

What is the next development?

Why Develop a "High Tech" Vacuum Furnace?

Heat Treating in the late 1950's and early 1960's was dominated by the use of salt bath furnaces, atmosphere (hydrogen, argon & nitrogen) retort and box furnaces, pack carburizing and coating processes in inert gas atmospheres and tempering furnaces. The expected stress and fatigue demands on the parts then were significantly less than today's parts, due to the advanced parts designs, higher operating temperatures and cyclical stresses. Different advanced alloy materials and basic processing capabilities have been developed to meet these higher demands. The world was changing rapidly and moved into the advanced aerospace age with greater demands in materials. So the parts could only evolve as the material and processing capabilities evolve.

Vacuum technologies were developing rapidly at this time to support the "Space Race" to the Moon. A very important issue for high quality braze and other processes is the need to heat the alloys in an oxygen free or very low oxygen level. A typical inert gas or hydrogen atmosphere is purged with high purity gases, but it is common to still have enough oxygen in the atmosphere around or in the parts to cause oxide formation. These oxides cause substantial metallurgical "defects" in the final product reducing the strength and durability of the parts.

The first vacuum furnace consisted of a simple insulated, heated vacuum vessel with no positive pressure quench capability providing substantially lower levels of oxygen. Additionally, the vacuum chamber provided some of its own insulation capability, because there is no atmosphere to circulate around the parts, allowing higher processing temperatures. This processing method substantially improved the quality, repeatability and strength of the brazed joint. This allowed the joining of high cost/temperature/strength materials to lower cost materials making the overall cost of the new hybrid part lower. This new hybrid part allowed fast development of high performance parts in high production to support the fast growth of the aerospace/space industry.

Compare Current "High Tech" to the First Vacuum Furnace

The first vacuum furnace used all metal radiant shielding, molybdenum elements and hearth components to maximize parts purity and high temperature operation. This first furnace used standard molybdenum which becomes crystalline and brittle after heating to braze temperatures. This single chamber had to cool in a vacuum, so cycle times were lengthy and thermally inefficient. Today, the all metal vacuum furnace is still the highest purity furnace design available for sensitive braze and alloy materials. Many less purity sensitive braze, age & high temperature anneal cycles do not require all metal hot zones and can benefit from the higher thermal efficiency, reliability, productivity, performance and lower cost of maintenance and operation using graphite materials. This gives the parts maker the advantage of potentially higher levels of profitability and productivity.

Today's "high tech" vacuum furnaces commonly utilize graphite materials for the insulation, elements and hearth. When these graphite materials were initially introduced in the 1960's & 1970's they were not durable or strong enough for long life operation. The graphite fiber, board, structural and element materials have been substantially improved from the early 1960/70's.

The some of the first improvements from the initial all metal vacuum vessel was the use of ceramic fiber with molybdenum hot face and elements. An internal heat exchanger and fan /motor with inert gas back fill capability was added to allow the fan motor to start with out arching. These advances yielded shorter cycles and higher productivity. Improvements in graphite materials allowed the use of graphite fiber and board insulation reducing the absorption of moisture, further reducing cycle times. Improvements came in the 2 point element support system allowing both molybdenum and graphite elements to float in their supports substantially reducing stresses and increasing element life. Graphite hearths use became common because the density and strength of the graphite improved over time. The hearth strength of graphite increases as the temperature rises in the furnace allowing heavier loads without hearth breakage. This is opposite to alloy and molybdenum hearths that are weaker at elevated temperatures.

Vacuum vessels were also improved with reduced number of penetrations and the use of silicon high temperature o-ring materials, K-F clamps, single vessel penetration points for vacuum/pressure sensors, better door clamping/locking ring systems. Thicker and stronger door & penetration flanges and hinges were used to reduce distortion of the flanges to reduce seal leaks especially with high pressure quench vessels up to 20 bar. Larger diameter vessels were used to provide a larger gap between the outside of the hot zone plenum to the inside of the vessel wall to allow for higher volume cooling gas flow with less flow restrictions and faster parts cooling. A big improvement came with the thicker water jacket allowing a flooded cooling flow reducing particulate and ionic erosion of the vessel, thereby substantially extending vessel life. Another substantial improvement came with the simple redesign of the vessel power feed through assemblies providing improved cooling, sealing and little maintenance over the life of the hot zone. All of these enhancements reduce maintenance and cost of operation with higher productivity.

Today's High Tech vacuum furnace computerized control systems allow very tight control with improved parts processing yielding higher quality products. These computerized controls provide many enhancements over the past programmable P.I.D. controller and chart recorder systems offered on vintage vacuum furnaces.

Some of the biggest enhancements are user friendly interface helping the operator to easily select the correct program based on a part number or description instead of a 2 digit program number, helping to reduce operator errors. Programs are password protected protecting critical recipes. All data is encrypted and cannot be changed, much like a strip chart recorder of old. Real time calibrated load & control thermocouples compensated for up to 10 offset values over the temperature range, providing very accurate temperature control for critical processes like brazing. Specialized 7 bands of P.I.D. control yielding faster ramps with no temperature overshoot at the set-point, custom band alarms settable during ramp and soak segments built into each segment of the 20 segment recipe.

Upgraded digital vacuum and pressure instruments and sensors with filters that are more resistant to process contamination, commonly eliminating the less reliable ion gage vacuum sensor replaced with cold cathode vacuum gages. Open/closed vacuum/pressure sensor valves further protecting the sensors from the contaminating process gas & high pressure quench gas. Monitor and control the diffusion pump oil temperature and place the oil temperature at an idle temperature during furnace idle conditions, saving energy and aide in process trouble-shooting. Integration of maintenance tracking, run time, process and furnace component alarm functions automatically in the computerize control system. Enhanced high volume flow of hydrogen, argon and nitrogen partial pressure control capabilities up to 10,000 micron or 10 torr. These are all very specialized and useful tools that are available on current "high-tech" vacuum furnaces with computerized controls.

Conclusion - Current "High-Tech" Vacuum Furnaces Produce Quality Parts at Lower Cost

The improvements in materials and the innovative applications of these materials have allowed the rapid development of today's high tech, high performing, high pressure quench vacuum furnaces.

Proceedings of the 26th ASM Heat Treating Society Conference
B.L. Ferguson, R. Jones, D.S. MacKenzie, and D. Weires, editors
https://doi.org/10.31399/asm.cp.ht2011p0305

Energy efficient vacuum solutions for industrial furnaces

Uwe Zöllig, Mario Vitale
Oerlikon Leybold Vacuum, Cologne, Germany
Uwe.Zoellig@Oerlikon.com

Introduction

Fighting against global warming and increasing greenhouse gas concentrations, the regimentation of CO2 emission are the justification for the European Commission to set up the European Directive 2005/32/EC. This and several amending directives describe a framework of eco-design requirements for energy using products asking for increased efficiencies. New standards and norms have been written and others have been updated which now need to be considered in new product engineering projects. Beside this energy costs are rising constantly and will continue to rise in the future.
Considering these aspects the design of modern furnaces has changed recently. New designed vacuum furnaces are operating much more efficient as older furnaces. In consequence, also the vacuum pumps and pump systems must be designed accordingly to support these energy saving attempts.
For sure first priority for the vacuum pumps and system is always to reliably provide the required vacuum level; trouble-free operation is a must. The need of energy saving does not allow uptime reduction of the furnace. High reliability can be reached by using traditional technologies as e.g. oil-sealed rotary piston or rotary-vane pumps, roots blowers and diffusion pumps, but next to these also more modern technologies such as dry screw pumps already have a proven track record to work most reliable even under harshest conditions. But, are all these pumps also optimized in view of energy consumption?
The presentation will demonstrate that modern vacuum solutions with dry-compressing screw type vacuum pumps and new innovative roots blower designs can help meeting the goals in energy saving. Further on it describes measures taken at conventional vacuum components as for example rotary-vane pumps or even diffusion pumps support fulfillment of the new energy-saving requirements.

Standard vacuum systems used on furnaces today

Oil-sealed vacuum pumps, combined with roots-type booster pumps are today's standard equipment for most industrial vacuum furnaces. These vacuum pumping systems do have relatively low investment costs and are still the norm for a broad field in heat-treatments.

Mostly depending on regional preferences, two concepts for oil-sealed fore-vacuum pumps compete against each other; rotary-vane pumps and rotary-piston pumps. Rotary piston pumps are for sure more robust as vane pumps, but they become more and more obsolete.
For applications such as annealing, hardening and tempering, which cause only a benign load with negligible impact on the vacuum system, oil-sealed vacuum pumps offer cost-effective and very reliable performance. Compared with piston pumps, rotary vane pumps offer the same sufficient process stability, have a lower capital cost and consume app. 5-10% less energy, simply as the rotating mass is lighter weighted. But, is there additional potential for even higher energy savings?

For more harsh applications as e.g. brazing, carburizing or sintering more and more end-users choose dry screw pumps as backing pumps, simply as these are more capable and reliable to withstand the higher process requirements as any oil-sealed pump. Tendentious, even in standard applications many oil-sealed pumps are substituted by dry screw pumps because of the significantly lower cost-of-ownership of these pumps.
But, today's dry pumps are typically more energy consuming as oil-sealed pumps of the same size. Is this a pig in a poke? Does the user have to accept a higher power demand as trade-in for the higher robustness?

Roots-blowers are used on nearly each furnace and have a more or less unchanged design since decades. Their power consumption is minor compared to fore-vacuum pumps but can their power demand even be further reduced by modern designs or intelligent controls?

Most high-vacuum furnaces use oil-diffusion pumps to reach pressure ranges below 10-3 mbar. These type of pumps are proven "working-horses" and very robust, but their power consumption is high and does not change even if the furnace is in idle mode. Can this also be changed by innovative designs or controls?

The following chapters will give answers to each of the above raised questions. It will be demonstrated that by using modern solutions higher efficiency and reduced power consumption can be reached. "The most environmental friendly and cost saving kWh is the one which is never consumed – it pays off"

Detail view on rotary-vane pumps:

The principle of an up-to-date rotary-vane pump is already quite optimal for low power consumption. Rotating masses (including oil) are minimized, internal friction is low and by principle the pumps do the most efficient adiabatic compression. The power consumption of such pumps is clearly less than the nominal motor power. A SOGEVAC SV630B for example is equipped with a 15 kW flange motor. The effective power consumption of this pump at a typical operation pressure for a furnace below 1 mbar is approximately 7,4 kW.

New pump deliveries will include IE2 motors, which meet latest standards. Motors declared IE2 have a higher efficiency factor, all kind of losses, such as iron- and copper-losses, are minimized to achieve a higher efficiency. Realistically additional energy saving can only be reached by operating the pumps motor with a matched variable speed drive (converter). Thus an optimum in performance can be achieved. Ramp up function and torque optimization are two features among other provided by modern variable speed drives. Next to this frequency drives offers extended possibilities for process control, e.g. a suction speed reduction during the boost phase of a carburizing process to reduce HC-gas consumption.

Picture: Energy efficient standard pump for benign applications: SOGEVAC SV300B

Detail view on screw-type dry vacuum pumps:

Today's standard dry pumps are screw pumps with variable pitch rotor design. Continues compression along the rotor length minimizes the energy demand. Older technologies with constant screw pitch or even dry pumps based on multiple stages of roots- or claw type rotors, have a significantly higher power consumption due to design and pumping principle.

But even the plurality of today's screw-pumps with variable pitch, differ a lot from each other. Most pumps of the 600 m³/h class demand app. 10 kW power at typical furnace operation pressures below 1 mbar of approximately 10 kW, which is a clearly higher value as those of a comparable rotary-vane pump.

During the development of DRYVAC, reduction of energy consumption was a mayor focus. Optimizing the mechanical rotor design, the electrical motor concept and by selection of a perfectly matched build-in frequency converter the achieved result is overwhelming.

The DRYVAC 650S does only consume 6,9 kW at 1 mbar, is even more energy efficient than a rotary-vane pump and by this become the new bench mark for power consumption in the market.

The build-in frequency converter also offers potential for additional savings and more process control. Many process steps do not require "full-power" suction speed, especially during operation at rougher pressures (e.g. during carburizing). Soft start and ramping functionality can be realized with the variable frequency drive. Chamber pressure can be controlled by varying the rotational speed. The customer can even realize a process specific "standby condition" considering certain valve positions to save for example the volume of supplied Nitrogen.

Next to these environmental issues, for sure the modern design of the DRYVAC does eliminate sensitive components as shaft-seals or couplings which clearly increase robustness and reliability of the pump.

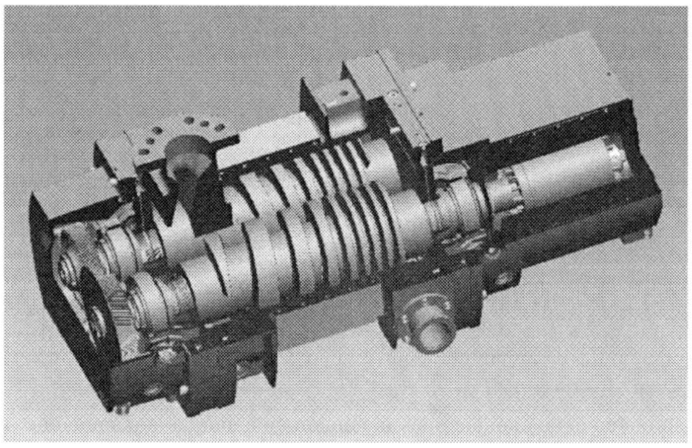

Picture: Layout of the hermetically sealed DRYVAC screw pumps featuring 2 rotors with progressive pitch profile and build-in, frequency-converter driven high-efficiency motor

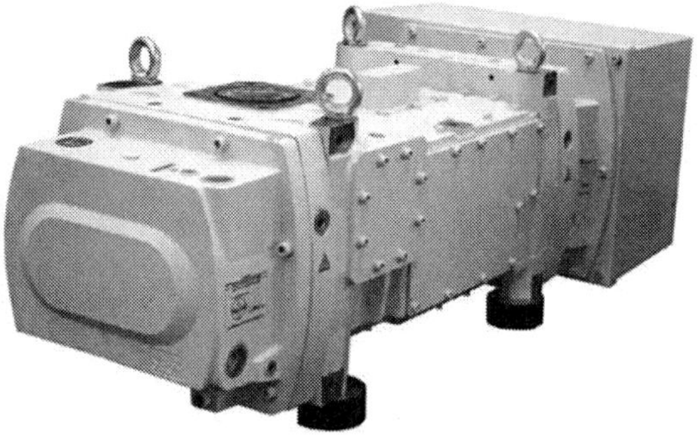

Picture: DRYVAC Enduro 650S, the new, most energy efficient standard for demanding heat-treatment applications

Detail view on roots-type vacuum blowers:

Most roots pumps used today are standard designs with flanged motors and lip-type shaft sealing. For most applications this design is sufficient, but it includes some weaknesses:

- Shaft-sealing, couplings and motor-bearings wear down over time
- Shaft-seals do not allow an operation at significantly increased rotary speed.
- The motor is mostly a standard motor not really matched to the pump for optimal usage

Picture: RUVAC WH4400, most compact and power saving design on market

The first two points could be addressed by usage of the well-known "canned-motor" pumps, such as the RUVAC WS2001 – since years available on the market. However, when looking at energy-efficiency this is a step backwards, as the canned motors typically operate by far less efficient than standard flange motors.

More promising are the most modern developments, using build-in, potted motor designs. Here the advantages of the canned-motor design are combined with reduced power consumption and offers improved compactness. In addition, the potted motors used for RUVAC WH roots-type booster pumps are high efficiency IE2 motors without external shaft seal or couplings. Further on the pumps are hermetically tight, a benefit for many processes.

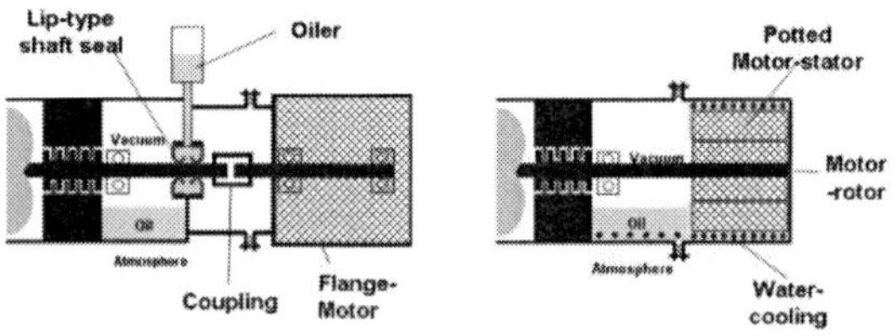

Picture: Motor concepts of a standard roots-pump (left) and a RUVAC WH roots-type booster pump (right).

At typical furnace operation pressures a RUVAC WH can be sped up by a frequency converter to rotary speeds up to 120 Hz (varying with pump-size). Thus a small pump can often substitute a much bigger pump, which reduces the installed and consumed motor power.

Detail view on diffusion pumps:

Diffusion pumps work in the high-vacuum field and require therefore a complete different pumping principle as the before described displacement pumps. Inside a diffusion pumps oil is evaporated, guided through multiple jet stages to build a steam "umbrella" which traps gas molecules flying into it. The oil steam is condensed when reaching the outer wall of the pump and recirculated.

Being a very simple, very robust and proven principle one the one hand side, a diffusion-pump on the other hand requires lots of energy. Most energy is consumed to evaporate the oil and not to compress the gas. All this evaporation energy is then removed by the coolant again as all oil is condensed again. Obviously, this is not a pump-principle with low energy demand.

Poorly the industry still need to stick to this energy wasting solution for most applications, as alternative high-vacuum pumps which are much more energy efficient (e.g. Turbomolecular pumps or Cryo pumps) are simply not robust enough to handle most furnace applications and next to this their initial costs are much higher.

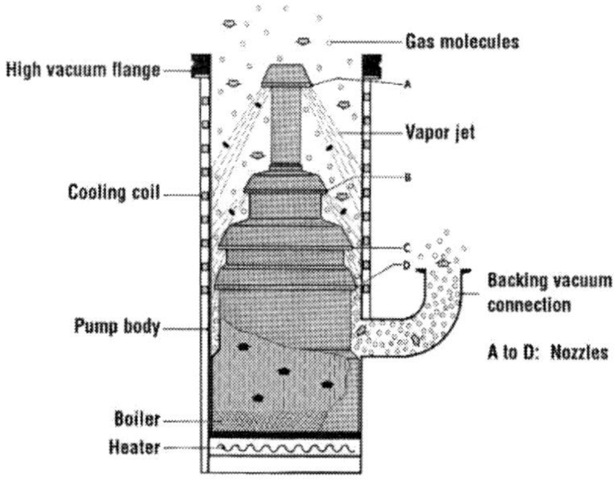

Picture: Pumping principle of a diffusion pump

But there are ways to minimize the power consumption of a diffusion pump. A major influence is given by the heater design. Most producers use heating-plates which are positioned below the oil-reservoir (boiler). This is not very energy efficient as there are high losses during the heat transfer into the oil, e.g. caused by the air-gap in between the heater element itself and the bottom of the diffusion pump. Better designs use heating elements positioned directly inside the oil-reservoir, by this clearly improving the heat transmission, bringing all heating energy directly into the oil.

Picture: Heater elements positioned inside the oil reservoir of a diffusion pump

Most energy is used during the heating phase of the diffusion pump. To shorten the heating time, it is important to minimize the mass of the pump which needs to be heated. Energy used to heat the pumps material is generally a wasted energy. Intelligent designed diffusion pumps as those from the DIP or LEYBOJET type have a total mass around 25% less as that of most competitors, therefore offering a much shorter heating-up time and lots of energy saving.
The full heater power is only necessary during the pumps heating phase. Compared with the total operation time of a typical furnace process, this time is relatively short. Most time the furnace is constantly maintained at high-vacuum or the diffusion pump is idling behind a closed valve. During all this time, a diffusion pump does not have such high power demand. Despite this fact, most diffusion pumps are continuously operated with full heating power, wasting lots of energy. Innovative control-systems can identify the actual power demand and regulate the power supply of the heaters accordingly. By these measures a reduction in power consumption of more than 30% can be realized.

Picture: Oil-diffusion pump LEYBOJET 630, improved for high throughput and lowest power-consumption.

Conclusions

There are many possibilities to safe energy. For sure vacuum pumps demand only a smaller part of the total energy consumption in a heat treatment process, but the responsible user should take all useful measures to reduce energy consumption, helping to reach our all environmental goals.

To use modern dry-compression screw type vacuum pumps or roots-pump clearly helps to minimize the environmental footprint of a vacuum pumps and pump-systems. Along with this, the implementation of latest vacuum pump designs, even for more traditional technologies as diffusion pumps or rotary-vane pumps will further reduce the CO^2 footprint. Intelligent controls will play a more and more important role and definitely help to meet the environmental goals.

Users should evaluate all options and not just stick to conventional solutions. Modern pump technology is robust and reliable, but most environmental friendly, too. They help to safe operation costs and to protect our green environment.

Proceedings of the 26th ASM Heat Treating Society Conference
B.L. Ferguson, R. Jones, D.S. MacKenzie, and D. Weires, editors
https://doi.org/10.31399/asm.cp.ht2011p0309

Multi-Purpose LPC+LPN+HPGQ 25 Bar N_2/He Single Chamber Vacuum Furnaces

Maciej Korecki, Józef Olejnik
Seco/Warwick S.A., Świebodzin, Poland
m.korecki@secowarwick.com.pl, j.olejnik@secowarwick.com.pl

Piotr Kula, Robert Pietrasik, Emilia Stańczyk-Wołowiec
Technical University of Lodz, Poland
piokula@p.lodz.pl, robert.pietrasik@p.lodz.pl, emilia.wolowiec@p.lodz.pl

Keywords: High pressure gas quench (HPGQ), low pressure carburizing (LPC), low pressure nitriding (LPN)

Introduction

The article describes the achievements and application of a new generation of HPGQ vacuum furnaces. Implementation of 25 bar quenching enables reaching hardening properties compared to the ones obtained with oil, while vacuum carburizing and nitriding additionally create a great potential for running different heat treatment and thermo-chemical processes as well as multi-processes combined in a single furnace cycle. Technical and technological aspects of the furnace exploitation are presented and operational costs reduction and energy saving are considered.

Despite the global economic difficult, the development of vacuum technologies and heat treating equipment continues in applications wherein vacuum is the basis for protective and technological atmosphere. This pertains especially to the applications based on single chamber vacuum furnaces equipped with high pressure gas quench systems (HPGQ). Intensification of gas quench enables heat treatment of not only alloy steels but also steel grades conventionally quenched in oil. There is continuous advancement in vacuum carburizing (LPC – FineCarb®, PreNitLPC®), which is becoming more and more competitive to traditional carburizing. Furthermore, first applications of vacuum nitriding have appeared (LPN - FineLPN®).

Those furnaces are fully automatic, computer controlled and equipped with technical support systems in the form of simulation software for vacuum carburizing treatments (SimCarb®, SimHard®) and quenching (G-Quench Pro®). Presently, the HPGQ single chamber vacuum furnaces are capable of handling a number of HT technologies such as: annealing, brazing, sintering, quenching, tempering, carburizing, nitriding, etc. These treatments may be run individually or grouped in a single treatment cycle, e.g. brazing + carburizing + quenching + tempering [1], or quenching + tempering + nitriding, etc. This makes the HPGQ furnace a versatile, flexible and multipurpose piece of heat and thermo-chemical treatment equipment which ensures high quality, repeatability and reliability at a minimal cost, while cutting down on process time and consumption of utilities and maintaining neutrality to the surroundings and the natural environment.

Single Chamber Vacuum Furnace HPGQ

The sophisticated vacuum furnace HPGQ is a unit featuring an internal quench system based on gas nozzles distributed evenly in the heating chamber around workloads or selectively depending on the shape of the workpieces and workload configuration. The outstanding effectiveness of the nozzle system comes from the fact that the nozzles aim the gas stream directly onto the workload and accelerate to the velocity of 100-150 mph. Such an intensive gas stream at high pressure results in very efficient quench and thorough penetration even through densely packed workloads [2, 3].

In the middle of the previous decade the single chamber HPGQ furnaces were only available in the pressure class of 10 – 12 bar, which enabled quenching of alloy steels, mainly tool and high speed steels (grades T and M), for cold and hot working (D2, D3, A1, O1, O2, H10, H11, H13), with a limitation imposed by the size of workpieces and workload density. Those furnaces achieved cooling efficiency expressed by heat transfer coefficient α at the level of 300-500 W/m2K. The current standard is class 15 furnaces which feature cooling efficiency of 400-700 W/m2K and thus have a wider application range which includes alloy steels for carburizing (9310, 4320, 8620), tool steels (O1, O2) and HSLA (high strength low alloy 4140, 4340) for workpieces of small cross-sections [3].

The next border was crossed in 2009 with the appearance of 25 bar single chamber furnaces for nitrogen and helium quench (Fig. 1).

These furnaces obtain impressive cooling rates in helium comparable to slow and medium oil quench, at the level of 1000-1500 W/m^2K, which enables quenching of a wide range of typical carburizing steels and HSLA [4] grades and even bearing steels (52100).

As far as tool steels are concerned, the cooling efficiency parameter is met even by the standard 10 bar furnaces. Using the example of hot working tool steel H13 and the NADCA [5] heat treatment requirements for dies, a minimum average cooling rate was determined at 50 °F/min in the temperature range of 1880 to 1000 °F. The tests following the requirements, done on a 16/16/16" steel block in standard HPGQ furnaces confirmed the effectiveness of the latter in die hardening (Fig. 2).The cooling rates obtained significantly exceeded the limit and, depending on the furnace working area, were respectively: for a 24/24/36" furnace approx.. 140 °F/min, 32/36/48" - 100 °F/min, and for 48/48/72" – 70 °F/min [4]. Therefore, class 15 bar furnaces may be expected to yield cooling rats higher by approx. 30% and the 25 bar He ones – even twice to three times higher [2].

Cooling rate is of key importance for impact strength and thus for the resistance of tools and dies to thermal fatigue cracking. For example, in the Charpy V-notch test for H13 steel, the cooling rate of 100 °F/min gave the impact strength of 17 J, the rate of 200 °F/min produced approx. 24 J, while a significantly lower cooling rate of 15 °F/min yielded a mere 8 J [6].

Fig. 1: Furnace HPGQ 25 bar N_2/He type 25.0VPT-4035/36 manufactured by Seco/Warwick S.A.

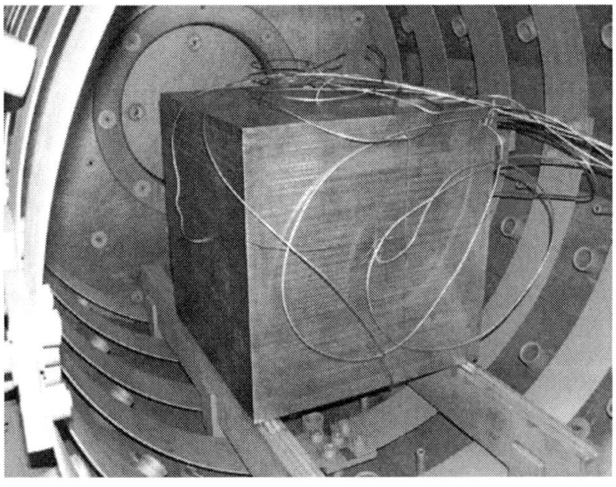

Fig. 2: Cooling rate test for dies in HPGQ furnace size 24/24/36" (Seco/Warwick)

In the course of hardening it is equally important to achieve uniform cooling from all sides and to disallow extensive temperature differences between the surface and the core of a workpiece as these may lead to major distortions due to thermal stress and, in extreme cases, cracking or damaging of the tool.

It is for such considerations that the HPGQ furnaces are equipped with a system of controlled quench which provides for the adjustment of cooling rate according to one of the thermocouples placed inside a workpiece. Apart from that, it is possible to run cooling based on the temperature difference between the workpiece surface and its core as well as interrupting the surface quench until the core reaches the temperature (interrupted quench, martempering, austempering). The controlled cooling options are provided by workload thermocouples interfaced with fan rotations, which directly influences the cooling gas flow rate. An example of die quench acc. to NADCA with surface and core temperature progress is presented in Fig. 3.

A major aspect of enhancing operational properties of the equipment is simulation software which enables prediction of treatment results in the given circumstances. The G-Quench Pro® software offered with HPGQ furnaces by Seco/Warwick provides for quench simulations of hot and cold working tool steels. The simulator takes into account a number of quenching parameters such as the type and size of furnace, the type and pressure of quench gas, the steel grade, the geometry of the workpieces and their loading density. Based on the above input, a cooling curve is plotted for a selected point from the surface to the core of the reference part. The outcome of the simulation is a CCT graph with cooling curve and expected hardness (Fig. 4).

Furthermore, the software facilitates on-line simulations in real time based on temperature readouts obtained from workload thermocouples directly during the quench.

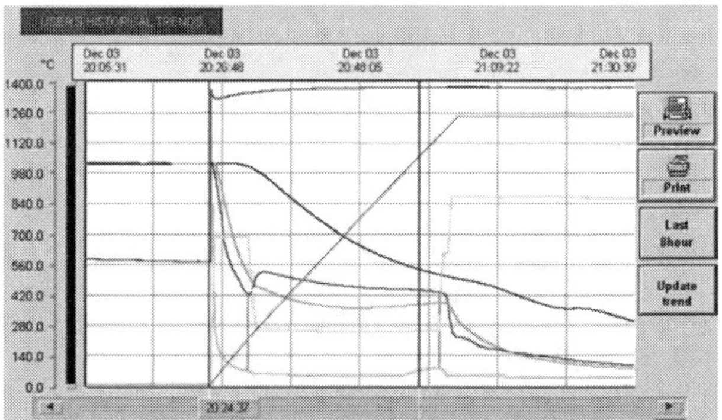

Fig. 3: The real trend of interrupted quench of the H13 tool

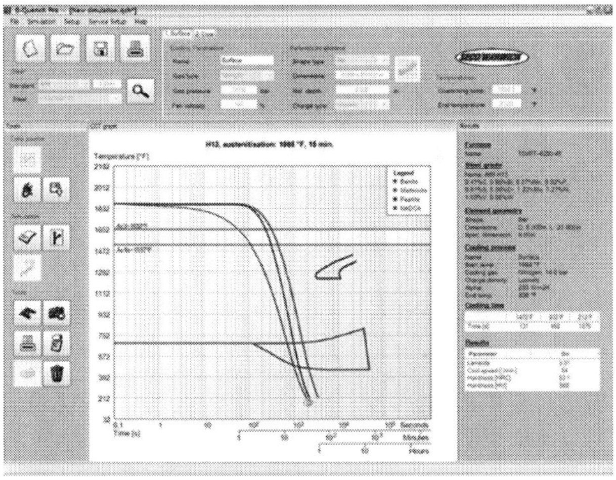

Fig. 4: The tool steel hardening simulator G-Quench Pro® (Seco/Warwick)

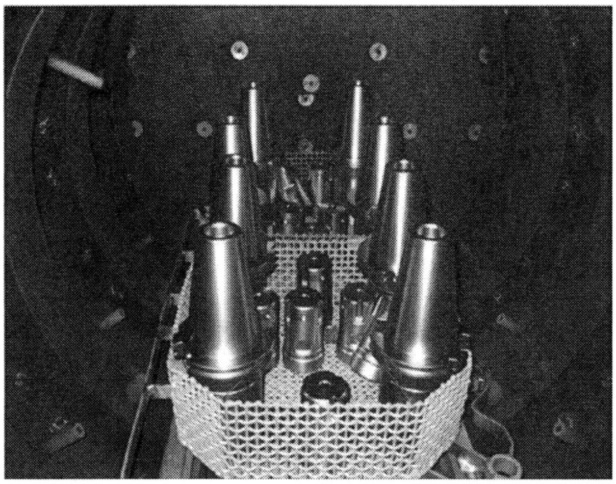

Fig. 5: Sample workload after carburizing and hardening in vacuum furnace LPC+HPGQ type 15.0VPT-4022/24 (Seco/Warwick)

Thermo-Chemical Treatment in HPGQ Furnaces

Vacuum carburizing by FineCarb® method has been introduced in over 70 industrial applications with single chamber HPGQ furnaces. It is based on a ternary mixture of carburizing gases (acetylene, ethylene, hydrogen) and an adequate manner of treatment, which ensure high efficiency, uniformity and purity [7, 8]. Combined with gas quench it provides an attractive alternative to conventional processes of case hardening carried out in atmosphere furnaces with oil bath [9] (Fig. 5).

A further sophisticated stage in the development of FineCarb® vacuum carburizing is the currently implemented method of carburizing preceded by nitriding - PreNitLPC®. This technology consists in feeding ammonia at the initial phase of treatment i.e. at heating for carburizing. The nitrogen introduced into the surface case in this way aids carburizing by accelerating carbon diffusion and lowering the tendency to build up carbides and, most importantly, by significantly limiting the growth of austenite grain (Fig. 6).

These advantages facilitate a considerable shortening of treatment time through an increase of the carburizing temperature. At the same time the case obtained features a proper microstructure and mechanical properties which are equal to those obtained in conventional treatments at a lower temperature [11]. For comparison, carburizing with the PreNitLPC® method at the temperature of 1900 °F is 4-5 times shorter than the one carried out at the conventional temperature of 1700 °F (Table 1).

Due to their disequilibrous nature, running vacuum carburizing treatments requires a computer assistance. The SimVaC® constitutes an integral part of the FineCarb® vacuum carburizing technology and of the expert system [12] which focuses on the development of the latter. It facilitates the design of processes of vacuum carburizing and hardening in high pressure gas as well as the analysis and optimization of treatments without the need

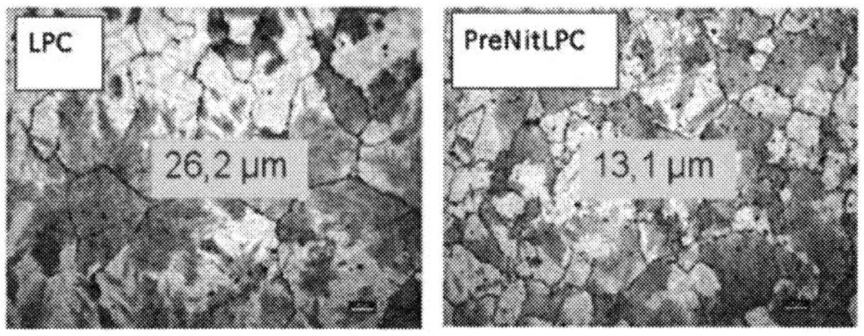

Fig. 6: The comparison of grain size after the LPC and PreNitLPC® treatment at 1830oF for steel 4320

Table 1: The influence of carburizing temperature acc. to PreNitLPC® on treatment time

	Carburising time for 8620					
Case depth, in.	1700°F	1740°F	1800°F	1830°F	1870°F	1900°F
0.02	1h23m	0h57m	0h39m	0h30m	0h24m	0h19m
0.04	5h30m	3h50m	2h35m	2h00m	1h35m	1h15m
0.08	22h00m	15h10m	10h20m	8h00m	6h10m	4h50m
	100%	69%	47%	36%	28%	22%

for real tests which usually are time-consuming and costly. The SimVaC® is a sophisticated simulation software consisting of a vacuum carburizing module SimCarb® and a hardening module SimHard® (Fig. 7).

It allows a high precision prediction of the results of real processes based on a process or an outcome simulation. The process simulation presents the effects of a preset process as a carbon profile and a case hardness profile. The outcome simulation prompts the treatment for the input case requirements. The system takes into account the steel grade, the shape and geometry of workpieces, their surface area, carbon concentration in the surface case, the case depth criterion, the carburizing temperature, the time sequence for boost and diffusion.

Other factors considered include precooling for hardening, the type and pressure of cooling gas and the size of the furnace. Apart from the carbon and hardness profiles, the simulation yields the demand factor for the mixture of carburizing gases (Fig. 8).

Similarly to vacuum carburizing, the HPGQ furnaces may be used for vacuum nitriding (already functioning in the PreNitLPC® method). The treatment consists in feeding ammonia to the vacuum furnace chamber at conventional nitriding temperatures. Currently research and tests are being carried out, chiefly on tool steels [13], aimed at mastering the process and working out simulation methods. While it appears obvious that the lengthy nitriding treatments in the HPGQ furnaces will not be justified, an interesting alternative might come from the relatively short, limited to several hours, nitriding of tool steels, applied as complementary to hardening and leading to a very hard and thin case which boosts the functional parameters of the tools. This would be particularly advantageous when the entire heat treatment is done at a single furnace cycle, without opening the furnace door and transferring the workload, by going through a sequence of: hardening, multiple tempering and final nitriding. Further advantages of the so conducted treatment are: excluding the chemical activation of surfaces before nitriding and obtaining a rapid and uniform increase of the nitrided case. This is due to heating in vacuum, which has strong reduction properties and cleans and activates the surfaces of the workpieces.

To confirm the above, a complete heat treatment of H11 steel tools was run in a HPGQ furnace type 15.0VPT-4022/24.

Austenitization was effected at the temperature of 1880 °F, followed by hardening in 12 bar nitrogen, then twice tempered at 1060 °F/2 h and finally nitrided at the temperature of 1000 °F for 4 h (Fig. 9).

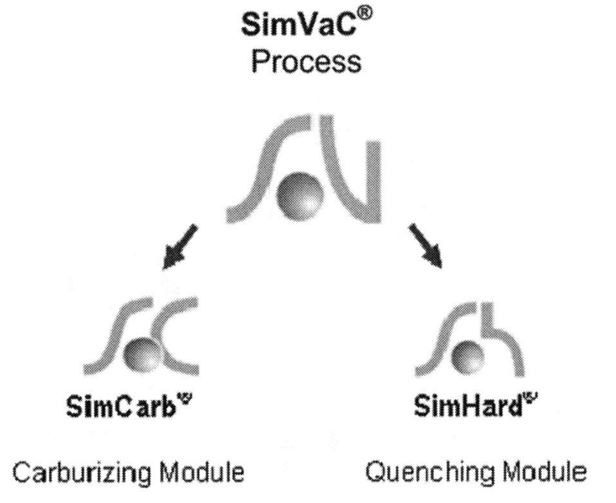

Fig. 7: The structure of SimVaC® simulator – vacuum carburizing and hardening (Seco/Warwick)

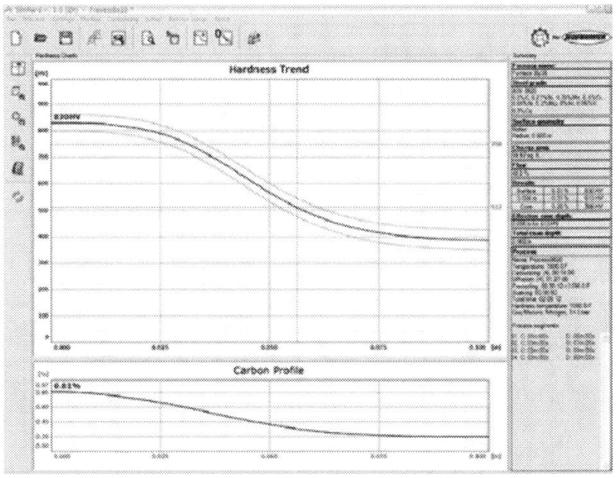

Fig. 8: The outcome of SimVaC® simulation of vacuum carburizing (LPC) and gas hardening (HPGQ) shown as a hardness profile based on carbon profile. During the 2 h treatment at the temperature of 1900 °F the case obtained was 0.055" for steel 8620

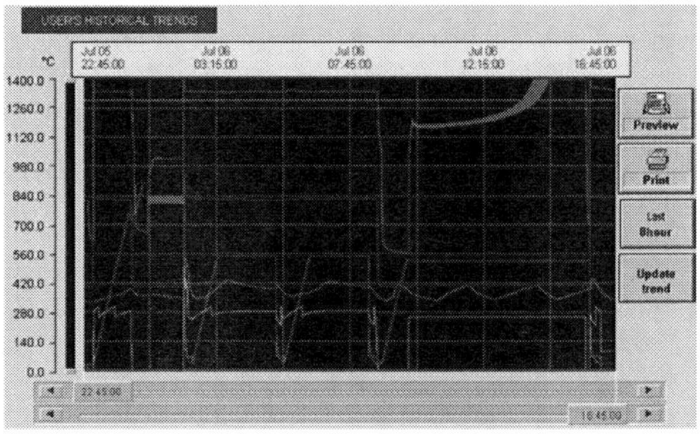

Fig. 9: The trend of complex heat treatment process of tool steel H11

The treatment resulted in a uniformly nitrided diffusion case of approx. 0,0055", surface hardness of approx. 900 HV and core hardness of 500 HV, respectively (Fig. 10).

Energy Efficiency

A reduction in energy requirements for the HPGQ furnace occurs at a few areas simultaneously. Among the basic areas are technical solutions which reduce thermal losses in the heating chamber and the application of electric receivers of increased energy efficiency. Very important is the optimization of the process, mainly its length. For that purpose a temperature monitoring in the workload is used as well as processes of high temperature carburizing (PreNitLPC®). As far as electrical power supply is concerned, it is essential to ensure a stable power demand not exceeding the maximum level while maintaining the highest possible power factor (P/S).

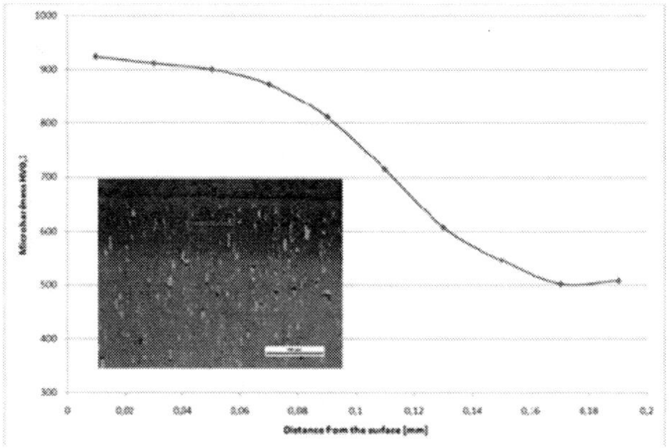

Fig. 10: The hardness profile and microstructure obtained after complex treatment for steel H11

The HPGQ vacuum furnaces are equipped with a power management system which ensures:

- Power demand limitation depending on temporary requirements.
- A substantial improvement of the power factor P/S in the heating and cooling phases.
- Start-up of fan motor without exceeding the rated currents (elimination of the starting current peak in the motor).
- Increased efficiency of the blower motor.

The HPGQ vacuum furnace comprises two main systems which use up most of the electrical energy (heating and cooling), each of which is conventionally equipped with individual power control systems. For heating, there are SCR controllers or a transducer to control the power of the resistance heating elements. For cooling, there are a soft-start or an inverter to control the blower motor. Since the furnace operating sequence does not provide for simultaneous heating and cooling, only one of the control systems may be activated while the other one is switched off. This dependency led to a search for a single system capable of alternate control of heating power or cooling intensity in the furnace. After theoretical analysis and testing an appropriate device was found – an inverter which, apart from controlling the work of an induction motor, may control the power of the resistance heating elements supplied through a transformer. What is more, such application enhances the operating features of the furnace and decidedly reduces its power consumption by increasing the power factor PF=P/S.

Figure 11 presents a comparison of electrical power requirement for active, passive and apparent power during the hardening treatment as exists between the traditional SCR controller and an inverter. In the given case the inverter control reduced energy consumption by 42 % and improved the power factor by 27 % (Table 2).

This method of heating and cooling with the aid of an inverter is protected with a patent [14] and has been successfully used in a few dozen HPGQ furnaces all over the world.

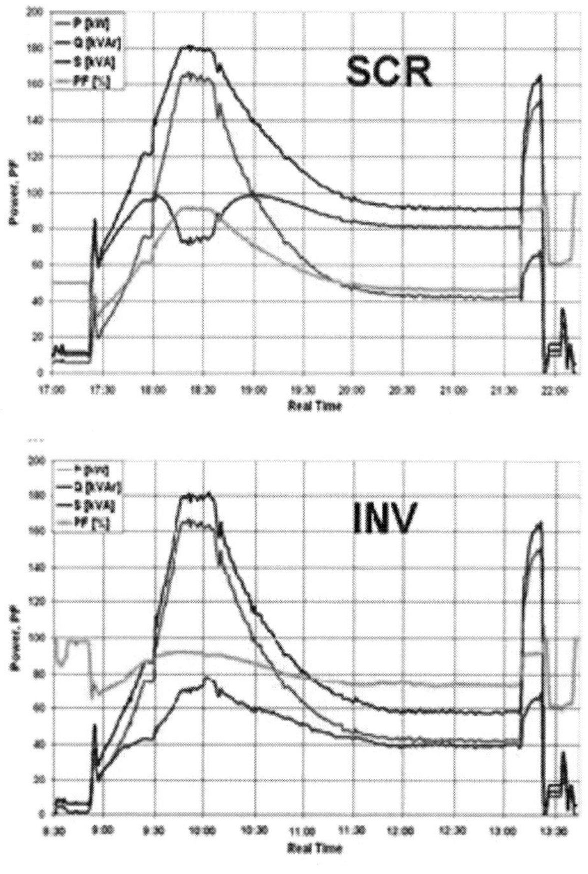

Fig. 11: The demand for active, passive and apparent electrical power during a hardening process depending on heating control mode: conventional thyristor (SCR), and inverter (INV)

Table 2: The energy comparison for reference treatment for thyristor (SCR) and inverter (INV) controllers

Energy consumption	SCR	INV	Safe	
Pt [kWh]	325	325	0	0
Qt [kVArh]	366	212	154	42%
St [kVAh]	489	388	101	21%
PF_{av}	0.66	0.84	0.18	27%

Summary

The single chamber vacuum furnaces HPGQ by Seco/Warwick provide a versatile and efficient tool for heat and thermo-chemical treatment. They are sophisticated devices which comply with the toughest quality, economic and environmental standards thanks to the following advantages:

- Multiple increase in cooling rate in the class of 15-25 bar N2/He furnaces enables heat treatment of steels conventionally hardened in oil.
- The exclusion of quench oil eliminates washing and utilization of washing means and the oil, which reduces the HT cost and makes the technology environmentally friendly.
- The single chamber furnace within which the workload is not moved enables application of workload thermocouples and complete temperature monitoring inside the workpieces, which in turn permits process optimization while simultaneously meeting the stricter requirements (the aviation industry).
- The function of isothermal and controlled cooling enables control of cooling rate and temperature distribution in the workload, thus reducing potential deformations and the risk of thermal fatigue cracking.
- The advanced high temperature carburizing acc. to the PreNitLPC® method permits a multiple shortening of treatment time, thus minimizing the costs.
- The possibility of vacuum carburizing and nitriding combined in a single cycle with hardening and tempering broadens the technological potential by including the multistage processes in one cycle.
- The simulation software for hardening and carburizing enable prediction of treatment results with high accuracy and eliminate the need for tests done on the treated workpieces.
- The electrical power management system facilitates economical and optimal use of energy.

References

[1] Korecki M., Adamek A., "Flexibility and Versatility of Heat Treatment under Vacuum", Furnace International, Part 1 – March/April 2005, pp. 6-8, Part 2 – May/June 2005, pp. 8-12

[2] Kowalski J., Korecki M., Olejnik J., "Next-Generation HPQ vacuum furnaces", Heat Treating Progress, September. 2008, pp. 39-44

[3] Korecki M., Siniarski D., "Seco/Warwick Single- and Multi-Chamber Vacuum Furnaces for Tools and Dies Heat Treatment", 1st Conference of IFHTSE for Tools and Dies, Pula, Croatia, June. 2005

[4] Korecki M., Olejnik J., Szczerba Z., Bazel M., "Single-Chamber HPGQ Vacuum Furnace with Quenching Efficiency Comparable to Oil", Industrial Heating, September. 2009, pp. 73-77

[5] Nord America Die Casting Association, "Recommended Procedures for H-13 Tool Steel", 2006

[6] Wallace J., Schwam D., Birceanu S., "The Effect of Cooling Rate on Thermal Fatigue Cracking and Impact Resistance of PG H13 Die Material", Die Materials and Technologies, 2009

[7] Kula P., Olejnik J., Heilman P., "Hydrocarbon gas mixture for the under-pressure carburizing of steel", U.S. Patent 7,513,958, 2009

[8] Kula P., Olejnik J., Heilman P., "Method for the under-pressure carburizing of steel work pieces", Patent No.: US 7,550,049, 2009

[9] Korecki M., Kula P., Reuss G, "Single-Chamber Vacuum Furnace as an Alternative for Case Hardening Technology", Heat Treatment Surface Engineering Conference, Jihlava, Czech Republic, November. 2009

[10] Kula P., Korecki M., Pietrasik R., Stańczyk-Wołowiec E., "FineCarb® – the Flexible System for Low Pressure Carburizing. New Options and Performance", 17th IFHTSE Congress, Kobe, Japan, October. 2008

[11] M. Korecki, P. Kula, R. Pietrasik, K. Dybowski, M. Krasowski, "Reducing Carburizing Time with PreNitLPC® Technology", 25th Heat Treating Society (ASM) Conference, Indianapolis, USA, September. 2009

[12] Stańczyk-Wołowiec E., Dybowski K., Atraszkiewicz R., "SimVaC® - Advanced Software for Supporting of Low Pressure Carburizing by FineCarb®", XII Seco/Warwick Seminar, Swiebodzin, Poland, September. 2008

[13] Kula P., Pietrasik R., Dybowski K., Stańczyk-Wołowiec E., Atraszkiewicz R., Korecki M., Olejnik J., "New Technological Pathways for Universal Vacuum Furnaces", 18th IFHTSE Congress, Rio de Janeiro, Brazil, July. 2010

[14] Korecki M., "Temperature control system for controlling heat treatment of metals, that is heating, soaking and cooling by a single frequency converter", US Patent 7,193,188, 2007

AUTHOR INDEX